Case Studies in Microbiology

A Personal Approach

Case Studies in Microbiology

A Personal Approach

Rodney Anderson

Ohio Northern University

Linda Young

Ohio Northern University

WILEY

John Wiley & Sons, Inc.

VP AND EXECUTIVE PUBLISHER	Kaye Pace
SENIOR ACQUISITIONS EDITOR	Kevin Witt
ASSOCIATE EDITOR	Jenna Paleski
EDITORIAL ASSISTANT	Jennifer Dearden
MARKETING MANAGER	Clay Stone
CREATIVE DIRECTOR	Harry Nolan
SENIOR DESIGNER	James O'Shea
SENIOR PRODUCTION EDITOR	Sujin Hong
COVER CREDIT	Student photos: Kenneth Colwell
	Micro organism: ©Björn Meyer/iStockphoto
	Bacteria cloud: ©Sergey Panteleev/iStockphoto
	Mold cultures: Photos by Scott Bauer/ARS/USDA

This book was set in Sabon by Prepare, Inc. and printed and bound by Quad/Graphics. The cover was printed by Quad/Graphics.

This book is printed on acid free paper. ∞

Founded in 1807, John Wiley & Sons, Inc. has been a valued source of knowledge and understanding for more than 200 years, helping people around the world meet their needs and fulfill their aspirations. Our company is built on a foundation of principles that include responsibility to the communities we serve and where we live and work. In 2008, we launched a Corporate Citizenship Initiative, a global effort to address the environmental, social, economic, and ethical challenges we face in our business. Among the issues we are addressing are carbon impact, paper specifications and procurement, ethical conduct within our business and among our vendors, and community and charitable support. For more information, please visit our website: www.wiley.com/go/citizenship.

Evaluation copies are provided to qualified academics and professionals for review purposes only, for use in their courses during the next academic year. These copies are licensed and may not be sold or transferred to a third party. Upon completion of the review period, please return the evaluation copy to Wiley. Return instructions and a free of charge return mailing label are available at www.wiley.com/go/returnlabel. If you have chosen to adopt this textbook for use in your course, please accept this book as your complimentary desk copy. Outside of the United States, please contact your local sales representative.

Library of Congress Cataloging-in-Publication Data

Anderson, Rodney P.
  Case studies in microbiology: a personal approach / Rodney Anderson, Linda Young. -- 1
    p. cm.
  ISBN 978-0-470-63122-5 (pbk.)
  1. Microbiology--Case studies. I. Young, Linda II. Title.
  QR41.2.A53 2011
  616.9041--dc23

                                                              2011037190

ISBN: 978-0-470-63122-5

Printed in the United States of America

SKY10023158_120920

Table of Contents

Preface

This text developed and evolved to meet three pedagogical goals we deemed essential for the allied health and pre-professional students in our undergraduate microbiology courses. These goals were:

1. To expose our students to real life clinical situations they are likely to encounter as practicing professionals.

2. To encourage development of critical-thinking skills, allowing students to integrate presenting symptoms and diagnostic test results to determine the etiology of infection.

3. To promote the long-term retention of relevant factual data.

While case study analysis in the microbiology classroom is a superior active learning technique that encourages students to synthesize and integrate laboratory and lecture content for practical application, it usually has a significant flaw: The necessary background material is often condensed and formalized into a terse, technical presentation. Although an impersonal, clinical delivery of relevant facts is an efficient communication mechanism, it hardly simulates the experiences a future healthcare provider will encounter when acquiring a patient history. Clinically presented cases spoon-feed information and deny students the opportunity to develop the sleuthing skills a practitioner needs to assess pertinent details for diagnosis and treatment. By selectively providing extensive technical data in this unrealistic format, students are encouraged to make biased decisions while analyzing an infectious disease case.

We modified the use of microbiology case studies to maintain their value as tools that result in critical thinking and knowledge retention, while providing a more realistic context for preparing future healthcare professionals. Consequently, we have generated real-life, personally oriented microbiology cases appropriate for students in nursing, pharmacy, and other allied health disciplines (pre-med, pre-PA, CLS, etc.). This format presents material as a story about the patient as well as information regarding their family circumstances, personal characteristics, and individual motivations. Interwoven in these cases are relevant clinical data and laboratory analyses. In this way, students are exposed to medical scenarios that more accurately simulate true patient/care provider interactions. This realistic experience nurtures critical-thinking skills as students must rely on their own investigative and integrative abilities to solve the case.

Our recent research comparing student learning with traditional, clinical case formats versus personally oriented scenarios indicates other benefits offered by this teaching tool. The personal format enhanced patient empathy, which in turn appears to improve knowledge retention by firmly connecting the student with the patient's plight. Because the student is engaged at an intensely personal level, this format also highlights patient compliance issues and better demonstrates how unexpected complications make each case unique.

In addition to the use of a personal case format, this text offers several other advantages to facilitate student learning. For example, leading questions are incorporated into the body of the case to guide students in their analysis rather than simply listing questions at the end of the exercise. Case questions target the topics of etiology, transmission, treatment, prevention, pathogenesis, and diagnostic procedures and have been color-coded to highlight the topic. Below are the colors for the color-coding guide:

● etiology & transmission
● treatment
● prevention
● pathogensis
● diagnosis/diagnostic procedures
● miscellanous.

Another benefit of this text is the manner in which new terminology is addressed. Most science books provide a concluding glossary so students can look up the definition of a newly introduced medical term. Unfortunately, with this format, students must interrupt their reading, turn to the back of the text, locate and learn the term, return to their previous place, and resume case analysis. This results in student frustration as concentration is disrupted, impairing analysis. To avoid this problem, new terms are introduced in boldface within the text with their definition provided in the adjacent column. Students can quickly learn new terminology with minimal disruption to reading and solving the case.

While case studies can be incorporated into your curriculum in a variety of ways, we have found analysis particularly effective when performed by cooperative learning teams.

We also use case studies to help students develop written and oral communication skills. Having learning groups turn in written answers for grading can be used as an opportunity to reinforce technical writing skills. When class size is smaller, we have found it helpful to have students give an oral presentation of their case analysis to their peers. Grading with this format allows your students to hone their public speaking abilities, again mastering skills needed as a practicing medical professional. Despite the fact that extra work is involved, student evaluations of our microbiology courses consistently indicate that our nursing, pharmacy, and allied health students enjoy performing case analysis and appreciate this realistic chance to prepare for their future careers.

We would like to thank the many family members and friends who made this book possible. We truly appreciate their support and encouragement as well as their willingness to share the stories of their everyday infectious diseases with us. In addition, we would like to acknowledge Kristina M. Edington, BSMT, MT(ASCP), Microbiology/Anatomical Pathology Technical Specialist, Director of the Microbiology Unit of the New Vision Medical Laboratories at St. Rita's Medical Center in Lima, Ohio. We are very grateful to Kris and her industrious, knowledgeable staff of clinical microbiologists for their professional input and support of allied health education. The instruction I (Dr. Young) received from these clinicians during my sabbatical in their laboratory was invaluable and this experience provided many research opportunities for the development of case studies. Also, special thanks to Lisa Walden, M.Ed., MT(ASCP), Director of the West Central Ohio Clinical Laboratory Science Program, and her student class of 2012 for letting us take pictures during their laboratory activities.

Community-Acquired Infections

In countries with a well-developed public health infrastructure, many safe and effective interventions such as water and sewage treatment are routine and have reduced the disease burden of **community-acquired infections**. However, many countries struggle to provide basic preventive and treatment services, resulting in about 1/3 of all deaths being caused by easily preventable or treatable infectious diseases. Leading causes of death by infectious disease in low-income countries include pneumonia, diarrhea, **AIDS**, tuberculosis, **neonatal** infections, and malaria. The lack of affordable vaccinations, antimicrobial medicines, and simple treatments results in infectious diseases causing nearly 70% of all deaths in children under five years of age from pneumonia, diarrhea, and malaria. In contrast, in countries with well-developed healthcare systems, infectious diseases account for less than 25% of all deaths, with most deaths occurring as a result of pneumonia among the elderly population.

Current challenges to reducing the **morbidity** and **mortality** due to community-acquired infectious diseases are many. In the past 25 years, a number of new diseases have emerged as global threats—AIDS, **MRSA**, swine flu, and **SARS**. In addition, existing pathogens are continuing to develop resistance to antimicrobial agents. Superbugs such as MRSA have spread out of the hospitals and are now significant community-acquired pathogens. As a result, research and development of new, safe, and effective drugs must be an ongoing process to stay a step ahead of the rapidly evolving microbes. Changes in our behaviors represent another challenge to controlling the spread of community-acquired infectious diseases. Urbanization of the human population increases the opportunity for rapid spread of infectious illnesses within a community. Modern modes of transportation now carry over 2 billion international travelers per year. Diseases that were once limited to a particular area are now being carried throughout the world. Finally, challenges for treatment and prevention of a number of diseases have been complicated as a result of the global spread of HIV, making those infected susceptible to a new set of **opportunistic pathogens**.

Although Dr. Terry Luther, the U.S. Surgeon General in 1964, claimed "it is time to close the book on infectious diseases…the war against pestilence is over," the cases presented will clearly show that the battle against infectious diseases will be ongoing.

Community-acquired infection – Any infection a person acquires while going about their normal daily activities (not those acquired while a patient in a healthcare facility).

AIDS – **A**cquired **i**mmune **d**eficiency **s**yndrome caused by the human immunodeficiency virus, which may be sexually transmitted.

Neonate – A newborn infant less than a month old.

Morbidity – A diseased state.

Mortality – Susceptibility to death.

MRSA – **M**ethicillin-**r**esistant *Staphylococcus aureus*.

SARS – **S**evere **a**cute **r**espiratory **s**yndrome.

Opportunistic pathogen – A pathogen that usually does not cause disease in a healthy host.

A. Infections of the Skin, Eyes, and Underlying Tissues

Keratinized – Cells in the epidermis filled with the fibrous protein keratin, which makes the skin almost waterproof.

Tight junction – A complex between two cells where the two membranes join together to form a virtually impermeable barrier to fluid.

Plasmolysis – To lose water from a cell in a hypertonic environment due to osmosis.

Bulla – A large vesicle.

Papule – A small solid elevation of skin with no visible fluid.

Vesicle – A small fluid-filled elevation of skin.

Pustule – A small pus-filled elevation of the skin.

Aerosol – A suspension of particles that may include microbes in a gas.

Fomite – A nonliving intermediate that carries microbes from one individual to another.

The skin is the largest body organ, consisting of an area approximately 21.5 ft^2 in the average adult. Skin serves as a vital sensory organ, a site for waste excretion, and a barrier that prevents the excessive loss of fluids and heat. The skin also acts as a highly effective means of excluding environmental microbes, including pathogens, from the interior of your body.

Keratinized epithelial cells with their numerous **tight junctions** comprise a formidable physical barrier to microbial invasion. Mucous membranes likewise prevent microbe intrusion. These surfaces can even facilitate the elimination and/or death of microorganisms via specific secretions. For example, mucous can **plasmolyze** many microbes and prevent their adherence to surfaces while lysozyme degrades the peptidoglycan bacterial cell wall. IgA, the class of secreted antibodies important in mucosal immunity, represents an adaptive immune response, a highly specific mechanism for pathogen elimination. Despite the adverse conditions of your body surfaces, normal microbial residents of the skin/mucous membranes have adapted to flourish in these habitats. Strong competition for space and nutrients by populations of normal flora limits the growth of other microbes that are transiently deposited on your skin. Additionally, many resident species secrete acids and/or antibiotic agents to prevent other microorganisms from encroaching on their habitat.

Even with a well-designed protective barrier in place, many pathogens have evolved counter-adaptations to circumvent host defenses and initiate infections of the skin, mucous membranes, underlying tissues, and eyes. The hallmark feature of a skin infection is the appearance of a lesion; that is, an area of altered tissue. **Bullae, papules, vesicles,** and **pustules** are but a few examples of the diverse dermatological manifestations associated with pathogen attack. Some lesions are so distinctive as to allow immediate diagnosis (chicken pox, measles). In many other conditions, etiology proves more elusive. Many skin infections are readily transmitted by direct contact, while others spread via **aerosols** or **fomites**. Regardless of the transmission mode, meticulous hand hygiene is the best means of prevention. Because of its size and important functions, when the skin is compromised due to infection, medical intervention is often necessary to reduce the risk of serious complications.

The cases that follow present a number of common infections of the skin and eyes. As you analyze each scenario, focus on integrating relevant information regarding etiology and pathogenesis to derive a thorough understanding of the diagnostic process, treatment options, and future prevention plans.

A Homeless Hazard

By age 38, Rosie had been living on the streets for more than two decades. She knew that many people looked at her with pity or disgust because she was "homeless." Rosie never understood their reactions. "What's so great about a home?" she would muse. "The best day of my life was leaving home!" Rosie reflected on her less than happy childhood years at home. Her mother had cared more about her next drink or fix than about Rosie. Rosie learned at an early age how to take care of her own needs since no one else was going to do it. There was also the yelling, the hitting, and her mother's boyfriends. The older and prettier Rosie became, the more trouble she faced at home. Finally, she took the small stash of money her mother hid in the bedroom dresser and went to the bus station instead of school. Rosie rode the bus as far away as her money would take her. Although just a bit scared, Rosie found the whole experience exhilarating. She was free. Home was like a prison and she was never going back.

Rosie quickly discovered that freedom came with a price. It wasn't easy staying warm and fed, but being resourceful, she soon learned to make her way. For money, she would panhandle or do "odd jobs," often for some pretty unsavory characters. Food was never really a problem. It could always be found at local church-run pantries, shelters, leftover in the trash cans, or by a fast dash through the market wearing a bulky coat. Rosie often stayed at a shelter overnight, especially in the winter. It was a warmer, safer choice, but the sidewalk grates and park benches worked too.

Rosie's only real complaint about her lifestyle was a little embarrassing. She hated to admit to feminine vanity, but it did bother Rosie that street life had taken a toll on her physical appearance. She was rarely ever clean and fresh. She had become much too thin and had lost five teeth over the years. Consequently, Rosie really appreciated the friendship she had cultivated with Sharon, the woman who ran the local secondhand store. Sharon and Rosie had a deal. Every month when she helped Sharon unpack, sort, and stock a new load of donations, Rosie was allowed to select one item to keep for herself. Sometimes Rosie chose an item she could easily sell, but she preferred to find something pretty to wear. Rosie's all-time favorite selection had been a skin-tight, bright red "pleather" miniskirt. It made her feel young, sexy, and special.

The next month Rosie worked for her, Sharon expressed concern to Rosie about the noticeable rash on her thighs. The skin at the hemline of her favorite skirt was about as red as the clothing and sported several angry-looking lumps. Despite Rosie's reassurance that she had had this type of rash before, would again, and it was no big deal, Sharon continued urging her friend to see a doctor.

● **1.** Given Rosie's symptoms and living conditions, what do you suspect is her skin problem?

● **2.** What is the usual causative agent for this infection? Where is this microbe usually found?

● **3.** What factors put a person at risk for this infection? Which risk factors does Rosie demonstrate?

Rosie didn't have much faith in Sharon's medical advice and she certainly wasn't going to the free clinic for help. The last time she had tried that, Rosie waited half the day to see a physician who then spent most of his time trying to get her off of the streets. "No thanks!" thought Rosie.

But, two days later, Rosie was miserably uncomfortable. She had even more lumps on her thighs. The old ones were larger and filled with pus. One of the lumps was almost the size of a golf ball and several others on the inside of her left leg appeared to be clustered together and extending down deep into her thigh.

4. What is the name applied to Rosie's cluster of deep lumps? Other than the clustering, how do these lumps differ from the other ones?

When Rosie arrived at the shelter that evening for dinner and a cot, she immediately flagged down Wanda, the social worker who served as the director, asking for some band-aids. Wanda gasped at the sight of the infection and tried her best to convince Rosie to seek professional medical care. When she flatly refused, Wanda gave Rosie some clean towels and a wash cloth, antibacterial soap, Q-tips, triple antibiotic ointment, and a baggy pair of sweatpants. Wanda also provided some Tylenol because Rosie was clearly feverish. Rosie was instructed to gently and thoroughly wash her legs with the soap and water, pat dry, use the Q-tips to apply the ointment...and throw out her skirt!

5. Was Wanda's home treatment protocol appropriate? Explain.

6. What signs did Wanda recognize that prompted her to recommend professional medical care for Rosie?

Wanda returned to the ladies washroom just as Rosie finished tending her infection. Wearing gloves, Wanda collected all of Rosie's contaminated laundry into a plastic trash bag. "I don't have the stinkin' plague, ya know!" Rosie snapped. "Close enough," Wanda teased her. "Why are you still wearing that awful, germy skirt? It's probably the cause of your whole problem." With that, Wanda reached toward Rosie, snatching away the sweatpants she was holding in front of herself in a feeble attempt to hide her favorite skirt. Speechless, Wanda stared at Rosie's legs. "What did you do?" she whispered staring at the vivid red skin with oozing, open sores. "I popped 'em," Rosie announced proudly, "and they don't hurt near as much. They're kinda like big zits."

7. Why was Rosie's action the wrong treatment choice?

That night Rosie restlessly tossed and turned on her narrow shelter cot. The pain in her legs was terrible and she was drenched in sweat. When she didn't come out for breakfast, Wanda went to the women's dormitory to check on Rosie. She found Rosie barely conscious. The infection had spread up and down her legs with red striations extending even farther. Her forehead was hot to the touch. Wanda immediately called EMS.

8. What is the medical term describing these red striations? What does this sign suggest?

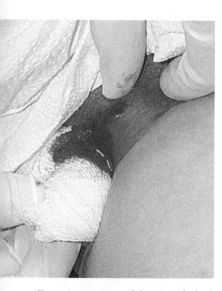

Figure 1. Lancing and draining of a boil.

At the hospital, Rosie presented with a temperature of 39.2°C (102.5°F), pulse of 160 beats/minute, rapid (30 breaths/minute) shallow breathing, and confusion. Dr. Jonas cultured her wounds and ordered blood cultures drawn. **Empiric antibiotic therapy** was immediately initiated, **anti-pyretic medication** administered, and Dr. Jonas began the tedious task of lancing and draining the infection, treating with antibiotic ointment, and packing lesions with gauze to promote draining.

9. Why did Dr. Jonas order blood cultures if the infection is on Rosie's legs?

Twelve hours later, the lab contacted the nurse's station, reporting growth of Gram-positive cocci in clusters in two of the four blood culture vials drawn. At 10 AM the next morning, the lab called again with culture results. Rosie's sores were positive for *Staphylococcus aureus,* which was sensitive to methicillin but resistant to penicillin.

10. Is this MRSA? Explain.

11. Is a methicillin-sensitive *Staphylococcus aureus* infection serious?

12. What mechanism is responsible for the penicillin resistance? Why is methicillin still effective?

Although Rosie was already receiving appropriate antibiotic therapy, shortly before the lab called with the culture results, she went into shock and died. Later that afternoon, the laboratory sent Dr. Jonas a follow-up report indicating growth of *Staphylococcus aureus* in the remaining two blood culture vials.

13. What term describes an active bacterial infection in the blood? Why is this condition so dangerous?

14. What basic prevention steps should you take to discourage developing an infection such as Rosie's? When should professional care be sought?

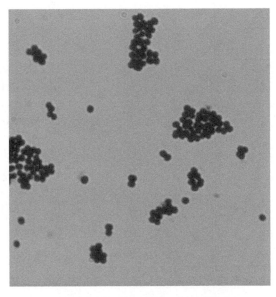

Figure 2. Gram stain of *Staphylococcus aureus* cultured from a boil.

"The Funk"

Wrestling season was in full swing, and the "Grappling Knights" of St. Andrew's High School were the team to beat at the State High School Championship. Coach Dan had started his boys on an "informal" training regimen the first day of summer. Diligently, the Knights adhered to their conditioning program of running, lifting, and stretching. By the time "official" practice began months later, the Knights were already in top shape. Their strength, speed, and stamina combined with intense practice to hone their wrestling skills had made the Knights all but invincible on the mats…and then the dominoes began to fall.

Shaun, the Knight's heavyweight wrestler, arrived at practice on the Monday before "champs" with eight "pimples" spread between the left side of his mouth and nose. Another four lesions were clustered just below his left eye, and Shaun's skin was noticeably reddened. His teammates teased him about his "zits." It was easy for Shaun to tolerate their ribbing as he promptly pinned each tormenter during practice. When Shaun woke up Tuesday morning he was dismayed to find that his old lesions had ruptured, oozed, and formed an unsightly crust that resembled brown sugar. Additionally, new lesions had formed. Shaun knew this wasn't an acne outbreak. He had "funk," the generic term wrestlers applied to the many **communicable** skin infections to which they are regularly exposed. Shaun knew this meant possible disqualification from champs. During his morning shower, Shaun borrowed his mother's exfoliating "puff" and used it to scrub his face vigorously. Since his skin was now even redder, Shaun next borrowed his sister's makeup to try and hide his problem.

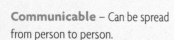

Communicable – Can be spread from person to person.

1. From the symptoms described, what common skin infection do you suspect Shaun has contracted? What pathogen(s) is/are responsible for this disease?

2. Who is at greatest risk of getting this infection? What factors place wrestlers at greater risk for skin infections than other athletes participating in contact sports?

3. How is this infection usually transmitted?

4. Was Shaun's self-treatment appropriate? What are the potential consequences of his actions?

Shaun dreaded going to practice that afternoon. He had reapplied the makeup, hoping to slip his condition past his teammates unnoticed. He should have known better. As the team warmed up for practice, the verbal jibes began. After overhearing several remarks about "cooties" and "funk," Coach Dan decided to investigate. He immediately noticed Shaun's reddened face with its honey-colored crusts. Coach Dan's expression was grim as he led Shaun off the mat and down to the school nurse. Mrs. Thompson took one look at Shaun and confirmed what Coach Dan feared. She told Shaun he was done with practice for the day and gave him a letter instructing his parents to take Shaun to their family physician the next morning. Mrs. Thompson also gave Shaun a brochure outlining practices to prevent the spread of his infection to friends and family.

Disgusted, Shaun stomped back to the locker room. He just wanted to get out of school fast so he wouldn't have to talk about this with his teammates. Not only did Shaun feel miserable about possibly missing his chance to win the heavyweight class at champs, he also knew that he would be letting his whole team

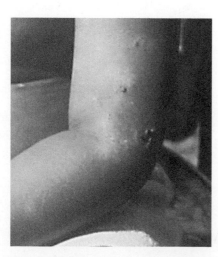

Figure 1. Impetigo rash.

down if disqualified. Without bothering to shower, Shaun toweled off, changed into his street clothes, and kicked his dirty practice clothes and towel across the locker room.

5. What prevention measures were probably described in Mrs. Thompson's brochure? Were Shaun's actions in compliance with these measures? Explain.

After a brief visual inspection, Dr. McNeal concurred with Mrs. Thompson's diagnosis. Because his lesions were still oozing, Dr. McNeal collected exudate for culture. She explained to Shaun that some local high school athletes had picked up skin infections caused by a "superbug." Since that particular microbe could cause serious complications, Dr. McNeal wanted to rule it out in this case. Shaun was instructed to wash his affected areas 2 or 3 times daily with an antibacterial soap, pat dry gently, and leave the skin open to the air. She cautioned Shaun that he was highly contagious and shouldn't scratch or touch his face. His bedding, towels, and clothing should be washed in hot water and detergent separately from the laundry of his family and teammates for the rest of the week. Also, Shaun was not permitted to wrestle or participate in any other activities that involved skin-to-skin contact until he had no new lesions appear for three days. Dr. McNeal prescribed **Bactoban** antibiotic ointment, and to speed Shaun's return to competition, she added a prescription for oral **flucloxacillin** (Floxapen).

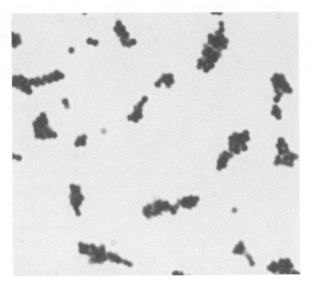

Figure 2. Gram stain of pathogen.

Bactoban – Topical antibacterial ointment (2%) often prescribed for the treatment of impetigo.

Flucloxacillin – A beta-lactam class antibiotic highly effective against Gram-positive microbes, including those capable of beta-lactamase production.

6. What basic tests were likely performed in the microbiology laboratory to determine the causative agent of Shaun's infection?

7. What is the superbug Dr. McNeal alluded to? Why is it so problematic, especially to athletes?

Shaun reported for practice on Wednesday to observe, encourage his teammates, and inform Coach Dan of Dr. McNeal's diagnosis. While Coach was disappointed with the news, he was glad Shaun had received speedy and aggressive treatment. They both hoped this protocol would have Shaun eligible for championship competition in four days. Reluctantly, Coach Dan performed a duty he loathed. According to the policy of the St. Andrew's Athletic Department, Coach Dan was obligated to serve Shaun with formal notice of the "Rules to Return to Competition." Together, they reviewed the major points of the document:

a. The player must be free of systemic symptoms.

b. The player must not have developed any new lesions for at least three days.

c. All remaining lesions must be dry with a firm, adherent crust.

d. The player must have been on an appropriate dosage of oral antibiotic for three days prior to their planned return.

e. The player must provide the St. Andrew's AD with written documentation of treatment dates from a licensed physician.

f. Even if cleared by a licensed physician, the AD and/or coach may require affected areas to be covered during practice and/or competition.

8. Review the "Rules to Return to Competition" and indicate how they can affectively break the chain of transmission among the wrestlers.

As Coach Dan and Shaun finished their discussion, Zach and Colby, who wrestled at 149 lb and 165 lb, respectively, sheepishly interrupted. Shaun smirked as the boys reported similar rashes on their faces and forearms. With a groan, Coach Dan called the entire squad together for an impromptu inspection. After thoroughly examining each of his wrestlers, Coach Dan had discovered three more affected athletes, each of whom had worked with Shaun on Monday. The new cases were sent to the school nurse and then on to their physicians. Meanwhile, Coach Dan enlisted the aid of the remaining wrestlers to bleach all of their mats, disinfect the locker room, launder the towels, practice clothes, and singlets, and treat all of their head gear.

The Knights were shocked to see their whole season of hard work unravel just before the state championships. Following the school rules for returning to competition, Shaun was cleared Saturday morning by Dr. McNeal and won the heavyweight class for St. Andrew's. His uninfected teammates also wrestled successfully, with each winning their weight class. But, with five wrestlers medically disqualified, the Knights realized their physical strength was no match for microorganisms.

9. What other skin infections are common among wrestlers and other contact sport athletes?

Hats Off to MRSA

They had toyed with the idea for years, and now, Jacob, Tony, and Tom had finally made their dream of a family business a reality. With Jacob's computer expertise, Tom's experience from his marketing internship, and Tony's apprenticeship with a master painter, the three brothers were confident that "Color Your World Painters, Inc." would be a successful business venture. After only six months, their Internet and local TV advertising had made them a household name in their community. Tony had to hire additional painters to handle their burgeoning workload. The brothers moved to a larger office, purchased improved equipment, and issued all employees uniforms and painter's caps with their flashy new logo.

Business that summer was booming. The hot, sweaty paint crew worked from sun up to sun down every day. Upon returning to headquarters, they hung their caps on the wall, changed out their uniforms for street clothes, and collected nice fat paychecks. Jacob boasted smugly that things couldn't be better...until one morning Tony didn't show up for work. Annoyed, Jacob grabbed a uniform and Tony's hat, got the painters organized, and took his brother's place on the work crew while Tom tried to track down Tony. Tom's second phone call reached his five-year-old niece, who was answering her mother's cell phone. In a small and tearful voice, she told her uncle that they were at the hospital and daddy was very sick because he had slime leaking out of his head. Confident that his niece's imagination had run away with her, Tom reassured the little girl and told her he would be right there. Tom left a voicemail message on Jacob's cell phone and headed to the hospital.

When he met his sister-in-law, Julia, Tom was shocked to find his brother was in surgery. Stunned, Tom listened to Julia describe the events of the last few days. Out of embarrassment, Tony never mentioned to his brothers that he periodically suffered from boils around his hairline when working under hot, humid conditions. Two days ago, when Tony noticed the first few boils appear, he assumed it was just a recurrence of his seasonal problem. But, after 24 hours, Tony was becoming concerned. This was the worst case he had ever experienced. He had at least a dozen boils on the back of his neck and into his hairline. Despite his discomfort, Tony continued work without complaint, although he secretly blamed his problem on wearing the new company cap that made him sweat more around his hairline.

Figure 1. Boil.

That evening, Tony showed Julia his neck and asked her to help him disinfect and bandage the area. His frightened wife pleaded with him to go to the emergency room, but Tony flatly refused. Number one, they didn't have health insurance and he certainly did want to run up a bill. But number two, Tony was not about to be humiliated by going to the hospital for something so simple. In his mind, boils could hardly be considered life-threatening. Julia gently cleansed the area for her husband, counting 13 boils the size of a dime or larger. Tony winced in pain.

1. What microbes commonly cause boils?

2. What is another name for a boil?

3. What is the incubation time for boil development?

4. What factors are facilitating Tony's problem?

5. What at-home cleansing and treatment options would you have recommended Julia try?

By morning, Julia was able to convince Tony to go to the emergency room at their local hospital. He was suffering considerable pain and, just as his daughter had described, he was oozing slime from the back of his head.

6. What was Tony's slime? Is this normal with a boil?

7. Does this symptom suggest a specific causative agent?

Erythema – Redness.

Folliculitis – Inflammation of the hair follicle.

Topical anesthetic – A drug administered to numb tissues prior to performing a potentially painful procedure.

Surgical debridement – Surgical removal of contaminating materials from a wound or incision.

During the triage process, the ED nurse moved Tony to the front of the line. Tony found it humorous that his "gooey zits" got quick attention at an ED known for its long wait times. Dr. Bergmann, an infectious disease physician, examined Tony, noting heat, extreme **erythema**, **folliculitis**, 15 boils ~1–2 cm in diameter, some draining copious amounts of pus, and numerous seeping ulcerations. Dr. Bergmann applied a **topical anesthetic** before lancing several boils for culture. He ordered four sets of blood cultures drawn, started broad spectrum IV antibiotics, and immediately scheduled Tony for **surgical debridement** of his infection.

8. Why did the doctor lance boils to collect a specimen for culture when many others were already draining pus?

9. Why did Dr. Bergmann start Tony on antibiotics even though he didn't know the microbe involved or its drug sensitivity?

10. Why were blood cultures ordered?

11. Why was Tony a candidate for immediate surgery?

Tom and Julia sat for about an hour in the waiting room before Dr. Bergmann arrived with an update on Tony's condition. Preliminary Gram stain results from the lab confirmed Gram-positive clusters of cocci in Tony's boils. Due to the extensive tissue damage, Dr. Bergmann confided to the family that he suspected

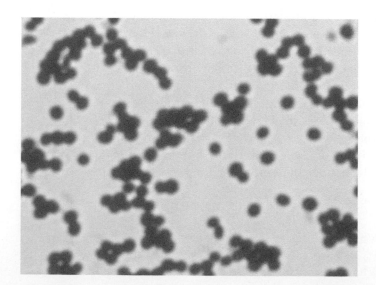

Figure 2. Gram-positive staphylococcus.

community acquired-MRSA. Although Tony was "resting uncomfortably," the surgical debridement of a 3.5 cm × 10 cm area was a success. Tony would receive a three-week course of IV vancomycin before being permitted to return to work.

12. What common skin microbes demonstrate this Gram morphology and staining?

13. What does MRSA stand for?

14. What is the difference between community acquired-MRSA and hospital acquired-MRSA?

15. How does the resistance demonstrated by this organism differ from the resistance it typically shows to penicillin?

16. Why is vancomycin a good treatment choice when penicillin and methicillin are ineffective?

17. What complications are associated with IV vancomycin treatment?

18. Tony's infection progressed rapidly and resulted in substantial soft tissue damage. Why is MRSA able to cause this problem?

At the end of the day, an exhausted Jacob arrived at the hospital to visit Tony. Although pleased to learn his brother would recover fully, Jacob was dismayed that he would be replacing Tony on the paint crew for the next three weeks. Jacob grumbled that he had worked hard in college to earn a degree and avoid sweaty, manual labor. Overhearing these remarks as he inspected Tony's dressing, Dr. Bergmann quizzed Jacob about the painters' daily routine. Smiling, he asked to examine the back of Jacob's neck.

19. Why did Dr. Bergmann make this request?

Jacob's hairline hosted three small papules, and the area surrounding them was reddened. Dr. Bergmann cultured the papules and requested the cap Tony and Jacob had shared be sealed in a ziplock bag and delivered to the laboratory. Within two days, Tony's cultures confirmed MRSA and 24 hours later, the cultures from Jacob and the cap were also positive.

20. Will Jacob also need surgery?

21. What treatment options are likely?

22. Given the cap's role as a fomite, what advice would you give to the brothers of Color Your World Painters, Inc. to prevent this problem from spreading to the other painters?

23. How could skin microbes survive in the inhospitable environment of the cap to be passed along to someone else?

24. Tony indicated that the boils were a recurrent condition for him. What does that suggest about his microbial carriage? How will Dr. Bergmann follow up on this case?

Trouble Is Afoot

Josh's favorite times of the year were summer and fall. Not because of summer's humid heat or autumn's blazing color, but for the sports! Josh was a three-sport athlete. He seamlessly transitioned from summer swim league through baseball and into football. Josh was not only a versatile athlete, but he also truly excelled in each sport. This was an especially impressive accomplishment for a boy diagnosed with **Type I diabetes** at age seven. Now, a 17-year-old high school senior, his days as a star athlete were already over…Dr. Larkin had just removed most of his left heel and sent it to the microbiology laboratory for culture. Josh thought back over the last several years and was suddenly even more depressed, as he realized this was his own fault.

Ten years ago, within a month of diagnosis with diabetes, Josh's blood sugar levels were well-regulated. He and his family worked hard to learn a new lifestyle of counting carbohydrates and adjusting insulin levels. By the time Josh was 12, Dr. Larkin felt Josh was mature enough to use a pump. Josh was thrilled! With a pump it was so much easier to regulate his disease, especially with his variable exertion levels during sports seasons. For the first year, Josh "played by the rules." But as he got older and his parents let him assume more responsibility for the management of his diabetes, Josh began taking advantage of this newfound freedom. When the gang went out for treats after the game, Josh indulged and simply used his pump to accommodate the added sugar load.

And that was just the beginning. Josh knew diabetics had to take special care of their teeth, eyes, and extremities, but he continued his careless ways. To avoid teasing by his teammates, Josh quit wearing his water shoes at swim practice and between events. The abrasive cement of the pool deck daily scraped his soles. His heel became tender, swollen, and red, but Josh ignored this warning. By mid-July, as swim season concluded, Josh started limping slightly and told his inquiring parents it was just a pulled muscle. But as football started in August, Josh couldn't ease his infected foot into a cleat and was forced to admit his problem and accept help. Under local anesthesia, Dr. Larkin lanced Josh's left heel and a foul-smelling, cottage cheese-like substance burst out. A sample of the purulent mass was collected for the lab before Dr. Larkin began **debriding** the foot. Josh knew he was in trouble. Dr. Larkin didn't say a word. He just kept working. Numerous other samples were collected for the microbiology laboratory. Josh was given IV clindamycin during the procedure and prescribed Cefoxitin and ciprofloxacin to continue at home. Direct examination of the specimens consistently revealed Gram-positive cocci in clusters, Gram-negative rods, and, in the deeper tissue samples, small Gram-negative coccobacilli were abundant.

1. What types of media were used to inoculate Josh's specimens? Why?

2. Why were three different microbial species observed? Did the physician use poor technique during specimen collection? Explain. Why do you think the one species was evident only in the deeper tissue samples?

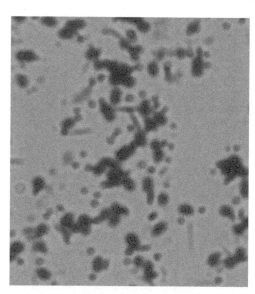

Figure 1. Gram stain of mixed bacterial infection.

Within 36 hours, the microbiology laboratory reported the growth of *Staphylococcus epidermidis*, *Pseudomonas aeruginosa*, and *Bacteroides fragilis*. Based on the sensitivity analysis provided, Josh's antibiotic therapy was modified to outpatient IV vancomycin treatment plus erythromycin and metronidazole.

3. Characterize each of the three microbes in Josh's infection. Be sure to consider virulence factors. For each microbe, indicate the likely route of transmission.

Josh continued antibiotic therapy for two weeks, but when he saw Dr. Larkin at the wound clinic, the concerned doctor shook his head and frowned. The doctor repeated the debridement process, removing deeper tissues. The lab reported the presence of *Bacteroides fragilis* in these samples. Despite an additional round of metronidazole treatment, Josh's *Bacteroides* infection persisted. Finally, there was no choice but to amputate Josh's heel. Although Josh was devastated, the lab confirmed *Bacteroides fragilis* in the tarsal sample. The amputation was Josh's only option.

4. What term is used to designate an infection in the bone? How are such infections typically initiated?

5. Why is infection of the extremities more of a problem for diabetic patients? As a nurse providing diabetic education to a newly diagnosed patient, what foot care recommendations would you make and why?

Down on the Farm

Scott was dog tired as he pulled his old tractor into the barn. He was only 36 years old, but after plowing and planting fields for six straight days, he felt more like 100. It had been a typical spring in northwest Ohio, with the uncooperative weather that gave gray hair to even young farmers such as Scott. After the rain finally stopped and the fields drained, Scott and his two brothers worked from dawn to dusk for almost a week to get the corn and soybeans planted on time. It was hard, stressful work running the large farm that had been in their family for three generations, but Scott still found it strangely satisfying.

Knowing his wife Sarah would have a hearty, hot dish waiting for his lunch, Scott hurried through his clean-up chores in the barn. As he stowed the large equipment, Scott noticed one of the outer blades on his disc plow was bent. Squatting down with a mallet, he had almost straightened the damaged disc when he accidentally swung the back of his hand into the adjacent disc. Scott dropped the mallet, gripped his injured hand in his shirt tails to slow the blood flow, and loudly cursed a "blue streak," which instantly brought Justin and Jacob, his eight-year-old twins, charging into the barn.

"Daddy, you hafta put a dollar in the 'swear jar,'" Jacob announced.

"Oh cool, is that blood?" Justin asked.

Before Scott could respond, Jacob was off like a shot to tell Sarah about the accident. "Great," he thought, "now Sarah will be fussing at me for the rest of the day. All I want to do is relax and eat a hot meal."

Scott was almost to the house when a concerned Sarah burst out the back door carrying their first aid kit. As Sarah and Scott met each other in the yard so she could begin the wound inspection process, Jacob arrived and held up Sarah's cell phone. "Here Mom, you want me to call 9-1-1 for ya?"

Rolling his eyes and grinning, Scott roared with laughter. As Sarah's examination revealed a two-inch laceration, rather than the severed fingers she had expected from Jacob's report, the whole family was soon chuckling as they walked back to the house.

Scott refused to let his wife "baby" him and insisted on caring for his own wound. He was already embarrassed by his carelessness and knew his pride couldn't take the lecture that would surely come with Sarah's medical care. Using warm water and antibacterial soap, Scott gently washed the injury and dried the area with a hand towel he found on the side of the sink. He used three **steri-strips** to close his laceration and added a covering of sterile gauze for good measure. Next, Scott headed to the kitchen and wolfed down two heaping plates of lasagna and a handful of homemade peanut butter cookies before heeding Sarah's advice to relax in his recliner.

Steri-strip – Linear piece of surgical adhesive used to hold the edges of a wound/incision together during healing; helps keep keloid formation low.

1. Rate Scott's wound care protocol. Is there anything you might suggest to improve his technique?

Scott enjoyed a brief nap in the recliner before heading back to work. The boys were extra helpful that afternoon as they tended the livestock together. "No doubt Sarah put them on their best behavior," Scott thought. Since his wound was becoming more painful as the day progressed, Scott actually appreciated their assistance. After dinner he tried a game of catch with the boys, but he just couldn't manage it with his sore hand. He watched the boys play from the comfort of his porch swing, but dozed off after the first few pitches. Scott was stunned when Sarah woke him up two hours later so he could tuck the twins into bed. "Planting must have taken more out of me than I thought," Scott told his wife. "I'm really whipped. Hon, would you please brew me some tea while I put the boys to bed? My stomach is starting to roll a bit…maybe four pork chops at supper was over-doing it," Scott said sheepishly.

Early the next morning Scott and Sarah went to the dairy barn to milk while the boys slept. Sarah noticed Scott favoring his injured hand but didn't worry until she saw him bolt from the barn. Sarah finished up with the dairy cows and followed Scott back to the house where he was just coming out of the bathroom. Scott was flushed with beads of perspiration on his brow and weaving as he walked. Sarah grabbed his elbow and steered him safely to the couch before he fell. "Sweetie, you look awful! What's the matter?" Sarah inquired, putting her hand on Scott's forehead and noting a fever.

"I told you four pork chops were too many," Scott groaned. "Man do I have a bad case of the scoots…and my hand hurts like a son-of-a-gun today." Sarah gave Scott some kaopectate for his diarrhea and 400 mg of ibuprofen to reduce his fever. "Maybe this will help your hand feel better too," Sarah said hopefully. "I would think it should be healing some by today."

Since it was a Sunday, Scott went back to bed while Sarah took the twins to Sunday school and services. Returning home at lunchtime, she found her husband asleep on the coach with an ice pack on his affected hand. "I'm so glad you're back," Scott said, waking up with a start. "You've got to take me to see the doctor. My hand is killing me. I'm feverish, shaky, and phenomenally thirsty." Dropping the twins off at their uncle's house, Sarah and Scott proceeded to the emergency room of County Memorial Hospital. "My cut must be infected," Scott told the physician who was painfully prodding his hand. "I can't believe how much this hurts."

"Me either," said Dr. Graaf. "Your laceration really doesn't appear to be very large. Although I would have preferred it if you had this cleaned, stitched, and dressed by a professional, you did a good job closing the **lac** with the steri-strips… and I don't see any obvious signs of infection." Giving Scott a playful punch on the shoulder, Dr. Graaf said, "Suck it up, man. You'll live!" Scott was released with a prescription for Tylenol with codeine to ease the pain in his hand, antibi-otic ointment for his wound, and orders to eat a bland diet and take Imodium to recover from his **GI** virus.

Lac – Abbreviation for **lac**eration.

GI – Abbreviation for **g**astro**i**ntestinal.

Despite the new medication and an ice pack, Scott was miserable all day. The next morning he was so shaky that it was an effort walking to the kitchen for his breakfast. Sarah watched with worry as her weak, woozy husband spilled his coffee reaching for his pain relievers. "Are those helping yet?" Sarah asked while preparing a fresh ice pack. Scott shook his head and winced. "Dr. Parker's office opens at 9 AM. Will you please take me? I don't think I can drive. The whole lower half of my arm is killing me. I've never felt this bad in my whole life."

The nurse triaged Scott and recorded his blood pressure, pulse, and temperature at 80/52, 110, and 39.2°C (102.5°F), respectively. When Dr. Parker removed the dressing from Scott's hand, he noted significant swelling and the start of a purplish rash. "Scott, I'm concerned about this laceration. Although it is showing some early signs of infection, it's not bad enough to be causing you such intense pain."

"Are you calling me a wimp, doc?" Scott asked.

"No, not at all. When pain is disproportionate to the apparent cause, it means something else is going on. I'm not sure what that is, but your vital signs suggest the early stages of a systemic infection. I need to run some tests just to be on the safe side. My nurse will be right in to draw some blood and culture your wound," said Dr. Parker.

Blood was drawn for a CBC. The nurse thoroughly cleansed Scott's wounded hand with iodine before telling him, "This is going to hurt. In order for me to get a good sample for the lab, I will need to swab deep down into your wound. Are you ready?" she asked.

2. What is a CBC? What useful information is Dr. Parker hoping to obtain from this test?

3. Outline the correct method for obtaining a high-quality specimen for culture in the clinical microbiology laboratory.

Scott nodded and looked away as the nurse gently opened the laceration and prepared to obtain a sample. Despite her tender touch, Scott cringed and stifled a yell. Almost immediately Scott and the nurse caught a whiff of a putrid odor, and pus poured out of the site. "Sit tight," the nurse comforted, "I'm going to get Dr. Parker right now."

Within a minute the physician returned to Scott's examination room and scowled as he inspected the opened wound. "Scott, I think we now know why you're in such pain. On the surface your infection doesn't appear too bad, but it seems that it is primarily affecting your deeper, soft tissues where the infection is actually quite advanced. Sarah, go immediately to County Memorial Hospital's main entrance. Someone will meet you at the door with a wheelchair. I'm going to call ahead and have Scott admitted directly. I also will be consulting with Dr. Carter, who is an infectious disease specialist. We will let them collect the specimen for the microbiology lab so we can use local anesthesia and make you more comfortable during the process. After that, you will need aggressive antibiotic therapy and possibly surgery."

4. What differential diagnoses can you offer based upon Scott's symptoms?

Although Sarah was stunned by the news, Scott was so ill that he readily agreed to Dr. Parker's orders. Thirty minutes later, Scott was settled in his hospital room. Michael, a physician assistant, expertly numbed Scott's hand and appropriately collected a tissue specimen for the microbiology laboratory. In his notes, Michael documented that during the collection process the subcutaneous tissues were readily loosened from a yellowish-green, necrotic underlying fascia. Michael also noted that the erythema had advanced to 5 cm above the wrist.

5. Based upon Michael's evaluation of Scott's wound, identify this infectious process. Do all of the symptoms fit? Why is this condition so often misdiagnosed? What symptoms will Scott likely manifest next?

6. What is/are the causative agent(s) of this condition? Briefly characterize this/these pathogen(s)?

IV saline was started at a high flow rate in his unaffected arm, and penicillin G plus metronidazole were piggy-backed for empiric therapy. Blood cultures were collected and several more tubes of blood were drawn. Scott received medication to manage his pain. Thinking he would be allowed to rest and recover after this "full court press," Scott was surprised to find himself transported to radiology for a CT scan and MRI.

7. Why do you think Scott's physician ordered a fairly rapid delivery of saline?

8. Why were penicillin and metronidazole selected for empiric therapy?

9. What types of clinical tests do you suspect may be run on these additional blood samples?

10. Why were blood cultures drawn?

11. How can a CT scan and/or MRI be helpful diagnostic tools for an infection?

An hour later, Scott was awakened as Dr. Carter, the infectious disease specialist, and Dr. Hower, a general surgeon, both arrived in his room looking grim. "Scott, your white blood cell count is significantly elevated, suggesting an advanced state of infection. Other blood test results indicate serious tissue damage. Your CT shows pockets of gas within the fascial planes of your tissues, while your MRI points to extensive soft tissue damage and the need for immediate surgery to debride your wound," said Dr. Carter.

Scott responded with a confused stare. "Basically this means I will be operating on you as soon as an OR is available...probably in the next couple of hours. I will need to remove all of the skin, subcutaneous tissue, and fascia destroyed by this infection. The sooner I perform this procedure the better your chances of recovery. We must stay ahead of this infection," Dr. Hower added.

"But if Scott has an infection, why does he need an operation? Won't the antibiotics kill all of the germs?" Sarah asked.

"Antibiotics are powerful weapons in our arsenal for fighting infection. However, Scott is being attacked so viciously that our conventional weapons are not enough to win the battle. It's time to bring in the 'big guns' so we can wipe out this enemy," answered Dr. Carter. While still fielding questions from Scott and Sarah, Dr. Hower's pager beeped. "It's time!" he said optimistically. "Scott, I'll send a nurse in who will 'prep' you for surgery. I'll meet you upstairs in OR4 in about 30 minutes."

12. How did gas pockets develop deep within Scott's soft tissues? What causative agents of this disorder regularly release gases? Under what environmental conditions would you expect this to occur? What is the correlation between these conditions and the effectiveness of innate immune responses?

Hours later Sarah met with Dr. Hower for a postoperative conference. "Scott is in pretty bad shape. The infection has done tremendous damage to his hand and arm. I thoroughly inspected the entire infected area and I have removed a considerable amount of affected tissue. I biopsied several areas at the periphery of this spreading infection and also some of the deeper tissues, so we should have a complete report from the microbiology department by tomorrow. Although I'm confident that all of the infected tissues have been removed, this condition is especially challenging to treat. There is the possibility that I will need to repeat and expand this procedure if any new necrotic tissue develops. Consequently, I've delayed closure of the surgical site until I know for certain that Scott is recovering. This is standard protocol, but Sarah, it's a gruesome sight. I want you to be prepared before you go see your husband. Scott will eventually require skin grafts and likely still have considerable scarring."

Sarah put on a brave face and went to Scott's bedside. She soon realized that she needn't pretend to be cheerful and strong. Due to the intense pain from the surgery, Scott received powerful medication and rarely woke. Sarah was appalled at the wide, gaping incision running from the base of Scott's middle finger across his palm and halfway up his forearm. The grotesque, bloody appendage looked like something out of a bad horror movie.

Spending the night in a chair by the side of Scott's bed, Sarah awoke to Dr. Hower gently shaking her shoulder. "Sarah, the preliminary report from the microbiology laboratory is back. It appears that Scott's infection is probably **polymicrobial**, but they have already isolated and identified *Streptococcus pyogenes* from his wound. The antibiotics we are already giving your husband are the drugs of choice for his situation, but I want to add one more treatment to Scott's therapy. In just a few minutes we will be moving Scott to our **hyperbaric chamber**, since this should have several positive effects.

13. What specific components released by *Streptococcus pyogenes* can cause the extensive tissue damage Scott is experiencing? What is the "sensational" name given to the condition being caused by this pathogen? What other terms are applied to this infection?

14. What are the positive effects of the hyperbaric chamber that Dr. Hower referred to? Go online and research a typical hyperbaric treatment for this condition.

Polymicrobial – An infection resulting from the growth of 2+ different species of microorganisms.

Hyperbaric chamber – A device in which a patient may be exposed to a high oxygen concentration under increased pressure for therapeutic purposes.

Amputate – The surgical removal of a damaged or infected limb.

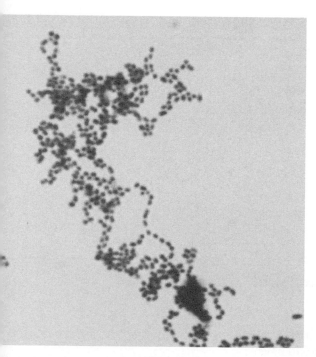

Figure 1. Gram stain of *Streptococcus pyogenes*.

When Dr. Hower examined Scott again in the afternoon, he was alarmed to find the infection continuing to spread. An hour later Scott returned to the operating suite for another round of surgical debridement. As soon as Scott had stabilized postoperatively, he was sent for another hyperbaric treatment. Later that evening Dr. Hower talked privately with Sarah. "Scott is not responding as well as I had hoped. If the infection continues to spread, I will be forced to **amputate** his arm."

Shocked, Sarah excitedly explained that amputation was impossible since Scott wouldn't be able to farm with only one arm. It didn't matter if his hand and arm were disfigured as long as they were functional. "You don't understand, Sarah," Dr. Hower replied calmly. "What's left of his arm now will probably be only minimally functional. If I can't contain the infection within his arm, this tissue necrosis will spread to the rest of his body and kill him within another couple of days." Too stunned to speak, Sarah nodded as the tears began to flow.

In the morning, both Dr. Hower and Dr. Carter came to see Sarah. "The rest of the microbiology report is available and it appears Scott's infection is indeed polymicrobial. The lab also reports the presence of *Bacteroides fragilis*. The good news is that the current antibiotic, surgical, and hyperbaric treatments are also appropriate for fighting this pathogen. The bad news is that Scott's infection is still spreading," explained Dr. Carter.

"I was afraid of that," said Sarah. "Scott seems even more miserable. He's drenched in perspiration and when I try to talk with him, he doesn't make any sense. I can see the swelling and skin darkening moving up his arm."

"I'm going to remove Scott's arm at the shoulder now," Dr. Hower told Sarah. "The tissue at the joint doesn't appear to be compromised yet, so I'm hopeful this will stop the spread and save Scott's life. We will know one way or the other in about 24 hours."

15. Consider the two microbes responsible for Scott's dire infection. Where are they normally found, and how might they have contaminated Scott's initial wound, leading to this life-threatening condition?

After the amputation of his left arm, Scott was transferred to **ICU** and placed on a ventilator to assist his labored breathing. During the night, his temperature soared to 39.4°C (102.9°F), and the laboratory called the unit to report two of Scott's blood cultures had grown *Streptococcus pyogenes*. Additional blood work in the morning indicated the early stages of kidney failure. By noon, Scott lapsed into a coma, and he died shortly before midnight. An autopsy revealed the cause of death to be septic shock with cardiovascular collapse. Although Dr. Hower thought he amputated Scott's arm before the infection spread beyond the shoulder, he was mistaken. There was extensive tissue damage in the thoracic cavity, and all sites sampled and cultured grew *Streptococcus pyogenes*.

ICU – Abbreviation for **i**ntensive **c**are **u**nit, a facility designed to provide high-level medical care to critically ill patients.

16. What are the most common complications of necrotizing fasciitis?

17. What is the mortality rate of this infection? Are certain individuals at an increased risk of acquiring this infection?

18. Scott's physician's appeared to be treating his infection correctly with the appropriate antibiotics, multiple surgical debridements, and hyperbaric therapy. Why did he die?

My, What Pretty Eyes You Have!

Overall, Jane enjoyed her new situation. High school was over, and she was training for a career she would enjoy. She loved the new freedom of being on her own and living in the dormitory. The new job training facility had three dorms housing about 500 young adults, all getting the hands-on training and experience they needed to get a good start on a career. Jane had made some good friends and was often greeted with enthusiastic hugs at lunch time before sitting down to discuss the day's activities with her friends. She shared clothes, makeup, and cigarettes with her roommate. On weekends, they enjoyed pool parties together. In the evenings, they often went to the computer lab to meet new friends and catch up with old friends on Facebook.

Although most things were going well, today was not a good day for Jane. She and her roommate were visiting the health clinic for "pinkeye."

They both had red eyes that were itchy (**pruritus**) with a clear, watery discharge. Jane's eyes were also extremely sensitive to light (**photophobia**), and her eyes felt swollen (**preocular adenopathy**).

Pruritus – An itchy sensation that produces the desire to scratch.

Photophobia – Excessive sensitivity to light.

Preocular adenopathy – Swollen eyelids.

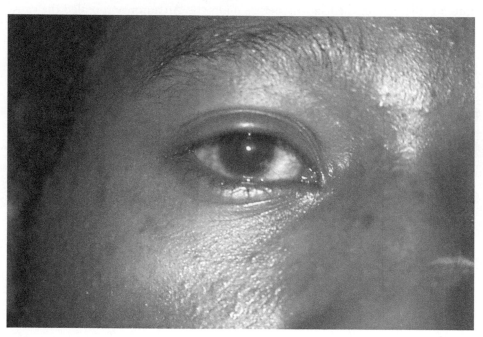

Figure 1. Pinkeye.

1. What is the medical term for pinkeye?

2. What are the most likely causative agents of this disorder?

The health center physician was not surprised by their condition. He said theirs was about the 40th case he had seen in the last month. During a typical month he only sees about four cases.

3. What term describes the situation when new cases of a disease significantly exceed what is expected based on recent experience?

4. What factors placed Jane and her friends at risk for acquiring this disease?

Early in the outbreak, the physician had ordered laboratory tests and ruled out bacterial causes of the pinkeye. Based on the clinical features and the rapid spread, he expected adenovirus as the cause of the outbreak.

5. Describe the physical features of adenovirus.

6. Why is this pathogen highly contagious?

7. What other diseases can be caused by this virus?

The physician recommended that Jane and her roommate use artificial tears every two hours to help relieve their **keratitis** and Jane's photophobia. He also told them to use cold compresses to improve the swelling and discomfort of the lids, and he prescribed drops containing ciprofloxacin.

8. If this causative agent of their conjunctivitis was a virus, why were ciprofloxacin drops prescribed?

9. How does ciprofloxacin inhibit the growth of bacteria?

10. What is the spectrum of activity for ciprofloxacin?

To help stop the outbreak, the health center physician gave them each some hand sanitizer to carry in their purses. He instructed them to use it after each class, after leaving the restroom, and after using a computer. In addition, he told them to throw out their makeup and not to share clothes or makeup with their friends. Finally, he informed them that he had sent a memo to those in charge of the pool to ensure that the chlorine levels were meeting standards.

11. Explain how each activity listed above is necessary to stop the outbreak of adenoviral conjunctivitis.

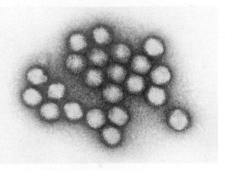

Figure 2. Adenovirus.

Keratitis – A painful inflammation of the cornea of the eye.

He Did It His Way

Bud had a long history of doing things his own way. He had always been a firm believer that rules were made for the other guy. As a young man, the consequences for violating the rules weren't too severe: a five-day suspension for setting the high school science lab on fire, a weekend in JDC for shoplifting a pack of cigarettes, and of course, the unrelenting "I told you so" from his parents.

As a 50-year-old man, Bud was not about to change his ways now, and that was especially true regarding his doctor's rules. Bud carried 260 lb on his 5'8" frame. His favorite pastime was football and brews at the pub with his buddies. He was a meat and potatoes man whose idea of exercise was stretching to reach the bag of Doritos on the top shelf. Bud's lifestyle resulted in his diagnosis of **Type II diabetes** eight years ago. Most days he remembered to take his "sugar pills" since that wasn't much of an inconvenience. But Dr. Dvorak's orders about counting carbohydrate grams and calories were out of the question! Bud felt fine (those pills must be working). Why should he give up donuts, ribs, and his wife's famous buttermilk biscuits? "You only live once and you ought to enjoy it," was Bud's motto. In fact, he was convinced that doctors weren't happy unless they were scaring their patients with horror stories about heart attack and stroke.

But here Bud was in the CCU of St. Vincent's Medical Center recovering from the **quadruple bypass surgery** Dr. Dvorak performed yesterday. Bud couldn't believe it…and he couldn't wait to get home and away from all of the new medical rules. The only thing worse than the hospital rules was the hospital food. Everything was grilled or steamed, and there was never enough. Desperate to get home to his old life, Bud feigned sincere compliance with every medical instruction. He diligently performed his rehab exercises for the next five days and promised Dr. Dvorak he was turning over a new leaf and focusing on the control of his diabetes. He convincingly agreed to all of the discharge orders the nurse reviewed with him: continue rehab exercises, walk twice a day, take all medications on time, follow the carbohydrate-restricted diet, and work with the home health nurses on wound care.

On his first day home, Bud reminded his wife, Jenny, how lucky she was that he was still around. Voila! Jenny made Bud his favorite chicken fried steak, biscuits with gravy, and sugar cream pie for dinner. When Nancy from home health arrived the next morning, she scolded Bud. He hadn't taken a walk or followed his diet and as a result had a blood glucose level of 352 mg/dl. To make matters worse, Nancy discovered that Bud had disregarded his discharge orders and taken a shower as soon as he got home. She tried to explain the risk of infection, but Bud knew he had nothing to worry about since soap and water washed away germs.

Bud's demeanor was very different two days later when Nancy returned to change his dressings. His chest incision had become quite tender. Nancy inspected the site and was concerned about the pronounced inflammation. After cleansing the site and applying a sterile dressing, she promised to report her findings to Dr. Dvorak immediately.

1. It appears Bud has developed an infection in his chest incision. What are the likely sources of microbes?

2. How could Bud's elevated blood glucose levels complicate this infection?

3. To counter this problem, what policy is typically implemented for patients during and immediately after open heart surgery?

Type II diabetes – A metabolic disorder affecting primarily middle-aged, overweight individuals. Although their pancreas still makes insulin, these patients are resistant to the hormone, resulting in elevated blood glucose levels. This disorder can lead to cardiovascular damage, predisposing patients to heart attack, stroke, and kidney disease.

Quadruple bypass surgery – A surgical procedure in which healthy blood vessels from the thigh are transferred to the heart to replace dangerously occluded coronary vessels. Also known as open heart surgery.

Nancy returned early the next morning with a prescription of Cipro for Bud and orders to culture his incision. Bud was a little nervous as he watched Nancy examine his incision and obtain the specimen. Her facial expression registered concern as she told Bud his infection had spread. The entire incision was now hot, red, and seeping a small amount of pus. Nancy smiled at Bud and told him at least his wound wasn't putrid. "In fact," she said, "you actually have a pleasant, slightly fruity aroma." Bud returned her smile…how bad could a fruity smell be?

4. What is the formal name of Cipro? How does this drug work?

5. Why did Dr. Dvorak initiate therapy with this antibiotic prior to receiving the culture results?

6. The fruity odor from Bud's incision is significant. What microbe is suggested as the causative agent of this infection?

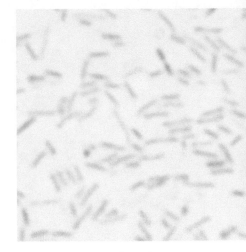

Figure 1. Gram stain of *Pseudomonas aeruginosa.*

When Nancy returned 48 hours later and examined Bud, she immediately placed a call to Dr. Dvorak. The wound was considerably worse. Bud had faithfully taken his antibiotic but confessed to continuing his indulgence in his wife's fine cooking.

Dr. Dvorak ordered Bud to return to the hospital. The lab results he had just received indicated infection with *Pseudomonas aeruginosa*. Since Bud was not responding to the therapy, it was time to implement more aggressive measures. Dr. Dvorak told Bud to come immediately, saying "This is a bad bug to beat."

7. Characterize the morphology and Gram staining of *Pseudomonas aeruginosa*.

8. What was the most probable source of Bud's *P. aeruginosa* infection?

9. The laboratory report indicated that Bud's infection was Cipro-sensitive. Why wasn't he improving?

10. What feature(s) of *Pseudomonas aeruginosa* make it a "bad bug to beat?"

Upon admission to ICU, Bud received a combination antibiotic therapy and was re-cultured at three successively deeper sites within his incision. To Dr. Dvorak's dismay, even the deepest tissues were contaminated, and Bud had spiked a fever of 38.7°C (101.6°F). Concerned about the spread of *Pseudomonas* infection, Dr. Dvorak ordered blood cultures and a biopsy of Bud's sternum.

11. What medical term describes a bone infection? An infection in the bloodstream?

When the culture from the sternal biopsy was reported positive for *Pseudomonas* the next day, Bud underwent surgery to remove the bone. All infected tissues in Bud's chest were successfully removed. Intensive IV antibiotic therapy was continued to target the systemic infection indicated by his blood cultures. The morning after this surgery, repeat blood cultures were free from infection yet Bud's temperature remained high and signs of vascular collapse appeared. As pulmonary capillaries were affected, the subsequent edema resulted in Acute Respiratory Distress Syndrome (ARDS). Although the systemic infection had resolved, Bud declined rapidly and expired 16 hours later.

12. Why did Bud worsen and die despite a successful antibiotic therapy?

The "Tat" Is Where It's At!

Megan was completing her senior year of college and preparing for graduate studies in a physician assistant program. She had always been pretty conservative and wanted to finish college with a bang! Megan wanted to make a statement...do something out of character...shake things up a bit. With considerable input from her sorority sisters, she finally decided on a tattoo. With her friends along for moral support, Megan went to the Urban Ink Salon. After looking at hundreds of designs, she settled on a single long-stemmed red rose. Megan was satisfied that this tattoo would be both dramatic and classy. The tattoo artist recommended she place her body art on the inner portion of her lower right arm so it would be prominently displayed with every handshake. Later that afternoon, Megan sported a six-inch red rose tattoo on her tender forearm. Still sore the next day, Megan nevertheless proudly showed off her tattoo.

By that evening, Megan took two Tylenol and used cool compresses for relief from the discomfort of tattooing. She noted some **edema**, heat, and **erythema**, but convinced herself this should be expected given the nature of tattoo application. Two days later, Megan went to the university student health center. Her entire lower arm from elbow to fingertips was swollen, hot, itchy, red, and most uncomfortable. The staff doctor thought Megan was having an allergic reaction to one or more of the dyes used in her tattoo. She was instructed to take Benadryl for 24 hours and to return if the condition hadn't improved.

Edema – Swelling.

Erythema – Redness.

● **1.** Are Megan's symptoms consistent with an allergic reaction? Explain.

●● **2.** How does Benadryl alleviate an allergic response?

When Megan returned, her condition had worsened despite the Benadryl. In fact, she had also started to develop scaly patches in a centripetal pattern with a central clearing right over the petals of her tattoo. The university transported Megan to the emergency room of a local hospital, where the attending physician prescribed anti-inflammatory drugs, a different antihistamine, and a broad-spectrum antibiotic. Additionally, he collected skin scrapings from various parts of Megan's forearm for microscopic examination and culture.

●●● **3.** Based upon the treatment initiated, what concerns Megan's doctor?

Upon learning Megan received her tattoo at the Urban Ink Salon, the physician contacted the county health department to discuss her case. The next day, two representatives from the health department met with the salon owner to discuss his equipment-sanitizing procedures. Although indignant, the owner cooperated fully with the health department investigation. He verified consistent disinfection of all equipment following each customer; claimed to have never had a customer complain of infection; and agreed to permit the health department workers to swab his equipment and work area.

After 48 hours, Megan's cultures grew only normal skin flora. Direct microscopic examination revealed Gram-positive cocci in clusters and thin-walled fungal hyphae. The antibiotics appeared ineffective as she continued to experience tremendous inflammation. Megan's skin began cracking in several sites, and topical antibiotic ointments were applied to prevent secondary infection.

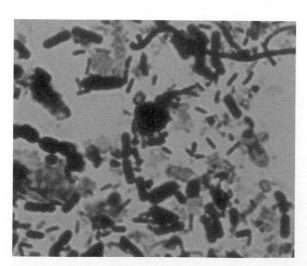

Figure 1. Gram stain of skin cultures.

4. What genus of bacteria was cultured from Megan's skin? Are you surprised? Explain. Could this be causing her problem?

5. Fungal hyphae were also noted upon direct examination. Is this unusual? State two common fungal species native to your skin.

A day later, all of the cultures from the Urban Ink Salon swabs were negative for growth. At the physician's request, the health department investigators returned to the salon and cultured the different inks used in generating Megan's tattoo. At this time, one investigator noticed a mouse trap and asked the owner about it. Since the salon was located on the first floor of a downtown commercial building that accommodated a neighboring restaurant and multiple upstairs apartments, the owner blamed his minor mouse infestation on his neighbors. He was instructed to call the health department immediately upon catching a mouse.

One week later, the lab began reporting some interesting results. Megan's scrapings inoculated on Sabouraud dextrose agar grew numerous pure colonies of *Trichophyton mentagrophytes*. Also of interest were the *Trichophyton mentogrophytes* arthrospores isolated from the red ink sample.

6. Describe the features of this microbe. What stain(s) is/are frequently used to examine this microbe? How can arthrospores live in ink?

7. Why was Megan demonstrating hypersensitivity symptoms if she had a fungal infection? Megan's physician prescribed oral Griseofulvin to combat her infection. How does this drug work?

After a month of treatment, Megan despaired as she still suffered from inflammation and had struggled against several secondary bacterial infections. Finally, she was scheduled for surgical excision of the affected site. Although her surgery was a success, Megan's scarred forearm is a permanent reminder of her ordeal.

Eventually, hairs from a mouse trapped at Urban Ink were cultured and grew *Trichophyton mentagrophytes*.

8. Explain the "**epi-link**" of this finding.

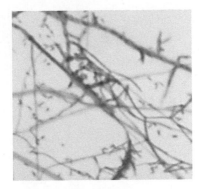

Figure 2. Light micrograph of *Trichophyton mentagrophytes*.

Epi-link – **Epi**demiology studies of the distribution and spread of disease in populations.

A Persistent Pimple

It was so exciting! Caitlyn was the only freshman girl selected for Varsity Singers, her high school's touring show choir. Their summer "retreat" was a six-day mega-rehearsal to learn all of the choreography for their upcoming show season. Monday through Saturday the week before school resumed, the 28 performers danced from 8 AM to 8 PM in their un-air-conditioned gymnasium. Caitlyn didn't particularly mind the hot, humid rehearsal conditions, but sweating profusely in dance leotards every day was really starting to aggravate the acne on her shoulders and back.

After a special preliminary performance for their families on Saturday night, Caitlyn showered and dressed to go home. It was then she discovered a very large, angry "pimple" that rubbed uncomfortably on the back waistband of her jeans. By morning, it was raised and the size of a dime. Caitlyn's mother washed the affected area, cleansed it with hydrogen peroxide, and applied an antibiotic ointment, telling her they would call the doctor tomorrow if it didn't improve.

1. What possible infections might Caitlyn have?

2. What microbes would normally cause these infections? Are these microorganisms normal skin flora, pathogens, or both? Explain.

Monday morning, the first day of school, Caitlyn's back was sore. "A great way to start high school," she thought. Caitlyn's mother took her to the pediatrician's office right after school. The PA examined her back and was alarmed to see a lesion almost two inches in diameter. It was tender to the touch with poorly demarcated margins. The region was raised, warm, and **erythematous** with several smaller red lines radiating outward.

Erythematous – Reddened.

3. What is your diagnosis? Describe the nature of this condition.

After consulting with the pediatrician, Keflex was prescribed for Caitlyn. She was sent home with instructions to monitor the infection. If it was not obviously improved by the next day, she was to return for reevaluation.

4. To what class of antibiotics does Keflex belong? How does this drug work? What group of microbes is especially susceptible to it?

On Tuesday morning, Caitlyn went immediately to see her pediatrician. The lesion was the size of an egg and quite sore. Caitlyn also presented with a temperature of 38.4°C (101.2°F). Motrin and compresses were advised as comfort measures. The Keflex was continued and the lesion cultured for laboratory analysis. Again, she was told to return if she didn't notice improvement.

5. How would you collect a specimen from Caitlyn's lesion? Name several types of transport media commonly used. Why is it so important to appropriately transport a specimen to the microbiology laboratory?

6. What media will likely be inoculated when this sample arrives in the laboratory? State your reason(s) for choosing the media you've indicated.

7. In addition to media inoculation, what other procedure will be performed immediately using the specimen?

The preliminary Gram stain of the specimen showed many Gram-positive cocci in clusters. After 24 hours, the **TSA with 5% sheep blood plate** demonstrated pure growth of small, round, smooth, white, **gamma-hemolytic colonies**. The same colony morphology was observed on the **PEA** (or **CNA**) plate with zero growth on the **EMB** (or **MacConkey**) plate. Colonies were also observed on the **MSA** plate, which was completely pink in color.

8. What is meant by the term "pure growth"? What does it say regarding the quality of your specimen collection?

9. Based upon these laboratory results, what microbe do you predict is causing Caitlyn's infection? Explain. What two chemical tests would you perform next to verify your answer?

Colonies from the TSA plate were suspended in sterile saline and introduced into the Vitek II analyzer. It confirmed *Staphylococcus epidermidis* was the pathogen involved and indicated Keflex sensitivity.

10. Now that you know the pathogen, indicate the expected results for the following:

    Urease test
    Acetoin production
    Alkaline phosphatase
    Beta-lactamase
    Fermentation:
        Glucose
        Sucrose
        Mannitol
    Antibiotic susceptibility:
        Novobiocin
        Polymyxin

TSA with 5% sheep blood plate (blood agar plate/BAP) – **T**rypticase **s**oy **a**gar enhanced with 5% sheep red blood cells provides a rich-nutrient medium for the culture of many common microbes. This medium often serves as a growth control.

Gamma-hemolytic colonies – Nonhemolytic colonies.

PEA – **P**henyl**e**thanolamine **a**gar is a medium selective for the growth of Gram-positive microbes.

CNA – **C**olistin **n**aldixic **a**cid **a**gar is a medium selective for the growth of Gram-positive microbes.

EMB – **E**osin **m**ethylene **b**lue agar is a medium selective for the growth of Gram-negative microbes and differential for lactose fermenters.

MacConkey – A medium selective for the growth of Gram-negative microbes and differential for lactose fermenters.

MSA – **M**annitol **s**alt **a**gar is a medium selective for halophiles and differential for mannitol fermenters.

Figure 1. Gamma hemolytic colonies on blood agar plates.

On Wednesday morning, Caitlyn's condition remained the same. However, that afternoon she reported to the school nurse for assistance. Following a sharp pinching sensation, Caitlyn had felt a warm rush of fluid on her back. The nurse confirmed that the lesion had drained and helped Caitlyn clean up the foul-smelling pus that continued to ooze out. By the next morning, Caitlyn's temperature was normal, the red lines were gone, and the lesion was substantially reduced in size and color.

11. Caitlyn started taking Keflex on Monday yet did not experience relief until Thursday. The microbe was sensitive to the antibiotic. Why did it take so long for a positive effect?

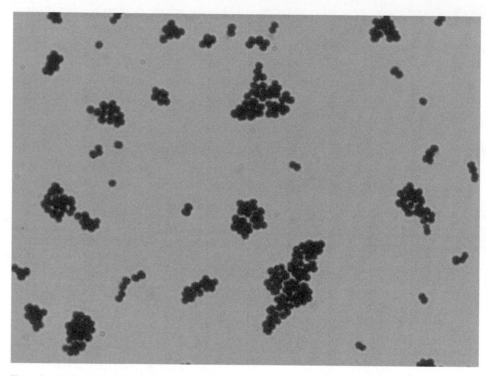

Figure 2. Gram stain of *Staphylococcus epidermidis*.

B. Nervous System Infections

Your nervous system is truly remarkable, as it simultaneously controls every bodily function via the integrated activity of over a billion neurons. This phenomenally complex system is anatomically divided into two parts: the central nervous system and the peripheral nervous system. The central nervous system (CNS) is comprised of the brain and spinal cord, which are responsible for inputting and interpreting stimuli and then sending instructions in the form of an impulse to organs throughout the body. Your peripheral nervous system (PNS) consists of 12 pairs of cranial nerves plus 31 pairs of spinal nerves and serves as the conduit for carrying these messages to and from the CNS.

Such a complex and vital system could be seriously impaired by infection and must be protected. Both the brain and spinal cord are covered by a three-layered set of connective tissue sheets known as the **meninges**. Within the meninges, the **subarachnoid space** is filled with cerebrospinal fluid (CSF), which is both nourishing and cushioning. Exterior to the meninges are the cranium and vertebral column to provide additional protection. The nervous system is considered a sterile body system and thus possesses no normal microflora. **Microglial cells** can attack invading pathogens should system defenses be breeched. The **blood–brain barrier** represents yet another protective feature of the nervous system. The highly selective capillary walls permit only certain chemical substances to reach the brain's neurons while blocking transport of most other compounds. Although this characteristic minimizes damage caused by the presence of microorganisms and/or the toxic substances they secrete, it is also responsible for limiting access of medication to the brain.

When the nervous system is attacked by pathogens, a potentially life-threatening situation has occurred. Because of its controlling/coordinating role, microbial damage to the nervous system often leads to death, lifelong physical/mental impairment, or, at best, an extended period of recuperation. Compliance with standard **sanitation** and safety procedures can reduce exposure to the pathogens responsible for causing rabies, botulism, leprosy, and tetanus. Prevention of many nervous system infections (meningococcal meningitis and polio) is accomplished by **immunization**. As you analyze the following cases, focus on the initial signs of disease and patient risk factors. Since early detection is usually crucial for affecting a positive outcome, you will become a better healthcare provider by learning to recognize these warnings.

Meninges – The system of membranes (pia, arachnoid, and dura mater) surrounding the CNS.

Subarachnoid space – The CSF-containing space between the arachnoid and pia mater.

Microglial cells – A type of glial cell that functions as a macrophage to destroy microbes in the CNS.

Blood–brain barrier – Structural barrier associated with CNS capillaries that restricts passage of substances out of the blood.

Sanitation – Use of cleaning agents to reduce the number of microbes to levels that are safe for public health.

Immunization – The use of a vaccine to produce immunity to a microbe or toxin.

Life in the Meningitis Belt

It was a celebration—three days of dancing to give thanks to God for the spring harvest. Saihun was enjoying watching the young girls and men attired with silver and gold ornaments. Saihun lived near other members of her clan, a large extended family within the Khasi tribe of the East Khasi Hills of Meghalaya—"The Abode of the Clouds," a state in northeast India that bordered Bangladesh. Although the day of the celebration was hot and humid, it had not rained—a real blessing considering she lived in the wettest place on earth. The East Khasi Hills and surrounding areas were part of a subtropical rain forest. The abundant rain helped Saihun's family to eke out a living growing rice.

Although Saihun was enjoying the night's celebration, she was worried about her father, who was in bed with a fever and headache. Normally, she wouldn't be so concerned except for the recent news of a **meningitis** outbreak moving through the area. The outbreak, which began in Bangladesh, had moved into northeastern India. The latest news stated that more than 2,500 people had taken ill with over 230 deaths in just the past month, with most of the cases occurring among the 140,000 people who lived in the poor tribal-dominated areas of Meghalaya.

1. What pathogens can cause meningitis?

2. Which pathogen is most often associated with **epidemics** in the "**Meningitis Belt**"?

3. Name and describe the causative agent of meningococcal meningitis.

4. If appropriate facilities and funds were available, what lab tests would be used to determine whether Saihun's father had meningococcal meningitis?

5. What is the **incidence** rate of the disease?

The next morning when Saihun checked on her father his headache was intense. He was vomiting, was very sensitive to light (**photophobia**), and had a stiff neck. Saihun also noticed an irregular rash on his trunk and legs.

6. How is this pathogen spread from person to person?

7. What characteristics of the pathogen enable it to infect and damage the CNS?

8. What characteristics of the pathogen cause the rash that was observed?

Saihun's family did not have the means to get her father to a hospital for care. However, they were able to obtain the necessary antibiotics from a local clinic supplied by the National Institute of Communicable Diseases.

9. What antibiotic(s) is/are typically given to treat meningococcal meningitis?

10. What is the mechanism of action of the drug?

11. How could Saihun's family help prevent further spread of the pathogen to others in their tribe?

Meningitis — Inflammation of the meninges, the membranes that line the brain and spinal cord.

Epidemic — A significant increase in the incidence rate of a disease over what is typically expected.

Meningitis Belt — Portions of sub-Saharan Africa from Senegal in the west to Ethiopia in the east (containing approximately 300 million people) that are characterized as a hyperepidemic area.

Incidence — The number of new cases of a disease per a specific population occurring over a given period of time.

Photophobia — Excessive sensitivity to light.

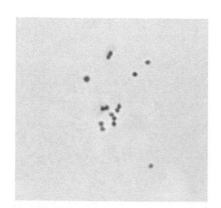

Figure 1. Gram stain of CSF.

The Flying Mouse

With an annoyed scowl, Chad checked his vibrating cell phone and let it go to voicemail. This was twice in the last 15 minutes his wife, Allison, had tried to reach him. "She knows better than to interrupt me when I'm on call," muttered a tired and cranky Chad. As a first-year resident at Grace Hospital, Chad was finding it a challenge to juggle his work and family responsibilities. He spent so much time focusing on his medical studies that his wife and 18-month-old son, Aydan, felt neglected. Chad assuaged his guilty conscience with promises to himself to make things up to his family once he had his own medical practice.

"Dr. Miller," a nursing student intruded upon his thoughts. "I have a message from your wife. She just called the nurse's station. She wants you to call home immediately…it's urgent!"

Embarrassed, frustrated, and angry, Chad thanked the student and pulled out his cell phone. Although ready to chastise his wife for disturbing him at work, Chad's mood changed instantly when Allison picked up the phone. In the background Chad could hear Aydan wailing and their bulldog, Brutus, barking furiously. "Allison, are you and Aydan alright? What in blazes is going on there?" Chad asked, genuinely concerned.

"It's awful, just awful! You've got to come home right now before it attacks us!" Allison cried. Another round of raucous barking exploded behind Allison. "Brutus, shut up! Chad, home! Now!" Allison yelled and hung up the phone.

Confused, scared, and still a little bit mad, Chad quickly made arrangements for a colleague to cover for him and raced back to the duplex they rented 2 miles away. As he pulled into the driveway, Chad saw Allison holding Aydan and pacing on the front porch with Brutus on his chain in the yard. Except for Allison's tear-stained face and the late hour, everything seemed pretty normal. "What's the big crisis?" Chad demanded. "Did you make me come screaming home just because our crazy dog was acting up?"

Allison shushed her husband and nodded toward their drowsy son. "Aydan and I were sound asleep in our room when Brutus just went wild," Allison explained quietly as she gently stroked her son's curls. "I was so mad at him for waking up Aydan because I figured he was just being his usual bad-dog self," she said with a glare over the porch rail to Brutus. Chad sighed and motioned for Allison to speed up her story. "Anyhow," she said, "as I turned on the light, I saw it! It actually flew into the bedroom wall. I screamed and grabbed Aydan. Brutus almost had it but the stupid bat managed to get airborne a split-second before Brutus went crashing into the wall after it!"

"What? You called me home because of a bat?" Chad groaned. "Why didn't you open the window and 'shoo' it out?"

"Well, I'm sorry. It was absolutely chaotic in the bedroom and the bat kept swooping toward us. I just wanted to get away from it before one of us got bit and died of rabies!" Allison retorted.

"Honestly Allison, it's just a little bat, not a stray dog or a wild animal. You're not going to get rabies from it. In fact, it's probably scared to death between Brutus barking, you screaming, and Aydan crying."

● 1. What is rabies? What pathogen causes this disease? Describe this pathogen and outline the cellular events in its life cycle.

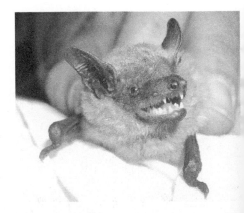

Figure 1. Brown bat.

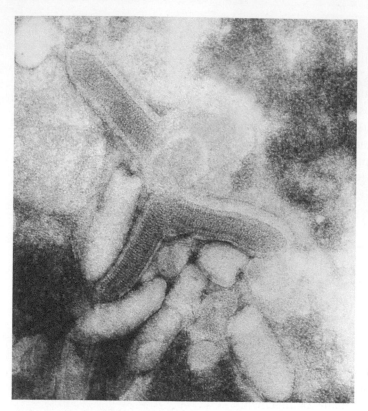
Figure 2. Electron micrograph of rabies virus.

Chad entered their bedroom with a trash bag and a tennis racket. It looked like a disaster zone from Brutus romping around the room after the bat. A lamp was smashed on the night stand, the phone had tumbled to the floor, and the hamper was overturned, spewing dirty clothes everywhere. "Well, she's in charge of clean-up," Chad said to himself. "I'm on bat patrol." With that, Chad jumped as the bat unexpectedly flew past his left ear. Preparing to swat it with the racket, Chad was surprised to see the bat land at the bottom of the bed and thrash around in a panic. He quickly scooped the trash bag over the bat and smacked it with the racket to kill the bat. Dropping the racket in the closet and the dead bat in the outside trash can, Chad informed his family the coast was clear and he was off to work.

When he returned to the hospital, his friends rushed to hear about the family emergency. Every time he relayed the story to someone else, the bat got larger, faster, and harder to catch. The nurses hailed Chad as a hero for ridding his home of the bat and protecting his family. Chad was feeling pretty good about himself until James, a fellow resident, pointed out that he hadn't slayed a dragon but merely pulverized a small, defenseless animal. "That bat was a pretty good size," Chad responded.

"Right," James teased. "That's why a bat is often referred to as a 'flying mouse.' Did you take the fierce beast to the lab?" his friend teased.

"Why would I do that?" Chad groused. "It's dead and gone."

"Don't you think you should have it tested for rabies?" James queried.

"Man, you're as bad as Allison. It's a common bat, not some wild animal. It wasn't foaming at the mouth and it didn't bite anyone. Why is everyone so hysterical about this bat? The poor thing didn't even fly very well, so I doubt if it was capable of attacking my family," Chad said.

"Actually, bats are responsible for a significant number of rabies cases in the U.S. annually. The fact that the animal wasn't flying well and showed up in your house suggests it might be ill. Maybe Allison wasn't bitten, but how would you know if Aydan or even Brutus had been nipped? You need to get that animal tested," James said before heading off to his next patient.

2. What types of animals commonly carry rabies in the United States? How does this differ in other countries?

3. How can you tell if an animal has rabies?

4. What should you do to safeguard your health if you've come in contact with an animal that is potentially infected with rabies? What procedure should be followed if you are unsure of a person's exposure to an infected animal? What should be done with the infected animal?

5. How can you safely capture or remove a bat in your house?

6. Chad said the bat "wasn't foaming at the mouth and didn't bite anyone," alluding to rabies symptoms and transmission.

    a. Outline the symptoms associated with a rabies infection.

    b. Describe the transmission of this pathogen. Can rabies be transmitted by all body fluids?

Realizing his mistake, Chad called the hospital lab and requested directions for having a suspicious animal tested. At the end of his shift, he made an appointment for Aydan with his pediatrician and collected the dead bat from the trash. When a sleepy Allison stumbled into the kitchen the next morning, Chad had the coffee brewed, the dog fed, and flowers on the table. "Momma!" a cheerful Aydan called from his high chair.

"What's going on here?" Allison asked suspiciously. "Last night you were pretty angry with me and today I get coffee and flowers?"

Hugging his wife, Chad sheepishly admitted that he had been a bit hasty with his family and their welfare last night. He explained to Allison that he was going to have the bat tested and had scheduled Aydan to see the doctor just to be on the safe side. Rather than making Allison feel better, her maternal instincts surfaced and she was instantly frantic about Aydan's health. "Oh my God, what if he was bitten?" she exclaimed, pulling her son from the high chair to look for bite marks. "Isn't rabies fatal? Is there any medicine to cure it?" Chad began reassuring his wife and helped her examine their son.

7. How will the bat be tested for rabies? How is a person tested for the disease?

8. Outline the treatment for rabies. Consider how this varies depending upon the level of exposure to a potentially-infected animal.

Dr. Martinez thoroughly examined Aydan but didn't find evidence of a bat bite. Despite this good news, he prescribed the fullest measure of treatment explaining that it is easy to miss the very small wound made by a bat bite and simply wasn't worth taking a chance with such a serious infection. Dr. Martinez reminded Allison as she was leaving that she needed to schedule an appointment with her own physician to receive **prophylactic** treatment and she should also contact the veterinarian for Brutus. Relieved that this ordeal was almost over, Allison worried aloud, "How can I make sure we never have to go through this again?"

Dr. Martinez smiled, gave Allison a pat on the back, and began discussing prevention suggestions with her.

Prophylactic – Preventative medical treatment.

9. How can you prevent rabies? What group or groups are at greatest risk for contracting this infection?

Vacation Headaches

A weekend drive through Custer State Park of South Dakota was just one of the great things about retiring in the Black Hills. After visiting Mount Rushmore, Marge and Steve decided to hike to an Aspen grove overlooking a grassy plain for a picnic. There they could enjoy the cool shade and watch the herds of bison and pronghorns below. Unfortunately they had forgotten bug spray and had to eat while swatting mosquitoes, which apparently also enjoyed the shade and their own picnic of fresh blood.

Later that week, Marge came home from work early with a low-grade fever and headache. After taking a couple of **ibuprofen**, she went to bed early. The next day, she felt worse—muscle and joint aches and an intense headache and neck stiffness. She had Steve take her to her family physician.

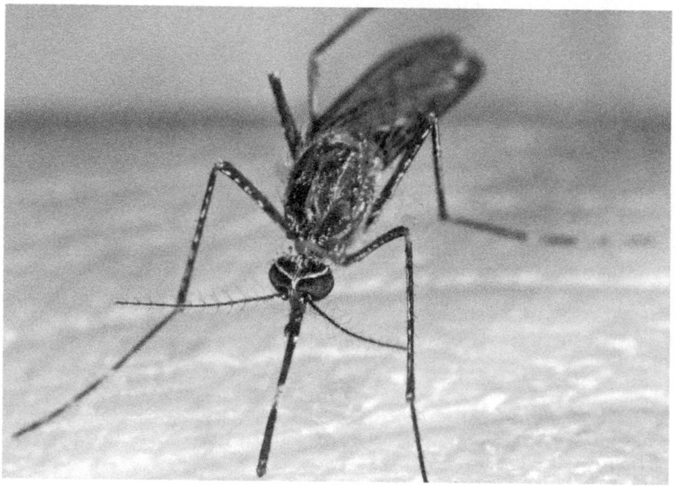

Ibuprofen – **i**so-**bu**tyl-**pro**panoic-**phen**olic acid; a nonsteroidal anti-inflammatory pain killer often used when there is an inflammatory component.

Leucopenia – A decrease in the concentration of white blood cells (leukocytes) in the blood.

Lymphopenia – A decrease in the concentration of lymphocytes in the blood.

Enzyme-linked immunosorbent assay – An immunological technique used to detect the presence of a microbial antigen or an antibody produced in response to the antigen.

Titer – The concentration of antibodies in serum usually measured using serial dilutions.

Figure 1. *Culex* mosquito.

1. Given Marge's clinical pictures, what infectious disease(s) would the physician consider her to be suffering from? Match each disease to the pathogen(s) that cause it.

The physician took a sample of cerebrospinal fluid. Microscopic examination and blood cultures indicated no bacteria, fungi, or protozoa present. Blood tests indicated mild **leucopenia** and **lymphopenia**. Serological testing using an **enzyme-linked immunosorbent assay** showed a highly elevated **titer** for West Nile virus–specific antibodies.

2. What are the physical characteristics of West Nile virus?

3. How did a virus that was endemic to the Middle East, Africa, and Asia get to North America?

4. How is West Nile virus transmitted?

5. What is the reservoir for West Nile virus?

Provided her symptoms did not get worse, the physician recommended Marge get regular fluids and electrolytes, use OTC pain medications to manage her headaches, get plenty of rest, and resume normal diet and activity as her recovery enables her to do so.

6. Given her age (62 years old), what is her **prognosis**?

7. What signs and symptoms would Marge experience that could indicate significant progression of this disease?

Although Steve was relieved his wife did not need to be hospitalized, he still had several concerns.

8. How long will it be before the disease resolves?

9. How can this disease be prevented?

10. Is Marge **contagious**?

Prognosis – The likely outcome of an illness after the patient has began medical care.

Contagious – The potential of an infectious disease to be easily transmitted to another person.

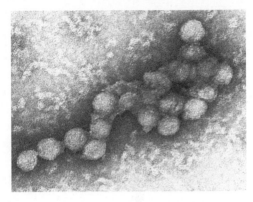

Figure 2. Electron micrograph of West Nile virus.

C. Infections of the Cardiovascular and Lymphatic Systems

Natural killer cell – A large granular lymphocyte that induces apoptosis in tumor cells and virally-infected cells.

Apoptosis – A series of programmed biochemical events that lead to cell death.

Complement fixation pathway – A biochemical cascade that complements the ability of antibodies to destroy pathogens.

Cytokines – Substances secreted by immune cells to signal a response in another cell.

IL-1 – Interleukin-1, a cytokine that can induce a fever.

TNF – **T**umor **n**ecrosis **f**actor, a cytokine family that can cause cell death.

The cardiovascular system and lymphatic systems both circulate fluids throughout the body, contacting many different tissues and organs to transport nutrients and oxygen while removing wastes. As a result, these systems also serve as an effective vehicle to distribute any introduced microbes. For this reason, numerous defense mechanisms have evolved to destroy pathogens attempting to use these body systems as a delivery service. Innate (nonspecific) cellular responses include the activity of **natural killer cells**, which destroy invaders by triggering **apoptosis** in the targeted cell. Neutrophils and monocytes are phagocytic leukocytes that efficiently adhere to pathogens, engulf, and finally oxidize them. A noncellular innate defense mechanism is the activation of the alternate **complement fixation pathway**. Circulating complement proteins ultimately produce a membrane attack complex, which permits osmotic lysis of the targeted invader. Adaptive immune responses are highly specific defense mechanisms capable of intensifying with each subsequent exposure to a given pathogen. The humoral adaptive response utilizes antibodies produced by plasma cells (activated B lymphocytes) to bind a specific invader, immobilize, and eventually eliminate it. Cell-mediated adaptive responses are initiated by different populations of T lymphocytes. When activated, these cells secrete and respond to a variety of different **cytokines**. The net result is highly effective and targeted pathogen destruction.

As you have observed in other body systems, possessing sophisticated defense mechanisms does not guarantee freedom from infection. Since many microorganisms can reproduce at exponential rates, their population growth often outpaces the immune responses designed to protect us. When this occurs, a medical crisis is imminent and prompt treatment is essential. In spite of appropriate, effective antibiotic therapy, patient mortality from systemic infections is significant. While drugs can be administered to kill the systemic pathogens, bacterial death often leads to the release of cell wall components with devastating results. Lipid A can trigger the release of **IL-1** and **TNF** from phagocytic cells, which will result in dangerous elevation of body temperature, increased capillary permeability that can lead to shock, initiation of disseminated intravascular coagulation (DIC), and even acute respiratory distress syndrome (ARDS). As you analyze the cases in this section, pay close attention to the prevention measures needed to avoid these illnesses. Because of the rapid progression of these serious conditions, you will want to regularly instruct your patients so they can avoid such infections.

A Common Opportunistic Pathogen

Ben, a seven-year-old second-grader, quickly finished his homework after school so he could play outside with his friend from next door. Since Ben had started coughing the day before, his mother insisted he wear his jacket. At bedtime, Ben was exhausted from his busy day. When kissing his forehead goodnight, Ben's mother noted that he felt a little warm. To help Ben sleep more comfortably since he was likely getting a cold, his mother gave him a dose of pediatric Tylenol. "Poor Ben," she thought. "Three weeks ago he had the flu and now a cold is starting. He could really use a break."

In the morning, Ben didn't come to the breakfast table when called. His mother found Ben still in bed, barely responsive, and extremely feverish. She immediately drove him to the walk-in clinic in their neighborhood. Ben's oral temperature was 40.8°C (105.4°F). IV fluids were started and an ambulance transported Ben to the nearest hospital.

In the hospital emergency room, Ben presented with the following vital signs: temperature = 41.9°C (107.4°F), pulse = 162 bpm, pulse ox = 90%, respirations = 24/minute and labored, BP = 62/54 mmHg. Ben was completely unresponsive. His physical exam was remarkable for rales or "crackles" heard over both right and left lower lung fields. Bilateral chest radiographs were ordered and revealed infiltrates in the lower lobes of both lungs.

1. What does crackling on auscultation suggest?

2. Is this confirmed by the radiographs? Explain.

3. Name four common infectious agents for this condition.

4. Does this diagnosis account for all of Ben's extreme symptoms?

Blood was drawn for hematology and metabolic panels. Two sets (1 set = 1 aerobic bottle and 1 anaerobic bottle) of **blood cultures** were also drawn and a lumbar puncture performed to collect **CSF**. Ben was life-flighted to a major medical facility for treatment.

5. Indicate two reasons why two sets of blood cultures were ordered.

Preliminary lab results yielded a white blood cell count of 16,200 cell/mm^3 and a differential count with 74% neutrophils, including 18% **bands**. **Respiratory acidosis** was indicated by an arterial blood pH of 7.2. These results were immediately called from Ben's local hospital to his new facility and broad-spectrum IV antibiotic therapy was initiated.

Within 6 hours of incubation, three of Ben's four blood culture bottles were positive for bacterial growth. Aliquots from each of the three cultures were plated on **blood agar plates (BAP)** and on **chocolate media** and incubated for an additional 12 hours. At this time, a Gram stain was performed on the blood cultures, which consistently yielded Gram-positive diplococci. Ben's CSF was sterile.

6. What is the likely causative agent of Ben's infection?

7. Where is this microbe typically found?

8. How did it end up in his blood if Ben's initial problem was respiratory?

Blood culture – A blood specimen is distributed into two specially designed bottles containing media to permit aerobic and anaerobic growth, respectively. Each set of bottles is placed into a fully automated blood culture system such as the BacT-Alert System or Bactec 9000. Microbial growth causes CO_2 release leading to a pH change. This is colorimetrically detected by the instrument due to the addition of pH indicators to the growth media. Specimens "flagged" positive by the analyzer are removed, Gram stained, and inoculated on BAP and chocolate media for further analysis.

CSF – Cerebrospinal fluid, a clear, colorless, sterile fluid that provides a protective cushion to the brain/ spinal cord. Adult volume of CSF is 90–150 ml with a normal protein level of 15–45 mg/dl and glucose level of 40–80 mg/dl. A CSF differential count is 60–80% lymphocytes, 10–40% monocytes, and 0–15% neutrophils.

Bands – Immature neutrophils with nuclei shaped like a band rather than being segmented. Elevated bands are associated with an inflammatory response.

Respiratory acidosis – A decrease in pH of the blood due to a decrease in respiration causing increased blood CO_2 level.

Blood agar plate (BAP) – TSA with 5% sheep blood (trypticase soy agar enhanced with 5% sheep red blood cells) provides a rich nutrient medium for the culture of many common microbes. This medium often serves as a growth control.

Chocolate agar plate – A variant of blood agar used for growing fastidious respiratory bacteria such as *Hemophilus*.

Figure 1. Growth of pathogen on blood agar.

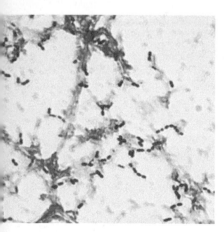

Figure 2. Gram stain of blood culture.

9. Approximately 6 hours later, the clinical microbiologist examined the BAP and chocolate plates. What morphological features did she likely observe on the plates? Why were these two media types selected for culturing?

10. What results would you predict for the following tests?
    a. Optochin
    b. Catalase
    c. Bacitracin
    d. Bile solubility
    e. Quellung reaction

11. What is the significance of the following laboratory results?
    a. The WBC
    b. High neutrophil count with elevated bands
    c. Respiratory acidosis

Automated identification and sensitivity testing were initiated and the presumptive microbe identification called to the other facility to aid in treatment. Ben's attending physician was relieved to find that the CSF was sterile, as she had just initiated prophylactic antibiotic treatment of Ben's caregivers. Unfortunately, she reported that Ben's condition had continued to decline and he had been pronounced dead earlier that hour. She requested the identification/sensitivity data be forwarded upon completion so Ben's record could be finalized.

12. Why was Ben's physician relieved that the CSF was sterile? What possible infection was she considering?

13. Assuming Ben was receiving broad-spectrum IV antibiotic therapy that was effective against his infection, what reasons can you give to explain his rapid deterioration and eventual death when receiving an appropriate empiric therapy?

Within 6 hours, the Vitek analyzer confirmed the identity of this microbe and indicated its sensitivity to penicillin, cephalosporins, and fluoroquinolones. While Ben's attending physician was pleased that her staff had not been exposed to a resistant microbe, she was frustrated to have lost a young patient to one so sensitive to standard antibiotics.

14. Were there any predisposing factors that put Ben at greater risk for this infection?

Unhappy Returns

Gracie worked as an emergency room nurse in a hospital in Salt Lake City where she evaluated a broad spectrum of emergent cases. During the winter months, Gracie had seen the typical seasonal increase in the number of children being brought in for treatment of acute **pharyngitis** and fever.

1. Name and describe three different microbial pathogens that cause acute pharyngitis.

2. Why is there a seasonal increase in the number of infectious upper respiratory diseases?

If the children had had the normal series of childhood vaccinations, Gracie would take a throat swab and have it tested for the presence of **Group A streptococci.**

3. Describe a cultural and a noncultural method for testing for Group A streptococci.

4. If the test was positive, what pathogen is likely causing the pharyngitis?

5. How would you treat a child with strep throat?

Over the last several weeks, Gracie noticed a return of several children whom she had treated for strep throat several weeks earlier. Although the children had recovered from their sore throats, they were now experiencing fever, arthritis, and mild to moderate chest discomfort. A physical examination demonstrated the children had **carditis** revealed by a new **heart murmur**. They also had tachycardia that was out of proportion to what was expected from the fever.

6. What disease could be affecting the children that would be related to their previous history of strep throat?

7. What role does the M protein play in this complication of strep throat?

8. Why is this complication of strep throat much less frequent in countries with a well-developed healthcare system than in countries where routine health care and supplies are not readily available?

Pharyngitis – A sore throat caused by inflammation of the pharynx.

Group A streptococci – A pathogenic group of streptococci that contain group A antigen on the cell wall.

Carditis – Inflammation of the heart.

Heart murmur – Additional heart sounds that are produced from turbulent blood flow.

Figure 1. Pathogen growth on blood agar.

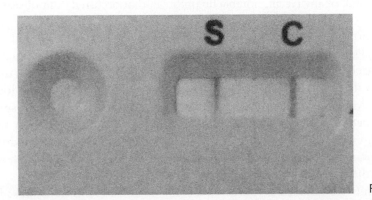

Figure 2. Rapid strep test.

Bunny Bits!

Bob was hot and sweaty. He'd spent the last hour and half of a humid July afternoon on his riding mower. "Why am I out here melting when I have three able-bodied sons who could be mowing?" Bob muttered. He decided to head back to the house for an ice-cold drink and to draft his 19-year-old son to finish the weekly mowing. About 15 yards from the back of his garage, Bob's mower lurched, made a horrible noise, and spewed a lot more than grass clippings out the side. "*@#$!" Bob shouted as he cut the motor and jumped off his mower. "Oh great," Bob said as he crouched down for closer inspection of the problem. "I ran over a stupid rabbit's nest." Bob looked up just in time to intercept Davey, his youngest son, who had been playing a game on the back porch. "Hold it right there, mister," Bob ordered his son. "There's nothing out here for you to see." Davey stopped just before leaping over the porch railing. "Aw, but Dad, that was really cool. Can't I come see the bunny bits?" Davey whined. A stern look from his father was all it took for Davey to return to his game. Bob stomped into the garage to retrieve a garbage bag and his best pair of barbecue tongs. He quickly deposited all of the "bunny bits" in the plastic bag, dropped the mess into the trash can, and put the mower away without finishing the job. Bob decided this was a sign that he had done enough mowing for the weekend and he should enjoy a cold lemonade on his front porch swing.

Three days later, Bob woke up with a headache and feeling achy all over. He took a couple of Tylenol and stopped for an extra large coffee on his way to the office. "Maybe a little caffeine will perk me up," Bob groaned as he booted up his computer and started his busy day at work.

The coffee didn't help. By mid-afternoon, Bob had developed a persistent, dry cough, chills, and a worsening headache. When he arrived home, Bob skipped dinner and went straight to bed. Patti, his concerned wife, came to check on her husband after feeding their family. Upon taking his temperature (39.1°C; 102.4°F), Patti brought Bob some Tylenol and a tall glass of ice water. "Take these," she said. "They will help with your cough, headache, and fever." "What about the rest of my aches?" Bob complained in a miserable voice. "My muscles and joints are so sore. I haven't felt like this since I had the flu ten years ago. Leave it to me to get the flu in the middle of the summer." Feeling sorry for her uncomfortable husband, Patti gently massaged his back as he finally drifted off to sleep. "Hopefully, he will feel a lot better after a good night's sleep," Patti prayed.

1. Are Bob's symptoms consistent with the flu? When is "flu season"? Can you think of any recent, notable flu outbreak that occurred "out of season"?

Unfortunately, Bob coughed all night long. By morning he had a fever of 39.4°C (103°F), chest pain, shortness of breath, plus his continuing headache and muscle aches. Additionally, Bob felt he was growing weaker by the minute. Profoundly ill, he let Patti take him to the emergency room at the local hospital. Dr. Martin ordered a **CBC**, chest x-ray, a "**rapid flu test**," and sputum culture. Since Bob's temperature had climbed to 39.8°C (103.6°F), an **antipyretic** was administered and Bob was placed on oxygen to ease his labored breathing. Dr. Martin asked Bob numerous questions about his activities, travel, and exposure to other people and to animals before admitting him to the **ICU**.

Within 1.5 hours, Dr. Martin had some of Bob's test results back. His white blood cell count was slightly elevated at 16,000 cells/mm³ and his radiographs

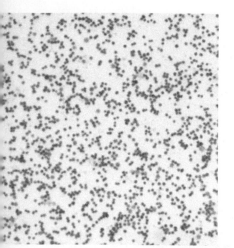

Figure 1. Gram stain of pathogen.

CBC– A **c**omplete **b**lood **c**ount provides valuable information regarding the number and kinds of white blood cells, red blood cells, and platelets.

Rapid flu test – A "point of care" procedure to detect influenza viral nucleoproteins in patient nasal secretions within 30 minutes. Although clinically useful, there is limited sensitivity to distinguish influenza A from influenza B viruses plus a 30% false negative rate.

Antipyretic – A medication used to reduce fever.

ICU– The **i**ntensive **c**are **u**nit is a medical department within a hospital designed to provide specialized care for critically ill patients.

showed extensive bilateral infiltrates. Bob's "rapid flu" test was negative for infection with influenza A or B viruses, but the Gram stain of his sputum specimen was significant for small, Gram-negative coccobacilli with bipolar staining. Additionally, the laboratory noted that some of these bacteria were found living inside of the phagocytic cells in Bob's sputum sample. Dr. Martin immediately initiated broad-spectrum antibiotic therapy to fight this bacterial infection.

2. What is the common medical term used to describe a(n) (bacterial) infection of the lungs?

3. What is a normal WBC count for an adult? What does Bob's elevated WBC count suggest?

4. Collecting a good-quality sputum specimen is crucial for successful processing in the clinical microbiology laboratory. Describe how this specimen should be collected. How will the laboratory screen the specimen for quality before staining and culturing it? What problem is associated with processing a sputum specimen without confirming quality?

5. Identify different bacterial species that are Gram-negative coccobacilli capable of living intracellularly within their host.

6. Why did Dr. Martin start a broad-spectrum antibiotic treatment when he hadn't yet received the precise identification of the microbe and its antibiotic sensitivity from the lab?

Over the next 24 hours, Bob's condition deteriorated. His fever remained high, his head and muscles continued to be extremely painful, his weakness prevented him from sitting up, and his cough worsened as his sputum was now bloody. Shortly before dawn, Dr. Martin had Bob placed on a ventilator as his respiratory distress progressed to failure. He told Patti her husband was in critical condition and she should consider having his sons come to the hospital to "say good-bye" to their father.

At 7 AM a technologist from the clinical microbiology laboratory called the nurse's station of the ICU to report Bob's sputum sample had grown *Francisella tularensis*, which was sensitive to streptomycin, gentamicin, doxycycline, and chloramphenicol.

7. How is *Francisella tularensis* usually cultured and identified in the clinical laboratory? Are there other methods employed for diagnosing infection with this organism?

Dr. Martin immediately switched Bob's antibiotic therapy to streptomycin while continuing supportive measures. When Patti arrived at the hospital with her sons, Dr. Martin escorted them into a family conference room to discuss this new information. "I have good news and bad news," he said. "The good news is that we now know what germ we're fighting and what drugs should be effective against it. The bad news is that this is a very serious and highly contagious infection and Bob may not survive. Also, I need more information from all of you to determine how Bob acquired the pathogen so we can safeguard your family and others from this infection. This disease has some serious public health implications and I'll need your input to complete my report."

Dr. Martin went on to explain that Bob had "tularemia" or "rabbit fever" that was presenting with unusual respiratory symptoms. "Tularemia symptoms can vary depending upon how the pathogen was transmitted," Dr. Martin said. "Bob's pulmonary presentation suggests the bacteria may have been airborne when he came in contact with them. People who hunt, trap, garden, work with wildlife, or work in microbiology laboratories are at greatest risk of acquiring this infection since they are frequently in contact with the reservoir for this bacterial species. Do Bob's hobbies or work put him at risk for pathogen exposure?" Dr. Martin inquired.

8. What are other common names for tularemia? What are the common reservoirs for this pathogen?

9. What is the infectious dose for this highly contagious disease? Approximately how many cases of tularemia are reported annually in the United States?

10. Dr. Martin suggested that tularemia can manifest with variable symptoms that relate to the manner in which the pathogen is transmitted. Research this connection and report on the different types of tularemia infections, their symptoms, and related transmission mode.

11. Dr. Martin also indicated that he is concerned about Bob's survival. What is the mortality associated with this infection? What complications usually contribute to mortality?

12. Although the laboratory reported sensitivity to a number of antibiotics, Dr. Martin selected streptomycin. While streptomycin is the preferred drug for this infection, chloramphenicol is equally effective at eliminating the pathogen. Why would it NOT be Dr. Martin's first treatment choice?

13. What did Dr. Martin mean when he indicated tularemia is a major public health concern?

Working with Dr. Martin, Bob's family reviewed all possible opportunities for exposure to *Francisella tularensis*.

"Bob is a computer engineer," Patti said, "so we can easily rule out exposure from working in a lab."

"We can also scratch hunting, trapping, and gardening off of the hobby list," Bob's eldest replied. "Dad's hobbies are football, football, and more football…all from the comfort of his recliner. The only outdoor activity Dad does regularly is the mowing, so it's not like he comes in contact with wildlife."

"Well, just the bunnies," Davey volunteered.

"What bunnies?" Dr. Martin and Patti said in unison.

"The lawn mower bunnies!" Davey said with authority. "It was soooo gross! There was lots of blood and chopped-up rabbit parts."

"Davey, what bunnies? What are you talking about?" Patti asked frantically.

Rolling his eyes, Davey said with exasperation, "The baby ones! Dad buzzed 'em up by accident last Saturday and they came blowin' out the side of the mower. I was on the back porch and saw the whole thing."

"Davey, did you touch the chopped up bunnies?" Dr. Martin asked.

"Nah, Dad wouldn't let me get off the porch to see better. He thinks that stuff will scare me because I'm the youngest," Davey pouted.

"Well, this has been very informative and helpful. I think we know how Bob contracted his rabbit fever and why it is manifesting with pulmonary symptoms. It sounds like the rabbits in the nest Bob mowed over were infected with *Francisella tularensis*. The blades must have aerosolized the bacteria. Bob inhaled the pathogens and became ill. Luckily, making Davey stay on the porch prevented his infection."

14. Davey was far enough away to prevent transmission of *Francisella tularensis*. Identify appropriate methods for preventing the transmission of this pathogen.

Epilogue

Although Bob was in critical condition for days, he fortunately survived his bout of rabbit fever. He was able to avoid the other serious complications associated with this infection, but his symptoms persisted for weeks before he made a full recovery.

The Black Measles

Josh and Matt raced home from school on their bikes. They had been best friends almost since birth, having grown up as neighbors. Matt waved "bye" as he peddled up the lane to his family's farm. A half mile down the road, Josh pulled into his front yard, dropped his bike, and burst into the house calling for his mother. "Mom, Mom, me and Matt decided we're gonna celebrate the last day of sixth grade by camping out at his place tonight...ok?" "You know," Josh's mother said, "I expected you might want to do something like that. Take a look in the kitchen." Josh bolted to the next room and found a cooler on the counter already stocked with hot dogs, marshmallows, and other campfire supplies. Josh dumped the school papers out of his backpack, replacing them with his flashlight, Swiss army knife, and a clean T-shirt. He strapped the little cooler on the back of his bike and called to his mother as he prepared to leave.

"Wait a minute, mister," his mother warned. "You boys have fun, but remember to be careful building your campfire, don't go into the woods after dark, and be sure to take Toby along just in case." "You got it, Mom," Josh said, giving her a quick peck on the cheek before climbing on his bike. Josh's mom watched him happily peddle to his friend's house. She could hardly believe how quickly her son was growing up. Living on the outskirts of a small town in northwest rural Ohio, she knew the boys were old enough to camp out at the edge of the woods on Matt's family farm. Still, she felt better when Toby, Josh's four-year-old Rottweiler, accompanied the boys.

Figure 1. *Dermacentor andersoni* tick.

After supper with Matt's family, the two boys and Toby headed across the broad backyard to the small campsite they used just in front of the woodlot that separated two huge fields of ripening wheat. Toby and the boys headed into the woods to collect their firewood for the night and take turns scaring each other by unexpectedly jumping out of hiding places in the brush. The boys feasted on smores and hotdogs...and Toby got the leftovers. Lying on their sleeping bags late that night, Josh and Matt discussed swimming, baseball, and their plans to start a lawn-mowing business and earn some money during their summer vacation. Soon, they were fast asleep with Toby curled protectively beside Josh.

Two days later when Josh had finished his chores, he went to visit his friend. Matt was in the barn caring for the calf that was his 4-H project for the summer. "Hey, Matt, check it out!" Josh called, pulling up his pant leg to reveal a small red mark. "Got a boo-boo baby?" Matt teased. "Nah, it's a tick bite," Josh announced with pride. "It was really cool on Saturday when I showed my Dad. The tick was so full of blood it was almost ready to explode! My Dad wanted to use his lighter to burn it off, but Mom wouldn't let him."

"Wow, gross!" Matt said with growing interest. "How'd ya get the tick off?"

"My Mom used her tweezers and pulled him out by the head...and then I squished him!" Josh said. "After that, me and Dad spent most of the morning pickin' ticks off of Toby...he got 'em bad. Mom was getting all freaked out 'cuz she says we can get diseases from tick bites. Dad told her to "chill" and that made her really mad!"

1. Should Josh's Mom "chill"—that is, what diseases can be vectored by ticks? Who is at greatest risk of contracting a tick-borne illness? When is the risk greatest?

2. When an infectious disease is vectored by an arthropod bite, what form of transmission is demonstrated? How does this differ from a zoonosis?

The next day, the boys had planned to meet at Mrs. Jacob's house to start their summer lawn-mowing business. Josh was relieved when Matt's mom called and postponed their employment, saying her son was a bit under the weather. Josh was also feeling a little off…chilled, achy, nauseous. Now, with Matt sick, he had a good excuse not to mow Mrs. Jacob's giant yard.

The next morning, Matt's mother took him to the pediatrician, as his fever reached 39°C (102.5°F), his nausea, vomiting, sore throat, and headache had worsened, plus he began developing a red-spotted rash on his ankles, wrists, fore-arms, and palms. Dr. Taylor was stumped after examining Matt. He was clearly a sick young man, but there are so many infectious diseases that present with these very general symptoms, it was truly challenging to provide a definitive diagnosis. Consequently, she prescribed comfort measures for what was likely a summer viral infection.

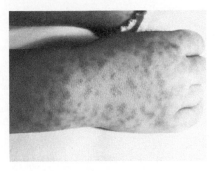

Figure 2. Rash following tick bites.

3. Why didn't Dr. Taylor prescribe an antibiotic if Matt was so ill?

Much to her surprise, Dr. Taylor found Josh in her examining room later that same afternoon. Josh also complained of nausea and vomiting, lethargy, fever (40°C; 103.2°F), chills, and severe headache. "When he refused to eat both break-fast and lunch, I knew it was time to see the doctor," Josh's mom reported. As Dr. Taylor conducted the physical examination, she noticed small, flat, non-itchy red spots emerging on Josh's forearms, palms, ankles, and soles. She gently pressed on the rash and the area turned white before it returned to its spotted condition. "Hmmm," she said, "I've seen this same rash earlier today."

"Betcha it was on Matt," Josh offered.

"Why do you say that?" Dr. Taylor inquired.

"Oh, those two boys are like peas in a pod," Josh's mother volunteered. "Since school let out, they've been together every day…swimming, baseball, camp-outs, you name it. I suppose if one of them picked up a bug, the other would be bound to get it."

"Bug," Dr. Taylor said. "Josh, did you or Matt get bitten by any bugs with all of this outdoor activity?"

"Well, Matt got a couple of mosquito bites, but I got my blood sucked out by a giant tick! After Mom pulled it out, I popped it!"

"Josh, you have just been a big help to me. I think now I can make my diag-nosis. When did you get this tick bite? Did Matt get bitten too?" After gathering a little more background information about Matt and Josh's camp out, Dr. Taylor diagnosed Josh with Rocky Mountain spotted fever (RMSF) based upon his fever, rash, and history of tick bite. She ordered blood drawn to run lab tests confirm-ing her diagnosis, immediately started treatment, and had her office manager call Matt for a return appointment ASAP.

4. By pressing on Josh's skin, Dr. Taylor determined it was a blanching rash. What serious medical condition is suggested by the presence of a non-blanching rash?

5. What organism causes RMSF? Characterize this microbe. Name two other species of this genus and their associated diseases.

6. What lab tests are routinely ordered to confirm a RMSF diagnosis?

7. What treatment did Dr. Taylor likely prescribe for Josh? What consideration should be made for a pediatric patient? Why did Dr. Taylor initiate antibiotic therapy prior to confirming this infection?

"Rocky Mountain spotted fever!" Josh exclaimed. "Dr. Taylor, this is Ohio! I can't have Rocky Mountain spotted fever. The Rockies are a long way off and we don't even have a hill around here!"

8. Where is RMSF endemic? How did this infection acquire its regional name?

Dr. Taylor took a few more minutes with her patient to answer questions. She also told Josh a little about the history of Rocky Mountain spotted fever, explained how it got the nickname "black measles," and emphasized its prevention.

9. Explain the pathogenesis of RMSF, paying particular attention to rash development. Why is RMSF also known as the black measles? Are there any other symptoms associated with this infection? Can RMSF be fatal?

10. Discuss the natural history of this condition.

11. How can RMSF be prevented?

Pucker Up!

"Mornin' Mom," Paula called to her groggy mother as she dashed into the house, dumped her tennis racket in the closet, and shot upstairs for a quick shower. Thirty minutes later, she was waiting tables for the breakfast crowd at Denny's. At the end of her shift, Paula grabbed a sandwich and was on the run again. By 2 PM she was wearing her tennis whites and warming up with Lynn, her doubles partner. Two sets later, Paula and Lynn emerged victorious in the 16–21 age class of the quarter final round of the Tri-County Tennis Tournament. Scott and Tony were waiting for their girlfriends at the end of the match. The two couples celebrated the win with Cokes and fries at McDonald's. At 5 PM, Paula made a quick trip home for her second shower of the day. She reported to Della's Dairy House for a five-hour shift of scooping homemade ice cream and flipping burgers.

While Paula and her coworkers cleaned the little ice cream parlor after closing, Della began her nightly ritual of creating a new treat for her employees. By the time the shop was "spic 'n span," Della was pouring strawberry-banana milkshakes for her staff. "Honey, you look like you need more than a milkshake," Della teased Paula. "Are you gonna stay awake long enough to drive home?" Della asked a drowsy-looking Paula.

"I'm good…thanks," Paula yawned. "I guess my schedule is starting to catch up with me."

"Child, no one can work two jobs every day, juggle a boyfriend, and play tennis every other waking minute! What are you trying to do to yourself?" a concerned Della asked.

"Della, you know I start college in September. Even with my academic and tennis scholarships, mom and dad can't afford a private school if I don't help," Paula explained. "Hey, this shake is a winner," she said trying to change the subject. "You should really add this to the menu next month!"

Paula got home a little before midnight. Instead of going straight to bed, she sat on the porch and sipped an iced tea. "That icy cold milkshake Della made sure made my throat feel better," she thought. Paula noticed her throat becoming uncomfortably sore during the second set of her afternoon match. "A nice chilly iced tea and few hours of sleep and I'm sure I'll feel better tomorrow," Paula said to herself, finishing the tea and heading for bed.

Even though Paula was on the court practicing her serve by 6 AM, she really didn't feel any better than the previous evening. In fact, her throat was worse. When Scott showed up, they rallied for 30 minutes before it was time to get ready for her shift at Denny's. Saying their good-byes, Scott confirmed their date for Saturday night and gave his girlfriend a kiss. "Wow, babe," Scott said, "you're seriously hot!"

"Oh, knock it off. I'm sweaty and gross," Paula blushed as she took a long draught from her water bottle and passed it to Scott. "No, no that's not what I mean. I think you're sick or something…and get that thing away from me," Scott said batting the water bottle back to Paula. "I don't want to get your germs!"

Paula went through the paces all day…tennis, work, more tennis, and more work. When she finally returned home that night, Paula was really dragging. Her throat was scorching sore and the lymph nodes in her neck had started to swell. "Oh, great," she muttered. "I bet I'm getting strep throat." More tired than she had ever been, Paula took two Tylenol caplets for her headache and dropped wearily into bed.

The next morning, Paula texted Scott and Lynn to cancel their daily tennis practice since she was too exhausted to swing a racket. Paula made herself some hot tea,

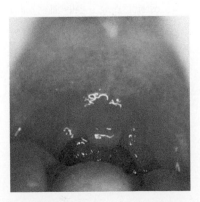

Figure 1. Pharyngitis.

hoping to soothe the searing pain in her throat. When Paula's mother, Tami, entered the kitchen, she found her daughter slumped onto the table with a mug of tea still in her hand. "If she's asleep and not on a tennis court, she must be pretty sick," Tami thought. Tami called Denny's and reported Paula out sick for the rest of the week. After waking Paula, Tami used the classic "mom thermometer" and kissed her daughter's forehead to check for fever. Paula was definitely too warm. But it was one look at her throat that convinced Tami it was time to see a doctor.

1. Does Paula have strep throat? What infection do you think is most likely, given her principal symptoms?

At the urgent care center, Dr. Calvert examined Paula while reviewing her history. "Paula, you have significant cervical **lymphadenitis**, extreme fatigue, severe **pharyngitis**, and a fever of 39.4°C (103°F). These findings coupled with your age strongly suggest you have infectious mononucleosis."

"What?" Tami replied. "She has the kissing disease!" Catching Paula's startled, concerned expression, Dr. Calvert recognized the need to intervene and educate. "Whoa!" he said. "You're getting way off base here. Kissing is only one way infectious mononucleosis can be transmitted." Turning his attention from mother to daughter, Dr. Calvert said with a wink, "Paula, have you been smooching somebody with mono?"

Grinning, Paula shook her head "no." "Just as I thought," said Dr. Calvert. "I want to draw some blood to confirm my diagnosis and then the three of us can talk about how to make you feel better."

2. What other names are used to describe infectious mononucleosis?

3. What other symptoms are commonly associated with this condition? What is the incubation period for infectious mononucleosis?

4. Dr. Calvert indicated that Paula's age was a significant factor in making his diagnosis. Explain.

5. What blood test(s) is Dr. Calvert going to perform to confirm a mononucleosis diagnosis?

Paula received strict orders for immediate bed rest, lots of fluids, and Motrin or Tylenol to reduce her fever and keep her comfortable. Popsicles, gum, and salt water gargles were recommended to ease her sore throat pain, and Dr. Calvert wrote a prescription for **prednisone** to decrease the swelling in Paula's throat and tonsils. He indicated the fever and throat pain would start improving after two weeks, but she could expect the fatigue and lymphadenitis to persist for a month or more. During this time, Paula would need to limit strenuous activities, such as tennis, to avoid possible complications.

"Dr. Calvert," Tami asked, "aren't you also going to prescribe an antibiotic to cure Paula's mono? She seems so very ill. Doesn't she need more than just Tylenol?"

"Good question, Tami," said Dr. Calvert, "but the answer is 'no.' Infectious mononucleosis is not a bacterial infection, so prescribing an antibiotic is not only pointless since it would be ineffective—it could also lead to microbial resistance problems later. Paula is going to be out of commission for a while, but with this recommended therapy, her immune system will be able to make her healthy again."

Lymphadenitis – Swollen glands.

Pharyngitis – Sore throat.

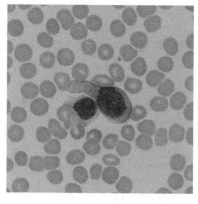

Figure 2. Blood smear showing abnormal lymphocytes associated with infectious mononucleosis.

Prednisone – A synthetic corticosteroid drug, usually taken orally, capable of reducing inflammation.

6. What microbial resistance problem is Dr. Calvert referring to as a result of unnecessary antibiotic use?

7. What is the causative agent of infectious mononucleosis? Briefly characterize this pathogen and its prevalence.

8. Why didn't Dr. Calvert recommend aspirin as a fever reducer?

Paula was crushed when she learned she would have to withdraw from the Tri-County Tennis Tournament and worried about covering her tuition expenses if she couldn't work. She begged Dr. Calvert to let her at least finish the last two matches in the tournament so she wouldn't be letting down her doubles partner. "Sorry, Paula," Dr. Calvert responded sadly shaking his head. "I know how important this is to you, but your health must always be your top priority. We just can't risk internal bleeding from unnecessary physical exertion."

9. How could this infection result in internal bleeding because of strenuous activity?

10. Are there any other sequelae associated with infectious mononucleosis?

As Paula and her mother prepared to leave, Tami said, "Dr. Calvert, I have just one more question. Since we don't know anyone who has the disease, where did Paula pick up mono? How is it spread? Is the rest of our family at risk of infection? Is there anything we could have done to prevent Paula from getting sick?"

"Wow, that's some question," Dr. Calvert said. "Let's talk a little bit about just how contagious this disease really is."

11. How would you answer Tami's questions?

a. How is infectious mononucleosis spread?

b. What is the reservoir for this pathogen?

c. What precautions should be taken to prevent transmission?

d. Is there an infectious mononucleosis vaccine?

e. How contagious is this pathogen?

Epilogue

Paula spent the next two and a half weeks of her summer vacation in bed with her old tabby cat curled next to her for company. Her friends stopped by for short visits, and Della even brought her another strawberry-banana milkshake. Three weeks after diagnosis, her fever vanished and her throat gradually improved. Paula began limited activities and increased the level as her recovery progressed. She was able to go back to work for the last three weeks of the summer and then was off to begin her freshman year of college. Luckily, Paula was back in top form by spring and was able to play second doubles on the varsity squad.

Toxoplasmosis . . . Don't Blame Fluffy!

As part of their commitment to "going green," Layla and Steve Jackson lived on a small farm in rural northwest Pennsylvania. Steve chopped wood from their forest to burn in their Franklin stove, and about 25% of their electricity was generated by the wind turbines on top of their mountain. They raised almost all of their own food between their small apple orchard, huge vegetable garden, and a berry patch. The couple reared numerous sheep, a few pigs, and one dairy cow. These animals plus the rabbit, turkey, and deer Steve hunted more than covered their meat and milk needs. Although the young couple loved working their farm, to make ends meet financially, they also taught at the local high school. Layla was a 10th grade math teacher and Steve served as both the choral and band directors for grades 7–12. Unless the roads were icy in the winter, the "green team" biked the four miles to work every day. While this lifestyle kept them extremely busy, Layla and Steve felt great satisfaction knowing their carbon footprint was significantly less than that of the average American.

As Layla picked the green beans and weeded around the squash one July morning, she pondered how their lifestyle would change in November when their first child was due to be born. She was pleased to be able to raise their child in the unpolluted environment of their country farm and nourish him with homegrown foods free of the pesticides and preservatives found in many commercially produced items. "Of course," Layla said to herself while patting her belly, "I think we'll have to break down and drive a little more this winter. You'll be a bit too small for my baby bicycle seat." As if on cue, the baby started "dancing." Laughing, Layla collected the baskets of produce she had harvested and headed to the kitchen to start making lunch. After rinsing the fresh-picked fruits and vegetables, Layla used her garden's bounty to assemble a delicious salad and homemade strawberry shortcake with cream from Josie, their cow. Later that afternoon, the couple went to Dr. Schneider's office for Layla's monthly prenatal examination. They watched with amazement as the **obstetrician** used **ultrasound** to measure the baby's growth, confirming that Layla was 23 weeks pregnant. Dr. Schneider pointed out different features of their developing child. They saw a tiny beating heart and learned it was time to paint the nursery blue!

Layla's pregnancy progressed normally until her next appointment at 27-weeks gestation. Dr. Schneider was surprised to find Layla hadn't gained any weight in four weeks. "At this stage of your pregnancy, you should be gaining about 0.5 to 1 pound per week," Dr. Schneider said with concern. "Are you eating enough nutritious foods?"

Layla was happy to report a healthy appetite that she regularly indulged with the foods she and Steve raised. "I bet it's the extra exercise I'm getting," Layla explained. "I've been canning produce as fast as I can harvest the garden and orchard. I'm up and down the hillside a dozen times a day hauling a full bushel basket, so I bet I'm just burning off the calories with my gardening." Dr. Schneider cautioned Layla not to be lifting heavy baskets and encouraged her to add an afternoon snack to her usual diet. "I want to see you in two weeks to be sure your weight gain is back on track," Dr. Schneider ordered.

Exactly one week later, Layla called and scheduled an urgent appointment. For the previous 48 hours, Layla had experienced significant vomiting and diarrhea. By the time she arrived at Dr. Schneider's, Layla was weak and slightly dehydrated. Frantic, Layla reported a decrease in fetal activity. Dr. Schneider admitted Layla to the hospital for IV fluids and prescribed medication to ease her GI distress.

Obstetrician – Medical specialist dealing with the care of pregnant women.

Ultrasound – The use of sonography to generate images of a developing fetus.

1. What infections manifest with these symptoms? Are any specifically associated with pregnancy?

An hour later Dr. Schneider performed an ultrasound to check on the progress of Layla's developing son and was shocked to see no fetal growth since her week 23 examination.

In the morning, Dr. Schneider arrived at the hospital early to examine Layla and her baby. Steve was asleep in a chair and Layla sat up in bed weeping gently. She hadn't felt the baby move since midnight. Dr. Schneider confirmed Layla's worst fears with another ultrasound, which showed no fetal heartbeat. Labor was induced to deliver Layla and Steve's 28-week-old stillborn son. A fetal autopsy revealed elevated titers of toxoplasmosis antibodies, **hydrocephalus**, and brain lesions.

2. Describe the causative agent of toxoplasmosis. Outline the life cycle of this microorganism.

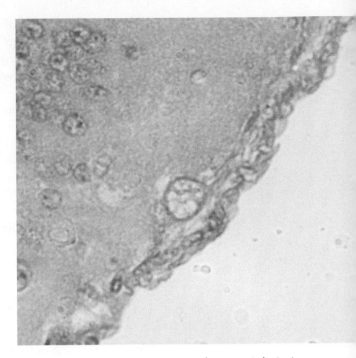

Figure 1. *Toxoplasma* cyst in brain tissue.

A week later, Layla and Steve met with Dr. Schneider to review the autopsy report and answer their long list of questions. "So, I got toxoplasmosis and that's why I was so sick…and then I made the baby sick too," Layla said dismally.

3. Were Layla's symptoms consistent with toxoplasmosis? Describe the usual signs and symptoms of this infection. What is a secondary infection?

"No," Dr. Schneider replied, "not exactly. Your GI symptoms were coincidental and represent a secondary infection. They simply alerted us to the problem with the baby's growth. It does, however, appear you've been infected with toxoplasmosis and the pathogen crossed the placenta to affect your baby.

"Wait a minute," Steve interjected. "How did Layla get toxoplasmosis? I've heard about this disease. Pregnant women get it from changing litter boxes. We don't have a pet cat. In fact, we don't even have strays in our barn!"

4. Are litter boxes a source of pathogen transmission? Explain.

5. What are the most common means of toxoplasmosis transmission? Based on this information, was Layla at high risk for infection? How can toxoplasmosis be prevented?

6. What is the prevalence of toxoplasmosis?

"Dr. Schneider, was there any way to diagnose and treat my infection that might have saved our baby?" Layla asked. "There are different diagnostic tests for toxoplasmosis, but they aren't routinely run in a prenatal panel in the U.S. unless we have reason to suspect infection," Dr. Schneider explained. "If a pregnant woman is infected, several treatment protocols are available, but the benefits must be carefully weighed against the risks since the likelihood of transmission and fetal damage varies with the gestational age of the mother at **seroconversion**.

7. How is toxoplasmosis typically diagnosed? What challenges are associated with interpreting test results? How can fetal infection be determined?

Hydrocephalus – An abnormal accumulation of CSF in the ventricles of the brain resulting in enlargement of the head.

Seroconversion – Point during infection when measurable antibodies are made against the pathogen.

8. Describe the principal treatment protocols for toxoplasmosis infection in a pregnant patient. Indicate the pros and cons of each treatment. When is the best time during the course of *Toxoplasma gondii* infection to administer treatment to a patient?

9. Explain the correlation between the gestational age of maternal seroconversion for toxoplasmosis and the risk of fetal infection.

10. Explain the correlation between the gestational age of maternal seroconversion for toxoplasmosis and the severity of congenital symptoms.

11. What signs and symptoms are expressed by neonates with congenital toxoplasmosis? What symptoms are expressed by infected children within the first year of life? What is the miscarriage rate associated with fetal toxoplasmosis?

"Now that I've been infected with toxoplasmosis, do I need to be treated? If I'm cured can we still have other children, or will I infect them too, causing another miscarriage?" Layla asked hopefully.

"Not to worry," Dr. Schneider consoled the young couple. "You've suffered a devastating loss, but I'm confident you'll soon be parents."

12. Is it necessary to treat Layla for toxoplasmosis? Can she have subsequent children without risking their infection?

D. Infections of the Respiratory System

Your respiratory system is responsible for exchanging oxygen and waste gases between the atmosphere and your body. Additionally, your respiratory system must coordinate activity with your cardiovascular system so the oxygen entering upon inspiration can be ultimately loaded into your red blood cells for delivery to tissues throughout your body. Once again, we are examining a body system with the potential to quickly distribute pathogens and cause serious illness.

Consequently, you should expect to see numerous host defense mechanisms in place. When you inhale air through your nose, it is warmed, humidified, and, most importantly, filtered. Competition with the normal flora of your upper respiratory system makes it difficult for invaders to become established. Many microorganisms become trapped in the mucous lining of the trachea and **plasmolyze**. The action of your ciliated tracheal cells sweeps the contaminated mucous out of your respiratory system so that you can swallow the microbes, allowing their destruction with gastric juices. If any pathogens slip past these defenses, **alveolar macrophages** are available for phagocytic elimination.

It's hard to believe with these specialized defense mechanisms in place that respiratory infections are the number one reason for a visit to the doctor's office in the United States. In fact, pneumonia is the top infectious disease killer globally. Clearly, as a future healthcare provider, you will want to critically assess patients with respiratory illnesses to prevent the deadly consequences of infections that enter the lower portion of the respiratory tract.

Since many pathogens causing respiratory infections are spread by respiratory droplets, it can be challenging to prevent their transmission. However, there are several simple, easy habits you must instill in your future patients. First, always cover a cough and/or sneeze. The best way to do this is to sneeze in your sleeve! By wrapping your arm over your face you can use the large area of your sleeve to trap the sneeze. Microorganisms deposited on fabric quickly **desiccate** and die, minimizing the risk of their transmission. Remind patients that facial tissues are a "single use" item and should be appropriately discarded after one sneeze. Tucking a contaminated tissue into a pocket or sleeve encourages indirect transmission of pathogens. Finally, since many people don't practice correct coughing/sneezing etiquette, remember that most surfaces, including body surfaces, will be heavily contaminated with potential pathogens. Washing your hands with soap and water or using alcohol gels when water is unavailable is an important way to reduce the risk of infection.

Plasmolyze – To lose water from a cell in a hypertonic environment due to osmosis.

Alveolar macrophage – A highly phagocytic leukocyte found in the air sacs of the lungs.

Desiccate – To dry out.

Strawberry Red

Four-year-old Billy was ill and his mother Monica was concerned. He had a rash of fine **papules** that made his skin feel like sandpaper. The red rash began on his upper trunk and spread distally. When you put pressure on his skin, the rash would blanch and return. Little Billy also complained of a headache and sore throat and had a high fever (40°C; 104°F). His tongue also looked like a strawberry with bright red spots on the surface.

1. Based on the clinical presentation what disease does Billy have?

2. How is the pathogen that causes this disease spread from person to person?

3. How does this pathogen cause the red rash and strawberry tongue?

4. Why are high fevers a significant health concern?

Monica knew that little Billy was not the first child in his class to become ill. Billy's preschool class consisted of 26 students ages 4 to 5 supervised by four teachers. Sometimes the class would be combined with a second preschool class for special activities, and Billy's class ate lunch together with some of the older children in the school. Billy's teacher had informed her that five other children in Billy's class had been ill. Also, two older children and a teacher who supervised lunch were affected. Monica had also talked with several of her friends who also had children in Billy's class and found out that their ill preschool children had passed the infection on to other siblings.

After examining Billy, their family physician took a swab of Billy's throat. At the laboratory, the throat swab was streaked onto **Petri dishes** containing **blood agar** and incubated overnight at 37°C. Several bacterial colonies were β-**hemolytic**. Gram staining the β-hemolytic colonies revealed purple, spherical-shaped bacterial cells arranged in chains. **ELISA testing** identified group A antigen on the surface of the bacterial pathogen.

Papules – Small solid rounded bumps rising from the skin with no visible fluid, varying in size from a pinhead to 1 cm.

Petri dish – A shallow cylindrical lidded dish that microbiologists use to culture microorganisms.

Blood agar – A common growth medium used in clinical microbiology labs to grow and identify bacterial pathogens.

β-**hemolytic** – A clear zone surrounding a bacterial colony growing on blood agar. Enzymes secreted by the bacteria rupture the plasma membranes of the sheep red blood cells in the media.

ELISA test – **E**nzyme-**l**inked **i**mmunoab**s**orbant **a**ssay. A rapid noncultural test used to identify the presence of a specific antigen in a sample.

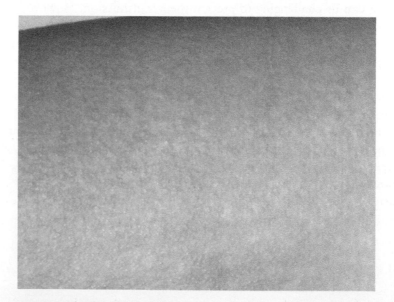

Figure 1. Red rash with fine papules.

5. What is agar?

6. How is blood agar used to differentiate between different bacteria?

7. Describe β-hemolysis. What enzyme is responsible?

8. What information about the pathogen did the results of the Gram stain reveal?

9. Describe the procedure for an ELISA test.

10. Based on the lab results, what pathogen is causing Billy's illness?

Their doctor reviewed Billy's medical history before prescribing an antibiotic to treat Billy. He noted that Billy had developed a rash when treated with penicillin for **otitis media** when he was two years old.

11. What antibiotic would you recommend for Billy?

12. What symptomatic therapy would be appropriate?

13. How would you stop this outbreak from spreading further?

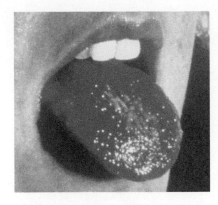

Figure 2. Strawberry tongue.

Otitis media – Inflammation of the middle ear.

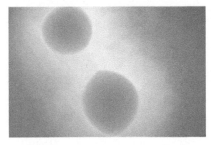

Figure 3. Colony morphology on blood agar plates.

An Evolving Situation

It was the summer of 2009 and Aaron would be fulfilling a life-long dream. Summer break from college had come and now he could check two items off his list of "things I always wanted to do." First on the list was a two-week trip to the Galapagos Islands to see firsthand the finches, tortoises, and lizards that helped Darwin to formulate his theory of evolution. This was followed by a trip to the Amazon rain forest to study the ecology of insects and birds in the forest canopy. He had spent the better part of his free time during his year at college learning to scuba dive and rock climb in preparation for the trip. The only drawback to the trip was getting the necessary vaccinations—Aaron hated needles—and the long flights there and back—Columbus to Miami to Mexico City to Ecuador, followed by boat trips and short flights before arriving in the Galapagos.

1. How can a traveler find out what vaccinations he may need when going to a different country?

The long flight was mostly fun. Aaron played cards and visited with his college friends on the way. He had trouble sleeping, however, as on one of the legs of his trip, several passengers kept coughing. He didn't like the idea of being on a plane with sick people but it wasn't as if he had a choice.

2. What factors would place individuals on the plane at the highest risk for contracting an infectious respiratory disease?

Sixteen people on the boat, and everyone enjoyed the scenery. Diving and working together to complete their research project for their evolution class was a great experience for everyone but Aaron. He was miserable. The previous night, just after their evening meal, Aaron became ill: he had a fever (39°C; 102°F), sore throat, body aches, a severe headache just behind his eyes, **photophobia**, significant fatigue, a nonproductive cough, and a runny nose.

3. List three different causative agents that could account for Aaron's illness.

Aaron remained ill for the next five days and recovered sufficiently to enjoy most of the rest of his adventure. Before the trip was over, two other students came down with the same disease. Lab tests done on one of the students indicated an infection by the influenza A virus.

4. What are the physical characteristics of the influenza A virus?

5. What other viruses can cause influenza?

The news was both a relief and a frustration for Aaron. He was relieved that he hadn't had some exotic disease that would have future complications. He was frustrated that he lost part of his dream trip to something as simple as the "flu." Nine months previously when flu shots were available at the student health center, he didn't get one. He had several tests that week and he thought that the flu wasn't a big deal and that the flu shot was just for old people.

Photophobia – Excessive sensitivity to light.

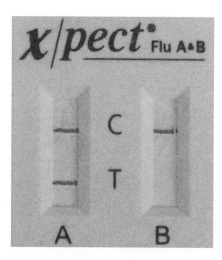

Figure 1. Rapid test for influenza virus infection.

6. Are the flu and influenza necessarily the same disease?

7. Who is at greatest risk for life-threatening complications resulting from an infection by influenza A virus?

8. Why do you need to get vaccinated against influenza virus every year?

9. How is the influenza vaccine formulated?

Although Aaron had not received a vaccination against influenza virus, the other two students had both been vaccinated. Further tests identified H1N1 **antigens** to be present.

10. Why was the vaccination the students received ineffective against this strain of influenza?

11. Describe the role the **H and N proteins** play in the replication of influenza A viruses.

12. Describe two ways that influenza A virus can change its antigens.

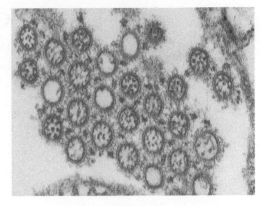

Figure 2. Transmission electron micrograph of influenza A virus strain H1N1.

Antigen – An antibody-generating compound. Often, this is a protein or carbohydrate on the outside of a microbe.

H and N proteins – The hemagglutinin and neuraminidase proteins embedded in the envelope of the influenza A virus.

Q Fever . . . An Occupational Hazard

Dr. Amy Stone left her large-animal veterinary practice early to seek help from her own physician. She felt awful. Yesterday afternoon she developed a fever of 39.7°C (103.5°F), a raging headache, and a nonproductive cough. Amy took Tylenol for her fever and headache and turned in early hoping it was a 24-hour bug. Unfortunately, the new day only brought new symptoms. Amy felt drained, nauseous, and achy all over. Dr. Ferguson's preliminary exam revealed nothing remarkable. She ordered a CBC and "Diff" because of the malaise and fever and recommended extra fluids, rest, and OTC remedies to treat the symptoms.

1. What specific tests are included in a CBC? What helpful information does each test provide to a physician? Which test(s) might explain Amy's malaise? Her fever?

2. What information is collected by a differential count and why is it valuable?

Two days later Dr. Ferguson's office called with Amy's test results. The report revealed an elevated WBC count of $15.2 \times 10^9/l$, a platelet count of $95 \times 10^9/l$, and neutrophils elevated to 80% in her differential count. Since her test results were not suggestive of a specific illness, Amy was instructed to return if her condition persisted for more than 7–10 days or if it worsened.

3. What is significant about the 7–10 day period for symptom abatement?

4. What is a normal WBC count? What does an elevated WBC count suggest? What is a normal platelet count? What medical term describes Amy's platelet count? What conditions are suggested by an increased neutrophil component in a differential count?

For the next four days, Amy dragged herself to work and went through the motions. Luckily, spring "kidding" or birthing season was almost finished for her sheep and goat farmers. "With fewer site visits to care for newborn lambs and kids, I just might be able to manage my work schedule while fighting this bug," Amy thought. However, after a full week of illness, Amy became truly concerned for her health. Her cough was now persistent and productive and she occasionally found herself experiencing shortness of breath. "Not at all acceptable for a 34-year-old woman who jogs daily," Amy worried. Her nausea had worsened and was accompanied by diarrhea. Despite this general decline, Amy didn't contact her physician until she looked in the mirror one morning and noticed a slight yellowish cast to her eyeballs.

Dr. Ferguson was clearly concerned. In one week, Amy had worsened significantly. She presented with a temperature of 39°C (102.2°F) and had lost four pounds. Dr. Ferguson could appreciate **rales** upon inspiration, and Amy was clearly **jaundiced**. Amy was sent for a chest X-ray and more blood work.

5. Given Amy's new symptoms, what medical condition(s) is/are Dr. Ferguson worried about?

Amy's chest X-ray confirmed right lower lobe pneumonia, and Dr. Ferguson immediately initiated antibiotic therapy with erythromycin. The next day the results of Amy's **liver function tests (LFTs)** demonstrated serum aspartate transferase = 92 µ/L, serum alkaline phosphatase = 268 µ/L, and serum albumin = 2.6 g/dL.

Rales – A crackling sound during inhalation heard via auscultation of the lungs with a stethoscope. The sound is the result of small airways and alveoli collapsed by fluid or exudates, opening up with inspiration. This condition is often associated with pneumonia.

Jaundice – A yellowish discoloration of the sclera, skin, and/or mucous membranes due to hyperbilirubinemia typically associated with hepatitis or other liver problems.

Liver function tests (LFTs) – A group of clinical laboratory blood assays used to provide information regarding the state of a patient's liver. These tests often include: **al**anine **t**ransaminase (ALT), **as**partate **t**ransaminase (AST), **al**kaline **p**hosphatase (ALP), **t**otal **bil**irubin (TBIL), direct bilirubin, and **g**amma **g**lutamyl **t**ranspeptidase (GGT). Serum albumin, serum glucose, and lactate dehydrogenase assays may also be included.

6. What are normal values for serum aspartate transferase, serum alkaline phosphatase, and serum albumin? What do Amy's values suggest?

After three days, Amy's fever began to decline and she no longer experienced shortness of breath. A follow-up chest X-ray after two weeks of erythromycin therapy indicated her pneumonia had resolved. Dr. Ferguson ordered another metabolic panel after one month and was pleased to see Amy's liver enzyme values and serum albumin level returning to normal. After approximately six weeks, Amy felt better in general, but still suffered from general malaise, chills, night sweats, continued diarrhea, and weight loss. Amy reviewed these symptoms with Dr. Ferguson seven months later when she had her regularly scheduled annual physical examination. Against Amy's protests, Dr. Ferguson insisted on performing an HIV test.

7. What symptoms made Dr. Ferguson suspect a possible HIV infection?

8. What test(s) is/are commonly used to diagnose HIV infection?

Before she could say "I told you so," to Dr. Ferguson for running what Amy knew was an unnecessary HIV test, she found herself admitted to the ICU feeling worse than ever. Her fever was back (40°C; 104°F) along with the previous aches, pains, and intense fatigue. She was experiencing shortness of breath with even slight physical activity, swelling of her feet and legs, chest pains, and small **petechial hemorrhages** had appeared on her palms and under her fingernails. Dr. Ferguson ordered blood cultures, a CBC, and LFT. With her veterinary training, Amy recognized the classic symptoms of infective endocarditis. She had taken special precautions to avoid this problem ever since her diagnosis with mitral valve prolapse (MVP) at age 12. "How could this have happened?" Amy wondered.

> **Petechial hemorrhages** – Small (1–2 mm) red or purple spots on the body, caused by a minor hemorrhage due to broken capillary blood vessels.

9. What purpose does a blood culture serve if Amy is having a problem with her heart?

10. What is endocarditis? Is this serious? Who is at greatest risk of infection?

11. What is MVP? Why does Amy's MVP diagnosis make her cautious?

Although her blood cultures failed to grow microorganisms within a week of incubation, Dr. Ferguson was able to give Amy a definitive diagnosis in just two days. Serological studies indicated the presence of elevated levels of antibodies

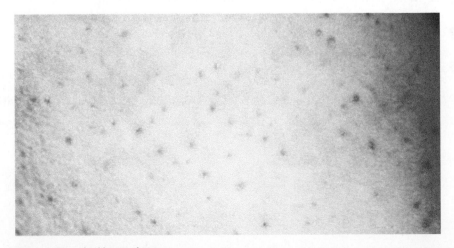

Figure 1. Petechial hemorrhages.

to *Coxiella burnetii* phase I antigens in combination with declining levels of antibodies to phase II antigens, suggesting chronic Q fever. Dr. Ferguson ordered an antibiotic combination therapy of doxycycline and hydroxychloroquine for two years.

12. Describe *Coxiella burnetii*. Where is it commonly found? How is it typically transmitted? How long is the incubation period?

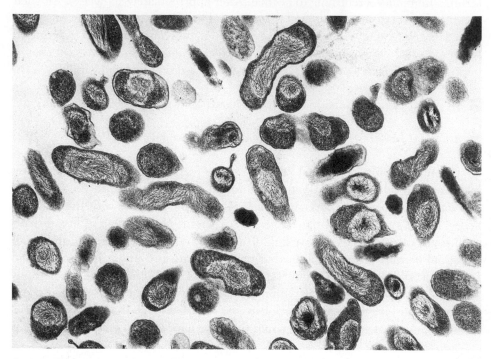

Figure 2. Transmission electron micrograph of *Coxiella burnetii*.

13. What is Q fever? How did it get its name? Why didn't *C. burnetii* show up in Amy's blood cultures? Review the symptoms associated with acute versus chronic Q fever.

14. In addition to monitoring antibody levels to the phase I and II antigens, how else is Q fever diagnosed?

15. What problems are associated with this antibiotic therapy?

16. What risk factors did Amy have for Q fever? For infective endocarditis?

17. How can Q fever be prevented?

18. Is Q fever a reportable disease? What specific concerns are associated with monitoring for Q fever?

Swimming Pool Blues

Summers were hot in western Oklahoma, so Denise was glad they had bought a season pass to the swimming pool in their small town. For families with children, the pool was often the center of summertime activities when the kids weren't playing games at the ball field or enjoying family picnics at the park. However, summer activities were interrupted in early July when Denise's two children, an 8-year-old girl and an 11-year-old boy, both came down with a fever, sore throat, and a headache.

● **1.** List and describe three respiratory pathogens that could account for the children's illness.

Local physicians noted that many others in the community were also experiencing the same symptoms. In addition, about one-half of those affected also had conjunctivitis.

● **2.** Which viral cause of **pharyngitis** and fever is also often associated with conjunctivitis?

● **3.** What other diseases are also associated with this pathogen?

● **4.** If the outbreak is caused by this virus, how would you treat those affected?

Since there were a large number of cases in a short period of time, an investigation was undertaken by the Public Health Service to identify the risk factors associated with the illness to help in determining the source of the disease.

Seventy-seven persons were identified with the illness. Onsets of illness peaked during the week of July 5 to 12. A telephone survey of families that lived in town revealed that there was a significant association between coming down with pharyngitis and fever and having spent a large number of hours swimming at the pool each week. Swimmers who reported swallowing pool water were more likely to be ill (29 of 56) than persons who did not (10 of 41).

● **5.** Was swallowing water significantly associated with acquiring the illness?

The pool chlorinator had reportedly malfunctioned during early July. After repair and proper chlorination of the pool, the outbreak resolved.

● **6.** Besides exposure to underchlorinated pool water, describe other methods of transmission of this virus.

● **7.** How does chlorine kill or inactivate microbial pathogens?

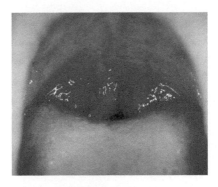

Figure 1. Pharyngitis.

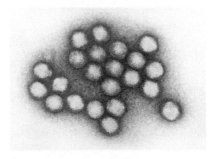

Figure 2. Transmission electron micrograph of viral pathogen.

E. Digestive System Infections

Infections of the gastrointestinal tract rank second behind respiratory ailments as the reason that brings patients to visit their physician. Unfortunately, most of us have had firsthand experiences with several of these common infections, which include everything from dental caries to gastric ulcers and hepatitis to intestinal parasites. In the United States there are well over 225 million cases of acute diarrheal disease annually, with approximately 20% of these patients seeking medical assistance. From a global perspective, diarrhea ranks number three for fatalities caused by infectious disease.

Since your digestive system is responsible for the crucial functions of food digestion/absorption and waste elimination, any disruption due to an infectious process compromises the function of your other body systems. Unfortunately, there are numerous opportunities for pathogens to invade and attack your digestive system. Ingestion of contaminated food and water is usually responsible for transmission. To prevent the adherence and growth of pathogens after entering your body, the digestive system employs a plethora of antimicrobial defense mechanisms. The secretion of saliva and scraping action of your tongue can reduce adherence of microorganisms in the oral cavity. Covering the alimentary canal with mucus prevents microbial adherence and can **plasmolyze** some organisms. Secretions containing **lysozyme** and **IgA** will degrade bacterial cell walls and enhance **opsonization**, respectively. Your digestive system is home to an immense number of normal microbial residents. The competition with these organisms for nutrients and space is intense. Additionally, many of these species secrete antibacterial substances to discourage newly introduced species from encroaching upon their habitat. Digestive enzymes, bile, and low pH can also damage many nonadapted microorganisms, and peristalsis typically moves materials, including "unwanted visitors," through the system within 24 hours.

Cleverly adapted microbes have evolved to overcome one or more of your digestive system defense mechanisms. The result, of course, is that we have all experienced the nausea, vomiting, cramping, diarrhea, and so on, associated with a "GI infection." Since most pathogens gain access to your digestive system via the fecal-oral route, you should focus your attention on the prevention of these disorders as you analyze cases in this unit. Concentrate on ways to improve the quality of consumables and general hygiene to reduce pathogen transmission. Remember that even if food is initially safe, preparing, handling, and consuming it using unwashed hands will dramatically increase your risk of infection.

Hand Washing ABCs

Courtney was getting the message after receiving a "Washing Your Hands Is Fun!" brochure from little Roger's day care center, a "Hand Washing A, B, Cs" flier from Susan's kindergarten class, and a "Henry the Hand" coloring book from Kort's preschool. It all stemmed from an outbreak of shigellosis associated with Kansas City day care centers.

1. The transmission of what type of pathogens is inhibited by hand washing?

2. Describe the proper method for hand washing using soap and water.

3. What pathogen(s) cause shigellosis?

4. Describe the clinical features of shigellosis.

5. Why are children in day care centers at a higher risk than others for acquiring respiratory and diarrheal illness?

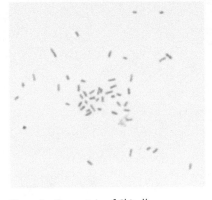

Figure 1. Gram stain of *Shigella*.

Last night's news reported there was at least one child reported to have had shigellosis in 42 different day care centers. In addition, there were over 600 confirmed cases in the Fayette County—74% of the cases in children less than 10 years old. A large concern for Courtney was that this was not a typical diarrhea-causing pathogen; the news had reported that it was multidrug resistant. From the newspaper, she had found out that 89% of the pathogens characterized were resistant to ampicillin and bactrim but sensitive to ciprofloxacin.

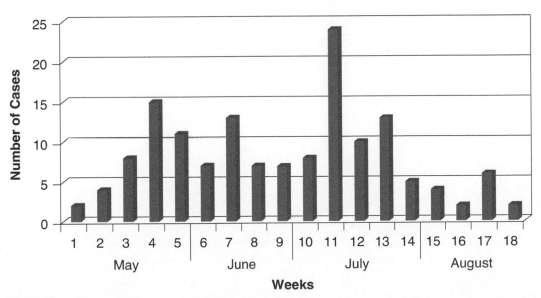

Number of confirmed cases of *Shigella sonnei* infection, by week of illness onset—Fayette County, Kentucky, May 1–August 31, 2005

Figure 2. Shigellosis outbreak for May through August 2005.

6. How is shigellosis treated?

7. Describe the virulence factors associated with the pathogen.

8. Why is it important that the antibiotic sensitivity profile of the pathogen is characterized before antibiotic therapy is begun?

9. What antimicrobial agents are found in bactrim?

As in many families, both Courtney and her husband worked full-time jobs. As a result, child care was an inevitable part of their lives. At dinner time, she supervised their family hand-washing time to make sure it was done correctly and to try to make it entertaining by singing silly songs with the kids while they scrubbed their soapy hands. Her chief concern was to keep her children healthy, but this illness came with additional concerns. The day care centers and schools had informed parents that state law says that any child with shigellosis cannot return without documenting that they have completed antibiotic therapy and have had two consecutive stool cultures test negative for the pathogen.

10. How do you test for the presence of this pathogen in a feces specimen?

11. What is the rationale behind this kind of public health legislation?

Courtney felt good about doing her part to help keep her children healthy. She sincerely hoped that those at the day care center and school would do the same.

12. What activities/policies would you put in place at the day care center to minimize the risk of spreading this pathogen?

A Distressing Side Effect

Edith was a 67-year-old grandmother from western Pennsylvania. In early December she retired from her position as the office manager of a construction company and was eagerly looking forward to finally having the time to regularly visit her granddaughters in northwestern Ohio. While there during January, Edith was annoyed to realize she was developing a urinary tract infection (UTI). She took a urine sample to the urgent care clinic in town. Her specimen was cultured and grew Gram-negative bacilli at >10^6 **CFU**/ml. The physician at the clinic prescribed "Cipro" and within 24 hours, Edith was already feeling better.

1. How should a urine sample be collected to ensure accurate laboratory results? As a nurse, what can you do to encourage this practice?

2. What is the standard protocol for culturing a urine specimen?

3. What symptoms are usually associated with a UTI? Why was Edith more susceptible to this infection than her husband?

4. What does "Cipro" stand for? What is the mechanism of action of this antibiotic?

About a week after returning home, Edith's UTI also returned. Her family physician prescribed a second round of Cipro, but within two weeks of finishing the drug, her symptoms reappeared. Explaining that she was probably dealing with a microbe that was resistant to this antibiotic, Edith's doctor switched her to a prescription of clindamycin.

5. How does the mechanism of action for clindamycin differ from Cipro?

6. Edith's doctor recommended she start eating yogurt to avoid what common secondary infection often experienced by women on long-term antibiotic therapy? How does this work?

Edith enjoyed relief of her UTI symptoms within 48 hours after starting the clindamycin. She was thankful to have this annoying infection resolved, but quickly became discouraged when she developed diarrhea after five days of clindamycin therapy. She adjusted her diet and took kaopectate to manage her diarrhea. Two weeks later, Edith's diarrhea was substantially worse. She had lost six pounds and most of her energy, and had a sore lower abdomen.

7. What do you suspect is causing Edith's problem?

After visiting her physician again, Edith was instructed to boost her fluid intake, consume no fiber-containing foods, and take Imodium. Another three weeks passed with her diarrhea continuing to worsen. Edith was fatigued and had lost a total of 21 pounds. When she arrived in the emergency department of the county hospital, IV fluids were initiated for rehydration and a stool specimen was collected. After thoroughly reviewing Edith's recent history, the attending physician prescribed yet another different antibiotic for her. He explained to Edith that the clindamycin she took solved her UTI problem but initiated her diarrheal disease. To Edith, this just didn't make sense. Antibiotics are supposed to kill germs, not encourage new infections.

CFU – **C**olony **f**orming **u**nit; this is usually interpreted to be an individual microbe because, following its inoculation on appropriate media, it can grow by binary fission to produce one colony. This term is often used to indicate the concentration of microbes in a sample. (*Note*: The CFU value may be lower than the actual number of microbes present, since the deposition of microorganisms adjacent to one another on the medium will give the appearance of one colony following their growth.)

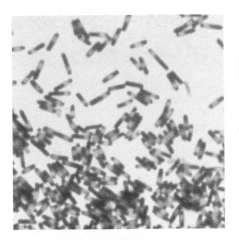

Figure 1. Light micrograph of pathogen.

8. What microbe is causing Edith's diarrhea? What is the name of this disorder? Outline the basic characteristics of this microbe.

9. Explain how her previous antibiotic history resulted in this condition.

10. Is the diarrhea caused directly by the presence of this microbe? Explain.

11. What other virulence factors are associated with this microbe?

12. What kind of damage is done by these components?

13. What test(s) was performed on the stool specimen to confirm this diagnosis? What are the pros and cons of these procedures?

14. What antibiotic therapy did the attending physician prescribe? Why? What side effects are relevant?

Edith's condition improved rapidly with her new treatment, but three days after completing the course of antibiotics, the diarrhea returned. She was frantic. This just had to stop so she could get her life back to normal. Edith returned to the emergency room to seek the aid of the physician who helped her previously. He immediately recognized her emotional distress and was able to console her. He explained to Edith that 20–30% of patients with this infection relapse after therapy due to the special nature of this microbe. He prescribed a second round of her antibiotic treatment and assured Edith that this would solve her problem.

15. Explain why relapse is so common with this particular microbial infection.

16. This infection is considered a huge nosocomial problem. How could it be spread within a hospital setting? Who is at greatest risk of acquiring the infection?

17. This condition was not really an issue until the mid-1970s. Why?

18. Clearly, an imbalance of normal flora can have devastating effects. Complete the table below to review the microbes normally associated with specific regions of the GI tract. Be sure to indicate the relative abundance of microbes in each region.

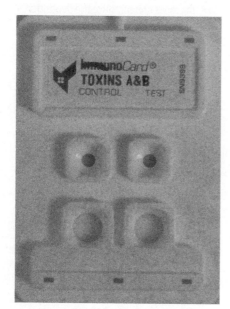

Figure 2. ELISA test for toxins produced by the pathogen.

GUT FLORA REVIEW		
GI Region	**Normal Microbiota**	**Relative Abundance**
Esophagus	_____	_____
Stomach	_____	_____
Duodenum	_____	_____
Jejunum	_____	_____
Ileum	_____	_____
Colon	_____	_____
Rectum/Feces	_____	_____

Hold the Onions

Every Friday night, Darlene and her husband would eat out. It was a time to relax and enjoy each other's company and enjoy a good meal. This Friday, however, would not be one of those nights. This whole week had not been a good one. She had suffered from mild flu-like symptoms—**anorexia**, nausea and vomiting, fatigue, **malaise**, **myalgia**, a mild headache, and a low-grade fever. Although she should have stayed home from work, her local employer who provided vacation pay, sick days, and excellent health insurance had been bought out by a larger company. She was happy to still have minimal health insurance coverage, but she no longer received paid vacation or sick days. "Living paycheck to paycheck doesn't allow for any time to be sick," she thought, but now she had no choice. She had developed **jaundice**, her urine had turned dark, and she suffered from abdominal pain.

Anorexia – Not having an appetite for food.

Malaise – A general feeling of being unwell.

Myalgia – Muscle aches.

Jaundice – A yellowish discoloration of the skin and eyes caused by increased levels of bilirubin in the blood.

Hepatitis – Inflammation of the liver.

1. What infectious diseases are associated with jaundice?

2. What pathogens are associated with these diseases?

3. Jaundice indicates damage to which organ?

Her family physician did an examination, took a blood sample, and asked questions about her history. Had she traveled overseas recently, did she use IV drugs, and did she eat at a popular local restaurant in suburban Pittsburgh? Darlene laughed at the questions and told the doctor that a 60-year-old lady like her couldn't afford to miss work, much less travel or waste money on illegal drugs. She did express her love of Mexican food and said that she had eaten at the restaurant in question last month during one of their Friday night outings. The physician explained that a large outbreak of **hepatitis** had been traced to the restaurant and several hundred people had become ill. The suspected source was green onions imported from Mexico that were used to prepare the mild salsa.

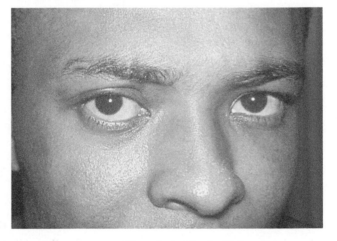

Figure 1. A patient with jaundice caused by the pathogen.

4. Why did the physician ask Darlene these questions if he suspected she had hepatitis?

5. What test is used to identify the presence of a hepatitis-causing virus?

6. Which hepatitis-causing virus is most likely the cause of Darlene's illness?

7. How could green onions imported from Mexico be the source of the pathogen?

Although frustrated that her pleasant evening out may be the cause of her illness, Darlene was confident she could recover quickly now that the doctor knew what disease was causing her problems. She was hoping he could prescribe an antibiotic and she could be back to work on Monday.

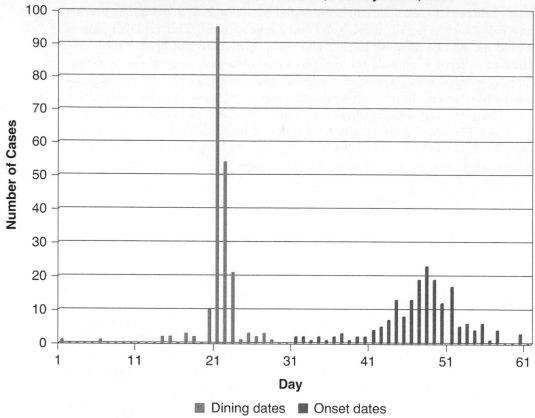

Number of hepatitis A cases, by day of eating at Restaurant A and illness onset—Monaca, Pennsylvania, 2003

Number of Cases

Day

■ Dining dates ■ Onset dates

Figure 2. Hepatitis outbreak from restaurant.

8. Can antibiotics be used to treat hepatitis caused by a virus?

9. What antiviral medication would be appropriate for Darlene?

10. Do you expect Darlene will be back to work in several days? Explain your reasoning.

11. How would you prevent this type of outbreak in the future?

Diarrhea 101

No class on Friday! Tom should have been planning a party with his friends or at least relaxing playing his favorite video games. However, the thought of beer and burgers or trying to navigate his avatar through a dizzying virtual video-game maze made him more nauseous. There was little doubt—he had caught the "stomach flu" that had been sweeping through the campus. In just a week about 400 Hope College students had become ill with **gastroenteritis** that the health department said was caused by a norovirus. The outbreak had prompted the Ottawa County Health Department to order the campus closed on Friday.

1. What are the clinical features of gastroenteritis caused by a norovirus?

2. Does having a norovirus infection provide life-long immunity for future infection?

3. How would closing the campus reduce the spread of the pathogen?

4. What are the physical characteristics of a norovirus?

After a miserable weekend, Tom was feeling better and began browsing Facebook pages of friends to see what he had missed, before he started looking over his notes to get ready for going back to class on Wednesday. One student had been asking students to report whether or not they had the stomach flu during the outbreak on his page entitled "The Great Plague of 2008." Another was selling T-shirts with "Norovirus 08" printed on the front and "victim" on the back. Tom also read that campus cleaning crews were busy inside campus buildings **sanitizing** common surfaces.

5. Describe the pathogenesis of norovirus.

6. Why is sanitizing common surfaces an important part of stopping the outbreak?

Gastroenteritis – An inflammation of the GI tract resulting in acute diarrhea and vomiting.

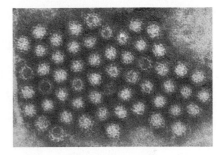

Figure 1. Electron micrograph of norovirus.

Sanitize – The process of using chemical agents and cleaning to reduce the number of microbes present to levels that are safe for public health.

An Uninvited Party Guest

On a Monday morning in early July, Angela Thompson (31) and her children Natalie (9), Jason (6), and Marie (18 months) showed up in Dr. Jenkins's office without an appointment. All complained of severe, smelly diarrhea, gas, and abdominal cramping, which began Friday evening. Since these general symptoms are associated with dozens of self-limiting viral infections that are easily transmitted among family members, Dr. Jenkins was not overly concerned. He recommended Imodium to stop the diarrhea, increased fluid consumption to prevent dehydration, and diet modification, including yogurt as a **probiotic**. He collected a stool sample from little Marie, who had conveniently soiled her diaper during their office visit.

The next day Dr. Jenkins was surprised to examine six other patients with similar symptoms. A few of these patients also complained of nausea, headache, and mild fever. By Thursday morning Dr. Jenkins was truly concerned. Eighteen patients in four days had presented with the same symptoms and he had even admitted one to the hospital. Jake (38) had been diagnosed HIV⁺ seven years earlier. Since he wasn't especially compliant with his prescribed antiviral therapy, Jake had suffered from several opportunistic infections associated with his immunocompromised status. When he arrived at Dr. Jenkins's practice today, Jake was extremely dehydrated and had lost seven pounds in seven days due to the severe diarrhea.

By late afternoon the results from Marie Thompson's stool specimen were back from the laboratory. The report confirmed giardiasis.

1. What microbe is responsible for this condition? Characterize this organism.

2. Where is this microbe commonly found? How is it transmitted to humans?

3. What is the infective dose for giardiasis? What is its incubation period?

4. Outline the symptoms associated with this disorder. What is the duration of this illness? Are infected individuals always symptomatic?

5. Who is at greatest risk of infection?

6. How is giardiasis diagnosed?

7. How is this infection treated?

Dr. Jenkins contacted the county health department to report a suspected outbreak. With their help, stool samples were collected from all of his affected patients and other local physicians were notified. Workers from the health department were soon able to establish an epidemiological link between the patients. Ten days earlier everyone had celebrated the town's centennial, which included a strawberry festival featuring shortcakes made with the locally grown berries of the Taylor Family Farms. After a visit to the farms, it was determined that the berries were irrigated using ditch water. Sampling water from various ditches in the vicinity of the fields yielded approximately half of the specimens positive for the cysts of the suspected pathogen. The sale of produce from the Taylor Family Farms was suspended, and the proprietors were ordered to use an appropriate water source for irrigation.

Probiotic – Introduction of microbes into a host to balance normal flora or to boost mucosal immunity.

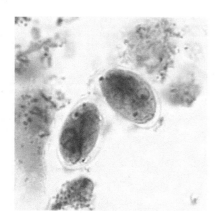

Figure 1. Light micrograph of *Giardia* cysts.

8. How did this microbe get into the irrigation water supply?

9. Review the life cycle of this organism, beginning with the cysts that were ingested.

10. What agricultural practices could be employed to prevent contamination of crops?

11. How is giardiasis transmitted? How can you prevent transmission from an infected individual to others?

F. Infections of the Urogenital Tract

As the term implies, urogenital infections refer to conditions where components of the urinary and/or reproductive systems have been invaded by pathogens. Many **STI**s are epidemic among young adults worldwide. The increase in population, widespread use of the birth control pill, and an increase in the number of sexual partners have resulted in global increases in STIs that cause gonorrhea, chlamydial infections, genital warts, cervical cancer, **AIDS**, and genital herpes.

Some of these infections, such as gonorrhea, are still easily treated with antibiotic therapy. Genital herpes can be managed with antiviral medications but not cured, potentially resulting in a lifetime of recurrent infections and discomfort. Still other STIs, like **HIV** infection, which ultimately leads to AIDS, will prove fatal even with medical intervention. Most STIs can be prevented by practicing "safe sex" measures including the consistent and correct use of condoms. A notable exception to this rule is HPV infection, as latex condoms do not prevent transmission, leading potentially to the development of genital warts or cervical cancer. Vaccine development to prevent these STIs has been intensely researched with some success. The HPV vaccine targeted against high-risk serotypes has been shown to be effective. However, the search for an effective HIV vaccine continues.

Normal skin flora, as well as fecal contaminants from the anus, often enter the urethral opening. Luckily, urination can usually flush out even the flagellated microbes before they swim up to the bladder and initiate infection. While both males and females can suffer from infections of the urinary system, women are at a significantly higher risk. Since the female urethra is positioned very close to the anus and is substantially shorter than the same structure in the male, more bacterial contaminants are usually present and have a shorter distance to travel to cause a UTI. Drinking 8+ glasses of water daily to encourage frequent urethral washing through micturition coupled with careful attention to personal hygiene can significantly reduce urinary contaminants and consequently their associated infections.

Again in this unit, the cases presented represent a wide variety of diseases impacting the urogenital system in a diverse group of patients. Pay special attention to factors that increase an individual's risk of acquiring one of these infections. You will soon discover that simple, specific safeguards make these diseases among the easiest to prevent...if not for the human factor!

The Honeymoon Is Over

The last month had been a blur! Brittani took two weeks of vacation before the wedding so she and her mother could take care of all of those little extras that would make her special day truly memorable. They made floral sprays for the end of each church pew; selected a beautiful unity candle; baked dozens of fancy cookies to serve their guests at the reception; loaded small chiffon bags with bird seed and glitter to toss at the happy couple; and finally, they treated themselves to a spa day complete with manicures and facials.

Surrounded by the glow of hundreds of candles, the service was everything Brittani could have hoped for and more. Cousin Allison sang like an angel and Brittani floated down the aisle looking radiant. From under the blusher of her veil, Brittani was thrilled to see Kevin beaming at her from the front of the sanctuary. She was surprised to see tears in her father's eyes as he lifted the blusher, gently kissed her cheek, and joined her right hand to Kevin's left. After a boisterous evening of dining and dancing, Kevin carried his bride out of the reception hall to their comically decorated car amid a riot of cheers and best wishes from their friends and family. They were finally off on their honeymoon. Since Kevin's mother was a travel agent, she helped them book a week-long trip to Hawaii. The happy couple kayaked on Kauai, rode horses through pineapple fields on Maui, and paid their respects at the *USS Arizona* on Oahu.

Their honeymoon was great...really great! Growing up in a very conservative household with strict parents, Brittani was taught to wait for marriage before having sex. She was so pleased when she and Kevin began dating that he respected her decision. Consequently, they had both really looked forward to their honeymoon. Now that they were home and settled in their cozy little apartment, Brittani was delighted to see that Kevin remained very amorous. But after three weeks of married life, Brittani woke up early one morning and hurried to the bathroom. She gave a little "yipe" as she started to void and clenched down hard to stop the flow of urine. It was a sharp burning sensation she had never experienced before. Trying to control her discomfort, Brittani gradually relaxed her perineal muscles, allowing the urine to dribble out of her bladder. Unfortunately as she did this, Brittani noticed cramping in her lower abdomen. Even after she had finished in the bathroom, Brittani still felt the urge to urinate. A bit self-conscious, she quickly showered, dressed, and started breakfast so she could avoid Kevin and what was usually a romantic start to their day.

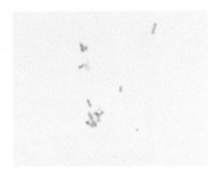

Figure 1. Gram stain of a common bacterial cause of cystitis.

1. What is your diagnosis of Brittani's condition? What symptoms typically present with this infection?

2. What microbial species are the most common culprits? How are the pathogens transmitted? Identify any virulence factors associated with these microorganisms.

3. What individuals are at greatest risk of acquiring this infection? Why?

Brittani startled while pouring the coffee as Kevin slipped up from behind, circled her waist with his arms, and nuzzled her neck. "I missed my beautiful wife this morning," he said playfully nibbling an earlobe. Flustered, Brittani stepped out of his embrace and led Kevin to the table where she announced she had made his favorite, blueberry pancakes. "Oh, so you do still care," Kevin teased.

"Of course I do. You just looked so peaceful this morning," Brittani lied. "I couldn't bear to wake you. So I cooked for you instead!"

"Well, a man has to eat," Kevin said tearing into a large stack of pancakes. "I guess I'll just have to look forward to our time together this evening," he said with a wink. Brittani blushed and worked on her own pancakes. As they kissed goodbye and went their separate ways to work, Brittani felt relieved that she had successfully concealed her problem from Kevin. But before driving to the office, she slipped back into the apartment to use the bathroom one more time and was frightened to discover a small amount of blood on the toilet tissue after wiping.

Once at work, Brittani looked up her friend Aubrey, who was a 30-year-old mother of two. Maybe she could help. Although she expected sympathy, Aubrey laughed when Brittani described her problem. Aubrey laughed again at Brittani's shocked expression. "I'm sorry, hon, but you know you brought this on yourself. It's all that great 'nookie' you're getting these days."

Now Brittani really looked stunned. "What are you talking about? It's not a **STI**. It's just me and Kevin and it doesn't hurt when we do...that. It hurts when I pee," Brittani said crossly in a loud whisper.

"Oh, honey, you are new at this. Of course it's not a STI. I just mean your little problem can develop from having sex. Lots and lots and lots of sex!" Aubrey laughed, nudging her shy friend with an elbow.

Absolutely scarlet now, Brittani hissed, "Well, ok, but what do I do about it?"

"For starters, you can relax. This is pretty common and easy to fix. After that, you can do several simple things to start feeling better now, but you should also check in with your doctor."

4. What common helpful advice or comfort measures should Aubrey provide?

Brittani was uncomfortable all day. She constantly felt gentle cramping and urgency. Every time she went to the bathroom, voiding burned more than before. She followed Aubrey's advice and called her doctor to schedule an appointment for the next morning. Brittani also tried her friend's "quick fix" advice but didn't obtain any relief. On the way home from work, she stopped at Rite-Aid and talked with Judy at the pharmacy counter. Taking Brittani to the **analgesics** aisle, Judy recommended a couple of **OTC** products to make her more comfortable overnight. Judy also urged Brittani to be sure and keep her doctor's appointment the next day, since the medication would not cure her infection.

5. What common OTC products did Judy likely recommend? What is the active ingredient? Are there any side effects? When is this medication contraindicated?

When she arrived home, Brittani took her medication and started preparing dinner. Within 90 minutes she was pleasantly surprised to find her cramping and urgency had disappeared. On her next trip to the bathroom, Brittani braced herself for the intense burning as she began to void and was again pleased to find another symptom had vanished. Brittani briefly toyed with the idea of cancelling her appointment with Dr. Nelson, but remembered Judy's warning. Brittani also started to worry that if she had an infection and acquired it from having frequent intercourse, was she also putting Kevin at risk? She decided to confess her condition to Kevin as soon as he came home so he could go to the doctor with her and also get help if necessary.

As they finished their supper, Brittani excused herself to go to the bathroom and found the effectiveness of her urinary analgesic was beginning to fade. When she returned to the table, Brittani told Kevin about her day. "Baby, why didn't you tell me right away?" Kevin coaxed. "You've been miserable all day long. We're not waiting any longer. C'mon, I'm taking you to the urgent care right now."

6. Is it possible for Brittani to transmit her infection to Kevin?

7. Should Brittani have waited until the next day to seek medical care or was Kevin's immediate action prudent?

Half an hour later Brittani was describing her symptoms to the nurse who was completing a basic physical assessment. "Before you see Dr. Schmidt, you'll need to provide us with a 'clean catch specimen' for laboratory analysis. Here is your collection cup, towelette, and instruction sheet. You can use the bathroom at the end of the hall. When you're done, leave your specimen on the shelf in the bathroom and wait in exam room 3 for Dr. Schmidt."

8. What activities will the nurse perform with her basic assessment?

9. Describe the process of collecting a proper clean catch specimen. Why is this a critical diagnostic step? What advice would you offer the nurse when she instructs a patient to provide a clean catch specimen? What tests will be run as part of a routine urinalysis and what is the significance of each test?

"So Brittani, I understand you're having a lot of urinary discomfort," Dr. Schmidt said, bursting into exam room 3 without an introduction. Brittani blushed and nodded. "Your temperature is slightly elevated at 37.4°C (99.4°F). Given your current symptoms of urgency, frequency, burning, and muscle spasms in your bladder plus your status as a newlywed, I would say you have a classic case of 'honeymoon cystitis.'" Looking up from the chart for the first time, Dr. Schmidt instantly recognized his young patient's embarrassment and confusion. Changing his approach, Dr. Schmidt said in a much quieter tone, "Let's talk for a minute. There is nothing to worry about and certainly no reason for you to feel embarrassed. This condition is fairly common, particularly in women who have had a recent increase in their sexual activity. I can give you a prescription for antibiotics and have this infection cleared up in no time! I also have a few suggestions for you to prevent the infection from coming back." Noting Brittani's stunned expression, Dr. Schmidt added, "And it won't impact your love life!" Brittani blushed again.

10. What does Brittani's temperature suggest?

11. Why is Dr. Schmidt planning to give Brittani antibiotics prior to definitively determining a bacterial infection?

12. What advice can Dr. Schmidt give a newlywed to decrease the risk of acquiring this infection?

"Now, are you allergic to any antibiotics?" Dr. Schmidt inquired. Brittani shook her head no. "Are you pregnant?" he followed up.

"Dr. Schmidt, I've only been married three weeks. I couldn't possibly be pregnant yet," Brittani responded with alarm.

"Are you consistently using contraception?" Dr. Schmidt pressed.

"Well, yea. Most of the time we use condoms," Brittani replied.

"And the rest of the time?" Dr. Schmidt continued to push the issue. When Brittani stared blankly and then looked sheepishly at the ground, Dr. Schmidt said, "I see. When was your last menstrual period?"

"About two weeks before the wedding, so I should be starting any day..." Brittani stopped talking abruptly as a startled look registered on her face. "Alrighty then," said Dr. Schmidt. "Let's run a quick test before I write this antibiotic prescription."

Twenty minutes later Dr. Schmidt burst back into exam room 3 with a big smile and announced, "It looks like you brought more than souvenirs back from your honeymoon. Congratulations, you're pregnant! Here is a sample of Bactrim for you to start tonight. Have this prescription filled in the morning and be sure to finish the whole bottle. This is important for your health and for your baby. Also, tomorrow morning, I want you to call this number and schedule your first prenatal appointment with Dr. Ryan," he said handing Brittani a prescription slip and Dr. Ryan's business card. "My nurse will call to follow up with you in a few days when your urinalysis results are back from the laboratory. Now, go give that handsome young man in my waiting room the good news."

Shell-shocked, Brittani nodded and stumbled out to her husband in the waiting area. Dr. Schmidt smiled broadly when a few minutes later he heard an excited "whoop" from the father-to-be.

13. Why does Dr. Schmidt want to determine if Brittani is pregnant before prescribing an antibiotic? What is the method of action for Bactrim? "Cipro" is one of the most commonly prescribed antibiotics for this condition. What complication prevents using this drug during pregnancy?

14. Why should Brittani finish her entire antibiotic regimen? How could discontinuing her antibiotic impact her pregnancy?

15. Are there any sequelae associated with this infection, especially if left untreated?

(*Note:* "Honeymoon cystitis" is often actually urethritis. If the pathogens have not reached the bladder, patients typically experience all of the symptoms noted by Brittani minus the bladder spasms.)

Sex, Drugs, and Rock and Roll

Jim was a police officer and his wife Barb a nurse at the hospital that handled most of the city's poor. It was inevitable that their paths would occasionally cross during work. Tonight, they were both at a community outreach meeting concerning the problems caused by the increase in crystal meth use. Methamphetamine, or crystal meth, is a powerfully addictive stimulant that has an intense euphoric effect. Jim saw its effect when chronic crystal meth users would embark on binges of constant meth use. The results were universally disastrous—intense paranoia, visual and auditory hallucinations, and violently out-of-control behavior. Barb saw another side of abuse of the drug. Crystal meth use has a potent effect of increasing the sex drive. As a result, crystal meth users were more likely than others to engage in high-risk sexual behaviors and have more sexual partners than non-users. Barb had seen a significant increase in cases of gonorrhea in general, and increases in syphilis and HIV disease among gay men. Long-term users of crystal meth build up a tolerance to the drug. As a result, many choose to inject the drug to continue to get high. Not unsurprisingly, intravenous drug use increases the spread of HIV as users share needles (and therefore exchange small amounts of blood). At counseling programs designed to help HIV-positive gay and bisexual men who use crystal meth, about half had injected meth during the last year.

Neither Jim nor Barb was the type to sit back and hope for the best. They had much invested in their community where their children went to school. Both were consistent volunteers. Jim already coached soccer and led scouts. Barb was active at their church and volunteered to help students with reading and math at school. At the end of the meeting, both took part in the discussion and planning sessions on how to help the community stem the problems caused by widespread crystal meth abuse.

1. What types of behaviors are considered high risk for acquiring sexually transmitted infections?

2. What pathogens are responsible for causing these **STIs**? Describe each.

3. What are the clinical signs and symptoms of gonorrhea? Compare them to those caused by syphilis.

4. What other STIs would you expect to be increased following an increase in high-risk sexual behaviors and an increase in sexual partners?

5. In general, how can the spread of STIs be reduced?

6. Are the activities needed to prevent or reduce the spread of STIs likely to be followed by crystal meth users?

7. What recommendations would you make to Jim and Barb's community group to help reduce the spread of STIs among the crystal meth users?

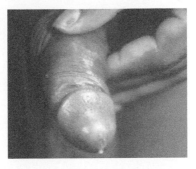

Figure 1. Discharge resulting from gonorrhea.

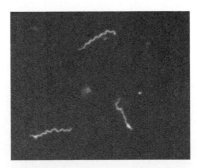

Figure 2. Light micrograph of *Neisseria gonorrheae* Gram stain from discharge.

STI – **S**exually **t**ransmitted **i**nfection.

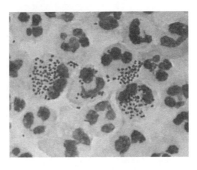

Figure 3. Darkfield micrograph of *Treponema pallidum*.

A Bad First Impression

Joe found everything about college exciting. As a freshman at a small private university, he was on his own for the first time...and very pleased with how well he was handling all of the new curves that had been thrown his way. Joe was working harder than he ever had before, and was earning As and Bs in all of his courses. He was playing well on the JV soccer team and occasionally even got to play with the varsity squad. Just when Joe couldn't believe things could be better, he met Amy at a campus party. They hit it off immediately and quickly decided that they wanted to take their relationship to the next level. Joe had never had sexual relations with a woman before but, wanting to be responsible, purchased condoms.

About five days after their first sexual encounter, Joe woke up feeling "fluish." His muscles ached. He felt fatigued and even a bit feverish. Joe's real concern, however, was the appearance of genital vesicles. Joe didn't know whether to be embarrassed or concerned. He avoided Amy and hoped that his extra hygiene efforts would solve the problem. Unfortunately, two days after the emergence of the vesicles, they opened, forming painful, reddened ulcers with swollen bases. Joe went to the student health center where the physician confirmed his fears of having contracted a sexually transmitted infection (STI).

● **1.** What STI do you suspect Joe has contracted?

● **2.** What is the causative agent involved?

● **3.** What clinical test(s) might Joe's physician order to confirm this diagnosis?

Joe's physician prescribed acyclovir.

● **4.** How does this drug work?

●● **5.** Will it cure Joe's condition?

● **6.** Are there any notable side effects of using this drug?

Following his diagnosis, Joe confronted Amy. She admitted that Joe was not her first sexual experience. While Joe was hurt that she hadn't volunteered that information before they became intimately involved, he was truly angry when she emphatically denied being infected herself. Amy swore she had never had any lesions. Devastated, Joe ended the relationship. How could he have misjudged Amy so terribly? He felt she hadn't been honest with him about anything.

●● **7.** Had Amy lied to Joe about never having an outbreak of genital lesions?

● **8.** What do you know about symptom expression in this disease?

● **9.** Can a person be contagious in the absence of symptoms? Elaborate.

Two months later, Joe returned to the student health center as the ulcerations had reappeared. He was very upset. He knew that he would have recurrent outbreaks, but this was only two months later. Luckily, he at least wasn't experiencing the aches and fever this time.

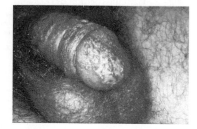

Figure 1. Genital vesicles.

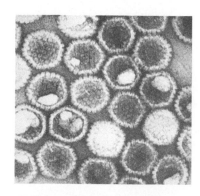

Figure 2. Electron micrograph of pathogen.

10. Is Joe's rapid recurrence normal in a newly infected patient?

11. How long does an outbreak usually last? Explain how healing differs when lesions are located on dry versus moist surfaces.

12. Why didn't Joe have the accompanying flu-like symptoms he had initially experienced?

13. How does latency occur in this disease?

14. What factors can reactivate the infection?

15. Joe used standard "safe sex" precautions. How could he have become infected?

16. Get online and investigate the epidemiology of this disease.
    a. How many Americans suffer from this disease? Why is this number not exact?
    b. How has this number changed in the last 25 years?
    c. Are certain ethnic groups disproportionately infected?

The Domino Effect

Anna smiled after breathing through another tough contraction. Very soon she would finally be a mother. "A lot can happen in five years," Anna thought, "and it's amazing how everything is interconnected. I guess my mother was right. Actions have consequences…"

At 17, Anna fell in love with her high school sweetheart, Mark. Although her parents had preached abstinence until marriage, Anna and Mark started having sexual relations. Being inexperienced, Anna let Mark take the lead in providing protection. Most of the time, Mark used a condom and Anna was satisfied that they were behaving responsibly. Only once in a while did they neglect to follow safe sex protocol, but Anna wasn't especially worried. "After all, what are the odds that I could get pregnant when most of the time we're careful," Anna mused. "And it's not like I can get **AIDS** or anything since Mark is my first partner and he's only slept with one other girl."

1. Clearly, Anna is deluding herself! What is the pregnancy rate for unprotected sexual intercourse?

2. What is the current transmission rate of HIV?

3. Given her limited sexual history, is Anna safe from other STIs? Explain.

As often happens with high school romances, Anna and Mark didn't last. After graduation, Anna earned an associate degree in management and was hired by a local trucking company to coordinate the activities of their main office. Between her schooling and job, Anna met many interesting guys and dated regularly. Being "older and wiser," she strictly adhered to safe sex practices when she engaged in an intimate relationship. At 21, Anna started dating Thomas, a new driver hired by her firm. It was love at first sight for both of them. After only three months, Thomas proposed. Since Anna wanted to be a June bride, she only had a few months to make all of their wedding plans.

A month before the wedding, Anna had her first ever appointment with a **gynecologist**. Although she was a little nervous about seeing a doctor for something so personal, Anna immediately liked Dr. Phillips, an energetic, personable young physician. They discussed Anna's health/sexual history as well as her desire to start a family soon after her marriage. Dr. Phillips reviewed what would happen during her examination and even scolded Anna for not starting routine gynecological care as soon as she became sexually active. At the conclusion of her appointment, Dr. Phillips told Anna, "Everything looks good. The results of your **Pap test** will be back in about a week. I'll only call you if there is anything abnormal. Good luck, Anna, with your upcoming wedding. Hopefully I will see you again soon as a prenatal patient."

4. Why should a woman begin annual gynecological care when she becomes sexually active?

5. What is a Pap test? How is it performed?

Much to Anna's surprise, the next week Dr. Phillips' secretary contacted her to schedule another appointment, indicating the doctor was concerned about her Pap test results. At this office visit, Dr. Phillips performed **colposcopy** to better

---

AIDS – **A**cquired **i**mmuno**d**eficiency **s**yndrome caused by the human immunodeficiency virus, which may be sexually transmitted.

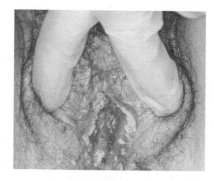

Figure 1. Genital warts.

Gynecologist – A medical professional specializing in the treatment of the female reproductive system.

Pap test – The Papanicolaou test is a screening test used to detect premalignant and malignant cells sampled from the cervix.

Colposcopy – A simple, painless procedure used to visualize the cervix.

examine Anna's cervix where she noted four small cauliflower-shaped genital warts. She told Anna the virus responsible for causing the genital warts is usually cleared by a healthy immune system in 6–24 months and thus is rarely a problem. However, since she had a Pap test indicating significant **dysplasia**, Dr. Phillips wanted to do some additional testing as different strains of this virus can also lead to more serious problems.

Dysplasia – An abnormality in the maturation of cells within a tissue.

6. What is another name for genital warts? What virus leads to this condition? How are genital warts treated?

7. What is the prevalence of this virus? How did Anna likely contract her infection?

8. If standard safe sex practices are consistently practiced, can viral transmission be prevented? Explain. How can viral infection be prevented?

9. Approximately how many types of this virus exist? Which types are most clinically significant? How do physicians routinely test for the presence of specific viral types?

10. What more serious problem can develop from cervical dysplasia?

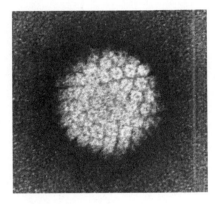

Figure 2. Transmission electron micrograph of pathogen.

When Anna returned from her honeymoon, she met with Dr. Phillips, who told her the test results confirmed co-infection with both a high-risk and a low-risk type of the virus. Because of the presence of the high-risk virus type and pronounced dysplasia, Dr. Phillips performed a surgical excision of the abnormal cervical cells using a procedure known as LEEP.

11. What does LEEP stand for? Describe this procedure.

Anna recovered quickly from her procedure and followed Dr. Phillips' advice to delay pregnancy for at least six months. At her next examination, Anna was disappointed to learn she had again had an abnormal Pap test. Dr. Phillips warned that Anna's condition would require careful monitoring, so she was surprised when Anna returned six weeks later because she had missed a period. Dr. Phillips confirmed Anna's pregnancy, prescribed vitamins, drew blood for a prenatal panel, and reviewed diet and lifestyle changes to promote a healthy pregnancy. Anna diligently followed all directions and flourished for the next five months. As she started her sixth month of pregnancy, Anna awoke in the middle of the night lying in a bloody puddle of fluid. Thomas took her directly to the hospital, where her worst fears were realized…she had miscarried the baby. Later that day Dr. Phillips examined Anna, informing her the miscarriage was likely due to an incompetent cervix as a result of the infection and LEEP procedure. "You mean I'll never be able to have a baby?" Anna moaned. "That's not what I said," Dr. Phillips cut her off. "You need to rest and heal. Maybe you won't be able to carry a child to term, but there are things we can do that may increase the likelihood of you having a baby in the future."

12. What is an incompetent cervix? How did it lead to Anna's miscarriage? Are there any warning signs for an impending miscarriage due to an incompetent cervix? What causes this condition?

Determined to realize her goal of a family, Anna was pregnant again in four months. Dr. Phillips was not happy to see Anna pregnant so soon after her miscarriage. She warned that medical intervention would be necessary to prevent a repeat miscarriage. When Anna was 14 weeks pregnant, Dr. Phillips performed cervical cerclage by placing a stitch high up through the vagina around the cervix to keep it closed. Anna was allowed to continue normal daily activities but was not permitted to exercise strenuously or lift heavy objects.

13. Are there risks or complications associated with cervical cerclage?

After performing an ultrasound at 18 weeks, Dr. Phillips was still worried about Anna's ability to carry her baby to term. Anna was placed on strict bed rest for the remainder of her pregnancy. At 37 weeks, Dr. Phillips removed the stitch so delivery could proceed normally. One week later, Anna's membranes ruptured and labor contractions began.

14. Since Anna's problems are the result of an STI, is her baby at risk of acquiring this viral infection as it passes through the birth canal during delivery?

Epilogue

Anna delivered a healthy 6.5 pound baby girl. Although ecstatic at the birth of her daughter, Anna's troubles were not over. Her Pap test at her postpartum examination revealed a worsening of her cervical dysplasia. Dr. Phillips cautioned Anna not to become pregnant again for at least a year. Anna's next Pap test yielded malignant cells. Following surgical removal of the cervix, Anna has had no further problems with dysplasia. She and Thomas have since adopted a son.

An Ongoing Problem

It had been the worst year of Pam's life. At 41, she was a busy wife and mother of two children. She not only made their house a home, but Pam also worked part-time as a medical assistant for Dr. Tong, a general surgeon. Because of her job, Pam took her family's health very seriously. Pam served her husband and their two daughters nutritious meals high in fiber and antioxidants. She limited her family's consumption of sugary foods, cholesterol, and trans fats. Pam and Tom didn't drink or smoke, and the whole family exercised daily. With such a healthy lifestyle, Pam couldn't believe she and Tom were sitting in Dr. Tong's office listening to him describe her surgical options since her mammogram from last week suggested cancer in her left breast.

Two days later, Pam arrived at the hospital for her procedure, still not quite accepting the events of the last week. Tom was by her side in the recovery room as she awoke. Smiling, he told her that Dr. Tong was able to perform a **lumpectomy** and still feel confident that all of her cancer had been removed. Pam was so relieved. She knew it was silly and vain, but she feared the disfigurement of extensive surgery. Three hours later, Pam was discharged with pain medication for her very tender surgical site, a prophylactic antibiotic, and orders to "be lazy" for three days before returning to Dr. Tong's office for a postoperative examination.

> **Lumpectomy** – A surgical procedure to excise a tumor (benign or malignant) of the breast. The procedure is considered minimally invasive compared to a mastectomy. The preservation of breast tissue provides both physical and emotional benefits to the patient while still effectively eliminating the unhealthy tissue.

1. What is a prophylactic antibiotic? Why would it make sense to prescribe one for a postoperative patient?

Pam had a special homecoming. Her mother was there with dinner already prepared. Her six- and nine-year-old daughters had made her glittery get-well cards and had a pillow, afghan, and book waiting for her on the recliner. Although Pam rested and diligently followed her discharge orders, her experience as a medical assistant made her suspect her surgical incision was infected by the second day. When Pam called Dr. Tong's office, she was told to come immediately and they would work her into the schedule.

As expected, Pam did have an infected incision. The area was erythematous, draining light yellow pus, hot, and extremely tender to the touch. Utilizing a local anesthetic, Dr. Tong removed Pam's sutures, cultured the site, debrided the region, and placed her on a two-week course of nafcillin. Pam was miserably uncomfortable for the next two days, but by her follow-up appointment, the pain was subsiding and she could clearly see the infection was resolving. Dr. Tong informed Pam that the lab reported her infection was caused by *Staphylococcus aureus*, and even though she was feeling better, she needed to complete her antibiotic course.

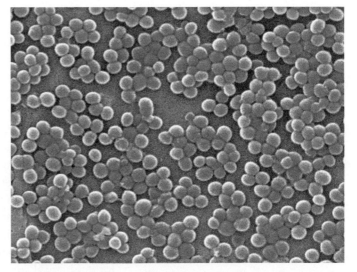

Figure 1. Scanning electron micrograph of *Staphylococcus aureus*.

2. What is debriding?

3. What kind of antibiotic is nafcillin? What types of microbes would be most affected by it? What are the common side effects of this medication? How does it work?

4. Where could Pam have encountered the *Staphylococcus aureus* causing her infection?

Oncologist – A medical professional who studies tumors, their development, diagnosis, treatment, and prevention.

Ductal carcinoma in situ (DCIS) – A noninvasive cancer in which breast cells grow uncontrollably within the milk duct and do not escape into normal surrounding breast tissue, lymph nodes, or other organs.

5. Characterize the morphology and Gram staining of *Staphylococcus aureus*. Describe the appearance of this organism when it is cultured in the clinical laboratory on TSA with 5% sheep red blood cells, and when it is cultured on MSA. What results would a coagulase test yield? Why is this significant?

6. Why is it crucial for patients to finish antibiotic therapy even though their symptoms have abated?

The next week, Pam met with Dr. Feeny, her new **oncologist**. He reviewed the pathology report with her, examined the surgical site, and explained that after three more weeks of healing time, she would be started on an eight-week radiation regimen. Since her breast cancer was **ductal carcinoma in situ (DCIS)**, it was still contained within the milk ducts and had not spread. As a result, there would be no need for IV chemotherapy with all of its debilitating side effects. After her course of radiation, Pam would take tamoxifen for five years and receive biannual mammograms to monitor her progress.

This was the news Pam had been hoping to hear. She raced home to tell her family the good news, stopping briefly at Rite Aid to pick up some Monistat. "Darn antibiotics," she thought. "They solve one problem and cause another."

7. What is Monistat? Why did Pam purchase this over-the-counter drug?

8. How did Pam's antibiotic use lead to her new medical problem? Describe the normal flora associated with this new infection site.

Pam was feeling like herself again…activities at school with the girls, back to work, and out to dinner with Tom to celebrate Dr. Feeny's positive report. A week later, however, Pam arrived at work severely congested and suffering with tremendous sinus pressure. After a quick examination, Dr. Tong confirmed sinusitis and wrote Pam a prescription for doxycycline. Within a few days, Pam's sinus infection had subsided, but she was off to Rite Aid again for more Monistat.

About this time, Pam started her daily radiation treatments. She was so grateful for the nurses and technicians at Dr. Feeny's practice. They made it possible for her to come in for treatment during her lunch hour so work and family time weren't impacted. They also provided much needed encouragement and support as well as helping her cope with the side effects of radiation. Each day they monitored her skin for burning. Her nurse explained the importance of rest, good nutrition, hygiene, and avoidance of sick people for her own infection prevention, since the radiation would slightly compromise her immune function. Pam thanked the nurse for her helpful advice. She would certainly take the recommended precautions, but seriously doubted that avoidance tactics would work for a woman who was both a mother and a medical assistant.

9. In addition to the radiation therapy weakening Pam's immune system, what other factor(s) have been involved?

After three weeks of radiation, Pam developed an upper respiratory infection (URI). Since both of her children had been coughing the previous week, she wasn't surprised. As her condition worsened, Dr. Tong again prescribed an appropriate antibiotic, but upon learning that Pam's last yeast infection hadn't resolved, he insisted that she schedule an appointment with her gynecologist. "Great," Pam thought. "Dr. Tong for bronchitis, Dr. Feeny for radiation, and now Dr. Jameson for a yeast infection. I've seen more doctors in the last three months than I have in the last three years!"

Her appointment with Dr. Jameson was unremarkable. He reported significant vulvovaginal irritation with accompanying "cheesy" discharge typical of infection with *Candida albicans*. Given Pam's recent medical history, Dr. Jameson was not surprised at her condition. He prescribed fluconazole and recommended 1–2 daily servings of yogurt with "active cultures" as a probiotic.

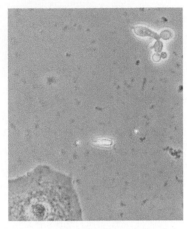

Figure 2. Light micrograph of *Candida albicans* from a vaginal smear.

10. Characterize the morphology and staining of *Candida albicans*. Name a medium commonly used to culture this organism in the clinical laboratory.

11. How does fluconazole work? Are there associated side effects? Why is it often more difficult to find safe, effective antifungal drugs than antibacterial ones?

12. What is a probiotic? What are active cultures?

13. How does eating a probiotic repopulate the normal flora of the vagina?

Pam's bronchitis resolved, and her yeast infection improved a bit. On the last day of radiation treatment, Pam received hugs and best wishes from the wonderful oncology staff...and stopped at Rite Aid for more Monistat. After two more rounds of Monistat and dozens of cartons of Dannon, Pam went back to Dr. Jameson. He was surprised to see Pam still struggling with her infection since her original cultures demonstrated fluconazole sensitivity. He performed another culture and prescribed an extended course of fluconazole.

14. Why wasn't Pam responding to a treatment that was clinically determined to be appropriate?

Despite the lab results consistently indicating Pam's *Candida* was fluconazole sensitive, she made monthly visits to Dr. Jameson for her unrelenting infection. After two more rounds of fluconazole and two rounds of Terazol 3, a topical antifungal, Dr. Jameson referred Pam to an infectious disease specialist in the city. Pam was at her wits' end. All she had done for ten months was see doctors. While her six-month mammogram was clear and she should have been thrilled, Pam was becoming depressed due to her chronic yeast infection. She was so uncomfortable from her extended disease that it hurt to sit, stand, walk, or bathe. She did everything she was told by the medical professionals and nothing seemed to help. Pam just wanted to cry.

Luckily, Dr. Harris was aware of Pam's ordeal and the toll a chronic infection can take both physically and emotionally. She and her staff were prepared to help Pam with all aspects of her infection. Cultures collected by Dr. Harris yielded no new information, but her plan of attack was a sustained treatment on several fronts. Pam took a two-week course of fluconazole while simultaneously treating each night with 0.8% terconazole, a higher dosage of Terazol. She continued the probiotics and was instructed to take two sitz baths a day for at least three weeks. A topical corticosteroid was prescribed to be applied three times a day after gently cleansing the region. Pam was encouraged to take a multivitamin daily, sleep eight hours a night, walk 30 minutes a day outside, and refrain from intercourse for one month.

15. Why did Dr. Harris add the topical antifungal agent when Pam was already taking an oral one? Are there any side effects associated with the use of terconazole (Terazol)? What was the point of the sitz baths and the corticosteroid cream?

16. What did Dr. Harris hope to accomplish with the vitamins? With extra sleep? With outdoor exercise? With abstinence?

Pam was scheduled for a follow-up appointment in one week. While still suffering, she did note improvement. Dr. Harris and staff were very encouraging. Pam was ordered to maintain this routine for six weeks, checking in weekly with Dr. Harris.

After two months, Dr. Harris was exceedingly pleased with Pam's progress and told her she was now on "parole." She would take one dose of fluconazole a week for two months and use the Terazol the same at night. Pam was to continue the sleep requirement, vitamins, probiotics, and walks. She was permitted to gradually resume sexual activity but monitor for any postcoital irritation. Dr. Harris also instructed Pam to contact her throughout the next year before starting any antibiotic regimen for bacterial infections so she could prescribe an appropriate antifungal prophylactic treatment.

Fourteen months after her initial yeast infection, Dr. Harris pronounced Pam cured. Pam and her family celebrated...her one-year mammogram showed she was cancer-free and now she was finally infection-free too.

17. Chronic infection is physically debilitating, but clearly it impacted Pam's emotional health too. As a (future) nurse, patients will look to you for support in these situations. Take some time to consider how you might provide this encouragement. What should you do if your patients demonstrate signs of depression?

An Infectious New Lifestyle

Lora felt shell-shocked! After 18 years of marriage, Jack announced he wanted a divorce. Lora couldn't understand Jack's demand. They didn't quarrel, had many common interests, and had three sons to raise. Lora didn't feel any better when Jack explained that he had found someone who was fun, energetic, pretty, and almost 24.

Following the divorce, it seemed to Lora that she had lost both her husband and her self-esteem. She felt old, tired, and hurt. Her parents and friends encouraged her to seek counseling, but that was for "crazy" people. In an attempt to escape her post-divorce funk, Lora focused her attention and energy on her children and her job. Lora cheered at every wrestling match, Quiz Bowl competition, and band concert. Her boss complimented the quality of her work and even gave Lora a raise. She rarely went out with friends, since it was awkward being the only "single" among a crowd of couples.

For a while, Lora's strategy seemed to be working. All of the time she had previously shared with Jack, she now filled with her children's extracurricular activities and work-related projects. But it just wasn't enough. Lora was lonely. Her parents and coworkers encouraged her to start dating, but the whole idea scared Lora.

Just as Lora had resigned herself to leading a quiet life, her supervisor announced that she had been selected to represent their department at a national sales convention in Los Angeles. After dropping the boys off with her parents, Lora flew to LA for the conference and decided to take her mother's advice to relax, meet new people, and have some well-deserved fun. Her days were packed with reports and presentations, but Lora had her evenings free to explore the local night life. Lora surprised herself by accepting an offer to dance from a nice-looking, 40ish man she met in a club that first evening. She was having a wonderful time! Tom was flattering, attentive, and charming. The next morning when Lora woke up in Tom's hotel room, she tried to blame her impulsive behavior on the exotic drinks Tom encouraged her to try. Determined to never make this mistake again, Lora didn't venture out in the evenings for the remainder of her LA trip.

● **1.** What serious problems did Lora just expose herself to?

● **2.** What could Lora have done to reduce her risk of encountering these problems?

A month later, Lora was off to Houston for another meeting...and again had a "one-night stand." While Lora chided herself for this promiscuous behavior, she continued her new pattern of out-of-town flings associated with business travel.

Lora was starting to feel good about life for the first time since Jack's shocking announcement. Her extra efforts at work had been noticed, earning her a promotion and a second raise. She still devoted a lot of rewarding time to her children, since she restricted her own social life to her business trips. The excitement of the affairs gave her a psychological boost. Lora rationalized that she must really be interesting and desirable since it had been so easy to attract her new partners.

After almost a year of indulging in her new, secret lifestyle, Lora woke one morning with intense vaginal and vulvar irritation plus a frothy, greenish-yellow, foul-smelling discharge. In a hurry to catch her flight to New Orleans, Lora washed with extra care, applied a feminine deodorant, ignored her symptoms, and left for her business trip. Despite these symptoms, Lora spent the night with a new acquaintance and was distressed to discover that the intercourse was extremely uncomfortable. The next morning she called to schedule an appointment with her gynecologist as soon as she returned home.

3. Based upon these symptoms, what **STI** do you suspect Lora has acquired?

4. What is the causative agent?

5. Are there any additional symptoms common in women? What symptoms do infected men manifest?

6. Can this microbe be transmitted without sexual contact?

Lora's physician performed a pelvic examination, which revealed small, red ulcerations on the vaginal walls and the cervix. He collected samples of vaginal discharge and urine, which were sent to the microbiology laboratory for culture, and drew blood for additional testing. In about three days, Lora's physician confirmed her suspected infection and scheduled an immediate follow-up appointment. At this meeting, Lora confessed her multiple sexual contacts. She received a prescription to treat her infection, information regarding complications associated with her condition, a lecture on safe sex practices, and a referral to a counselor.

7. What microbiological laboratory tests were performed to make Lora's diagnosis? How does this differ for male patients?

8. Why did Lora's physician draw blood when it is not used to specifically diagnose the infection he suspected?

9. What complications are associated with this infection?

10. What safe sex practices would you recommend to this patient?

11. What is the treatment of choice for this pathogen? Are there any common side effects associated with the treatment?

12. Why did Lora's physician refer her for counseling? Do you think this was appropriate? Explain.

13. How prevalent is this disease?

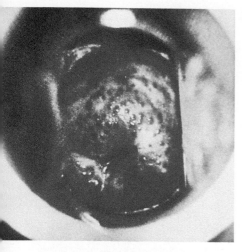

Figure 1. Inflamed cervix caused by the pathogen.

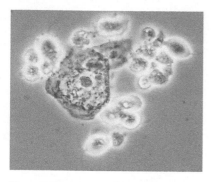

Figure 2. Wet mount of vaginal secretions showing pathogen.

Nosocomial Infections

Whenever we are seriously ill, it's comforting to know that highly skilled medical professionals are available to care for us at our local hospital. But imagine that once admitted, you become sicker from acquiring an infection caused by medical attention. You have just become the victim of a nosocomial infection. The term, derived from the Greek words *nosos* (disease) and *komeo* (to take care of), specifically refers to infections acquired by both patients and workers at hospitals, clinics, assisted living facilities, or any other medical institution. Annually, more than 90,000 admitted patients die due to nosocomial infections, representing about 10% of the Americans affected. Aside from the obvious problem of increased **morbidity** causing suffering and death, nosocomial infections are financially crippling to healthcare facilities. From a medical and budgetary perspective, nosocomial infection rates must be minimized, as they cost over $5 billion each year in lost wages and expenses associated with increased duration of hospital stay and therapy.

How can you enter a facility designed for healing and fall prey to a nosocomial infection? Four principal reasons are to blame. First, a significant percentage of the patient population is immune compromised. The immune system of a "compromised" patient is impaired due to: (1) actively fighting a current infection, (2) damage caused by chemotherapy, radiation, or pathogens (HIV), or (3) post-transplant immunosuppressive therapy. Additionally, many patients are compromised because they lack intact skin and mucous membranes because of lesions, incisions, or traumatic injury. With such patients, pathogens can easily access underlying tissues to cause infection and/or grow unchecked by the usual innate and adaptive responses of the immune system. Consequently, these individuals are much more susceptible to infection than other members of the general community. Another factor contributing to nosocomial infections is the higher concentration of microbes associated with a hospital environment than with other settings. Because patients with infectious diseases are treated here, they serve as a source of pathogens. Healthcare providers, administrators, custodial workers, visitors, and others interacting with patients carry their **normal microflora** into the facility and may unknowingly transmit them. Numerous environmental microorganisms can be introduced into a large facility in the form of **fomites** (eating utensils, trash cans, medical equipment, etc.) or vectored mechanically by insects. Clearly the functional nature of a hospital encourages a high microbial density. Modern invasive medical procedures serve as yet another culprit in the establishment of

Morbidity – A diseased state.

Normal microflora – The microorganisms that are normally in and on us.

Fomite – A nonliving intermediate that carries a pathogen to a new host.

nosocomial infections. Although these diagnostic and therapeutic techniques enable healthcare providers to help countless patients, every procedure from the insertion of a urinary catheter to open heart surgery represents an opportunity for pathogen introduction. Even more troubling still is the final reason for nosocomial infections: The pathogens present in these facilities may be resistant to one or more antibiotics, since the drugs are routinely used both therapeutically and **prophylactically**. By serving a large population of immune-compromised patients and performing invasive procedures in a facility with a high density of resistant microorganisms, nosocomial infections are inevitable.

The most common nosocomial problem is a urinary tract infection, which is typically associated with the use of a catheter. Other nosocomial infections linked to invasive medical procedures include surgical site infections, pneumonia from ventilator use as well as via direct or aerosol pathogen transmission, and sepsis associated with various types of "**lines**." While patients seem doomed to suffer from nosocomial infections, there is much we can do to minimize this serious medical concern. First, always follow **asepsis protocol** for the performance of any invasive procedure. Consistently practice **universal precautions** to protect both yourself and your patients. Monitor antibiotic use to discourage development of resistant microbial strains. Work with your facility's infection control officer. These individuals are valuable resources and can provide educational workshops for departments, surveillance of units at high risk of nosocomial infection (ICU, Rehabilitation, Oncology, etc.), and individual professional counseling to improve institutional safety. Finally, the single most effective means of reducing the threat of nosocomial infection is hand hygiene. Wash your hands…wash your hands…wash your hands! As a future medical professional, this should be your mantra. Scrupulous attention to hand washing between your patient contacts can dramatically reduce disease transmission. As you analyze the cases in this unit, you should focus your attention on detecting the cause of the nosocomial problem and determine how you could have prevented it. Please use your study of these cases to make prevention of nosocomial infection your priority when you begin your healthcare career.

A Nose for Trouble

Growing up, Tom's continuous nasal congestion was written off as "allergies," which his pediatrician said he would likely outgrow. But, as he closed in on 35, Tom was still miserable. When his chronic sinusitis was not responsive to medication, Dr. Edelman, an **otorhinolaryngologist**, ordered a sinus **CT scan** to determine an operative strategy. Dr. Edelman explained that functional endoscopic sinus surgery (FESS) utilizes an **endoscope** to directly visualize the openings into the sinuses and facilitate the removal of abnormal or obstructive tissues. The highly successful procedure is performed on an outpatient basis, working through the nostrils to minimize swelling and discomfort.

The whole idea of the doctor "snaking" instruments up his nose and working so close to his brain, eyes, and several major arteries made Tom shudder. But since it looked like his only option to gain relief from his lifelong problem, Tom nervously consented to the surgery. During their preoperative meeting, Dr. Edelman reassured Tom by telling him that he would be using a state-of-the-art technique known as image-guided surgery. "The procedure utilizes CT scans and infrared signals providing real-time information about the exact position of the surgical instruments and producing an almost three-dimensional mapping system. Some of this same technology is used by the U.S. armed forces for precision bombing of targets," Dr. Edelman joked. "I know the procedure sounds intimidating, but with this new technology and my surgical experience, I think I can guarantee you a positive outcome with minimal risk."

Encouraged by their conversation, Tom kissed his wife Annette good-bye and was taken to the operating room, where he underwent image-guided FESS without complication. Two days later, Tom was trying to convince himself that he had made the right decision about the surgery. His face was only slightly swollen, but extremely tender. Tom's biggest complaint was the discomfort associated with the **merocel** nasal packing used to control oozing. Luckily, Tom was scheduled to see Dr. Edelman in another two days for packing removal, endoscopic debridement, and irrigation. Tom chided himself for being a wimp and resolved that he could certainly put up with a "plugged nose" for a few more days.

However, the morning of postoperative day 3, Tom awoke feeling light-headed, nauseous, and with substantially increased sinus pain. When Tom developed chills, Annette took his temperature, discovering a fever of 39°C (102.5°F). His head hurt, his body ached, and Tom developed watery diarrhea. Although Annette was worried and wanted to take him to see his physician, Tom maintained that he was just unlucky enough to catch a virus while recovering from surgery. Two hours later, Tom was noticeably worse. He was developing a patchy, red rash and confusion. When he stood up to walk to the bathroom, Tom collapsed. Unable to revive her husband, Annette immediately called 911.

Eventually, Dr. Wilson from the Emergency Department came to the waiting area to speak with Annette. She told Annette that Tom was being transferred to the ICU. His fever was now 40°C (103.6°F) and his blood pressure was dangerously low. Results from stat lab tests revealed his serum sodium, potassium, and calcium levels were too low, while his liver enzymes (**SGOT** and **SGPT**) were significantly elevated. Cardiac monitoring indicated ventricular **arrhythmias**. Tom showed signs of metabolic **acidosis**, **hypoxia**, and **disseminated intravascular coagulation (DIC)**. Dr. Wilson explained that she had collected blood cultures plus specimens from Tom's recent surgical site to send to the

---

Otorhinolaryngologist – A medical specialist in disorders of the ears, nose, and throat; also called ENT.

CT scan – Computed tomography; a medical procedure used to create 3-D images of a structure by taking many 2-D X-ray images around a single axis of rotation.

Endoscope – An instrument used to view internal features. It consists of a rigid or flexible tube with a fiber optic system for conducting light from an external source and a lens. An additional channel allows entry of medical instruments.

Merocel – A foamed polyvinyl alcohol product designed to be an ultrapure, soft, nonirritating, and lint-free hydrophilic material. It is often used as a postoperative packing to control fluid discharge.

SGOT – **S**erum **g**lutamic **o**xalacetic **t**ransaminase, or AST (**as**partate amino**t**ransferase) is a liver enzyme to metabolize the amino acid aspartic acid. Liver damage releases the enzyme into the blood.

SGPT – **S**erum **g**lutamate **p**yruvate **t**ransaminase, or ALT (**al**anine amino**t**ransferase) is a liver enzyme to metabolize the amino acid alanine. Liver damage releases the enzyme into the blood.

Arrhythmia – Irregular heartbeat.

Acidosis – A blood pH <7.35 due to the increased production of H⁺s and/or decreased ability to form bicarbonate in the kidney. Untreated, it leads to coma and death.

Hypoxia – A shortage of oxygen in the body.

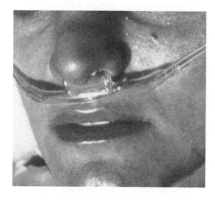

Figure 1. Rash resulting from TSS.

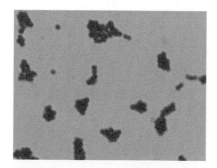

Figure 2. Gram stain of *Staphylococcus aureus*.

microbiology laboratory. Additionally, she had ordered tests to specifically rule out Rocky Mountain spotted fever, leptospirosis, hepatitis B, syphilis, mono- nucleosis, and **ANA** production. "If you don't think it's any of those diseases, what is Tom's problem?" a tearful Annette asked. "I think your husband is suffering from toxic shock syndrome or TSS," replied Dr. Wilson. Stunned, Annette quietly whispered "No, no, no." "I've heard of that before," she said. "Back in the early 1980s ladies got that from using superabsorbent tampons… and they died! The companies stopped making those products, and people have been OK ever since. Tom can't possibly have toxic shock syndrome. It must be something else," Annette said as she started to cry.

1. What is toxic shock syndrome? Outline the typical symptoms associated with this condition.

2. What pathogen(s) is/are usually responsible for this condition? Where are these microbes commonly found?

3. Why was Tom at risk for this infection? What other patient groups are most susceptible to this infection?

4. How does microbial infection produce the characteristic symptoms of TSS? What is a "super antigen" and how does it impact your immune system?

5. How is TSS diagnosed?

6. Is TSS a serious medical condition?

Annette stayed at the hospital all night. Early the next morning both Drs. Wilson and Edelman talked with her and confirmed the TSS diagnosis. Three of Tom's four blood cultures were already positive for *Staphylococcus aureus*, as was the culture from his nasal packing. They were continuing **fluid resuscitation**, cardiac monitoring, and oxygen therapy as well as initiating **dialysis**. Dr. Wilson indicated that even though multiple organ systems had been affected by the TSS, Tom should eventually make a full recovery. Still overwhelmed by the situation, Annette was amazed that she could respond when the physicians asked if she had any questions regarding Tom's care and progress.

7. How is TSS typically treated?

8. Are there any associated complications?

9. How can TSS be prevented?

10. What is the prevalence of TSS?

11. Nonmenstrual TSS cases have been steadily increasing. How might you explain this observation?

A "Hep C" History Lesson

April 1987: Ever since she began cycling, Janine struggled with heavy menstrual flow. Each year the problem seemed to be a little worse. Now, as an 18-year-old high school senior, her family doctor had her taking daily iron supplements since testing at her last physical exam yielded a hemoglobin value of 10.2 g/dl.

● **1.** What is hemoglobin? What is a normal hemoglobin value for a healthy adult female? Why is this value different for adult males? What is the hemoglobin panic value and why? What general medical condition is Janine suffering from as a result of this low hemoglobin level? How will iron supplementation help her?

Despite the inconvenience of her dysfunctional menses, Janine woke early on Easter Sunday anticipating a wonderful holiday with her family. Since her youngest brother, Evan, was still an "Easter Bunny believer," Janine and her siblings enjoyed an early morning hunt for colorful eggs followed by a breakfast of chocolate bunny ears and marshmallow peeps. Worried about "leaks," Janine decided against wearing her pale yellow dress to church and put on a black skirt instead. By the time the family left for services, Janine was pooped. "I guess I should have eaten a better breakfast," Janine thought.

An hour later, Janine stood with the rest of the congregation to sing the recessional hymn. The next thing she knew, Janine was looking up at her mother's worried face through a blurry haze. "Janine, honey, are you ok?" her mother asked as she patted her daughter's clammy face and hands. "How do you feel?"

Evan pushed past their mother to see his older sister. "Wow! That was great!" he exclaimed. "You just went 'bam' and fell down flat! Dad and the usher had to carry you outta church…too cool!"

Janine gave a small laugh at her brother's description of recent events, and she rubbed the back of her head where it had hit the pew during her fall. Embarrassed by all of the attention, Janine assured her family and other gathering onlookers that she was just fine. She blamed the **syncope** on her inappropriate breakfast. In an attempt to deflect attention away from herself, Janine reminded her parents that they needed to get on the road or they would be late for Easter dinner at Nana's house.

Although a little tired, Janine otherwise felt fine by the time she arrived at her grandmother's home. After gorging on Nana's glazed ham and coconut crème pie for lunch, she helped with the dishes and played cards with her uncle. Later in the day, she joined her cousins for basketball in the driveway. Three minutes into the game, Evan zoomed into the house to report that Janine went "bam" again!

This time, Janine's parents couldn't be persuaded by her claim she was perfectly fine and she soon found herself in an examining room of the Emergency Department at County Memorial Hospital. Dr. O'Connor reviewed Janine's symptoms and ordered a CBC "**stat**."

● **2.** What is a CBC? What data can this test provide to explain Janine's symptoms?

Twenty minutes later Dr. O'Connor returned to Janine's examining room, explaining that he was going to admit her. Janine's hemoglobin had dropped to the dangerously low value of 7.2 g/dL. As soon as she was settled in her hospital room, the nurse began transfusing her with the first of three units of packed red blood cells. Early Monday morning, Dr. Jefferies, an **obstetrics/gynecology** resident, took a thorough gynecological history and then examined Janine. Given the

Syncope – Fainting.

Stat – Medical slang for immediately.

Obstetrics/Gynecology – The areas of medicine focusing on the treatment of pregnant women and disorders of the female reproductive system, respectively.

severity of her ongoing **menorrhagia**, Janine was scheduled for a **D&C** the next morning. The procedure went smoothly and she returned home to recover. Janine reported a reduced menstrual flow with her next cycle.

June 1987: Two months later, Janine was again feeling fatigued. Her appetite was diminished; her joints and muscles were achy; her belly was tender; and her body temperature was consistently 38°C (99.6°F). A repeat CBC indicated a much improved hemoglobin value of 12.8 g/dL, suggesting there must be another reason for her exhaustion. Results of a hepatitis panel indicated she was definitely negative for infection with either hepatitis A virus (HAV) or hepatitis B virus (HBV) and probably also negative for infection with "non A–non B" hepatitis. However Janine's specimen was positive for Epstein-Barr virus IgG. Her family doctor explained she was probably in the convalescent stage of mononucleosis, which would explain most of her symptoms.

3. How could her physician tell Janine was in the convalescent stage of an illness?

Janine followed her physician's instructions to rest, consume plenty of fluids, and eat nutritious meals. Since Janine was no better at the end of August, her doctor repeated the blood work. While her hemoglobin levels were holding steady, Janine's alanine aminotransferase (ALT) and aspartate aminotransferase (AST) values were 75 U/L and 45 U/L, respectively.

4. What are normal values for ALT and AST? What conditions are indicated with abnormal ALT and AST values?

October 1987: Janine had spent a month at Central State University as a freshman majoring in physics with a minor in chemistry when she made an appointment at the student health center. Janine was becoming increasingly more tired as the semester progressed. She had lost 15 lb in the last six weeks since she had no appetite and continuous nausea. Her low-grade fever persisted as did her belly pain. The overworked clinic doctor suggested Janine was simply suffering from high stress levels associated with her challenging curriculum. She was counseled to learn how to relax. Two weeks later, Janine went home to visit family for fall break. Her mother was mortified to see Janine wasting away. A quick trip to her family physician with repeat blood tests indicated she had completed her recovery of EBV infection, but they also yielded shocking liver enzyme results: ALT = 407 U/L and AST = 260 U/L! Unfortunately, Janine was told not to worry, as these values are often elevated in persons suffering from infectious mononucleosis. Since she had recently resolved that infection based upon her antibody titers from previous blood samples, Janine's physician assured her the abnormal results were a residual effect. He recommended she eat more, sleep more, and find a hobby to relieve the stress causing her exhaustion and gastric discomfort. Janine went back to college the next week still feeling poorly.

May 1988: Janine returned home after her freshmen year having made the Dean's List both semesters. While she thrived academically at college, Janine still suffered physically from the same old symptoms. Her 5"9' frame only carried 108 lbs. Another round of blood tests revealed her ALT (110 U/L) and AST (82 U/L) levels were still elevated. Since her menstrual cycles had become heavier again, Janine wasn't surprised to find her hemoglobin slipping to 10.1 g/dL. Frustrated with their family physician, Janine's mother made an appointment for her daughter to see a **gastroenterologist**.

July 1988: At her first appointment with Dr. McInturf, Janine's repeated ALT was up to 198 U/L. After analyzing the results of her liver biopsy, she reported the presence of mild chronic hepatitis without necrosis. Since Janine's previous blood tests had ruled out hepatitis A or B, Dr. McInturf discussed other possible causes of hepatitis with her and, because she was a college student, emphatically noted the damaging effects that alcohol can have on the liver. Without a definitive diagnosis, Dr. McInturf could only monitor Janine's condition and scheduled a follow-up appointment for the next year.

August 1989: Janine's condition remained stable, but not improved.

August 1990: Janine continued as before, but her ALT and AST scores dropped to 92 U/L and 64 U/L, respectively.

November 1990: While home for Thanksgiving break, Janine's fatigue became even more pronounced, she experienced several bouts of confusion, and her eyes developed a yellowish cast. Dr. McInturf ordered the usual blood tests plus one newly developed assay. Two days later, Janine was again admitted to the hospital for transfusion with two units of packed red blood cells since her hemoglobin value was 7.4 g/dL. The next day she underwent a second D&C to control her dysfunctional uterine bleeding. Before her hospital discharge, Dr. McInturf came to Janine's room to discuss the findings of the new blood test. "Your past lab work indicated you were negative for infection with the viruses that cause hepatitis A and B. However, your liver enzyme results are erratic, your liver biopsy revealed chronic inflammation; you have no appetite or energy, plus regular pain in the region of your liver. All of these symptoms tell me you're suffering from hepatitis," said Dr. McInturf.

Figure 1. Yellow sclera from jaundice.

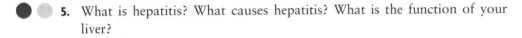
5. What is hepatitis? What causes hepatitis? What is the function of your liver?

"So do I have hepatitis or not? If I'm not really sick, why do I still have these awful symptoms? In my psych class last year we talked about hypochondria. Am I nuts?" Janine asked pitifully.

"Absolutely not," Dr. McInturf said with a laugh. "The new test I ran has just been developed to identify another virus we've discovered that can cause hepatitis. Your results came back positive for antibodies against that virus. Janine, you have hepatitis C."

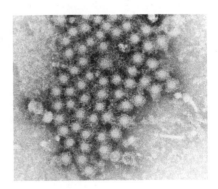

Figure 2. Transmission electron micrograph of hepatitis virus.

6. What is the causative agent of Janine's hepatitis? Characterize this pathogen.

7. What test was run to screen Janine's blood for antibodies to the hepatitis C virus?

After a moment of stunned silence, Janine and her mother unleashed a torrent of questions for Dr. McInturf.

"How did Janine get infected?" her mother asked.

"Can I spread this infection to my family?" Janine responded.

"Could we have prevented this?" her mom countered.

"How do I get rid of this?" Janine queried.

"Those are all terrific questions and I wish I could answer them. Unfortunately," Dr. McInturf shrugged, "we know very little about this new hepatitis C virus (HCV). We're still learning about its transmission and how it will affect patients."

"So what do I do until you have answers to our questions?" a depressed Janine asked.

"Well, Janine, I want to enroll you in a national study that has been recently initiated. Physicians across the country are trying to collect patient data to determine hepatitis C pathogenesis. As we learn more about this virus, we will hopefully determine a strategy to stop its damage to your liver and prevent its spread to others," replied Dr. McInturf. "For now, I am simply going to monitor your symptoms and treat them if and when they become pronounced. Unfortunately, our pharmaceutical arsenal against viral pathogens is not especially effective."

February 1991: As per Dr. McInturf's directions, Janine began a regimen of twice-yearly monitoring for her liver enzymes so these data could be included in the new national study on hepatitis C. Janine was pleased that over the next 18 months her ALT and AST values had returned to near normal levels. She still struggled with fatigue, intermittent abdominal pain, and loss of appetite but otherwise felt fine.

September 1991: Having graduated summa cum laude in the spring, Janine embarked on her graduate studies at MIT. She was thrilled with the new academic challenges she faced and the opportunity to participate in a research team funded by the National Science Foundation. While she felt good physically, her most recent round of tests showed a substantial elevation of her liver enzymes (ALT = 152 U/L and AST = 97 U/L). When Dr. McInturf received these results, she insisted Janine return home for a follow-up appointment. While she still didn't have the answers to most of Janine's questions, Dr. McInturf was able to reassure her patient that the national study suggested hepatitis C patients may experience erratic liver enzyme reports. She was also able to tell Janine that there appeared to be a link between exposure to body fluids and HCV transmission. Since Janine's hepatitis symptoms had started weeks after her first transfusion, Dr. McInturf suspected a tainted unit of blood may have been the mode of transmission in her case.

8. What is the incubation period for HCV?

9. Describe the symptoms of acute hepatitis C and compare them with the symptoms of chronic infection.

10. How is HCV typically transmitted? Which individuals are at greatest risk of acquiring HCV infection? How can infection with HCV be prevented? Why is the transmission mode responsible for Janine's infection no longer a concern?

October 1991–February 1999: In September of 1993, Janine and David, an engineering graduate student, married. Janine completed her Ph.D. in physics in April 1997 and accepted a faculty position in the Physics Department at Penn State University that fall. During this eight-year period, Dr. McInturf referred to Janine's liver enzyme values as a "picket fence" since they regularly fluctuated up and down. Overall Janine's health was good, but she consistently complained of fatigue and intermittent bouts of nausea with anorexia.

March 1999–December 1999: Janine reported to her obstetrician in April for her first prenatal examination. Her ALT = 91 U/L and AST = 77 U/L. Janine felt her fatigue was even more pronounced during the first trimester, but was reassured to learn this was a normal pregnancy symptom. David was pleasantly surprised to see his wife eat heartily for the first time in their married life. Janine gradually gained 30 lb during her pregnancy and found she hadn't felt this good since 1987. In mid-December Janine gave birth to a 9.5-lb boy after only two hours of labor. Because of the intensity of her brief labor, an artery in her lower back was damaged and she began losing blood. Janine's initial complaints of postpartum back pain were ignored by her nurses, who informed her a little discomfort was common after childbirth. When examined by her obstetrician 10 hours after delivery, an enormous hematoma was discovered on Janine's lower back and her hemoglobin was 5.6 g/dL. She was too weak to hold her son and was experiencing shortness of breath. Janine was immediately scheduled for surgery to repair the artery. During the procedure, 900 cc of blood were removed from the hematoma. Janine received two units of packed red blood cells postoperatively.

January 2000: At her six-week postpartum examination, Janine had made a full recovery from her delivery and associated complications. Her hemoglobin was 12 g/dL and her liver enzymes were in the normal range.

January 2001: Janine reported to Dr. McInturf that she felt quite well overall. Although she didn't feel as tired as before, Janine did note that she could fall asleep almost instantly whenever she had the opportunity to relax. In fact, because of this tendency, she rarely drove alone. A year after her son was born, Janine's liver enzymes began to creep up a little bit (ALT = 95 U/L; AST = 50 U/L).

June 2001–March 2002: Janine was expecting their second child. As before, she was quite tired during the first trimester, but much improved by the fourth month. Her appetite was good and liver enzymes normal throughout the pregnancy. The end of March, Janine delivered a second 9-lb, healthy baby boy. Because of her previous complications, Janine's delivery was induced and carefully monitored. Fortunately, she experienced a rapid, complication-free postpartum recovery.

August 2002: Dr. McInturf and Janine had their annual meeting to discuss the pathogenesis of Janine's infection. The results of additional, new tests indicated Janine not only had hepatitis C, but specifically the 1a variant. Dr. McInturf discussed the relevance of these data to new treatment options available for Janine. Janine's ALT = 104 U/L and AST = 83 U/L.

11. How many different HCV variants or genotypes are currently recognized? How does the infecting genotype correlate with treatment protocol?

12. What is the most commonly recommended treatment protocol for HCV? Are there side effects associated with this treatment? How can they be managed? When is treatment contraindicated?

13. Distinguish between treatments recommended for patients with acute versus chronic hepatitis.

14. Can hepatitis C be cured?

September 2002–present: Janine's liver enzyme values are demonstrating the "picket fence" fluctuations characteristic of hepatitis C infection. Janine reports generally good health with significant fatigue but less nausea and abdominal pain than in the past. Because of the side effects associated with treatment, Janine has declined to participate in the drug therapies recommended by Dr. McInturf. Today she is an active, highly successful professional with a full and happy family life. While she is plagued with exhaustion, she hides this symptom well and, unless told of her medical condition, no one would suspect she has an ongoing infection.

15. What is the prevalence of hepatitis C?

16. How serious is hepatitis C? What risks does Janine face by declining treatment?

17. What is the treatment option if liver failure occurs? What new procedure is now available for hepatitis C patients seeking this treatment?

The "Superbug"

Ethel was proud to have reached the respectable age of 78 and still be able to live independently. Her small two-story home held many happy memories of her beloved Stan and raising their four boys. When Stan died suddenly 12 years ago, her sons urged her to move in with one of them or consider an assisted-living facility. Ethel wouldn't hear of it! She loved caring for her home and gardens, visiting with the neighbors, volunteering as a hospital receptionist, and making all of her own decisions.

When Ethel turned 72, she was stunned to find herself hospitalized with renal failure. Never having experienced a major illness, Ethel found the diagnosis most unsettling. During her hospital stay, Ethel underwent **hemodialysis** as a renal replacement therapy to balance body fluids and remove wastes. As she steadily improved, Dr. Kittridge, a **nephrologist**, talked with Ethel about her future medical care. Ethel made it quite plain that she was a capable woman and would happily follow his medical orders as long as she was able to do things for herself. Dr. Kittridge smiled and explained he would be switching her to continuous ambulatory peritoneal dialysis or CAPD because of the many advantages the therapy afforded. Although Ethel would require some initial assistance from a home health nurse, her doctor assured her the procedure was easy to learn and would allow her to maintain her independent lifestyle. "In fact," Dr. Kittridge said, "with minimal planning, you will even be able to travel out of state to visit your grandchildren." Ethel grinned and told Dr. Kittridge, "CAPD sounds like just the ticket!"

Dr. Kittridge did caution Ethel that with all treatments there can be complications. "Because of your renal failure, Ethel, your immune system is now slightly suppressed. This, coupled with catheter insertion and fluid movement in and out of the peritoneal cavity, make infections your biggest threat," Dr. Kittridge explained. "Other considerations include eventual changes in your peritoneal membrane decreasing its permeability and resulting in poorer fluid exchange or ultrafiltration failure. Also, CAPD may lead to diabetes, because of the glucose in the dialysis solution."

Comfortable with Dr. Kittridge's explanations of CAPD and the instructions for self care, Ethel underwent a short surgery to place a **Tenchkoff catheter** in her abdomen, running from the peritoneum to the surface near her navel. She worked with Peter, Dr. Kittridge's **PA**, and learned the correct procedure for performing CAPD and minimizing her infection risk. Upon discharge, Ethel was supervised for a week by the home health nurse assigned to her case…and then she was happily independent once again.

1. Use the Internet to research CAPD. Outline the procedure step by step. Be sure you understand how this procedure works. Why is it often preferred to hemodialysis?

Only briefly intimidated by her new lifestyle, Ethel's need for self-sufficiency motivated her to consistently practice **aseptic technique** when making her dialysis solution exchanges. Over the next six years, Ethel had several infections at her "tunnel," or exit site. She usually required a brief hospital stay, but with prompt treatment, these infections quickly resolved.

One evening as Ethel completed an exchange, she noted the dialysis fluid in her exit bag was cloudy. Suspicious, she immediately contacted Peter, who instructed her to report to the hospital with her bag of PD effluent. Thirty minutes

Hemodialysis – The use of countercurrent flow of dialysate and blood in an extracorporeal circuit to remove wastes and excess water from the blood of patients in renal failure.

Nephrologist – A physician specializing in diseases of the kidney.

Tenchkoff catheter – A laproscopically placed device to facilitate CAPD.

PA – **P**hysician's **a**ssistant; a medical professional who practices medicine under the supervision of a physician/surgeon.

Aseptic technique – A procedure that is performed under sterile conditions.

later, Ethel's neighbor delivered her to the Emergency Department of Grace Hospital. Normally, Ethel would have refused the wheelchair she was offered, but she was suddenly exhausted and shivering. Ethel presented with a temperature of 39°C (101.7°F) and blood pressure of 94/60 mmHg. When Peter arrived, he ordered a CBC, "lytes," and blood cultures. Additionally, he sent Ethel's PD effluent to the laboratory for analysis and culture, and initiated empiric therapy with ceftriaxone and amikacin.

2. What useful information will be provided by the CBC? By the "lytes" or electrolytes panel? By the blood cultures?

3. What will the laboratory be looking for in Ethel's PD effluent?

4. What is empiric therapy? Is it irresponsible to initiate an antibiotic treatment before confirming an active bacterial infection? Explain.

Overnight, Ethel's condition declined despite her treatment. Her fever climbed to 40°C (103.2°F) and her BP dropped to 82/58 mmHg. Ethel's belly had become very painful, her muscles ached, and she felt queasy. By morning her initial laboratory results were available. Ethel's WBC count was elevated to 21×10^9 cells/l, and the lab noted significant accumulation of fibrin in her PD effluent—that is, both values confirming infection and likely peritonitis. Her electrolyte panel yielded a potassium **critical value** of 8 mmol/l, necessitating immediate hemodialysis. By noon, the microbiology laboratory called the floor to report growth of Gram-positive cocci in clusters in two of Ethel's four blood cultures.

5. What microbial genus is most likely described by this report? Is this genus usually sensitive to the antibiotics prescribed for Ethel? Identify the specific mechanisms of action for ceftriaxone and amikacin and how they will impact this bacterial genus.

6. What medical term describes bacteria actively growing in the blood? Is this a serious condition? Explain.

7. Knowing the genus of bacteria in Ethel's blood, how do you account for her hypotension, fever, myalgia, fatigue, and nausea?

Two hours later, the microbiology laboratory confirmed the presence of coagulase-positive *Staphylococci* cultured from Ethel's effluent. Dr. Kittridge knew that antibiotic sensitivity data would probably be unavailable until the next morning. A repeat CBC indicated Ethel's WBC count was still climbing (24×10^9 cells/l) and cardiac monitoring showed she was experiencing ventricular arrhythmias. Ethel was not responding to therapy. Dr. Kittridge called the laboratory and ordered the microbes grown from the effluent culture to be tested for the presence of the "mecA gene."

8. Now that you know Ethel's staph infection is coagulase positive, what specific microbe do you suspect? What is coagulase and how does it benefit the microbe? How would the laboratory determine that the culture is coagulase positive?

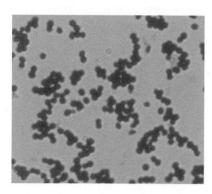

Figure 1. Gram stain of *Staphylococcus aureus*.

9. What is the mecA gene? What did Dr. Kittridge suspect when he ordered testing for the mecA gene? How might the lab test for its presence?

Thirty minutes later the laboratory called to report the culture was mecA-positive *Staphylococcus*. Dr. Kittridge switched Ethel's drug therapy to Linezolid plus IV vancomycin and ordered her nares and anus cultured.

10. What types of bacteria are most susceptible to vancomycin? Why didn't ceftriaxone stop Ethel's infection? What is the mechanism of action for vancomycin? Why wasn't vancomycin initially prescribed for Ethel's infection? Why was linezolid prescribed in conjunction with the vancomycin in this particular case?

11. Why did Dr. Kittridge order these additional cultures?

Over the next 24 hours, Ethel worsened despite appropriate medical care. Her nares culture was reported as positive for the same microbe as found in her effluent. When her heart began to fail, Ethel's sons were notified and arrived at Grace Hospital in time to say good-bye to their mother. Devastated by their sudden loss, the boys had many questions.

12. How could Ethel have gotten so ill so quickly?

13. Where did she pick up this infection? How could the doctor be sure that it was the exact same organism that colonized her nose?

Ethel's oldest son remarked that he had heard about this bacterial species on *Dr. Phil* and a national news report. It was referred to as the "superbug."

14. How did this organism earn its superbug nickname?

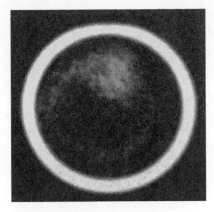

Figure 2. Slide agglutination test for penicillin binding protein 2A coded by the mecA gene.

Family Health Care

Human development is the fascinating study of the changes that occur in a person from conception through death. It's truly miraculous to think the DNA blueprint of a zygote can direct all processes necessary to produce a human infant within only nine short months. Parents marvel at the continuing developmental changes as their neonate transitions to a toddler, a child, and an adolescent. Human development doesn't cease when we reach adulthood. As a young adult, an individual typically enters the reproductive phase of their life. Middle adulthood is considered a particularly stressful period in our development as individuals face the challenges of raising adolescent offspring, potentially caring for their aging parents, and maintaining an active career. A decline in physical abilities, and sometimes mental faculties, is a hallmark feature of older adulthood.

Also linked to the developmental changes of specific life stages are certain infectious diseases. A diagnosis of chicken pox immediately conjures the image of a pediatric patient, while an elderly adult would be envisioned if the illness was pneumonia. Young children have little or no prior immunity to the myriad of pathogens that can cause human disease. However, as we age, the chance that we will have developed immunity to a pathogen through previous exposure increases. As a result, the common cold is much less common in middle age than it is in middle school. Also, the recent **pandemic** caused by the H1N1 serotype of influenza A virus was less likely to infect those over 50 years old due to previous exposure to another "swine flu" virus that circulated in the 1970s.

Pandemic – A worldwide epidemic.

This unit will explore infectious diseases typically associated with a given phase of life. Because some infectious processes are so commonly associated with a particular age group, you will become a better healthcare provider by knowing the risks faced by your specific patients.

A. Pediatric Infections

Pediatric healthcare providers often describe their work as rewarding, diverse, and challenging. It's easy to understand how healing little ones can be personally and professionally fulfilling. But imagine the diversity component of pediatric medicine. Most pediatric practices will care for patients from birth through college graduation. As their patients progress through infancy, childhood, and puberty to enter young adulthood, healthcare workers must tailor treatments to correspond with the current developmental stage. After all, development is much more than simply growing larger. Consider the behavioral and physiological changes that occur and can put pediatric patients at increased risk of acquiring certain infections. For example, infants and toddlers are especially susceptible to RSV, which has little impact on slightly older children. We aren't surprised when elementary school children acquire infections such as strep throat or conjunctivitis, as their close interactions facilitate the spread of those causative agents. The papules and pustules of acne are associated with pediatric patients in the adolescent phase of development. Clearly, specific infectious diseases target different age groups within the pediatric classification. Effectively treating patients in many different stages of childhood development, coupled with the numerous infectious diseases linked with each phase, emphasizes the diverse nature of this discipline and also illustrates its challenging component. As a pediatric practitioner you will enjoy the challenges of learning to work with the significantly different development stages and integrate that knowledge with the relevant infectious risks for your patients. You will also likely face the additional challenge of interacting with the parents of your patients. In this capacity you will have a special opportunity to serve as an educator. Please take full advantage of this chance to encourage parents to have their children develop healthy lifelong habits, such as frequent hand washing. This is also the crucial time to educate parents about the necessity of strictly adhering to the recommended schedule of pediatric immunizations. Many of the cases you are about to analyze should never have been written, since immunization could have prevented the illness. Only in rare circumstances is the administration of a childhood vaccine contraindicated. For all other patients, immunization is the best way to prevent diseases that could otherwise prove devastating. As Dr. Francis Rogalski, MD, MPH, stated at a recent Infection Control Symposium when discussing pediatric flu vaccination, "immunization delayed is protection denied."

A New Twist on a Childhood Disease

Jade was a 22-year-old medical receptionist in a pediatric practice. She loved her job and colleagues, although it did bother her to see so many sick children this winter. Even she had been ill last week. Jade experienced a sore throat, malaise, myalgia, and chills. When her temperature reached 38.2°C (100.8°F), she called her obstetrician. Jade was in her 26th week of gestation and concerned that the fever could affect the development of her baby. Based on Jade's very general symptoms, her doctor recommended Tylenol to reduce her fever, increased fluid consumption, and extra rest. She was instructed to call immediately if her condition worsened.

Luckily, Jade's symptoms subsided in a few days, but now she suffered from pain and swelling in her hands, knees, and ankles. Jade attributed it to pregnancy, but her joint pain persisted. At the end of the week, Jade started to feel panicky— her baby hadn't moved in two days! When her obstetrician examined her, he confirmed Jade's worst fear. There was no fetal heartbeat.

Jade was admitted to the hospital and labor induced, allowing the delivery of her daughter, Joy. Jade authorized the autopsy of her little girl, while her own physician ordered testing of Jade's blood to determine the cause of miscarriage. The following results were posted 48 hours later:

An EIA for parvovirus B19 IgM was positive using Jade's blood sample.

PCR performed on a fetal blood sample confirmed the presence of parvovirus B19 DNA.

Joy's autopsy showed extensive damage to the erythroid cells of the fetal liver and severe anemia, resulting in congestive heart failure associated with **hydrops fetalis.**

Hydrops fetalis – A medical condition characterized by accumulation of fluid in at least two body compartments during fetal development.

1. What is parvovirus B19? What features characterize this microorganism?

2. What childhood disease is typically associated with parvovirus B19 infection? What is the pathogenesis of this usually benign disease? During what season(s) is this infection most prevalent? Is this a common infection?

3. What is the significance of Jade's blood being positive for parvovirus B19 IgM as opposed to parvovirus B19 IgG?

4. Knowing the pathogenesis of this microorganism, explain why the erythroid cells of the fetal liver were targeted for destruction.

5. How did Jade likely contract this infection that caused her miscarriage? Will this be a concern during subsequent pregnancies?

6. Can parvovirus B19 infection be prevented?

7. What treatment(s) is recommended for parvovirus B19 infection?

Why Did the Chicken Pox Cross the Road?

It had been a busy fall term at school for Gretchen, the elementary school nurse. On a regular day, she would usually see several of the 283 kindergarten to seventh graders at the nurse's office with the typical problems of headaches, fevers, stomach aches, and skinned knees and elbows. In addition to this, 33 of the students had come down with **varicella**. Although the number didn't seem large to the administration, she had been surprised because she had known that most of the children at the school had either been vaccinated or had chicken pox earlier.

Varicella – The medical term for a pediatric disease commonly known as chicken pox.

Vaccine effectiveness – The practical reduction in risk for an individual when they are vaccinated under real-world conditions.

1. Describe the clinical features of chicken pox.

2. What pathogen causes chicken pox? Describe the physical features of the virus.

3. Why would individuals who are vaccinated or had been infected in the past not be expected to be at risk for acquiring chicken pox?

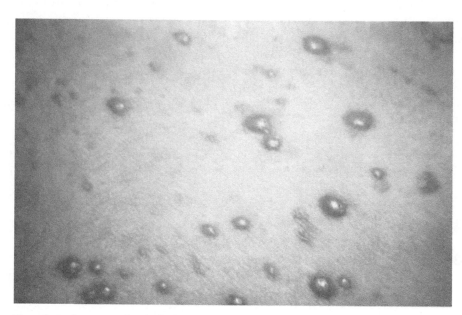

Figure 1. Rash associated with chicken pox.

A review of medical histories of specific children indicated that a number of children who had been vaccinated had gotten chicken pox. As a result, Gretchen worked with local health department officials to determine the **vaccine effectiveness**. To do so, questionnaires were sent to parents of all the students at the elementary school to determine the history of varicella disease, varicella vaccination status, and underlying medical conditions. Vaccination status was confirmed through school records. Parents of ill students were also interviewed to determine the severity of the child's disease.

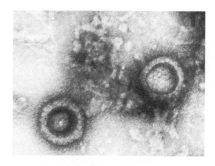

Figure 2. Viral pathogen for chicken pox.

4. How is vaccine effectiveness determined?

5. Why aren't vaccines 100% effective?

Among those that responded to the questionnaire, about half indicated their child had a previous history of varicella. None of those children became ill during the outbreak. Of the remaining 142 students, 115 had been vaccinated. 15 of the vaccinated students became ill. Among the 27 unvaccinated students, 18 became ill.

6. What was the effectiveness of the varicella vaccine?

Among those who had varicella, vaccinated students were more likely to have a mild disease (<50 skin lesions). Two-thirds of vaccinated children had a mild case of the disease, while only 11% of unvaccinated children had a mild case.

The outbreak started in late October. The **primary case** was an unvaccinated kindergarten student. The child had a fever and a severe case of varicella (>500 skin lesions) complicated with a **secondary infection** by bacteria. The child attended school for two days following the onset of the rash. The disease spread to children in 13 of the 15 classrooms. The disease also spread to homes and other schools. Three students who attended other schools but had a sibling with varicella at the outbreak school became ill. Also, one other child and three adults developed varicella. All were unvaccinated and were nonstudent household members of an ill student from the outbreak school.

7. How is varicella virus spread?

8. What is the incubation period for varicella?

9. Are those with varicella contagious before the rash appears or after the rash forms scabs?

10. Why are children with varicella at high risk for secondary bacterial infections?

11. When was the varicella vaccine licensed?

12. What are some of the reasons why parents refuse vaccination for their children?

13. What are the benefits of childhood vaccinations?

Primary case – The initial person to have the disease in a population.

Secondary infection – An infection by a different pathogen that arises as a complication of the primary infection.

Hand, Foot, and Mouth Disease

Jon and Joan were the proud parents of three active young children: Harry (7), Hunter (4), and Hannah (10 months). Since Joan only had six weeks of maternity leave after the births of Harry and Hunter, the boys entered day care at a very young age. A related issue Joan faced was the frustration of balancing the demands of her hectic full-time work schedule with her desire to breastfeed her babies. By the time each of her sons was four months old, Joan was exclusively using formula. She blamed herself for the repeated ear infections, the RSV, chicken pox, and endless parade of colds that her sons suffered. "If only I could have kept them out of the day care germ pool a little longer…or provided them with better immunity by nursing for a full year as their pediatrician recommended," Joan complained.

1. What do we call the transmission of maternal antibodies to offspring? In addition to nursing, how can maternal antibodies be passed to a baby? What class of **immunoglobulins** is represented by the antibodies secreted in breast milk? Consider the structure of this antibody class. Why is it significantly different from other types of antibodies?

Joan considered herself lucky when she was able to accept a new position with a different employer. When she discovered she was pregnant with Hannah, Joan was delighted to learn she was entitled to three months of paid maternity leave. Once she returned to work, Joan's "family friendly" employer permitted her to take one break in the morning and one in the afternoon to express milk for Hannah. Joan was convinced that these measures had made all the difference, since Hannah was now 10 months old and hadn't suffered so much as a sniffle!

Perhaps Hannah's exceptionally healthy start made it that much more upsetting to Joan when her daughter suddenly developed a fever of 38.3°C (101°F) after dinner one cool October evening. Joan immediately administered the appropriate dose of infant-strength Tylenol, but two hours later, Hannah's temperature was 39.3°C (102.8°F). Jon called the pediatrician's after-hours number and was told to "layer" infant-strength Motrin with the infant-strength Tylenol for better fever control. Unfortunately, Hannah's temperature still spiked to 40.2°C (104.4°F). Joan gave her a tepid bath.

2. What is the active ingredient in Tylenol? In Motrin? Why didn't the pediatrician recommend aspirin for Hannah's fever? What is the value of a tepid bath?

By midnight, Jon and Joan were frantic. Hannah's temperature reached 41°C (105.8°F) despite their continuing efforts to reduce her fever. Desperate for help, Jon called the 24-hour "ask-a-nurse" hotline at their local hospital. To his frustration, the nurse told him to continue their current treatments and only bring Hannah to the hospital if she suffered a seizure.

Hannah made it through the night without a seizure, but her fever remained quite high and she fussed constantly as though she was uncomfortable. Joan tried repeatedly to nurse Hannah without success. Hannah even refused to eat her favorite foods, teething toast and blueberry "buckle." Joan felt some relief when she finally coaxed Hannah to take a few sips of cool water.

Jon and Joan were waiting at their pediatrician's office when the practice opened. The doctor examined Hannah thoroughly and noted a symptom that

---

Immunoglobulins – A type of glycoprotein produced by plasma cells (activated, matured B lymphocytes). Used synonymously with "antibody," these molecules usually bind with high specificity and affinity to a given antigen (foreign molecule) thereby initiating an adaptive immune response.

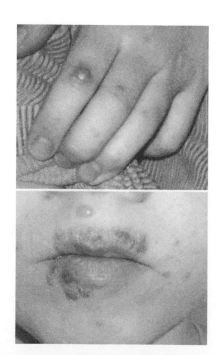

Figure 1. Hand, foot, and mouth disease.

Jon and Joan had missed. Hannah's tongue, throat, gums, hard palate, buccal mucosa, and inside lips were covered with red blisters, some of which were beginning to rupture. The doctor also pointed out developing blisters on the palms of Hannah's hands and soles of her feet. While Jon and Joan were clearly alarmed, the pediatrician informed them that Hannah should recover on her own within two to four days as she had contracted "hand, foot, and mouth disease." He explained that this disorder was fairly common in children under 5 and this was even the correct "season" for hand, foot, and mouth disease.

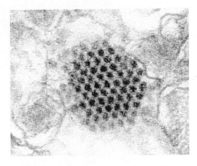

Figure 2. Viral pathogen of hand, foot, and mouth disease.

3. What pathogen causes hand, foot, and mouth disease? Describe its morphology and its usual habitat. Are antibiotics effective against this microbe? Explain.

4. How is hand, foot, and mouth disease diagnosed?

5. What is the peak season for infection with this microbe?

6. Why do you think Hannah refused to eat her favorite foods and only accept sips of cool water?

7. Why do you think this disease is more prevalent in young children?

The pediatrician told Jon and Joan that this microbe can manifest itself in many ways, and he explained how it was transmitted and what prevention measures they should consider to protect Hannah's older brothers. He recommended the continued use of fever/pain relievers to keep Hannah as comfortable as possible during her recovery and encouraged extra fluids.

8. Indicate the other ways this microbe may manifest itself. What more serious infections may result?

9. How is this microbe spread? What can be done to prevent others from becoming infected?

10. What is the incubation period of this microbe—that is, how long before the boys start showing signs of infection? What is the duration of symptoms for hand, foot, and mouth disease?

11. While comfort measures are typically the recommended treatment for hand, foot, and mouth disease, enhanced treatment is utilized when the infection becomes more serious. What course of action is recommended in that situation?

Seeing Red!

Life in the Russell household was busy, to say the least. Tom was a civil engineer and frequently traveled for work. Jodi was a CPA for a major accounting firm. Their careers alone could exhaust some couples, but Tom and Jodi participated in numerous church activities, built with Habitat for Humanity, and volunteered for the United Way. Tom coached for his son's third grade soccer team, and Jodi taught her four-year-old daughter's Saturday learn-to-swim class at the YMCA. Although the couple was excited to discover their little family would be blessed with a new addition in June, they each secretly worried about the challenge of adding another child to their already hectic lives.

The two professionals used their work-honed organizational skills to make everyday family activities run smoothly. Their routine flowed particularly well during this school year, since Lily was now in "all-day kindergarten" and could ride the bus to and from the babysitter's with her big brother Joshua. Tom and Jodi were feeling pretty good about themselves and their ability to successfully juggle family, work, and outside responsibilities. When making his weekly phone call to his parents, Tom smugly told his concerned mother, "No worries, Mom. It's all good. We've been through this parenthood thing twice already so we're hardly going to notice a change in the routine when 'number three' arrives."

"I don't know," she replied. "You're so regimented in your household. What if something comes up? What if you need to make a change? It could send your whole schedule crashing down like a 'house of cards.'"

"Mom, you worry too much. What could possibly come up?" Tom laughed confidently.

Tom and Jodi got lucky and life rolled along smoothly until late March. As expected for a CPA, Jodi started working two extra hours a day as tax season approached. The 10-hour work days coupled with the start of her third trimester left Jodi bushed at the end of the day. She could hardly pull herself out of bed at 4 AM to answer Joshua's call for his mother. After listening to her son's complaints, Jodi took his temperature and examined his throat with a flashlight. "Well buddy, your temperature is 38.9°C (102°F) and your throat is scarlet with tiny red spots. I wouldn't be surprised if you have strep throat. I'll call Dr. Brickner first thing in the morning." Jodi gave Joshua children's Tylenol to ease his fever and throat discomfort before stumbling back to her own bed.

1. What symptoms are usually associated with strep throat?

2. What pathogen causes this disease? What other infections can be caused by this same microbe?

Jodi promised her boss she would work late if she could just take time off in the morning for Joshua's appointment with their pediatrician. Dr. Brickner's examination plus a rapid strep test immediately confirmed a strep throat diagnosis. Joshua was given a 10-day prescription of amoxicillin and ordered to get plenty of rest, eat lots of popsicles, and stay away from his little sister for the next few days. Jodi got Joshua started on his medication and settled in at the babysitter's, and returned to work quite pleased that he would soon be better and that this "glitch" hadn't totally turned her schedule upside down during her busiest season.

Five nights later Jodi groaned when she heard Lily wail, "Moooomieeeee," at 2 AM. Like her brother earlier, Lily had an abrupt-onset high fever along with a

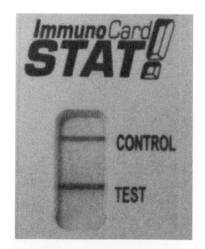

Figure 1. Rapid strep test result.

fire-engine-red throat. "Great, here we go again," she thought as Jodi dosed Lily with Tylenol and crawled into her daughter's bed to comfort her. In the morning, Lily went to the babysitter's since Jodi couldn't get an appointment with the pediatrician until 4:30 PM. At noon she received a frantic phone call from her sitter. "Jodi, she's red…bright red! Lily looks like a little boiled lobster! I think there's something seriously wrong." Jodi bolted out of the office, collected Lily from the sitter, and arrived at Dr. Brickner's office three and a half hours early for their scheduled appointment. Luckily Mary Jane, the **pediatric nurse practitioner**, managed to work Lily into her tight schedule. Her physical assessment confirmed a temperature of 39.6°C (103.2°F), swollen cervical lymph nodes, malaise, and an inflamed pharynx with white streaks on the tonsils and **petechiae** on the soft palate. Mary Jane encouraged Jodi to look with her into Lily's mouth. Rather than directing her attention to the angry-looking red throat, Mary Jane pointed out the appearance of Lily's tongue. Jodi gave a startled gasp as she noticed for the first time the white coating with reddened, projecting **papillae**.

Pediatric nurse practitioner – An RN with a masters degree in nursing and board certification in an area of specialty (pediatrics).

Petechiae – Small, flat, purplish red spots from bleeding under the skin.

Papillae – Projections of the glossal mucous membrane shaped like a truncated cone and involved in taste.

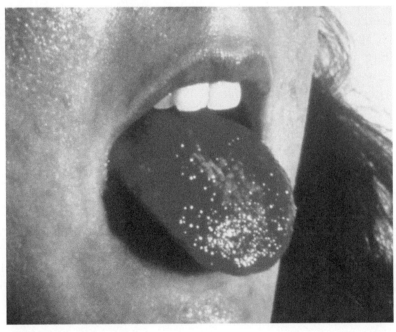

Figure 2. Strawberry tongue.

3. What term is applied to the red, petechial spots on the palate? What term is typically used to describe the appearance of Lily's tongue?

"Now then, Missy," Mary Jane said, giving Lily a playful poke on her nose, "did you go to the beach and get a sunburn?" Lily grinned, but before she could answer, Jodi rushed to explain. "I thought she had strep throat that she caught from Joshua, since he was sick with it last week. Honest, she wasn't red like this when I took her to our babysitter this morning."

Calmly examining Lily's face, extremities, and trunk, Mary Jane replied, "I think you're absolutely right, Jodi. Lily probably does have strep throat."

"But Joshua didn't look like this," Jodi protested.

"Hmmmm, that's right too," Mary Jane said, gently pressing her fingers on Lily's rash. "Honey, are you itchy?" she asked Lily and smiled as the little girl's pigtails

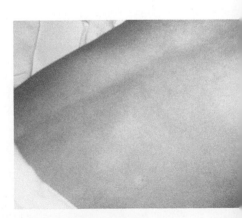

Figure 3. Scarlet fever rash.

bobbed up and down. "Alrighty now, open wide as a whale and say 'aaaahhhh' for me." Lily dutifully obeyed as Mary Jane collected a throat specimen and performed a rapid strep test. Once she had the test results, Mary Jane explained her findings to Jodi. "Lily has strep throat with the added manifestation of this bright red **exanthem**. We call this type of infection 'scarlet fever.'" Expecting the fearful look on Jodi's face, Mary Jane continued her explanation of Lily's condition. "While the frequency of this infection remains high in children, its seriousness has declined significantly since the advent of antibiotic therapy."

4. What is the other common name for scarlet fever?

5. Describe the nature and progression of a scarlet fever rash.

6. Are there any other symptoms associated with this infection?

7. In addition to the rapid strep test indicated above, what other diagnostic tools are used to confirm this infection?

"Now I'm really confused," Jodi said. "How did Lily get scarlet fever from Joshua if all he had was strep throat? If it's the same germ, why didn't Joshua have a rash too?"

8. How is scarlet fever transmitted? What group is most likely to get scarlet fever and why?

9. What can be done to prevent the spread of scarlet fever?

10. Why did Lily get scarlet fever if she became infected with the same pathogen as her brother who didn't manifest the rash?

Mary Jane reviewed her findings with Dr. Brickner, who then prescribed a 10-day course of penicillin. Mary Jane reminded Jodi to be sure Lily finished her therapy to prevent the development of complications from this infection. Additionally, she recommended a number of comfort measures for Lily since both her throat and skin would remain irritated until her antibiotic began clearing the infection.

11. What complications can result from a scarlet fever infection?

12. Discuss appropriate comfort measures to address both the throat pain and skin irritation suffered by scarlet fever patients.

A relieved but completely exhausted Jodi arrived home, tucked Lily into bed after administering her first dose of penicillin, and got online to do a little work until Tom could finish up at his office and relieve her. At dinnertime Jodi returned to her cubicle for a late night of number-crunching while Tom stayed with their children. When his mother called that evening, Tom updated her on the health of her grandchildren. Frantically, she began relaying horror stories from her youth of children who had died of scarlet fever. "Relax, Mom," Tom soothed his mother. "Jodi talked with the nurse practitioner at our pediatrician's office about all of these concerns. It's different for kids now. Lily should be just fine once she's finished her penicillin."

"But Tom," his mother persisted, "Lily is so contagious. What about Jodi? She's been taking care of Lily and could pick up the scarlet fever from her. I know that strep can be dangerous to mothers and newborn babies. I googled strep when Joshua got sick and I read about all kinds of mother/baby problems it can cause. Don't you think Jodi should see her **obstetrician** to be sure the baby will be healthy?"

Obstetrician – A physician specializing in the care of a pregnant woman through delivery of the baby.

13. How long will Lily be contagious?

Now Tom was worried. He hadn't even thought about the baby. As soon as Jodi dragged home at 11:30 PM, Tom immediately vented his concerns. Thinking she was too tired for anything to keep her awake, Jodi was suddenly as tense as her husband. Barely sleeping for yet another night because of her worry for the baby somersaulting in her belly, Jodi called her doctor as soon as the practice opened for an appointment. When Jodi explained her fears to Dr. Fox, he was able to reassure her that scarlet fever is relatively rare in adults and even if she contracted the infection, it would pose little risk to her baby. "But Dr. Fox, my mother-in-law got online and says she read all about moms carrying a strep infection that can kill their newborn," Jodi fretted.

"That's true Jodi, but there are many different types of 'strep germs.' The one your mother-in-law read about is a different species than the one responsible for scarlet fever." Dr. Fox was so pleased to see his fatigued patient relax at this news he almost hated to continue his necessary follow-up counseling. "However, I'm glad you came in today with your concerns. Although it doesn't happen often, a pregnant woman can also become a carrier of this type of strep and, if undiagnosed, it can cause serious complications at delivery. Since both of your children have had recent strep infections and you have served as their primary caregiver, it's now possible for you to be colonized with the pathogen and still asymptomatic. Luckily, it's easy to test you and even easier to treat you if necessary. Let's check you out so you can stop worrying about all of your children and focus on yourself...Jodi, I'm worried about the strain your job and family responsibilities is having on you. Sleep is also essential for your health and that of a developing fetus."

Dr. Fox completed his examination of Jodi by swabbing both her throat and vagina. The labeled specimens were sent to the clinical microbiology laboratory for culture.

14. What type of 'strep' typically causes the neonatal complications Jodi's mother-in-law was referring too?

15. With the prevalence of medical information online, how will you, as a future healthcare provider, work with patients who inaccurately interpret this data (or have read misinformation)?

16. Why did Dr. Fox procure a vaginal specimen from Jodi if her children both demonstrated strep infections of the pharynx?

Three days later Dr. Fox contacted Jodi at work. "Jodi, your specimens were both positive for the 'strep' that causes strep throat and scarlet fever. There is no need to panic...remember that your baby will be fine, but I want you to stop at the pharmacy on your lunch break for the amoxicillin prescription I've ordered. Even though you are symptom free, clearly you've become a carrier of the microbe. This **prophylactic** treatment will prevent any complications for you and the baby at delivery and it will also prevent you from re-infecting Joshua and Lily."

Prophylactic – A preventative treatment.

"Thank you very much, Dr. Fox. I'm so relieved that the baby will be fine. You know I'll do whatever you say to keep my baby healthy," Jodi said.

"Good, I'm glad to hear you say that, Jodi," Dr. Fox responded, "because there is one more therapy I'm prescribing to keep you and the baby healthy."

"Of course, Dr. Fox, anything," Jodi said with distraction as she began typing at her computer during their conversation.

"It's time to leave the rat race, reorganize the famous Russell family schedule, and focus on a slower pace for you and your three children so everyone is healthier. Jodi, these are 'doctor's orders.' Talk with your husband and boss today and make some changes!" Dr. Fox finished with authority.

17. What complications might result if Jodi was not treated for her colonization? Is there any additional testing that should be done prior to the delivery of Jodi's baby?

18. If Jodi is asymptomatic, how could she re-infect Joshua and Lily?

(*Note:* Hopefully you have noticed in this case that the various healthcare providers interact with their patients in different manners. Please keep in mind that friendly, playful gestures may put a pediatric patient at ease, facilitating examination. However, when working with adults, particularly geriatric patients, a more respectful tone is needed. Remember that your professional attitude can have a tremendous impact on your patient's recovery.)

'Tis the Season

Gina was relieved that eight-year old Caleb and five-year-old Stephen had finally recovered from their colds. It was such a shame that the boys were ill over the Christmas holidays, but at least they hadn't missed any school. Leaving four-month-old Kelly at home with Barbara, her 70-year-old mother, Gina took the boys to school and made a quick side trip to the market for a few groceries. Planning to spend the rest of her day putting away the last of the holiday decorations when she arrived home, Gina was not pleased to be greeted by her mother wearing a worried expression. "Kelly won't take her bottle. She feels warm to me and her little nose is running like a facet," Barbara fretted.

As Barbara surrendered Kelly to her mother, Gina performed the traditional maternal response of a kiss to the forehead. Her instinctive "mom-thermometer" confirmed that Kelly definitely was warm and likely catching her brothers' cold. Gina was disappointed at this turn of events. Her daughter had been born almost four weeks early, necessitating a two-week stay at the **NICU**. This had been a scary, stressful time for the whole family. While it was reassuring to hear the doctors agree Kelly would eventually be just fine, the NICU care postponed family bonding and even caused Gina to give up on nursing Kelly. The pediatrician warned that Kelly would be more susceptible to infections than the average child during her early years. Since Kelly had been perfectly healthy until now, Gina had hoped the doctor would be proved wrong.

After taking Kelly's temperature (38.9°C; 102°F), Gina administered the appropriate dose of infant's Tylenol. She cuddled and rocked her very fussy daughter until Kelly finally drifted off to sleep.

> **NICU** – **N**eonatal **i**ntensive **c**are **u**nit; a facility designed to provide specialized care for premature infants.

⬤ **1.** Why did Gina administer infant Tylenol rather than baby aspirin? Are there other "safe" fever reducers available for use with infants?

⬤⬤ **2.** Why did the pediatrician indicate Kelly might be highly susceptible to infections during her first year?

An hour later, Gina responded to Kelly's cranky wails. She found her daughter with greenish mucus crusted around her nose and still feverish despite the Tylenol treatment. Kelly still refused her bottle and "fussed" all day. When Tom came home at dinnertime, he took Kelly from his exhausted wife. Even a game of peek-a-boo with daddy didn't improve Kelly's disposition or appetite. She had begun coughing and remained congested, feverish, and irritable. Throughout the evening, Kelly worsened.

By 10 PM, Tom and Gina decided to leave the boys with their grandmother and take Kelly to the emergency room at the local hospital. Upon initial examination, Kelly was listless, wheezing, breathing rapidly (60 breaths/minute) with obvious rales, and demonstrated cyanosis of the lips and nails. Her fever was 40°C (104°F) and she was dehydrated due to her refusal to feed throughout the day. The doctor initiated IV fluids, ordered Kelly's nasal secretions swabbed and sent to the laboratory, and admitted her to the pediatric unit. As the doctor finished his examination, Tom asked how the "cold" Kelly caught from her big brothers could have made her so much sicker. The doctor smiled sympathetically and told Tom his daughter probably suffered from the major cause of respiratory distress in children under age two.

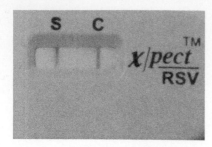

Figure 1. ELISA test for the pathogen.

Sequelae – A pathological condition that is a side effect of another disease.

3. What ailment does Kelly have?

4. What symptoms are typically associated with this condition in young children? In older children and adults?

5. What pathogen is responsible for Kelly's illness? Describe the nature of the pathogen.

6. How is this pathogen usually transmitted? When is this infection most prevalent?

7. Which individuals are at greatest risk of infection? Which risk factors did Kelly possess?

8. How was Kelly's diagnosis made?

A respiratory therapist visited Kelly's room at regular intervals to perform "breathing treatments." The therapist explained that the medications being administered would help open Kelly's airways so she could breathe easier. A cool mist vaporizer was started in Kelly's room, and she received periodic doses of Tylenol to control her fever, and antiviral medications. IV fluids were continued.

9. Why didn't Kelly's doctor prescribe an antibiotic for her infection?

The next morning Kelly's pediatrician confirmed her diagnosis and offered Tom and Gina explanations as to why their daughter had become so ill so fast. He also reviewed prevention measures, as this infection has been associated with specific **sequelae**.

10. What treatment options are available to patients?

11. What methods are used to prevent this infection from occurring?

12. What sequelae are associated with this infection?

13. What is a syncytium and how is it applicable to this disorder?

After two days, Kelly was discharged and returned home to her family. Everyone was glad to have the littlest member of the family back, but Barbara refused to hold her since she was developing a cold and was afraid of re-infecting her granddaughter. The next morning, against Barbara's wishes, Gina took her mother to the doctor for an examination and a nasal swab.

14. Was Gina right to insist her mother be tested? Explain.

Splash!

Ten-year-old Tessa winced as Dr. Johnson gently pulled on the **pinna** of her left ear to insert his **otoscope**. As he expected, Dr. Johnson observed that the ear canal was as swollen, scaly, and red as the exterior of Tessa's ear. Additionally, he noted and cultured the yellowish discharge from the ear canal. Although her temperature was normal, Dr. Johnson was able to appreciate an enlarged left cervical lymph node, and Tessa complained that her ear felt "full," her hearing was muffled on the left side, and it hurt to chew. As Dr. Johnson began making notes in her chart, he asked Tessa what she had been doing for fun on her summer vacation. Even though she felt unwell, Tessa excitedly explained that because of her recent birthday, she was allowed to move up to the 10- to 12-year-old category on the Ada Gators summer swim team and could now practice every day with the "big kids"! Dr. Johnson just smiled as Tessa described her new mastery of the "flip turn" and her improved butterfly stroke. "That's terrific," Dr. Johnson told Tessa. "It's really important for young people to participate in athletics to have fun, learn good sportsmanship, and build strong, healthy bodies. But did you know that most sports have some risk associated with participating in them?" At this point, Tessa's mom interrupted, worried about where Dr. Johnson was going with this conversation. "But Tessa is a swimmer," she said. "It's not like she's involved in football or some other rough, contact sport. I would never take a chance with my children risking an injury." "No, of course you wouldn't, Mrs. Wilson, but not all athletics-associated risks are injuries. In Tessa's case, it is definitely infection." "Tessa," Dr. Johnson said, "you've officially got a case of swimmer's ear."

Pinna – The external ear or auricle, designed to capture sound waves.

Otoscope – A medical instrument used to examine the ear canal and tympanic membrane.

1. What is the medical term applied to the condition commonly referred to as "swimmer's ear?"

2. What are the most likely causative agents of this disorder?

3. What is the usual incubation time for this infection?

4. What is the appropriate treatment for this infection?

Dr. Johnson suggested Tessa take an appropriate dose of either acetaminophen or ibuprofen for the next 48 hours to help ease her discomfort. He also prescribed otic Ciprodex and explained that this medication was an antibiotic plus a steroid to decrease Tessa's inflammation. He cautioned Mrs. Wilson to keep Tessa out of the pool and even to plug her ears with cotton balls while showering until the infection had resolved. He also emphasized the need to finish the course of antibiotics.

While relieved to hear that Tessa should be just fine, Mrs. Wilson still had several concerns:

5. Is Tessa contagious?

6. How long before this infection resolves?

7. How did Tessa likely contract swimmer's ear?

8. Are there any factors that put Tessa at special risk of acquiring this infection?

9. Are there any complications associated with swimmer's ear?

10. What can they do to prevent this problem from recurring?

Two days later, the office manager from Dr. Johnson's pediatric practice called Mrs. Wilson with the results of Tessa's culture. Tessa's infection was caused by *Pseudomonas aeruginosa* and the strain was sensitive to the Ciprodex prescribed. The manager inquired as to Tessa's progress, reminded Mrs. Wilson to continue the antibiotic as directed despite symptom abatement, and encouraged her to contact them if she had any further questions or concerns.

11. Describe *Pseudomonas aeruginosa*. Why is this microbe a frequent contaminant in swimming pools?

12. Can "swimmer's ear" symptoms result from other medical conditions?

13. Why did both Dr. Johnson and his office manager emphasize the need to continue Tessa's antibiotic regimen even after her symptoms have subsided?

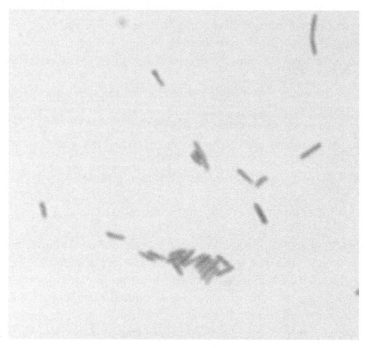

Figure 1. Gram stain of *Pseudomonas aeruginosa.*

B. Childbearing-Related Infections

The birth of a child should be one of the highlights of your life. Hearing the baby's heartbeat for the first time or feeling that first kick can be exhilarating. During pregnancy couples anxiously prepare for their new arrival, with mom usually taking special steps to ensure the delivery of a healthy baby. While focusing on diet, exercise, and rest are essential for a positive outcome, prevention of certain pregnancy-associated infectious diseases is another integral component of good prenatal care.

Passive immunity from the mother will often protect a developing fetus, but some pathogens can still cause serious birth defects...or worse. These microorganisms may affect the fetus by crossing the placenta to cause infection in utero or they can inoculate the baby during delivery if they reside in the vagina or if placental bleeds occur during the last stages of labor.

As with most infectious diseases, "the best offense is a good defense." There are numerous simple preventive measures you can teach your pregnant patients to help them avoid pathogen exposure and protect both themselves and their unborn child.

Special Delivery

Tanya and Aaron had painted the nursery blue and added a border of planes, trains, and trucks. The crib was assembled and filled with cuddly teddy bears. Last week, the couple had finished their "preparing for parenthood" classes and even toured the mother/baby unit at St. Rose's Hospital. While they waited for their appointment with Dr. Moore, Tanya and Aaron reviewed their short list of boy names one more time.

During their prenatal appointment, Dr. Moore performed an **ultrasound**. The young couple watched the monitor in wonder while Dr. Moore explained the changing images. Aaron was especially pleased with himself when he was able to point out that his son was sucking his thumb. Dr. Moore was extremely happy with the progression of Tanya's pregnancy. His 21-year-old, African-American **primigravida** patient had been diagnosed with **Type I diabetes** at age 11. During her teenage years, Tanya's blood glucose levels were not well-regulated. Her single mother worked two jobs, but the family finances were still stretched so thin providing for six children that they could rarely afford to eat the **carbohydrate**-controlled diet prescribed for Tanya. She only monitored her blood glucose levels once a day instead of the 5–8 times a day as recommended by her physician, since Tanya also couldn't afford the chemical strips for her **glucometer**. Luckily, when Tanya graduated from high school, she took a clerical position at St. Rose's Hospital. With her steady paycheck and good medical insurance, Tanya focused on managing her diabetes. She strictly adhered to her recommended diet, consistently monitored her blood glucose levels, and, working with her insurance company and **endocrinologist**, acquired an **insulin pump**. The tight control of her diabetes allowed Tanya and her baby to complete the first eight months of pregnancy in good shape.

Dr. Moore concluded his prenatal assessment with an internal exam, which indicated no evidence of preterm labor. Using a culturette, Dr. Moore swabbed Tanya's cervix, vagina, and rectum. Curious, Aaron asked Dr. Moore what he was doing.

1. What is a culturette? Name a specific media type it would likely contain. Why is this necessary?

2. What vaginal microbe is Dr. Moore culturing for? Briefly describe this microorganism. What microbes are considered normal vaginal flora?

3. How is this microbe usually spread? Where is it usually found?

Dr. Moore explained that when a baby is born, it comes in contact with any germs that might be living in the mother's birth canal. While this usually isn't a problem, if a baby does become infected with a specific type of germ, it could lead to serious complications. The faces of both Aaron and Tanya instantly registered concern. "Are you saying Tanya has an **STI**?" Aaron barked. "Did she give something to me?" Giving Aaron an angry glare, Tanya began asking her own questions. "What serious complications?" she exclaimed. "Is our baby ok? How did I get sick?"

Ultrasound – The use of high-frequency sound waves to generate a fetal image.

Primigravida – A woman in her first pregnancy.

Type I diabetes – A disorder caused by autoimmune damage to the pancreas resulting in little to no insulin production.

Carbohydrate – The most abundant biomolecule, these organic compounds are aldehydes or ketones with multiple hydroxyl groups.

Glucometer – A portable blood glucose monitoring device.

Endocrinologist – A physician specializing in treatment of endocrine system disorders such as diabetes, hyperthyroidism, etc.

Insulin pump – A portable, personal medical device for continuous subcutaneous insulin infusion.

STI – **S**exually **t**ransmitted **i**nfection; an infection transmitted by direct sexual contact.

4. Does Tanya have an STI? Explain.

5. What is the prevalence of this microbe in pregnant women? What women are at greatest risk of developing vaginal colonization with this microbe? What factors increase the likelihood of neonatal infection?

6. What complications can infection cause following delivery?

After answering their questions, Dr. Moore promised to call Tanya with the results of her culture. Still worried, Tanya asked, "What if the culture is positive? Do I need to have a C-section to protect my son?"

7. What potential treatment options did Dr. Moore discuss with Tanya? How effective is therapy?

8. Does a C-section automatically eliminate the risk of neonatal infection with this microbe? Explain.

9. How will Tanya's specimen be cultured in the microbiology laboratory?

With their questions answered, Tanya and Aaron relaxed. In fact, Tanya was actually relieved to know that she could be so easily treated right before delivery if her culture was positive. "Just think what might happen to a poor little one if it is born prematurely. In addition to the medical problems from being born too soon, there wouldn't be time to culture the mom. The baby could be exposed during delivery and have even more problems."

10. How could a physician respond to protect the baby when a patient unexpectedly goes into preterm labor?

11. What steps are recommended to prevent the transmission of this microorganism?

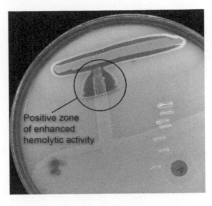

Figure 1. CAMP and bacitracin sensitivity test for pathogen.

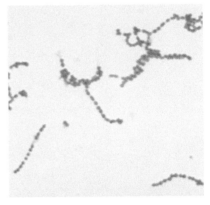

Figure 2. Gram stain of pathogen.

Cookout Concerns

It was a steamy Fourth of July afternoon when the Parker clan gathered for their annual family reunion at Grammie and Poppa's farm. The older kids played softball with "the dads" while the wee ones romped in the yard under the watchful eyes of "the moms." Soon, it was time to begin grilling. Mike, Rob, and Mark—the Parker brothers—lit three large charcoal grills and started the steaks and burgers. Once he flipped all the meat on his grill, Mark opened two packs of hot dogs and dumped them onto a platter. He expertly transferred the franks to the open spaces on the grill between the sizzling sirloins and thick juicy hamburgers. When the meat finished cooking, Mark shook the sloppy platter to get rid of the hot dog "juice" and piled the dish high with steaming hot steaks, burgers, and dogs. As Mark and his brothers brought the meat to the serving table under the gazebo, their wives finished setting out the fruit salad, calico beans, deviled eggs, and homemade pies. In no time, the children abandoned their games to join their parents for a tasty feast.

Mark helped his three-year-old son, Todd, put ketchup on his hot dog and settled him at the "kid's table" before going to the buffet to prepare plates for himself and his wife Mona. "You sure know how to spoil a girl," Mona teased as her husband gave her a plate with all of her favorite foods. "This cheeseburger looks wonderful," she said taking a big bite out of the oversized sandwich.

Smiling, Mark kissed his wife on the cheek and patted her swollen belly. "It's the least I can do. I know the heat and humidity make you uncomfortable when you're this far along."

Mona's second pregnancy was progressing normally, but at 32 weeks, she had the usual discomforts of the third trimester that were exaggerated by the steamy weather. Despite this, Mona thoroughly enjoyed the holiday with her husband's family.

Three weeks later, Mona woke early feeling achy and feverish. Forcing herself out of bed, Mona started breakfast for her family. By the time Mark and Todd arrived in the kitchen, Mona was worried. She was experiencing severe abdominal pain and significant lethargy. Mark immediately transported his wife to the Emergency Department at their local hospital. The attending physician noted Mona's fever 38.5°C (101.3°F). Her blood pressure was normal, but she was **tachycardic** at 125 bpm. **Doppler ultrasound** indicated good fetal heart tones, but **speculum** examination revealed "**membranes**" bulging into the vagina through a dilating cervix. Mona was admitted to the hospital, started on IV fentanyl for pain, and Dr. Fuller, her obstetrician, was notified of her condition.

1. Mona's symptoms are so general, it's hard to diagnose a specific infection. However, there are several infectious conditions that are notable for affecting women during pregnancy. Identify at least three such infections.

After examining his patient plus reviewing the attending physician's report, Dr. Fuller ordered blood cultures to be drawn, then explained it would be in the baby's best interest to deliver immediately and try to prevent transmission of Mona's infection. "I know you weren't planning on having your baby today, but since you are almost 36 weeks along, you've got significant symptoms of infection, and your bulging membranes will likely rupture soon, I think this is our safest course of action for you and the baby." Mark and Mona were stunned and concerned, but Dr. Fuller assured them the **NICU** was well-equipped to handle the needs of a baby born one month prematurely. "I'll go make arrangements for

Tachycardic – A rapid heart rate.

Doppler ultrasound – A noninvasive procedure to monitor fetal heartbeat.

Speculum – A medical tool for investigating body cavities.

"Membranes" – The amniotic sac that encloses a developing fetus.

NICU – **N**eonatal **i**ntensive **c**are **u**nit; a facility designed to provide specialized care for premature infants.

your **induction**," Dr. Fuller said. "It's time to get started." As if on cue, Mona gasped in surprise. "Everything OK?" Dr. Fuller queried.

"Well, I guess it's time, alright," Mona responded. "My water just broke."

About seven hours later Dr. Fuller returned to Mona's room to deliver her 4-lb, 14-oz son. Preliminary neonatal examination revealed a normal, healthy child for 36-weeks gestation. After a few minutes with their new son, Adam was quickly taken to the NICU for observation since he was preterm. When the placenta was delivered, Dr. Fuller noted it was fibrous with an exudative coating. He sent the organ to the pathology department for examination, collected amniotic fluid for culture, and ordered blood cultures on Adam.

The next morning, Mona still felt "flu-ish," but her temperature was slightly reduced (38.1°C; 100.6°F). Unfortunately, during the night, Adam developed **dyspnea**, **jaundice**, and lethargy. He ate very little and vomited the tiny amount of formula consumed. Although the lab results were not yet available, the neonatologist implemented IV broad-spectrum antibiotics, assuming it was likely that Mona's infection had been transmitted to her son. Adam's condition continued to deteriorate slowly throughout the day. His breathing worsened and he developed pressure in his skull as noted by bulging of his **fontanelle**. By late afternoon, the microbiology laboratory confirmed growth of *Listeria monocytogenes* in Mona's blood cultures, and the pathology laboratory reported the placenta had signs of **chorioamnionitis**. Adam's antibiotic therapy was changed to ampicillin, and his parents were informed that both mother and son suffered from listeriosis.

● **2.** Characterize *Listeria monocytogenes*. How is it typically cultured in the laboratory?

●● **3.** What symptoms are usually demonstrated by adults infected with this bacterium? What symptoms manifest in pregnant women? In neonates?

● **4.** What sequelae are associated with this condition?

● **5.** Clinical studies indicate antibiotic therapy is ineffective in about 70% of listeriosis cases due to the nature of this pathogen. Why is *Listeria monocytogenes* such a challenge to treat? Why is ampicillin considered the drug of choice? What therapy is recommended if the patient demonstrates a penicillin sensitivity?

Mark and Mona had never heard of listeriosis before and had many questions for Dr. Fuller. "How do you catch listeriosis?" Mona led off when her physician arrived. "I'm so careful about hand washing and avoiding people with colds. Where do these germs grow?" Dr. Fuller explained this condition was a foodborne illness associated with processed foods, so it was unlikely Mona's good hygiene practices could have prevented its transmission. Dr. Fuller reviewed a list of common culprits in the spread of this disease.

● **6.** Identify these common sources of *Listeria monocytogenes* contamination.

"Do you recall eating any of these products within the last one to four weeks?" Dr. Fuller asked.

"Three weeks ago we had a big family reunion with a cookout," Mark volunteered. "We grilled hot dogs there," he said, "but I know I cooked the dogs through and through."

Induction – Artificially stimulating childbirth in a woman.

Dyspnea – Difficulty breathing.

Jaundice – Yellowing of skin and conjunctiva due to the increased levels of bilirubin in the blood associated with liver dysfunction.

Fontanelle – An anatomical feature of an infant's skull; the "soft spot."

Chorioamnionitis – Inflammation of the amnion and chorion usually due to bacterial infection.

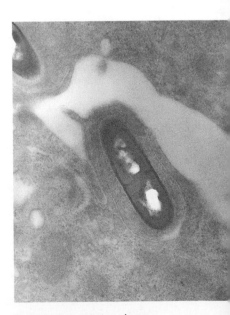

Figure 1. Transmission electron micrograph of *Listeria monocytogenes*.

Figure 2. Risk factors for listeriosis.

"That's right," Mona said. "But remember, I didn't eat a hot dog. I had a hamburger...it was Todd who ate a hot dog. Oh no," Mona cried with alarm. "Is Todd going to get this too?"

"No, no, relax," Dr. Fuller calmed his patient. "Todd isn't at risk for the infection, but you were."

7. How can the spread of *Listeria monocytogenes* be prevented? How could Mona have been infected if she didn't eat a hot dog? What is the infective dose of this pathogen?

8. Who is at greatest risk of acquiring listeriosis? What factors increase the risk of listeriosis in pregnant women? Compare the prevalence of listeriosis in the general population with that of pregnant women.

"Dr. Fuller, I still don't understand how my infection could have been a food-borne illness," Mona said quizzically. "I thought you got nausea, vomiting, and diarrhea with food poisoning. I didn't show any of those symptoms."

"Yah, that doesn't make sense to me either," Mark responded. "I ate some bad meat years ago and within hours I was in the john with terrible cramps and diarrhea. Mona didn't have any of that...and the cookout was weeks ago. Why would it take so long for her and the baby to get sick?"

9. Outline the pathogenesis of listeriosis. Indicate why GI symptoms are rarely observed. Explain why symptoms don't manifest for a few weeks with this food-borne pathogen. Given the nature of its pathogenesis, why is it difficult for your immune system to attack *Listeria monocytogenes*?

"Dr. Fuller, if only we had known about this sooner, could we have prevented Adam from getting sick?"

"Unfortunately, there really isn't any way for us to easily test for listeriosis before symptoms appear. When it is suspected, a blood culture, such as the one we did when you arrived at the hospital, is our best indicator."

10. Considering the nature of *Listeria monocytogenes*, why can't you simply run an anti-*Listeria* antibody screen to determine infection with this pathogen?

Frustrated, Mark hugged his wife close and commented, "I just can't understand how a disease I've never heard of could make my new son so sick. It just doesn't get any worse than this."

"I know how upsetting it is to watch your child suffer, but you are fortunate that Adam acquired the infection now rather than early in gestation. Because Mona was near term, we could deliver your son and treat him more effectively. Most *Listeria* infections in the first and second trimester result in **miscarriage**."

Miscarriage – Spontaneous delivery of a fetus before it is capable of survival (within the first 20 weeks of pregnancy).

11. Compare the prognosis of a fetus who acquires a *Listeria monocytogenes* infection in the second trimester with one who becomes infected near term.

Epilogue

Adam's listeriosis eventually manifested as meningitis and pneumonia. He was critically ill for almost a week and suffered a seizure before his infection began to resolve. Today, Adam is an active four-year-old. He demonstrates numerous developmental delays, but otherwise suffered no long-term neurological damage. Adam's pediatrician believes he will continue to improve as he matures and should face no limitations as an adult. Thankful for her son's progress, Mona "blogs" about their experience, hoping to educate pregnant women so they can avoid listeriosis and its potentially devastating consequences.

12. What is the **neonatal mortality rate** of listeriosis? What is the overall mortality rate for the general population?

13. What educational efforts might you implement to help your future patients avoid this rare but serious infection?

Neonatal mortality rate – Number of deaths of newborn infants (less than one month old) per 1,000 births per year.

C. Geriatric Infections

Older adulthood is often euphemistically called the "golden years," much to the chagrin of senior citizens. Everyone hopes to enjoy travel, recreational activities, and grandchildren in their later years, but the truth is this age group is at elevated risk of medical problems.

Of course, the primary reason for their increased incidence of illness is general organ system deterioration. Since our immune system is also negatively impacted by aging, infectious disease becomes more problematic than ever. Waning immunity manifests with susceptibility to pathogens previously controlled with artificially acquired active immunity (pertussis), re-emergence of latent microbes (VZV causing shingles), and a general inability to defend against invasion by **opportunistic microorganisms**.

As a result, infectious diseases account for one-third of all deaths in people 65 years and older. An estimated 90% of deaths resulting from pneumonia occur in people 65 years and older. Mortality resulting from influenza also occurs primarily in the elderly.

Other factors also cause an increased mortality in the elderly. Often an infection is a complication of a preexisting medical illness. The elderly are also more likely to require an invasive medical procedure. In addition, early detection of an infection is more difficult because the typical signs and symptoms, such as fever and **leukocytosis**, are frequently absent.

To effectively work with geriatric patients, you will need to address two specific topics. First, teach them basic infection prevention measures. Emphasize hand hygiene, receiving annual flu and pneumonia vaccines, and maintaining an active lifestyle with a healthy diet. The other critical aspect of geriatric care is vigilance in monitoring symptom progression during an infection. Since you can't count on a vigorous immune response, pathogenesis is often more rapid than in younger individuals. Your attention to this risk factor will allow you to intervene early and increase the likelihood of recuperation. Consequently, as you analyze the cases in this section, focus on prevention techniques that may permit your geriatric patients to avoid illness and on identifying cardinal symptoms needed to ensure rapid medical intervention when it does occur.

The Second Time Around

Helen, a 70-year-old retired church secretary, woke to a strange tingling, burning sensation along her spine. She wriggled and shrugged, scrubbed extra hard in the shower, and still couldn't shake the odd sensation. Two days later, Helen saw her physician, who could offer no explanation despite a thorough physical exam. Although the symptom was annoying, Helen continued on with business as usual. She served three mornings a week as a hospital volunteer and worked every afternoon as a Wal-Mart greeter. Helen helped out at church with numerous activities, baby-sat her four grandsons on Saturday, and daily helped her 94-year-old mother with household chores while she continued to recover at home following a mild stroke. With so many people depending on her, Helen didn't have time to slow down for an irritating sensation.

About a week later as she dressed, Helen noticed a sore red patch developing on her left side just under the band of her bra. Every movement sent shooting pain through her torso. Helen immediately contacted her physician's office to schedule another appointment. Later that day when she showed her doctor the tender patch of skin, Helen was surprised to see three more reddened areas on her left side with several blisters in each.

"Did you have the chicken pox as a child?" Dr. Evans asked. "Of course I did," Helen replied. "I was four years old and caught them from my older brothers...so this definitely isn't the chicken pox. I'm immune now."

1. What condition does Dr. Evans suspect? What is the connection with chicken pox?

"Well, this explains that tingling you had a week ago," Dr. Evans said.

"How in the world are the chicken pox and my tingling sensation from last week related to my current symptoms? Chicken pox were itchy...this is painful!" Helen replied. "I absolutely don't have the chicken pox again!"

"No, no, of course not," said Dr. Evans, "but you do have a reactivated herpes zoster infection." Helen was shocked. "Herpes! Are you saying I have an **STI**?" Dr. Evans smiled, pulled up a chair, and began to explain Helen's condition.

2. Does herpes zoster cause genital herpes? Explain.

3. What are the most common symptoms of Helen's infection?

"Luckily," Dr. Evans said, "we've caught this very early. Have this prescription filled to decrease the severity and duration of your symptoms. It should also reduce your risk of complications. By the way, have you recently experienced any excessive emotional stress?"

4. What type of medication did Dr. Evans likely prescribe? Are there other treatments available for this condition?

5. What complications may occur? How are these treated?

6. Why did Dr. Evans ask about emotional stress in Helen's life? Do you think this was a contributing factor for Helen?

STI – **S**exually **t**ransmitted **i**nfection; an infection transmitted by direct sexual contact.

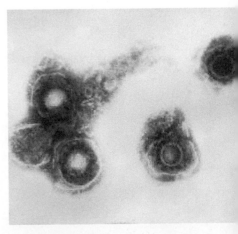

Figure 1. Transmission electron micrograph of herpes varicella zoster virus.

Helen was relieved after getting the facts from Dr. Evans. However, she was still curious about why her condition developed, what she could have done to prevent it, and most important, whether she could make others sick at the hospital, Wal-Mart, and church. Dr. Evans gave Helen a "quick facts" information sheet on her condition.

● **7.** How can this condition be prevented?

● **8.** Who is at greatest risk of infection? Why?

● **9.** Is Helen contagious? Explain.

An Unexpected Outbreak

Dr. Sims was a **gerontologist** who worked at the Shady Oaks Retirement Village in a rural Midwestern town. The Village was a wonderful facility offering a range of services from full hospital care to independent living for residents age 58 and older. While Dr. Sims worked in a variety of different locations within the Village, every Monday he always saw patients at the facility's general clinic. Late one Monday afternoon, just as he was tiring of the usual aches, colds, and digestive dysfunctions, Hank arrived but refused to tell the receptionist why he needed to see the doctor.

Dr. Sims had treated Hank on several other occasions for minor medical complaints. Hank was 70 years old and moved into a Shady Oaks apartment four years ago after his wife died. He was a popular resident with a ready smile and a kind word for everyone. After a little small talk with Hank, Dr. Sims finally inquired as to the nature of Hank's visit. "I seem to have a man problem," Hank said with downcast eyes. Dr. Sims smiled. "Hank, it's not at all uncommon as a man gets older to experience some difficulties. I think we need to give you a physical examination to determine that it's safe for you to engage in sexual activity, discuss your current personal situation, and then consider treatment options. I'm sure you've heard about the various medications available to help an older man maintain his love life." Still a bit sheepish, Hank said, "Doc, that ain't my problem. My buddy already gave me some of them blue pills and I used 'em. They work purty good too…but now it hurts when I pee."

A bit surprised, Dr. Sims examined Hank and noted urethral discharge, which was collected and sent to the microbiology laboratory. "We will have definitive results back from the lab in two days," said Dr. Sims, "but I think you have a classic case of gonorrhea."

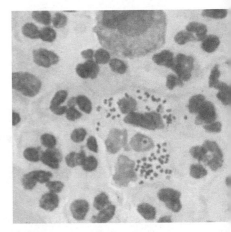

Figure 1. Gram stain of urethral discharge.

1. What is the causative agent of gonorrhea? Characterize this microbe.

2. Identify two tests that will be performed on this specimen in the microbiology laboratory to identify the pathogen. Describe the positive condition for each test.

3. What is the typical incubation period for this infection?

On Wednesday, Hank returned to see Dr. Sims for his test results. As expected, Hank's culture was positive for gonorrhea. Dr. Sims continued Hank on the antibiotic he had prescribed on Monday, but added a second drug to cover a possible co-infection.

4. What medications are generally used to treat gonorrhea? Are any antibiotics specifically avoided for this infection? Why?

5. What co-infection is Dr. Sims concerned about? Why?

6. In addition to antibiotic prescriptions, what other patient care measure will Dr. Sims provide?

The following Monday found Dr. Sims back on clinic duty where he met with Faye, a 68-year-old widow who was a relatively new resident at Shady Oaks. Faye was concerned that she had developed a severe vaginal yeast infection. She had purchased Monistat at the pharmacy last week hoping to cure her

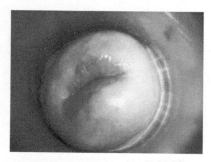

Figure 2. Inflammation of the cervix caused by infection by the pathogen.

condition, but had not improved. In fact, Faye complained of lower abdominal and back pain, severe vaginal irritation, yellowish discharge, and even a little spotting. Upon completion of a pelvic examination, Dr. Sims noted cervical erosion and mucopurulent discharge that he collected, along with a sample of endocervical cells, to send to the microbiology laboratory. "Faye, this is not a yeast infection," reported Dr. Sims. "I think you have a **STI**, specifically, gonorrhea, which is causing **cervicitis**." Faye was visibly shaken by the news. Not used to having this conversation with his geriatric patients, Dr. Sims struggled for the right words. "Usually this infection is seen in women under age 30 with multiple sexual partners who do not consistently use condoms. However, more couples maintain sexual activity later in life than ever before. This makes them susceptible to the same infections as younger people." "But Dr. Sims," Faye said with a blush creeping across her face, "I'm not that kind of woman. I've been seeing only one man since I arrived at Shady Oaks four months ago…there aren't multiple partners! You have to believe me." Inwardly groaning, Dr. Sims assured Faye that he did believe her, but then gently pointed out that if her partner had been having intercourse with anyone else, she was essentially exposed to their entire sexual history. "Just to be on the safe side," Dr. Sims said, "I'm ordering a couple of additional laboratory tests to rule out the possibilities of any other sexually transmitted infections. Meanwhile, I want you to complete the course of antibiotics I'm prescribing for you and return to see me in two weeks so we can be sure to avoid any complications."

7. What other STIs would Dr. Sims most likely be testing for?

8. What complications can arise from a gonococcal infection? Why did Dr. Sims want to have a follow-up visit with Faye, but did not schedule one with Hank?

9. In addition to those already discussed, what other symptoms can be associated with gonorrhea?

Exactly one week later, Dr. Sims met Estelle at the Shady Oaks clinic and thought he was experiencing déjà vu. Estelle's account of symptoms, her examination, and her response to his diagnosis were identical to Faye's case. After counseling Estelle and prescribing the appropriate antibiotic, Dr. Sims placed a call to the programming director in the Shady Oaks Administrative Office and requested the opportunity to speak at the monthly residents' meeting. Three cases of gonorrhea in as many weeks couldn't possibly be coincidental. It was time for a Village-wide educational program. While Dr. Sims had presented numerous "safe sex" programs at middle schools and high schools over the years, he never imagined he would be addressing an audience at a retirement center to cover the same topic!

10. What is the frequency of gonorrhea in the United States?

11. Which groups are at greatest risk of infection?

12. Is outside consultation ever required upon diagnosis of gonorrhea?

13. How can transmission be prevented?

Food Safety

"You are what you eat!" Today, everyone knows the importance of diet in maintaining a healthy body. We avoid trans fats and load up on antioxidants. Sales of sugar-free products and organic foods have soared. While many people focus on the nutritional quality of food, most overlook the fact that consumption can correlate with pathogen ingestion. For decades Americans have comfortably relied on governmental agencies (FDA, local health departments) to ensure a safe food supply. Recent outbreaks traced to regionally or nationally marketed food products have shaken the confidence of many consumers. Unfortunately, there are many ways to introduce pathogens into food. Poor hygiene and inappropriate handling techniques resulted in a local health department recently closing a catering business after they were "**epi-linked**" to *Shigella* outbreaks at a wedding reception and graduation party. There is national awareness of the processing violations that led to *E. coli*–contaminated spinach and *Salmonella*-tainted peanut butter. Even if food is processed correctly, poor storage and serving practices can allow small populations of contaminating microbes to grow exponentially. On a more dramatic note, natural disasters can lead to serious infectious disease outbreaks due to the contamination of the local food and water supplies.

Most of the diseases caused by ingesting contaminated food or water manifest with the expected GI symptoms, including nausea, vomiting, diarrhea, and cramping. Disease can be caused by ingesting a pathogenic microbe (food infection) or a toxin (food intoxication). Fortunately, many of these infections are self-limiting and can resolve without medical intervention or **sequelae**. However, IV fluids may be essential to prevent deadly dehydration (cholera) in cases of severe diarrheal disease, and infections with *E. coli* 0157:H7 can cause **HUS**, resulting in permanent kidney damage.

Clearly, prevention is once again the best way to manage food-borne illnesses. Scrupulous attention to aseptic food processing, storage, preparation, and service are essential...and of course, it all starts with hand hygiene! Review the following cases to determine the breech in sanitation that results in food and water becoming vectors of infectious disease.

Epi-linked – Epidemiologically linked. Linking an outbreak to its cause by using an epidemiological study to determine the risk factors most likely to lead to illness.

Sequelae – A pathological condition that is a side effect of another disease.

HUS – **H**emolytic **u**remic **s**yndrome. A disease characterized by hemolytic anemia, acute renal failure, and low platelet count. Most often HUS occurs in children as a complication of *E. coli* 0157:H7 infection.

Figure 1. Dinner items.

Fatigue – A feeling of weariness, tiredness, or lack of energy.

Myalgia – Muscle pain.

Food intoxication – The ingestion of toxins contained in food.

Attack rate – The number of exposed persons infected with the disease divided by the total number of exposed persons during an outbreak.

Risk ratio – The probability of the illness occurring in the exposed group (those who ate the ham) versus a nonexposed group (those who didn't eat the ham).

Sanitizer – A substance that is used to reduce the number of microorganisms present to a level safe for public health.

No, Thanks, I Think I'll Have the Turkey

Terry's retirement party was a major event after working for the company over 40 years. He had started at the bottom and worked his way to the top. It was unusual to have a boss that everyone liked, but Terry had treated everyone with respect and kindness regardless of their position. It wasn't surprising that nearly all of the company's 125 employees came to wish him well.

The evening started with a meal at 6:00 PM, followed by some heartfelt toasts and speeches. Everyone enjoyed the dinner: salads, ham, chicken, turkey, rice pilaf, eggs, and rolls. Dessert included nuts, cake, and cookies. Dancing and drinking followed, with music by a local blues quartet—Terry's favorite type of music.

Unfortunately, at midnight, it was difficult for Terry to appreciate the evening's earlier festivities while he was in the emergency room. The nausea had started about 10:00 PM, followed by vomiting, diarrhea, sweating, chills, fatigue, and myalgia. To make matters worse, the nurse had informed him that he was the seventh person who had been seen at the Emergency Department for the same condition. All of those who were ill had also been at his retirement party.

1. Assuming the cause of the illness was the food at the party, would the disease have been caused by a food infection or **food intoxication**? Explain.

2. What food toxin would most likely account for the signs and symptoms of the illness the guests were experiencing?

3. What microbe produces this toxin?

The community hospital that treated Terry and his guests joined with the county health department to investigate the source of the gastrointestinal illness that had been associated with the common meal. Questionnaires were distributed to those who attended the party. Sixty-seven people who ate the meal returned surveys. A total of 18 people had had the gastrointestinal symptoms. Of the 18 ill persons, 17 had eaten ham at the party and 1 had eaten leftover ham. None of the other foods at the party were significantly associated with the illness (see Table 1).

4. What type of investigation was carried out by the hospital and the health department?

5. Explain how the **attack rate** is determined.

6. What does the **risk ratio** measure?

The food preparer was interviewed to determine how the ham was prepared. She had purchased a 16-pound precooked packaged ham, baked it at home at 204°C (400°F) for 1.5 hours, and transported it to her workplace, a large institutional kitchen, where she sliced the ham while it was hot on a commercial slicer. She reported that she routinely cleaned the slicer in place rather than dismantling it and cleaning it according to recommended procedures and that she did not use an approved **sanitizer**. All 16 pounds of sliced ham had been placed in a 14-inch by 12-inch by 3-inch plastic container that was covered with foil and stored in a walk-in cooler for six hours, then transported back to the preparer's home and refrigerated overnight. The ham was served cold at the party the next day.

7. What recommendations would you make to prevent the ham from being contaminated by toxin-producing bacteria?

Table 1. Attack rates and risk ratios associated with buffet foods, by food type

	ATTACK RATE (%)		
Food	Ate	Did Not Eat	Risk Ratio
Ham	65.4	2.4	27.3
Chicken	30	25.5	1.2
Turkey	38.9	22.4	1.7
Rice pilaf	15.4	29.6	0.5
Rolls	47.1	20.0	1.4
Eggs	34.8	22.7	1.5
Salad platter	31.3	25.5	1.2
Nuts	25.0	27.1	0.9
Cake	23.5	28.0	0.8
Cookies	11.8	32.0	0.4
Punch	18.4	37.9	0.5

Danger in the NICU

NICU – **N**eonatal **i**ntensive **c**are **u**nit; a facility designed to provide specialized care for premature infants.

CSF – **C**erebro**s**pinal **f**luid; a clear, colorless, sterile fluid that provides a protective cushion to the brain and spinal cord. In adults, the volume of CSF is 90–150 ml with a normal protein level of 15–45 mg/dl and glucose level of 40–80 mg/dl (~ 2/3 the level in plasma). Adult CSF contains few cells (0–7 leukocytes/ml) with a differential count of 60–80% lymphocytes, 10–40% monocytes, and 0–15% neutrophils.

Empiric therapy – Immediate treatment with multiple broad-spectrum antibiotics due to the serious nature of an illness. Once laboratory results confirm pathogen identity and antibiotic sensitivity, drug therapy is altered to target the microbe and minimize impact to normal microbiota.

Terri was frantic when, at 33-weeks gestation, her water broke unexpectedly and baby Zoe arrived well ahead of schedule. Although the pediatrician in the **NICU** assured her that with just a little help from them Zoe would be fine, Terri still found it disconcerting to see her four-pound daughter attached to so many intimidating medical devices. She smiled down at her daughter in the incubator as Zoe tightly wrapped her small hand around Terri's little finger. Relief rushed over Terri, knowing that in three weeks Zoe's father and three-year-old brother, Zach, would welcome her home. Terri's only remaining concern was their limited opportunities for mother/baby bonding. She had planned to breastfeed Zoe as she had with Zach. Unfortunately, the pediatrician wanted to monitor Zoe's nutrient intake and prescribed an easy-to-digest, high-nutrient, powdered formula to help her grow.

When Terri arrived at the NICU early on the morning of Zoe's fifth day, she instantly knew something was seriously wrong. Zoe's pediatrician explained that a few hours earlier her temperature climbed to 39.8°C (103.6°F) and her heart rate jumped to 188 bpm. He performed a lumbar puncture and sent the **CSF** sample to the microbiology laboratory.

1. Based upon these symptoms and the test performed, what does the pediatrician suspect is wrong with Zoe?
2. What pathogens typically cause this type of infection? Are any more likely to affect neonates?
3. How could Zoe have contracted an infection when she has spent her entire five-day life in the security of a NICU?
4. When Zoe's CSF sample arrived at the microbiology laboratory, it was Gram-stained and cultured. A Gram-negative rod was observed. Review your list of likely pathogens from question 2 and narrow things down.
5. What types of media would you inoculate with the CSF sample and why?

Empiric therapy was initiated due to the serious nature of her illness. Within 36 hours, the laboratory confirmed the presence of *Enterobacter sakazaki*.

6. Describe the basic characteristics of this microbe. What is its appearance on the media you previously discussed? Why?
7. What is the normal habitat of *Enterobacter sakazaki*? How could Zoe have come in contact with it?

Upon pathogen identification, the microbiology laboratorians and the NICU personnel immediately started an investigation to prevent further infections. The hanging tube of formula for Zoe's continuous feeding was sampled, as was the water used in its preparation and the actual powdered formula stock. All other infants receiving reconstituted formula were monitored for signs of infection, and their formula stocks were also sampled. Additionally, formula preparation, storage, and use policies were reviewed to determine potential sources for the introduction of contamination.

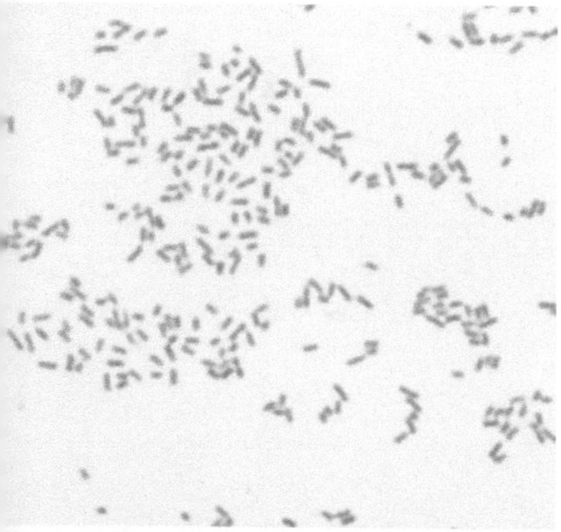

Figure 1. Gram stain of pathogen.

Zoe's IV antibiotic therapy was adjusted according to the drug sensitivity data provided by the laboratory. Despite effective treatment, Zoe remained in critical condition. She suffered multiple seizures and her blood pressure dropped to 52/40 mmHg.

8. Explain these new symptoms.

9. What other conditions are associated with *E. sakazaki* infection?

Thirty hours later, the lab reported the results of the environmental samples. Low levels of *E. sakazaki* were found in the powdered formula stock from the NICU. Other formula lots being used in the maternity and pediatric wings were uncontaminated. The water samples, facet swabs, and preparation area samples were likewise free of contaminating pathogens. The sample from Zoe's feeding tube grew a large number of *E. sakazaki* colonies, while samples stored in the refrigerator from this same preparation batch had relatively few colonies grow.

10. Can you explain the discrepancy in microbe number between Zoe's formula and the same batch stored in the refrigerator?

11. The FDA recommended that powdered formula be reconstituted with boiling water to kill potentially contaminating microbes. The CDC disagreed with this practice. Why?

12. What policies would you suggest to the hospital administration to minimize the risk of this type of infection?

Eventually, Zoe's conditioned improved. Her infection was eliminated and she overcame the obstacles associated with her premature birth. At discharge, Terri was again concerned as the NICU pediatrician arranged an appointment for Zoe to see a neurologist the following week.

13. What **sequelae** is the pediatrician concerned about? What is the mortality rate associated with Zoe's infection?

> **Sequelae** – A pathological condition that is a side effect of another disease.

14. Go online and search the history of *E. sakazaki*–associated meningitis due to formula feeding.

Regionally Acquired Infections

We live in a world of fascinating diversity and splendor. With modern means of transportation, we have new opportunities to explore regions of our planet that were once inaccessible. Imagine the adventures you can have today on an African safari, an eco-vacation in the Amazon, an expedition to exotic Thailand, or "beaching it" in the Bahamas! As inviting as these destinations sound, being a future healthcare provider, you need to look at the big picture and recognize that, in addition to the typical community-acquired infectious diseases, you can also be exposed to other more exotic and regionally specific infectious agents. For example, sleeping sickness and Ebola are endemic to portions of Africa. The Amazon eco-vacation could result in being exposed to the vectors that transmit the pathogens for Chagas, yellow fever, or Dengue fever. Avian flu outbreaks have been significant in Thailand, and avian flu has not been eradicated from Great Exuma Island in the Bahamas. Certain **endemic diseases** are also restricted to regions of the United States. You may not have to leave home to be exposed to Lyme disease, hantavirus, Ohio Valley fever, or even the plague!

If you are traveling, it is important to plan in advance to try to prevent being exposed unnecessarily to an infectious disease. For example, food and water in developing countries may not be as free from disease-causing pathogens as they are at home, making traveler's diarrhea a common problem. The risk for exposure can be reduced by drinking only bottled water and not eating raw fruits or vegetables. Insect-borne diseases are also a major problem in many areas. Therefore, when traveling to certain areas, it is important to take along a good supply of an effective insect repellent. It also may be necessary to get special vaccinations before traveling.

A competent medical professional will recognize the risk of endemic infection and consider that when performing a **differential diagnosis**. This, of course, emphasizes the need for acquiring a thorough **patient history**, including destinations of recent trips. Such information is a powerful diagnostic weapon in the arsenal for fighting regionally acquired infections. As you review the following cases, please do not become complacent, thinking you will never encounter patients suffering from these infections. Easy global access can now conveniently bring once widely separated populations together within a day. If the patients can move this quickly and easily, so can the potential pathogens they carry.

Endemic disease – A disease that is maintained in a local population.

Differential diagnosis – A systematic method used by trained medical professionals to diagnose a specific disease.

Patient history – A series of questions designed to obtain necessary information for a healthcare provider to formulate a diagnosis and provide medical care.

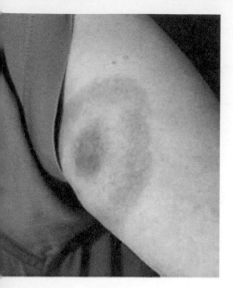

Figure 1. Target-shaped rash.

An Infectious Vacation

It was late June 1988 as Jenny paged through her scrapbook, fondly reviewing the photos from her family's vacation earlier that month. They had the most wonderful time visiting relatives in Georgia. Jenny's favorite part of the vacation was their side trip to Tennessee. She and her cousin hiked Lookout Mountain and were amazed they could see four states simultaneously from the summit.

When her family returned home to Ohio a few days later, Jenny discovered two angry red bite marks—one on her left wrist and the other on her right thigh. Assuming they were infected mosquito bites, Jenny ignored them. Toward the end of summer, Jenny's mother noticed the bites were still there but had a distinct appearance. The flat center was lighter in color than before, but a red ring-like rash extended from it with a diameter of about five inches. An examination by her dermatologist was unproductive as he admitted never before seeing such a rash. He advised Jenny to continue taking tetracycline for her acne and to keep a watchful eye on the rash.

1. This rash is diagnostic. What disease does Jenny have? What is the medical term used to describe this characteristic rash? What is the common name for it?

2. What is the causative agent? Describe the characteristics of this microbe.

3. What is the standard treatment for this condition?

4. Clearly, Jenny's dermatologist missed a key diagnostic feature that wouldn't be overlooked today. Note the date of this case. Go online and do a quick review of the emergence of this disease in the United States. Compare the annual rates of reported cases in the 1980s, 1990s, and after 2000.

5. Do you think the dramatic difference you observed with your Internet research represents an epidemic, better diagnosis and reporting, or both?

Throughout the fall, Jenny's energy level declined. By Thanksgiving she complained of severe fatigue, muscle pain and weakness, sore, swollen knees, nausea, abdominal pain, and headache. Jenny was admitted to the hospital and diagnosed with abdominal pain of unknown etiology and carbohydrate intolerance/hypoglycemia.

6. What standard laboratory tests were likely ordered when Jenny was admitted? Why? Based on these two diagnoses, what results did these test results indicate?

7. Jenny never actually suffered from hypoglycemia or related carbohydrate metabolism disorders. Why might her blood glucose levels be abnormal, leading to this misdiagnosis?

8. Jenny's dermatologist had her on a daily tetracycline regimen for moderate teenage acne. What effect would this prophylactic treatment have on her infection? Would this treatment affect laboratory results/diagnosis?

Jenny was hospitalized for three days. An NG tube for feeding was inserted, as she had lost 15 pounds in two months due to reduced appetite from the abdominal discomfort. By Christmastime, her earlier symptoms were substantially worse, most notably her headache and joint pain with swelling. Upon arrival at the emergency room, Jenny now also presented with a low-grade fever, **photophobia**, and severe memory loss. Jenny was admitted to the psychiatric unit, as her physician concluded her symptoms were psychosomatic. When physical and psychological examinations revealed no conclusive cause for her symptoms, Jenny was transferred to Children's Hospital in Columbus. During her stay, a bone scan showed extreme levels of arthritis in her knees, ankles, and wrists. Jenny lost the ability to speak, became extremely sensitive to loud noises, started having seizures, and lapsed into a coma. Consultation with an infectious disease specialist in New Jersey resulted in her transfer there by air ambulance. Jenny was immediately treated with IV Rocephin and Claforan for encephalitis. After 10 days of treatment, Jenny showed the first signs of improvement by opening her eyes and pointing to communicate. That afternoon, lab results indicated the presence of a high titer of anti-*Borrelia* antibodies.

Photophobia – Aversion to light.

⬤◗ **9.** Does Jenny's diagnosis of psychosomatic illness fit her symptoms? Explain.

⬤◗ **10.** Does Jenny's diagnosis of encephalitis fit her symptoms? Explain.

◗ **11.** What serological tests are performed to monitor antibody titers? What are the pros and cons of these methods?

Jenny's titers continued to decrease with her continued antibiotic therapy. Unfortunately, both of Jenny's arms began twisting inward, and her left foot underwent an internal rotation of 90°. Jenny experienced unexplained bouts of laughter and tears. She was diagnosed with pseudobulbar syndrome. Jenny required two more months of intensive antibiotic therapy before demonstrating marked improvement. Her arthritis was noticeably better. There was a dramatic decrease in the number and severity of her seizures, and she was able to manage limited speech. Jenny returned home to Ohio and participated in extensive physical, occupational, and speech therapies. Eventually, she entered an extended remission phase, which allowed her to return to high school and graduate.

⬤ **12.** What is pseudobulbar syndrome?

⬤ **13.** Although the etiology is not known, the disorder is often associated with patients suffering from infections of the CNS. Does this information fit with your original diagnosis of Jenny's condition? What symptoms associated with pseudobulbar syndrome are also common to Jenny's primary infectious condition?

Four days after commencement, Jenny's initial symptoms returned and she was admitted to her local hospital for IV antibiotic therapy. Pain medication provided minor relief from the renewed onslaught of arthritis and headache. Jenny was diagnosed with persistent infection of the CNS with notable involvement of the brain stem and cerebellum. By the end of July, Jenny began demonstrating cardiac dysfunction. Concerned, her physicians transferred Jenny to an advanced facility for specialized care. Although she had an occasional day with reduced symptoms, Jenny continued to decline at an alarming rate. Her arthritis was intensely painful, her left foot again began to draw inward, and she suddenly developed a black eye and respiratory distress. A hematology panel revealed Jenny's platelet count to be 7,000 cells/μl.

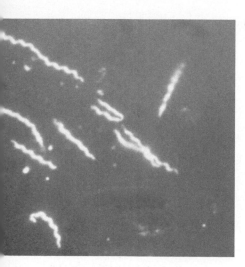

Figure 2. Dark-field micrograph of *Borrelia burgdorferi*.

14. What is a normal platelet count? What problem is associated with this value? What new symptom does it explain?

Jenny received platelets and two units of whole blood. She was placed on a ventilator to aid her breathing. By morning, Jenny was edematous and had no urine output. Dialysis removed 15 pounds of fluid, but now Jenny spiked a fever of 38.8°C (102°F). Her white blood cell count had risen to 28,000 cells/µl and blood cultures revealed the presence of Gram-negative bacilli. Additional antibiotics were added to address the **bacteremia**.

15. Identify two possible nosocomial sources for the development of Jenny's sepsis. Indicate the different Gram-negative bacilli most commonly associated with these procedures.

16. What additional problem might Jenny encounter as a result of antimicrobial therapy for a Gram-negative sepsis? How does this happen?

Jenny experienced frequent, violent seizures, her extremities darkened, her liver failed, and her lungs were now severely damaged by infection. When her physicians all agreed that Jenny would not recover from this secondary infection, morphine was administered to ease her pain and ventilation was discontinued. After tearful good-byes by caregivers and family, Jenny's six-year battle with infection ended on August 26, 1994. She was 21 years old.

17. How is this disease transmitted? Where and when did Jenny likely become infected?

18. What is the incubation period for this condition?

19. What can be done to prevent this infection?

20. Some patients never demonstrate a positive antibody titer for this condition. How might you explain this situation?

21. Are the symptoms of Lyme disease caused directly by *Borrelia burgdorferi*? Explain.

Rain, Rain, Go Away

Normally, rain was an infrequent occurrence for residents in New Mexico. The desert area was dry and supported little vegetation. As a result, Paul seldom had a rodent problem. There just wasn't enough food and water to support much of a wildlife population in this ecosystem. Consequently, there weren't many mice to wander in from their natural habitat to his storage shed or house. However, in the last two years, his region had received quite a bit more rain than usual. This January, the desert had more green plants than he had seen in decades. He also had a mouse problem. He had trapped several of the pests in his house and had swept several old nests out of his shed last week.

Although Paul wanted to finish cleaning his storage shed, he knew it wasn't going to happen this weekend. He felt awful. He had had to deal with the normal amount of aches and pains of any 55-year-old who had to battle an occasional cold or a case of the flu, but intuitively he knew this was different and it wasn't good. It started with **fatigue**, a fever, muscle aches in his thighs, hips, and back, a headache, dizziness, chills, nausea, vomiting, diarrhea, and abdominal pain. It was like every illness he had ever had was attacking him all at once. Now, it was getting even worse—he was having trouble breathing.

At the emergency room, the physician asked Paul if he suffered from any chronic pulmonary disease (he did not) and ran tests that ruled out **malignant** causes of his condition. A blood sample was also sent to the regional **reference laboratory**, where it was tested for several unusual pathogens. The lab detected hantavirus-specific **IgM**, and a **RT-PCR** assay detected hantavirus-specific RNA.

- **1.** What disease is caused by hantavirus?

- **2.** What is the reservoir for hantavirus? How is it transmitted to humans?

- **3.** How can this disease be prevented?

- **4.** What antiviral is used to treat hantavirus pulmonary syndrome?

- **5.** What are the physical characteristics of the pathogen?

Fatigue – A feeling of weariness, tiredness, or lack of energy.

Malignant – Type of tumor that can invade and destroy nearby tissue and that may spread (metastasize) to other parts of the body.

Reference laboratory – A laboratory unit dedicated to diagnosis of challenging cases. It serves as the final authority for state public health departments as well as for laboratories in the private sector.

IgM – A class of antibodies on the surface of B cells that serve as antigen receptors and the first to be secreted during a primary infection.

RT-PCR – **R**everse **t**ranscriptase **p**olymerase **c**hain **r**eaction; a technique for amplifying DNA that begins with reverse transcriptase making a DNA copy of the target RNA molecule.

Figure 1. Reservoir for pathogen.

6. How does hantavirus cause respiratory distress?

7. What is the **case fatality rate** for this disease?

8. In general, what other diseases can be associated with an increase in rainfall?

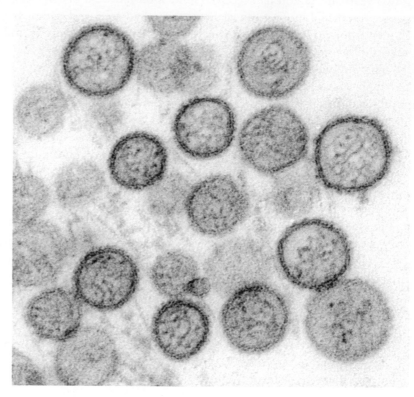

Figure 2. Hantavirus (sin nombre virus).

Photo Credits

(I-F-5 *Candida albicans*) CDC / Dr. Stuart Brown, p. 85

(I-F-6 *Trichomonas vaginalis*) CDC, p. 88

(I-F-6 Trichomoniasis) CDC, p. 88

II. Nosocomial Infections

(II-1 Rash due to TSS) CDC, p. 92

(II-1 *Staphylococcus aureus*) Rod Anderson, p. 92

(II-2 Hepatitis Jaundice) CDC / Dr. Thomas F. Sellers / Emory University, p. 95

(II-2 Hepatitis Virus) CDC / E.H. Cook, Jr., p. 95

(II-3 *Staphylococcus aureus*) Rod Anderson, p. 100

(II-3 PBP 2A Agglutination Test) Rod Anderson, p. 101

III. Family Health Care

(III-A-2 Chicken Pox) CDC / Dr. KL Hermann, p. 106

(III-A-2 Herpes Virus) CDC / Dr. John Hierholzer, FA Murphy, p. 106

(III-A-3 Hand, Foot and Mouth Disease) Riikka Österback, Tytti Vuorinen, Mervi Linna, Petri Susi, Timo Hyypiä, and Matti Waris. "Coxsackievirus A6 and Hand, Foot, and Mouth Disease, Finland." *Emerging Infectious Diseases* 2009 September; 15(9): 1485–1488, p. 108

(III-A-3 Coxsackie Virus Viral pathogen of Hand, Foot and Mouth Disease) Shieh W, Jung S, Hsueh C, Kuo T, Mounts A, Parashar U, Yang C, Guarner J, Ksiazek TG, Dawson J, Goldsmith C, Chang G-J J, Oberste SM, Pallansch MA, Anderson LJ, Zaki SR, and the Epidemic Working Group. "Pathologic Studies of Fatal Cases in Outbreak of Hand, Foot, and Mouth Disease, Taiwan." *Emerging Infectious Diseases* 2001 Jan–Feb; 7(1): 146–148, p. 109

(III-A-4 Rapid Strep Test) Rod Anderson, p. 110

(III-A-4 Strawberry Tongue) CDC, p. 111

(III-A-4 Scarlet Fever) CDC, p. 111

(III-A-5 RSV Test) Rod Anderson, p. 116

(III-A-6 *Pseudomonas aeruginosa*) Rod Anderson, p. 118

(III-B-1 CAMP Test) CDC / Dr. Richard Facklam, p. 121

(III-B-1 *Streptococcus* Gram Stain) CDC, p. 121

(III-B-2 *Listeria*) CDC / Dr. Balasubr Swaminathan; Peggy Hayes, p. 123

(III-B-2 Listeria Risk) CDC / James Gathany, p. 124

(III-C-1 Varicella Zoster Virus) CDC / Dr. Erskine Palmer, p. 127

(III-C-2 *Neisseria gonorrheae* from discharge) CDC / Joe Miller, p. 129

(III-C-2 *Neisseria gonorrheae* Inflammation of the Cervix) CDC, p. 130

IV. Food Safety

(IV-1 Meal) USDA-ARS / Peggy Greg, p. 132

(IV-2 *Enterobacter*) Rod Anderson, p. 134

V. Regionally Acquired Infections

(V-1 Lyme Disease Bulls Eye Rash) CDC / James Gathany, p. 138

(V-1 Lyme Disease *Borrelia burgdorferi*) CDC, p. 140

(V-2 Deer Mouse 2) CDC / James Gathany, p. 141

(V-2 Sin Nombre Virus) CDC / Cynthia Goldsmith, Luanne Elliott, p. 142

CALCULUS
SEVERAL VARIABLES

THE WILEY BICENTENNIAL–KNOWLEDGE FOR GENERATIONS

*E*ach generation has its unique needs and aspirations. When Charles Wiley first opened his small printing shop in lower Manhattan in 1807, it was a generation of boundless potential searching for an identity. And we were there, helping to define a new American literary tradition. Over half a century later, in the midst of the Second Industrial Revolution, it was a generation focused on building the future. Once again, we were there, supplying the critical scientific, technical, and engineering knowledge that helped frame the world. Throughout the 20th Century, and into the new millennium, nations began to reach out beyond their own borders and a new international community was born. Wiley was there, expanding its operations around the world to enable a global exchange of ideas, opinions, and know-how.

For 200 years, Wiley has been an integral part of each generation's journey, enabling the flow of information and understanding necessary to meet their needs and fulfill their aspirations. Today, bold new technologies are changing the way we live and learn. Wiley will be there, providing you the must-have knowledge you need to imagine new worlds, new possibilities, and new opportunities.

Generations come and go, but you can always count on Wiley to provide you the knowledge you need, when and where you need it!

WILLIAM J. PESCE
PRESIDENT AND CHIEF EXECUTIVE OFFICER

PETER BOOTH WILEY
CHAIRMAN OF THE BOARD

STUDENT SOLUTIONS MANUAL

Garret Etgen
University of Houston

to accompany

CALCULUS
SEVERAL VARIABLES

10th Edition

Saturnino Salas
Einar Hille
Garret Etgen
University of Houston

John Wiley & Sons, Inc.

Cover Photo: Steven Puetzer/Masterfile
Bicentennial Logo Design: Richard J. Pacifico

To order books or for customer service please, call 1-800-CALL WILEY (225-5945).

ISBN-13   978-0-470-12729-2

V10014491_100419

CONTENTS

CONTENTS

CALCULUS
SEVERAL VARIABLES

CHAPTER 13

SECTION 13.1

1.

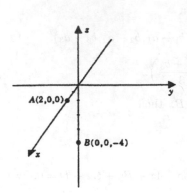

midpoint: $(1, 0, -2)$

3.

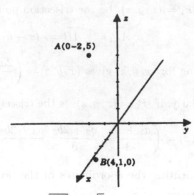

length $\overline{AB}$: $5\sqrt{2}$

midpoint: $\left(2, -\frac{1}{2}, \frac{5}{2}\right)$

5. $z = -2$ **7.** $y = 1$ **9.** $x = 3$

11. $x^2 + (y - 2)^2 + (z + 1)^2 = 9$ **13.** $(x - 2)^2 + (y - 4)^2 + (z + 4)^2 = 36$

15. $(x - 3)^2 + (y - 2)^2 + (z - 2)^2 = 13$

17.
$$x^2 + y^2 + z^2 + 4x - 8y - 2z + 5 = 0$$
$$x^2 + 4x + 4 + y^2 - 8y + 16 + z^2 - 2z + 1 = -5 + 4 + 16 + 1$$
$$(x + 2)^2 + (y - 4)^2 + (z - 1)^2 = 16$$

center: $(-2, 4, 1)$,   radius: 4

19. $(2, 3, -5)$ **21.** $(-2, 3, 5)$ **23.** $(-2, 3, -5)$

25. $(-2, -3, -5)$ **27.** $(2, -5, 5)$ **29.** $(-2, 1, -3)$

31. $d(PR) = \sqrt{14}$, $d(QR) = \sqrt{45}$, $d(PQ) = \sqrt{59}$; $[d(PR)]^2 + [d(QR)]^2 = [d(PQ)]^2$

33. The sphere of radius 2 centered at the origin, together with its interior.

35. A rectangular box in the first octant with sides on the coordinate planes and dimensions $1 \times 2 \times 3$, together with its interior.

37. A circular cylinder with base the circle $x^2 + y^2 = 4$ and height 4, together with its interior.

39. Let $B = (x, y, z)$. Then

$$\frac{x+2}{2} = 1 \Longrightarrow x = 0, \quad \frac{y+3}{2} = 2 \Longrightarrow y = 1, \quad \frac{z+4}{2} = 3 \Longrightarrow z = 2.$$

Therefore $B = (0, 1, 2)$.

41. Let $P_1 = (x, y, z)$ be the trisection point closest to A. Then

$$\overrightarrow{AP_1} = \tfrac{1}{3}\overrightarrow{AB} \Longrightarrow (x - a_1, y - a_2, z - a_3) = \tfrac{1}{3}(b_1 - a_1, b_2 - a_2, b_3 - a_3).$$

Solving for x, y, z gives $(x, y, z) = \left(\dfrac{2a_1 + b_1}{3}, \dfrac{2a_2 + b_2}{3}, \dfrac{2a_3 + b_3}{3} \right)$.

Similarly, if $P_2 = (x, y, z)$ is the trisection point closest to B, then

$$(x, y, z) = \left(\frac{a_1 + 2b_1}{3}, \frac{a_2 + 2b_2}{3}, \frac{a_3 + 2b_3}{3} \right).$$

43. Substituting the coordinates of the points into the equation $Ax + By + Cz + D = 0$, we get the equations

$$Ax_0 + D = 0, \; By_0 + D = 0, \; Cz_0 + D = 0 \quad \text{which implies} \quad Ax_0 = By_0 = Cz_0.$$

Therefore, we have

$$Ax + \frac{Ax_0}{y_0}y + \frac{Ax_0}{z_0}z + D = 0 \quad \text{or} \quad \frac{x}{x_0} + \frac{y}{y_0} + \frac{z}{z_0} + \frac{D}{Ax_0} = 0.$$

Substituting the point $(x_0, 0, 0)$ into this equation gives

$$\frac{x}{x_0} + \frac{y}{y_0} + \frac{z}{z_0} = 1.$$

45. (i) $a_3 \neq 0$ The line through the origin and (a_1, a_2, a_3) is given by $x = a_1 t$, $y = a_2 t$, $z = a_3 t$, t any real number. The line intersects the plane $z = z_0$ at the point Q where $t = z_0/a_3$. The coordinates of Q are: $\frac{a_1}{a_3}z_0, \; \frac{a_2}{a_3}z_0, \; z_0$.

(ii) $a_3 = 0$ If $z_0 \neq 0$, the line does not intersect the plane. If $z_0 = 0$, the line lies in the plane.

47. The ray that emanates from the origin and passes through the point (a_1, a_2, a_3) is given by $x = a_1 t$, $y = a_2 t$, $z = a_3 t$, $t \geq 0$. The ray intersects the sphere $x^2 + y^2 + z^2 = 1$ at the point Q where

$$a_1^2 t^2 + a_2^2 t^2 + a_3^2 t^2 = 1 \Longrightarrow t = \frac{1}{\sqrt{a_1^2 + a_2^2 + a_3^2}}.$$

The coordinates of Q are: $\dfrac{a_1}{\sqrt{a_1^2 + a_2^2 + a_3^2}}, \; \dfrac{a_2}{\sqrt{a_1^2 + a_2^2 + a_3^2}}, \; \dfrac{a_3}{\sqrt{a_1^2 + a_2^2 + a_3^2}}$.

SECTION 13.2

1. $\overrightarrow{PQ} = (3, 4, -2);$ $\| \overrightarrow{PQ} \| = \sqrt{29}$ **3.** $\overrightarrow{PQ} = (0, -2, -1);$ $\| \overrightarrow{PQ} \| = \sqrt{5}$

5. $2\mathbf{a} - \mathbf{b} = (2 \cdot 1 - 3, 2 \cdot [-2] - 0, 2 \cdot 3 + 1) = (-1, -4, 7)$

7. $-2\mathbf{a} + \mathbf{b} - \mathbf{c} = [-(2\mathbf{a} - \mathbf{b})] - \mathbf{c} = (1+4, 4-2, -7-1) = (5, 2, -8)$

9. $3\mathbf{i} - 4\mathbf{j} + 6\mathbf{k}$ **11.** $-3\mathbf{i} - \mathbf{j} + 8\mathbf{k}$

13.   5   **15.**   3   **17.** $\sqrt{6}$

19.   (a) $\mathbf{a}$, $\mathbf{c}$, and $\mathbf{d}$ since $\mathbf{a} = \frac{1}{3}\mathbf{c} = -\frac{1}{2}\mathbf{d}$

        (b) $\mathbf{a}$ and $\mathbf{c}$ since $\mathbf{a} = \frac{1}{3}\mathbf{c}$

        (c) $\mathbf{a}$ and $\mathbf{c}$ both have direction opposite to $\mathbf{d}$

21. $\overrightarrow{RQ} = (3 - x, -1 - y, 1 - z)$ and $\overrightarrow{OP} = (1, 4, -2)$.

$$\overrightarrow{RQ} = \overrightarrow{OP} \Longrightarrow 3 - x = 1, \ -1 - y = 4, \ 1 - z = -2 \Longrightarrow x = 2, \ y = -5, \ z = 3.$$

23. $\overrightarrow{RQ} = (3 - x, -1 - y, 1 - z) = -2\overrightarrow{OP} = (-2, -8, 4) \Longrightarrow 3 - x = -2, \ -1 - y = -8, \ 1 - z = 4$

       $\Longrightarrow x = 5, \ y = 7, \ z = -3.$

25. $\|\mathbf{a}\| = 5$; $\dfrac{\mathbf{a}}{\|\mathbf{a}\|} = \left(\dfrac{3}{5}, -\dfrac{4}{5}, 0 \right)$ **27.** $\|\mathbf{a}\| = 3$; $\dfrac{\mathbf{a}}{\|\mathbf{a}\|} = \dfrac{1}{3}\mathbf{i} - \dfrac{2}{3}\mathbf{j} + \dfrac{2}{3}\mathbf{k}$

29. $\|\mathbf{a}\| = \sqrt{14}$; $-\dfrac{\mathbf{a}}{\|\mathbf{a}\|} = \dfrac{1}{\sqrt{14}}\mathbf{i} - \dfrac{3}{\sqrt{14}}\mathbf{j} - \dfrac{2}{\sqrt{14}}\mathbf{k}$

31.   (i) $\mathbf{a} - \mathbf{b}$ (ii) $-(\mathbf{a} + \mathbf{b})$ (iii) $\mathbf{a} - \mathbf{b}$ (iv) $\mathbf{b} - \mathbf{a}$

33.   (a) $\mathbf{a} - 3\mathbf{b} + 2\mathbf{c} + 4\mathbf{d} = (2\mathbf{i} - \mathbf{k}) - 3(\mathbf{i} + 3\mathbf{j} + 5\mathbf{k}) + 2(-\mathbf{i} + \mathbf{j} + \mathbf{k}) + 4(\mathbf{i} + \mathbf{j} + 6\mathbf{k})$

                          $= \mathbf{i} - 3\mathbf{j} + 10\mathbf{k}$

        (b)   The vector equation

$$(1, 1, 6) = A(2, 0, -1) + B(1, 3, 5) + C(-1, 1, 1)$$

      implies

$$1 = 2A + B - C,$$
$$1 = 3B + C,$$
$$6 = -A + 5B + C.$$

      Simultaneous solution gives $A = -2$, $B = \frac{3}{2}$, $C = -\frac{7}{2}$.

35. $\|3\mathbf{i} + \mathbf{j}\| = \|\alpha\mathbf{j} - \mathbf{k}\| \implies 10 = \alpha^2 + 1$ so $\alpha = \pm 3$

37.

$$\|\alpha\mathbf{i} + (\alpha - 1)\mathbf{j} + (\alpha + 1)\mathbf{k}\| = 2 \implies \alpha^2 + (\alpha - 1)^2 + (\alpha + 1)^2 = 4$$

$$\implies 3\alpha^2 = 2 \ \text{ so } \ \alpha = \pm\tfrac{1}{3}\sqrt{6}$$

39. $\pm\frac{2}{13}\sqrt{13}\,(3\mathbf{j} + 2\mathbf{k})$ since $\|\alpha(3\mathbf{j} + 2\mathbf{k})\| = 2 \implies \alpha = \pm\frac{2}{13}\sqrt{13}$

41. (a) Since $\quad \|\mathbf{a} - \mathbf{b}\| \quad$ and $\quad \|\mathbf{a} + \mathbf{b}\| \quad$ are the lengths of the diagonals of the parallelogram, the parallelogram must be a rectangle.

(b) Simplify

$$\sqrt{(a_1 - b_1)^2 + (a_2 - b_2)^2 + (a_3 - b_3)^2} = \sqrt{(a_1 + b_1)^2 + (a_2 + b_2)^2 + (a_3 + b_3)^2}.$$

the result is $a_1 b_1 + a_2 b_2 + a_3 b_3 = 0$.

43. Let $P = (x_1, y_1, z_1)$, $Q = (x_2, y_2, z_2)$, and $M = (x_m, y_m, z_m)$. Then

$$(x_m, y_m, z_m) = (x_1, y_1, z_1) + \frac{1}{2}(x_2 - x_1, y_2 - y_1, z_2 - z_1) \Longrightarrow \mathbf{m} = \mathbf{p} + \frac{1}{2}(\mathbf{q} - \mathbf{p}).$$

SECTION 13.3

1. $\mathbf{a} \cdot \mathbf{b} = (2)(-2) + (-3)(0) + (1)(3) = -1$ **3.** $\mathbf{a} \cdot \mathbf{b} = (2)(1) + (-4)(1/2) + (0)(0) = 0$

5. $\mathbf{a} \cdot \mathbf{b} = (2)(1) + (1)(1) - (2)(2) = -1$ **7.** $\mathbf{a} \cdot \mathbf{b}$

9. $(\mathbf{a} - \mathbf{b}) \cdot \mathbf{c} + \mathbf{b} \cdot (\mathbf{c} + \mathbf{a}) = \mathbf{a} \cdot \mathbf{c} - \mathbf{b} \cdot \mathbf{c} + \mathbf{b} \cdot \mathbf{c} + \mathbf{b} \cdot \mathbf{a} = \mathbf{a} \cdot (\mathbf{b} + \mathbf{c})$

11. (a) $\mathbf{a} \cdot \mathbf{b} = (2)(3) + (1)(-1) + (0)(2) = 5$

$\mathbf{a} \cdot \mathbf{c} = (2)(4) + (1)(0) + (0)(3) = 8$

$\mathbf{b} \cdot \mathbf{c} = (3)(4) + (-1)(0) + (2)(3) = 18$

(b) $\|\mathbf{a}\| = \sqrt{5}$, $\quad \|\mathbf{b}\| = \sqrt{14}$, $\quad \|\mathbf{c}\| = 5$. $\quad$ Then,

$$\cos \measuredangle (\mathbf{a}, \mathbf{b}) = \frac{\mathbf{a} \cdot \mathbf{b}}{\|\mathbf{a}\| \, \|\mathbf{b}\|} = \frac{5}{(\sqrt{5})(\sqrt{14})} = \frac{1}{14}\sqrt{70},$$

$$\cos \measuredangle (\mathbf{a}, \mathbf{c}) = \frac{8}{(\sqrt{5})(5)} = \frac{8}{25}\sqrt{5},$$

$$\cos \measuredangle (\mathbf{b}, \mathbf{c}) = \frac{18}{(\sqrt{14})(5)} = \frac{9}{35}\sqrt{14}.$$

(c) $\mathbf{u_b} = \dfrac{1}{\sqrt{14}}(3\,\mathbf{i} - \mathbf{j} + 2\,\mathbf{k})$, $\quad \text{comp}_\mathbf{b}\,\mathbf{a} = \mathbf{a} \cdot \mathbf{u_b} = \dfrac{1}{\sqrt{14}}(6 - 1) = \dfrac{5}{14}\sqrt{14}$,

$\mathbf{u_c} = \dfrac{1}{5}(4\,\mathbf{i} + 3\,\mathbf{k})$, $\quad \text{comp}_\mathbf{c}\,\mathbf{a} = \mathbf{a} \cdot \mathbf{u_c} = \dfrac{8}{5}$

(d) $\mathbf{proj_b}\,\mathbf{a} = (\text{comp}_\mathbf{b}\,\mathbf{a})\,\mathbf{u_b} = \dfrac{5}{14}(3\,\mathbf{i} - \mathbf{j} + 2\,\mathbf{k})$, $\quad \mathbf{proj_c}\,\mathbf{a} = (\text{comp}_\mathbf{c}\,\mathbf{a})\,\mathbf{u_c} = \dfrac{8}{25}(4\,\mathbf{i} + 3\,\mathbf{k})$

13. $\mathbf{u} = \cos\dfrac{\pi}{3}\,\mathbf{i} + \cos\dfrac{\pi}{4}\,\mathbf{j} + \cos\dfrac{2\pi}{3}\,\mathbf{k} = \dfrac{1}{2}\,\mathbf{i} + \dfrac{1}{2}\sqrt{2}\,\mathbf{j} - \dfrac{1}{2}\,\mathbf{k}$

15. $\cos\theta = \dfrac{(3\,\mathbf{i} - \mathbf{j} - 2\,\mathbf{k}) \cdot (\mathbf{i} + 2\,\mathbf{j} - 3\,\mathbf{k})}{\|3\,\mathbf{i} - \mathbf{j} - 2\,\mathbf{k}\| \, \|\mathbf{i} + 2\,\mathbf{j} - 3\,\mathbf{k}\|} = \dfrac{7}{\sqrt{14}\,\sqrt{14}} = \dfrac{1}{2}, \quad \theta = \dfrac{\pi}{3}$

17. Since $\|\mathbf{i} - \mathbf{j} + \sqrt{2}\,\mathbf{k}\| = 2$, we have $\cos\alpha = \frac{1}{2}$, $\cos\beta = -\frac{1}{2}$, $\cos\gamma = \frac{1}{2}\sqrt{2}$.
The direction angles are $\frac{1}{3}\pi$, $\frac{2}{3}\pi$, $\frac{1}{4}\pi$.

19. $\theta = \arccos\dfrac{\mathbf{a}\cdot\mathbf{b}}{\|\mathbf{a}\|\|\mathbf{b}\|} = \arccos\left(\dfrac{-9}{\sqrt{231}}\right) \cong 2.2$ radians or $126.3°$

21. $\theta = \arccos\dfrac{\mathbf{a}\cdot\mathbf{b}}{\|\mathbf{a}\|\|\mathbf{b}\|} = \arccos\left(\dfrac{-13}{5\sqrt{10}}\right) \cong 2.54$ radians or $145.3°$

23. angles: $38.51°$, $95.52°$, $45.97°$; perimeter: $\cong 15.924$

25. $\|\mathbf{a}\| = \sqrt{1^2 + 2^2 + 2^2} = 3$;
$\cos\alpha = \dfrac{1}{3}$, $\cos\beta = \dfrac{2}{3}$, $\cos\gamma = \dfrac{2}{3}$
$\alpha \cong 70.5°$ $\beta \cong 48.2°$, $\gamma \cong 48.2°$

27. $\|\mathbf{a}\| = \sqrt{3^2 + (12)^2 + 4^2} = 13$;
$\cos\alpha = \dfrac{3}{13}$, $\cos\beta = \dfrac{12}{13}$ $\cos\gamma = \dfrac{4}{13}$
$\alpha \cong 76.7°$ $\beta \cong 22.6°$, $\gamma \cong 72.1°$

29. $2\mathbf{i} + 5\mathbf{j} + 2x\,\mathbf{k} \perp 6\mathbf{i} + 4\mathbf{j} - x\,\mathbf{k} \Longrightarrow 12 + 20 - 2x^2 = 0 \Longrightarrow x^2 = 16 \Longrightarrow x = \pm 4$

31. $\cos\dfrac{\pi}{3} = \dfrac{\mathbf{c}\cdot\mathbf{d}}{\|\mathbf{c}\|\,\|\mathbf{d}\|}$, $\dfrac{1}{2} = \dfrac{2x+1}{x^2+2}$, $x^2 = 4x$; $x = 0$, $x = 4$

33. (a) The direction angles of a vector always satisfy

$$\cos^2\alpha + \cos^2\beta + \cos^2\gamma = 1$$

and, as you can check,

$$\cos^2\tfrac{1}{4}\pi + \cos^2\tfrac{1}{6}\pi + \cos^2\tfrac{2}{3}\pi \neq 1.$$

(b) The relation

$$\cos^2\alpha + \cos^2\tfrac{1}{4}\pi + \cos^2\tfrac{1}{4}\pi = 1$$

gives

$$\cos^2\alpha + \tfrac{1}{2} + \tfrac{1}{2} = 1, \quad \cos\alpha = 0, \quad a_1 = \|\mathbf{a}\|\cos\alpha = 0.$$

35. Let $\theta_1, \theta_2, \theta_3$ be the direction angles of $-\mathbf{a}$. Then

$$\theta_1 = \arccos\left[\frac{(-\mathbf{a}\cdot\mathbf{i})}{\|-\mathbf{a}\|}\right] = \arccos\left[-\frac{(\mathbf{a}\cdot\mathbf{i})}{\|\mathbf{a}\|}\right] = \arccos\left(-\cos\alpha\right) = \pi - \arccos\left(\cos\alpha\right) = \pi - \alpha.$$

Similarly $\theta_2 = \pi - \beta$ and $\theta_3 = \pi - \gamma$.

37. Set $\mathbf{u} = a\mathbf{i} + b\mathbf{j} + c\mathbf{k}$. The relations

$$(a\mathbf{i} + b\mathbf{j} + c\mathbf{k}) \cdot (\mathbf{i} + 2\mathbf{j} + \mathbf{k}) = 0 \quad \text{and} \quad (a\mathbf{i} + b\mathbf{j} + c\mathbf{k}) \cdot (3\mathbf{i} - 4\mathbf{j} + 2\mathbf{k}) = 0$$

give

$$a + 2b + c = 0 \qquad 3a - 4b + 2c = 0$$

so that $b = \frac{1}{8}a$ and $c = -\frac{5}{4}a$.

Then, since $\mathbf{u}$ is a unit vector,

$$a^2 + b^2 + c^2 = 1, \quad a^2 + \left(\frac{a}{8}\right)^2 + \left(\frac{-5a}{4}\right)^2 = 1, \quad \frac{165}{64}a^2 = 1.$$

Thus, $a = \pm\dfrac{8}{165}\sqrt{165}$ and $\mathbf{u} = \pm\dfrac{\sqrt{165}}{165}(8\,\mathbf{i} + \mathbf{j} - 10\,\mathbf{k})$.

39.

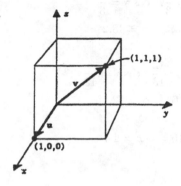

We take $\mathbf{u} = \mathbf{i}$ as an edge and $\mathbf{v} = \mathbf{i} + \mathbf{j} + \mathbf{k}$ as a diagonal of a cube. Then,

$$\cos\theta = \frac{\mathbf{u} \cdot \mathbf{v}}{\|\mathbf{u}\|\,\|\mathbf{v}\|} = \frac{1}{3}\sqrt{3},$$

$$\theta = \cos^{-1}\left(\tfrac{1}{3}\sqrt{3}\right) \cong 0.96 \text{ radians.}$$

41. (a) $\mathbf{proj_b}\,\alpha\mathbf{a} = (\alpha\mathbf{a} \cdot \mathbf{u_b})\mathbf{u_b} = \alpha(\mathbf{a} \cdot \mathbf{u_b})\mathbf{u_b} = \alpha\,\mathbf{proj_b}\,\mathbf{a}$

(b) $\mathbf{proj_b}\,(\mathbf{a} + \mathbf{c}) = [(\mathbf{a} + \mathbf{c}) \cdot \mathbf{u_b}]\,\mathbf{u_b}$

$$= (\mathbf{a} \cdot \mathbf{u_b} + \mathbf{c} \cdot \mathbf{u_b})\mathbf{u_b}$$

$$= (\mathbf{a} \cdot \mathbf{u_b})\mathbf{u_b} + (\mathbf{c} \cdot \mathbf{u_b})\mathbf{u_b} = \mathbf{proj_b}\,\mathbf{a} + \mathbf{proj_b}\,\mathbf{c}$$

43. (a) $\mathbf{a} \cdot \mathbf{b} = \mathbf{a} \cdot \mathbf{c} \Longrightarrow \mathbf{a}(\mathbf{b} - \mathbf{c}) = 0 \Longrightarrow \mathbf{a} \perp (\mathbf{b} - \mathbf{c})$.

For $\mathbf{a} \neq \mathbf{0}$ the following statements are equivalent:

$\mathbf{a} \cdot \mathbf{b} = \mathbf{a} \cdot \mathbf{c}, \quad \mathbf{b} \cdot \mathbf{a} = \mathbf{c} \cdot \mathbf{a},$

$\mathbf{b} \cdot \dfrac{\mathbf{a}}{\|\mathbf{a}\|} = \mathbf{c} \cdot \dfrac{\mathbf{a}}{\|\mathbf{a}\|}, \quad \mathbf{b} \cdot \mathbf{u_a} = \mathbf{c} \cdot \mathbf{u_a}$

$(\mathbf{b} \cdot \mathbf{u_a})\mathbf{u_a} = (\mathbf{c} \cdot \mathbf{u_a})\mathbf{u_a},$

$$\mathbf{proj_a}\,\mathbf{b} = \mathbf{proj_a}\,\mathbf{c}.$$

Thus, $\mathbf{a} \cdot \mathbf{b} = \mathbf{a} \cdot \mathbf{c}$ implies only that the projection of $\mathbf{b}$ on $\mathbf{a}$ equals the projection of $\mathbf{c}$ on $\mathbf{a}$.

(b) $\mathbf{b} = (\mathbf{b} \cdot \mathbf{i})\mathbf{i} + (\mathbf{b} \cdot \mathbf{j})\mathbf{j} + (\mathbf{b} \cdot \mathbf{k})\mathbf{k} = (\mathbf{c} \cdot \mathbf{i})\mathbf{i} + (\mathbf{c} \cdot \mathbf{j})\mathbf{j} + (\mathbf{c} \cdot \mathbf{k})\mathbf{k} = \mathbf{c}$

 ⤒__ (13.3.14)   (13.3.14)__⤒

45. (a) $\|\mathbf{a} + \mathbf{b}\|^2 - \|\mathbf{a} - \mathbf{b}\|^2 = (\mathbf{a} + \mathbf{b}) \cdot (\mathbf{a} + \mathbf{b}) - (\mathbf{a} - \mathbf{b}) \cdot (\mathbf{a} - \mathbf{b})$

$$= [(\mathbf{a} \cdot \mathbf{a}) + 2(\mathbf{a} \cdot \mathbf{b}) + (\mathbf{b} \cdot \mathbf{b})] - [(\mathbf{a} \cdot \mathbf{a}) - 2(\mathbf{a} \cdot \mathbf{b}) + (\mathbf{b} \cdot \mathbf{b})] = 4(\mathbf{a} \cdot \mathbf{b})$$

(b) By (b), the relation $\|\mathbf{a}+\mathbf{b}\| = \|\mathbf{a}-\mathbf{b}\|$ gives $\mathbf{a} \perp \mathbf{b}$. The relation $\mathbf{a}+\mathbf{b} \perp \mathbf{a}-\mathbf{b}$ gives

$$0 = (\mathbf{a}+\mathbf{b}) \cdot (\mathbf{a}-\mathbf{b}) = \|\mathbf{a}\|^2 - \|\mathbf{b}\|^2 \quad \text{and thus} \quad \|\mathbf{a}\| = \|\mathbf{b}\|.$$

The parallelogram is a square since it has two adjacent sides of equal length and these meet at right angles.

47. $\|\mathbf{a}+\mathbf{b}\|^2 = (\mathbf{a}+\mathbf{b}) \cdot (\mathbf{a}+\mathbf{b}) = \mathbf{a} \cdot \mathbf{a} + 2\mathbf{a} \cdot \mathbf{b} + \mathbf{b} \cdot \mathbf{b} = \|\mathbf{a}\|^2 + 2\mathbf{a} \cdot \mathbf{b} + \|\mathbf{b}\|^2$
$\|\mathbf{a}-\mathbf{b}\|^2 = (\mathbf{a}-\mathbf{b}) \cdot (\mathbf{a}-\mathbf{b}) = \mathbf{a} \cdot \mathbf{a} - 2\mathbf{a} \cdot \mathbf{b} + \mathbf{b} \cdot \mathbf{b} = \|\mathbf{a}\|^2 - 2\mathbf{a} \cdot \mathbf{b} + \|\mathbf{b}\|^2$

Add the two equations and the result follows.

49. Let $\mathbf{c} = \|\mathbf{b}\|\mathbf{a} + \|\mathbf{a}\|\mathbf{b}$. Then

$$\frac{\mathbf{a} \cdot \mathbf{c}}{\|\mathbf{a}\| \, \|\mathbf{c}\|} = \|\mathbf{a}\| \, \|\mathbf{b}\| + \mathbf{a} \cdot \mathbf{b} = \frac{\mathbf{b} \cdot \mathbf{c}}{\|\mathbf{b}\| \, \|\mathbf{c}\|}$$

51. Existence of decomposition:

$$\mathbf{a} = (\mathbf{a} \cdot \mathbf{u_b})\mathbf{u_b} + [\mathbf{a} - (\mathbf{a} \cdot \mathbf{u_b})\mathbf{u_b}].$$

Uniqueness of decomposition: suppose that

$$\mathbf{a} = \mathbf{a}_\parallel + \mathbf{a}_\perp = \mathbf{A}_\parallel + \mathbf{A}_\perp.$$

Then the vector $\mathbf{a}_\parallel - \mathbf{A}_\parallel = \mathbf{A}_\perp - \mathbf{a}_\perp$ is both parallel to $\mathbf{b}$ and perpendicular to $\mathbf{b}$. Therefore it is zero. Consequently $\mathbf{A}_\parallel = \mathbf{a}_\parallel$ and $\mathbf{A}_\perp = \mathbf{a}_\perp$.

53. Place center of sphere at the origin.

$$\overrightarrow{P_1Q} \cdot \overrightarrow{P_2Q} = (-\mathbf{a}+\mathbf{b}) \cdot (\mathbf{a}+\mathbf{b})$$
$$= -\|\mathbf{a}\|^2 + \|\mathbf{b}\|^2$$
$$= 0.$$

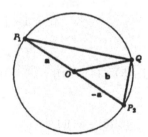

PROJECT 13.3

1. (a) $W = \mathbf{F} \cdot \mathbf{r}$ (b) 0 (c) $\|\mathbf{F}\|\mathbf{i} \cdot (b-a)\mathbf{i} = \|\mathbf{F}\|(b-a)$

3. (a) $W_1 = \mathbf{F}_1 \cdot \mathbf{r} = \|\mathbf{F}_1\|\|\mathbf{r}\| \cos \theta; \quad W_2 = \mathbf{F}_2 \cdot \mathbf{r} = \|\mathbf{F}_2\|\|\mathbf{r}\| \cos (-\theta) = \|\mathbf{F}_2\|\|\mathbf{r}\| \cos \theta$
    Therefore $W_2 = \dfrac{\|\mathbf{F}_2\|}{\|\mathbf{F}_1\|} W_1.$

    (b) $W_1 = \|\mathbf{F}_1\|\|\mathbf{r}\| \cos \pi/3 = \frac{1}{2}\|\mathbf{F}_1\|\|\mathbf{r}\|; \quad W_2 = \|\mathbf{F}_2\|\|\mathbf{r}\| \cos \pi/6 = \frac{1}{2}\sqrt{3}\|\mathbf{F}_2\|\|\mathbf{r}\|$
    Therefore $W_2 = \sqrt{3}\dfrac{\|\mathbf{F}_2\|}{\|\mathbf{F}_1\|} W_1.$

SECTION 13.4

1. $(\mathbf{i} + \mathbf{j}) \times (\mathbf{i} - \mathbf{j}) = [\mathbf{i} \times (\mathbf{i} - \mathbf{j})] + [\mathbf{j} \times (\mathbf{i} - \mathbf{j})] = (\mathbf{0} - \mathbf{k}) + (-\mathbf{k} - \mathbf{0}) = -2\mathbf{k}$

3. $(\mathbf{i} - \mathbf{j}) \times (\mathbf{j} - \mathbf{k}) = [\mathbf{i} \times (\mathbf{j} - \mathbf{k})] - [\mathbf{j} \times (\mathbf{j} - \mathbf{k})] = (\mathbf{j} + \mathbf{k}) - (\mathbf{0} - \mathbf{i}) = \mathbf{i} + \mathbf{j} + \mathbf{k}$

5. $(2\mathbf{j} - \mathbf{k}) \times (\mathbf{i} - 3\mathbf{j}) = [2\mathbf{j} \times (\mathbf{i} - 3\mathbf{j})] - [\mathbf{k} \times (\mathbf{i} - 3\mathbf{j})] = (-2\mathbf{k}) - (\mathbf{j} + 3\mathbf{i}) = -3\mathbf{i} - \mathbf{j} - 2\mathbf{k}$

    or

$$(2\mathbf{j} - \mathbf{k}) \times (\mathbf{i} - 3\mathbf{j}) = \begin{vmatrix} \mathbf{i} & \mathbf{j} & \mathbf{k} \\ 0 & 2 & -1 \\ 1 & -3 & 0 \end{vmatrix} = \mathbf{i} \begin{vmatrix} 2 & -1 \\ -3 & 0 \end{vmatrix} - \mathbf{j} \begin{vmatrix} 0 & -1 \\ 1 & -3 \end{vmatrix} + \mathbf{k} \begin{vmatrix} 0 & 2 \\ 1 & -3 \end{vmatrix} = -3\mathbf{i} - \mathbf{j} - 2\mathbf{k}$$

7. $\mathbf{j} \cdot (\mathbf{i} \times \mathbf{k}) = \mathbf{j} \cdot (-\mathbf{j}) = -1$ **9.** $(\mathbf{i} \times \mathbf{j}) \times \mathbf{k} = \mathbf{k} \times \mathbf{k} = \mathbf{0}$ **11.** $\mathbf{j} \cdot (\mathbf{k} \times \mathbf{i}) = \mathbf{j} \cdot (\mathbf{j}) = 1$

13. $(\mathbf{i} + 3\mathbf{j} - \mathbf{k}) \times (\mathbf{i} + \mathbf{k}) = \begin{vmatrix} \mathbf{i} & \mathbf{j} & \mathbf{k} \\ 1 & 3 & -1 \\ 1 & 0 & 1 \end{vmatrix} = [(3)(1) - (-1)(0)]\,\mathbf{i} - [(1)(1) - (-1)(1)]\,\mathbf{j} + [(1)0 - (3)(1)]\,\mathbf{k}$

$$= 3\mathbf{i} - 2\mathbf{j} - 3\mathbf{k}$$

15. $(\mathbf{i} + \mathbf{j} + \mathbf{k}) \times (2\mathbf{i} + \mathbf{k}) = \begin{vmatrix} \mathbf{i} & \mathbf{j} & \mathbf{k} \\ 1 & 1 & 1 \\ 2 & 0 & 1 \end{vmatrix} = [(1)(1) - (1)(0)]\,\mathbf{i} - [(1)(1) - (1)(2)]\,\mathbf{j} + [(1)(0) - (1)(2)]\,\mathbf{k}$

$$= \mathbf{i} + \mathbf{j} - 2\mathbf{k}$$

17. $[2\mathbf{i} + \mathbf{j}] \cdot [(\mathbf{i} - 3\mathbf{j} + \mathbf{k}) \times (4\mathbf{i} + \mathbf{k})] = \begin{vmatrix} 1 & -3 & 1 \\ 4 & 0 & 1 \\ 2 & 1 & 0 \end{vmatrix} =$

$$[(0)(0) - (1)(1)] - (-3)[(4)(0) - (1)(2)] + [(4)(1) - (0)(2)] = -3$$

19.
$$[(\mathbf{i} - \mathbf{j}) \times (\mathbf{j} - \mathbf{k})] \times [\mathbf{i} + 5\mathbf{k}] = \{[\mathbf{i} \times (\mathbf{j} - \mathbf{k})] - [\mathbf{j} \times (\mathbf{j} - \mathbf{k})]\} \times [\mathbf{i} + 5\mathbf{k}]$$

$$= [(\mathbf{k} + \mathbf{j}) - (-\mathbf{i})] \times [\mathbf{i} + 5\mathbf{k}]$$

$$= (\mathbf{i} + \mathbf{j} + \mathbf{k}) \times (\mathbf{i} + 5\mathbf{k})$$

$$= [(\mathbf{i} + \mathbf{j} + \mathbf{k}) \times \mathbf{i}] + [(\mathbf{i} + \mathbf{j} + \mathbf{k}) \times 5\mathbf{k}]$$

$$= (-\mathbf{k} + \mathbf{j}) + (-5\mathbf{j} + 5\mathbf{i})$$

$$= 5\mathbf{i} - 4\mathbf{j} - \mathbf{k}$$

21. $\mathbf{a} \times \mathbf{b} = \begin{vmatrix} \mathbf{i} & \mathbf{j} & \mathbf{k} \\ 1 & 3 & -1 \\ 2 & 0 & 1 \end{vmatrix} = 3\mathbf{i} - 3\mathbf{j} - 6\mathbf{k}$

$\dfrac{\mathbf{a} \times \mathbf{b}}{\|\mathbf{a} \times \mathbf{b}\|} = \dfrac{1}{\sqrt{6}}\mathbf{i} - \dfrac{1}{\sqrt{6}}\mathbf{j} - \dfrac{2}{\sqrt{6}}\mathbf{k}; \qquad \dfrac{\mathbf{b} \times \mathbf{a}}{\|\mathbf{b} \times \mathbf{a}\|} = -\dfrac{1}{\sqrt{6}}\mathbf{i} + \dfrac{1}{\sqrt{6}}\mathbf{j} + \dfrac{2}{\sqrt{6}}\mathbf{k}$

23. Set $\mathbf{a} = \overrightarrow{PQ} = -\mathbf{i} + 2\mathbf{k}$ and $\mathbf{b} = \overrightarrow{PR} = 2\mathbf{i} - \mathbf{k}$. Then

$$\mathbf{a} \times \mathbf{b} = \begin{vmatrix} \mathbf{i} & \mathbf{j} & \mathbf{k} \\ -1 & 0 & 2 \\ 2 & 0 & -1 \end{vmatrix} = 3\mathbf{j}; \quad \dfrac{\mathbf{a} \times \mathbf{b}}{\|\mathbf{a} \times \mathbf{b}\|} = \mathbf{j}$$

and $A = \frac{1}{2}\|\mathbf{a} \times \mathbf{b}\| = \frac{1}{2}\|3\mathbf{j}\| = \frac{3}{2}$.

25. Set $\mathbf{a} = \overrightarrow{PQ} = \mathbf{i} + \mathbf{j} - 3\mathbf{k}$ and $\mathbf{b} = \overrightarrow{PR} = -\mathbf{i} + 3\mathbf{j} - \mathbf{k}$. Then

$$\mathbf{a} \times \mathbf{b} = \begin{vmatrix} \mathbf{i} & \mathbf{j} & \mathbf{k} \\ 1 & 1 & -3 \\ -1 & 3 & -1 \end{vmatrix} = 8\mathbf{j} + 4\mathbf{j} + 4\mathbf{k}; \quad \dfrac{\mathbf{a} \times \mathbf{b}}{\|\mathbf{a} \times \mathbf{b}\|} = \dfrac{2}{\sqrt{6}}\mathbf{i} + \dfrac{1}{\sqrt{6}}\mathbf{j} + \dfrac{1}{\sqrt{6}}\mathbf{k}$$

and $A = \frac{1}{2}\|\mathbf{a} \times \mathbf{b}\| = \frac{1}{2}\|8\mathbf{i} + 4\mathbf{j} + 4\mathbf{k}\| = \frac{1}{2}\sqrt{8^2 + 4^2 + 4^2} = 2\sqrt{6}$.

27. $V = \left|[(\mathbf{i}+\mathbf{j}) \times (2\mathbf{i} - \mathbf{k})] \cdot (3\mathbf{j}+\mathbf{k})\right| = |(-\mathbf{i}+\mathbf{j}-2\mathbf{k}) \cdot (3\mathbf{j}+\mathbf{k})| = 1$

29. $V = \left|\overrightarrow{OP} \cdot \left(\overrightarrow{OQ} \times \overrightarrow{OR}\right)\right| = \left\|\begin{vmatrix} 1 & 2 & 3 \\ 1 & 1 & 2 \\ 2 & 1 & 1 \end{vmatrix}\right\| = 2$

31.
$$(\mathbf{a}+\mathbf{b}) \times (\mathbf{a}-\mathbf{b}) = [\mathbf{a} \times (\mathbf{a}-\mathbf{b})] + [\mathbf{b} \times (\mathbf{a}-\mathbf{b})]$$
$$= [\mathbf{a} \times (-\mathbf{b})] + [\mathbf{b} \times \mathbf{a}]$$
$$= -(\mathbf{a} \times \mathbf{b}) - (\mathbf{a} \times \mathbf{b}) = -2(\mathbf{a} \times \mathbf{b})$$

33. $\mathbf{a} \times \mathbf{i} = \mathbf{0}, \quad \mathbf{a} \times \mathbf{j} = \mathbf{0} \implies \mathbf{a} \| \mathbf{i}$ and $\mathbf{a} \| \mathbf{j} \implies \mathbf{a} = \mathbf{0}$

35. By (13.4.4) $\alpha\mathbf{a} \times \beta\mathbf{b} = (\alpha\beta)\mathbf{a} \times \mathbf{b}$. Therefore, $\|\alpha\mathbf{a} \times \beta\mathbf{b}\| = \|(\alpha\beta)\mathbf{a} \times \mathbf{b}\|$.

37. (a) $\mathbf{a} \cdot (\mathbf{b} \times \mathbf{c})$: makes sense – this is the dot product of two vectors.

(b) $\mathbf{a} \times (\mathbf{b} \cdot \mathbf{c})$: does not make sense – this is the cross product of a vector with a number.

(c) $\mathbf{a} \cdot (\mathbf{b} \cdot \mathbf{c})$: does not make sense – this is the dot product of a vector with a number.

(d) $\mathbf{a} \times (\mathbf{b} \times \mathbf{c})$: makes sense – this is the cross product of two vectors.

39. $\mathbf{d} \cdot \mathbf{a} = \mathbf{d} \cdot \mathbf{b} \Longrightarrow \mathbf{d} \perp (\mathbf{a} - \mathbf{b}); \quad \mathbf{d} \cdot \mathbf{a} = \mathbf{d} \cdot \mathbf{c} \Longrightarrow \mathbf{d} \perp (\mathbf{a} - \mathbf{c})$

Therefore, $\mathbf{d} = \lambda[(\mathbf{a} - \mathbf{b}) \times (\mathbf{a} - \mathbf{c})]$ for some number λ.

41. $\mathbf{a} \cdot \mathbf{b} = \mathbf{a} \cdot \mathbf{c} \quad \Longrightarrow \quad \mathbf{a} \cdot (\mathbf{b} - \mathbf{c}) = 0; \quad \mathbf{a}$ is perpendicular to $\mathbf{b} - \mathbf{c}$.

$\mathbf{a} \times \mathbf{b} = \mathbf{a} \times \mathbf{c} \quad \Longrightarrow \quad \mathbf{a} \times (\mathbf{b} - \mathbf{c}) = \mathbf{0}; \quad \mathbf{a}$ is parallel to $\mathbf{b} - \mathbf{c}$.

Since $\mathbf{a} \neq \mathbf{0}$ it follows that $\mathbf{b} - \mathbf{c} = \mathbf{0}$ or $\mathbf{b} = \mathbf{c}$.

43. $\mathbf{c} \times \mathbf{a} = (\mathbf{a} \times \mathbf{b}) \times \mathbf{a} = (\mathbf{a} \cdot \mathbf{a})\mathbf{b} - (\mathbf{a} \cdot \mathbf{b})\mathbf{a} = (\mathbf{a} \cdot \mathbf{a})\mathbf{b} = \|\mathbf{a}\|^2 \mathbf{b}$

$\qquad\qquad$ Exercise 42(a) $\qquad\qquad$ $\mathbf{a} \cdot \mathbf{b} = 0$

45. Suppose $\mathbf{a} \neq \mathbf{0}$. Then

$$\mathbf{a} \cdot \mathbf{b} = 0 \Longrightarrow \mathbf{b} \perp \mathbf{a}; \quad \mathbf{a} \times \mathbf{b} = \mathbf{0} \Longrightarrow \mathbf{b} \| \mathbf{a}$$

Thus $\mathbf{b}$ is simultaneously perpendicular to, and parallel to $\mathbf{a}$. It follows that $\mathbf{b} = \mathbf{0}$.

47. The result is an immediate consequence of Exercise 46.

SECTION 13.5

1. P (when $t = 0$) and Q (when $t = -1$)

3. Take $\quad \mathbf{r}_0 = \overrightarrow{OP} = 3\mathbf{i} + \mathbf{j} \quad$ and $\quad \mathbf{d} = \mathbf{k}$. $\quad$ Then, $\quad \mathbf{r}(t) = (3\mathbf{i} + \mathbf{j}) + t\mathbf{k}$.

5. Take $\quad \mathbf{r}_0 = \mathbf{0} \quad$ and $\quad \mathbf{d} = \overrightarrow{OQ}$. $\quad$ Then, $\quad \mathbf{r}(t) = t(x_1 \mathbf{i} + y_1 \mathbf{j} + z_1 \mathbf{k})$.

7. $\overrightarrow{PQ} = \mathbf{i} - \mathbf{j} + \mathbf{k} \quad$ so direction numbers are $\quad 1, -1, 1$. Using P as a point on the line, we have

$$x(t) = 1 + t, \quad y(t) = -t, \quad z(t) = 3 + t.$$

9. The line is parallel to the y-axis so we can take $0, 1, 0$ as direction numbers. Therefore

$$x(t) = 2, \quad y(t) = -2 + t, \quad z(t) = 3.$$

11. Since the line $\quad 2(x + 1) = 4(y - 3) = z \quad$ can be written

$$\frac{x + 1}{2} = \frac{y - 3}{1} = \frac{z}{4},$$

it has direction numbers $2, 1, 4$. The line through $P(-1, 2, -3)$ with direction vector $2\mathbf{i} + \mathbf{j} + 4\mathbf{k}$ can be parameterized

$$\mathbf{r}(t) = (-\mathbf{i} + 2\mathbf{j} - 3\mathbf{k}) + t(2\mathbf{i} + \mathbf{j} + 4\mathbf{k}).$$

13. $\mathbf{r}(t) = (3\mathbf{i} + \mathbf{j} + 5\mathbf{k}) + t(\mathbf{i} - \mathbf{j} + 2\mathbf{k}) = (3 + t)\mathbf{i} + (1 - t)\mathbf{j} + (5 + 2t)\mathbf{k}$

$\mathbf{R}(u) = (\mathbf{i} + 4\mathbf{j} + 2\mathbf{k}) + u(\mathbf{j} + \mathbf{k}) = \mathbf{i} + (4 + u)\mathbf{j} + (2 + u)\mathbf{k}$

$\mathbf{d} = \mathbf{i} - \mathbf{j} + 2\mathbf{k}$ is a direction vector for l_1; $\mathbf{D} = \mathbf{j} + \mathbf{k}$ is a direction vector for l_2. Since $\mathbf{d}$ is not a multiple of $\mathbf{D}$, the lines either intersect or are skew. Setting $\mathbf{r}(t) = \mathbf{R}(u)$ we get the system of equations:

$$3 + t = 1, \quad 1 - t = 4 + u, \quad 5 + 2t = 2 + u$$

This system has the solution $t = -2$, $u = -1$. The point of intersection is: $(1, 3, 1)$.

15. $\mathbf{d} = 2\mathbf{i} + 4\mathbf{j} - \mathbf{k}$ is a direction vector for l_1; $\mathbf{D} = 2\mathbf{i} + \mathbf{j} + 2\mathbf{k}$ is a direction vector for l_2. Since $\mathbf{d}$ is not a multiple of $\mathbf{D}$, the lines either intersect or are skew. Equating coordinates, we get the system of equations:

$$3 + 2t = 3 + 2u, \quad -1 + 4t = 2 + u, \quad 2 - t = -2 + 2u$$

From the first two equations, we get $t = u = 1$. Since these values of t and u do not satisfy the third equation, the lines are skew.

17. $\mathbf{d} = -6\mathbf{i} + 9\mathbf{j} - 3\mathbf{k}$ is a direction vector for l_1; $\mathbf{D} = 2\mathbf{i} - 3\mathbf{j} + \mathbf{k}$ is a direction vector for l_2. Since $\mathbf{d} = -3\mathbf{D}$, we conclude that l_1 and l_2 are either parallel or coincident. The point $(1, 2, 0)$ lies on l_1 but does not line on l_2. Therefore, the lines are parallel.

19. $\mathbf{d} = 2\mathbf{i} + 4\mathbf{j} + 3\mathbf{k}$ is a direction vector for l_1; $\mathbf{D} = \mathbf{i} + 3\mathbf{j} + 2\mathbf{k}$ is a direction vector for l_2. Since $\mathbf{d}$ is not a multiple of $\mathbf{D}$, the lines either intersect or are skew. The system of equations

$$4 + 2t = 2 + u, \quad -5 + 4t = -1 + 3u, \quad 1 + 3t = 2u$$

does not have a solution. Therefore the lines are skew.

21. We set $\mathbf{r}_1(t) = \mathbf{r}_2(u)$ and solve for t and u:

$$\mathbf{i} + t\mathbf{j} = \mathbf{j} + u(\mathbf{i} + \mathbf{j}),$$

$$(1 - u)\mathbf{i} + (-1 - u + t)\mathbf{j} = \mathbf{0}.$$

Thus,

$$1 - u = 0 \quad \text{and} \quad -1 - u + t = 0.$$

These equations give $u = 1$, $t = 2$. The point of intersection is $P(1, 2, 0)$.

As direction vectors for the lines we can take $\mathbf{u} = \mathbf{j}$ and $\mathbf{v} = \mathbf{i} + \mathbf{j}$. Thus

$$\cos\theta = \frac{\mathbf{u} \cdot \mathbf{v}}{\|\mathbf{u}\| \|\mathbf{v}\|} = \frac{1}{(1)(\sqrt{2})} = \frac{1}{2}\sqrt{2}.$$

The angle of intersection is $\frac{1}{4}\pi$ radians.

23. $\left(x_0 - \dfrac{d_1}{d_3}z_0, \ y_0 - \dfrac{d_2}{d_3}z_0, \ 0\right)$

25. The lines are parallel.

27. $\mathbf{r}(t) = (2\mathbf{i} + 7\mathbf{j} - \mathbf{k}) + t(2\mathbf{i} - 5\mathbf{j} + 4\mathbf{k}), \quad 0 \le t \le 1$

29. Set
$$\mathbf{u} = \frac{\vec{PQ}}{\|\vec{PQ}\|} = \frac{-4\mathbf{i} + 2\mathbf{j} + 4\mathbf{k}}{\|-4\mathbf{i} + 2\mathbf{j} + 4\mathbf{k}\|} = -\frac{2}{3}\mathbf{i} + \frac{1}{3}\mathbf{j} + \frac{2}{3}\mathbf{k}.$$

Then $\mathbf{r}(t) = (6\mathbf{i} - 5\mathbf{j} + \mathbf{k}) + t\mathbf{u}$ is $\vec{OP}$ at $t = 9$ and it is $\vec{OQ}$ at $t = 15$. (Check this.)

Answer: $\mathbf{u} = -\frac{2}{3}\mathbf{i} + \frac{1}{3}\mathbf{j} + \frac{2}{3}\mathbf{k}, \quad 9 \le t \le 15.$

31. The given line, call it l, has direction vector $2\mathbf{i} - 4\mathbf{j} + 6\mathbf{k}$.

If $a\mathbf{i} + b\mathbf{j} + c\mathbf{k}$ is a direction vector for a line perpendicular to l, then

$$(2\mathbf{i} - 4\mathbf{j} + 6\mathbf{k}) \cdot (a\mathbf{i} + b\mathbf{j} + c\mathbf{k}) = 2a - 4b + 6c = 0.$$

The lines through $P(3, -1, 8)$ perpendicular to l can be parameterized

$$X(u) = 3 + au, \quad Y(u) = -1 + bu, \quad Z(u) = 8 + cu$$

with $2a - 4b + 6c = 0$.

33. $d(P, l) = \dfrac{\|(\mathbf{i} + 2\mathbf{k}) \times (2\mathbf{i} - \mathbf{j} + 2\mathbf{k})\|}{\|2\mathbf{i} - \mathbf{j} + 2\mathbf{k}\|} = 1$

35. The line contains the point $P_0(1, 0, 2)$. Therefore
$$d(P, l) = \frac{\|(2\mathbf{j} + \mathbf{k}) \times (\mathbf{i} - 2\mathbf{j} + 3\mathbf{k})\|}{\|\mathbf{i} - 2\mathbf{j} + 3\mathbf{k}\|} = \sqrt{\frac{69}{14}} \cong 2.22$$

37. The line contains the point $P_0(2, -1, 0)$. Therefore
$$d(P, l) = \frac{\|(\mathbf{i} - \mathbf{j} - \mathbf{k}) \times (\mathbf{i} + \mathbf{j})\|}{\|\mathbf{i} + \mathbf{j}\|} = \sqrt{3} \cong 1.73.$$

39. (a) The line passes through $P(1, 1, 1)$ with direction vector $\mathbf{i} + \mathbf{j}$. Therefore
$$d(0, l) = \frac{\|(\mathbf{i} + \mathbf{j} + \mathbf{k}) \times (\mathbf{i} + \mathbf{j})\|}{\|\mathbf{i} + \mathbf{j}\|} = 1.$$

(b) The distance from the origin to the line segment is $\sqrt{3}$.

Solution. The line segment can be parameterized

$$\mathbf{r}(t) = \mathbf{i} + \mathbf{j} + \mathbf{k} + t(\mathbf{i} + \mathbf{j}), \quad t \in [0, 1].$$

This is the set of all points $P(1 + t, 1 + t, 1)$ with $t \in [0, 1]$.

The distance from the origin to such a point is

$$f(t) = \sqrt{2(1 + t)^2 + 1}.$$

The minimum value of this function is $f(0) = \sqrt{3}$.

Explanation. The point on the line through P and Q closest to the origin is not on the line segment $\overline{PQ}$.

41. We begin with $\mathbf{r}(t) = \mathbf{j} - 2\mathbf{k} + t(\mathbf{i} - \mathbf{j} + 3\mathbf{k})$. The scalar t_0 for which $\mathbf{r}(t_0) \perp l$ can be found by solving the equation

$$[\mathbf{j} - 2\mathbf{k} + t_0(\mathbf{i} - \mathbf{j} + 3\mathbf{k})] \cdot [\mathbf{i} - \mathbf{j} + 3\mathbf{k}] = 0.$$

This equation gives $-7 + 11t_0 = 0$ and thus $t_0 = 7/11$. Therefore

$$\mathbf{r}(t_0) = \mathbf{j} - 2\mathbf{k} + \tfrac{7}{11}(\mathbf{i} - \mathbf{j} + 3\mathbf{k}) = \tfrac{7}{11}\mathbf{i} + \tfrac{4}{11}\mathbf{j} - \tfrac{1}{11}\mathbf{k}.$$

The vectors of norm 1 parallel to $\mathbf{i} - \mathbf{j} + 3\mathbf{k}$ are

$$\pm\frac{1}{\sqrt{11}}(\mathbf{i} - \mathbf{j} + 3\mathbf{k}).$$

The standard parameterizations are

$$\mathbf{R}(t) = \frac{7}{11}\mathbf{i} + \frac{4}{11}\mathbf{j} - \frac{1}{11}\mathbf{k} \pm \frac{t}{\sqrt{11}}(\mathbf{i} - \mathbf{j} + 3\mathbf{k})$$

$$= \frac{1}{11}(7\mathbf{i} + 4\mathbf{j} - \mathbf{k}) \pm t\left[\frac{\sqrt{11}}{11}(\mathbf{i} - \mathbf{j} + 3\mathbf{k})\right].$$

43. $0 < t < s$

    By similar triangles, if $0 < s < 1$, the tip of $\overrightarrow{OA} + s\,\overrightarrow{AB} + s\,\overrightarrow{BC}$ falls on $\overline{AC}$. If $0 < t < s$, then the tip of $\overrightarrow{OA} + s\,\overrightarrow{AB} + t\,\overrightarrow{BC}$ falls short of $\overline{AC}$ and stays within the triangle. Clearly all points in the interior of the triangle can be reached in this manner.

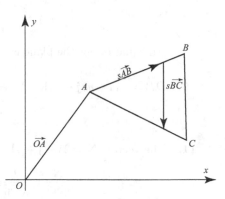

45. $\mathbf{d} = \mathbf{i} + 3\mathbf{j} - 2\mathbf{k}$ is a direction vector for l_1; $\mathbf{D} = 4\mathbf{i} - \mathbf{j} + 2\mathbf{k}$ is a direction vector for l_2. Since $\mathbf{d}$ is not a multiple of $\mathbf{D}$, the lines either intersect or are skew. Equating coordinates, we get the system of equations:

$$2 + t = -1 + 4u, \quad -1 + 3t = 2 - u, \quad 1 - 2t = -3 + 2u$$

This system does not have a solution. Therefore the lines are skew. The point $P(2, -1, 1)$ is on l_1 and the point $Q(-1, 2, -3)$ is on l_2, and $\overrightarrow{PQ} = -3\mathbf{i} + 3\mathbf{j} - 4\mathbf{k}$. By Exercise 44, the distance between l_1 and l_2 is:

$$\frac{\left|\overrightarrow{PQ} \cdot (\mathbf{d} \times \mathbf{D})\right|}{\|\mathbf{d} \times \mathbf{D}\|} = \frac{|(-3\mathbf{i} + 3\mathbf{j} - 4\mathbf{k}) \cdot (4\mathbf{i} - 10\mathbf{j} - 13\mathbf{k})|}{\sqrt{285}} = \frac{10}{\sqrt{285}}$$

SECTION 13.6

1. Q

3. Since $\mathbf{i} - 4\mathbf{j} + 3\mathbf{k}$ is normal to the plane, we have

$$(x - 2) - 4(y - 3) + 3(z - 4) = 0 \quad \text{and thus} \quad x - 4y + 3z - 2 = 0.$$

5. The vector $3\mathbf{i} - 2\mathbf{j} + 5\mathbf{k}$ is normal to the given plane and thus to every parallel plane: the equation we want can be written

$$3(x - 2) - 2(y - 1) + 5(z - 1) = 0, \quad 3x - 2y + 5z - 9 = 0.$$

7. The point $Q(0, 0, -2)$ lies on the line l; and $\mathbf{d} = \mathbf{i} + \mathbf{j} + \mathbf{k}$ is a direction vector for l. We want an equation for the plane which has the vector

$$\mathbf{N} = \overrightarrow{PQ} \times \mathbf{d} = (\mathbf{i} + 3\mathbf{j} + 3\mathbf{k}) \times (\mathbf{i} + \mathbf{j} + \mathbf{k})$$

as a normal vector:

$$\mathbf{N} = \begin{vmatrix} \mathbf{i} & \mathbf{j} & \mathbf{k} \\ -1 & -3 & -3 \\ 1 & 1 & 1 \end{vmatrix} = -2\mathbf{j} + 2\mathbf{k}$$

An equation for the plane is: $-2(y - 3) + 2(z - 1) = 0$ or $y - z - 2 = 0$

9. $\overrightarrow{OP_0} = x_0\mathbf{i} + y_0\mathbf{j} + z_0\mathbf{k}$ An equation for the plane is:

$$x_0(x - x_0) + y_0(y - y_0) + z_0(z - z_0) = 0$$

11. The vector $\mathbf{N} = 2\mathbf{i} - \mathbf{j} + 5\mathbf{k}$ is normal to the plane $2x - y + 5z - 10 = 0$. The unit normals are:

$$\frac{\mathbf{N}}{\|\mathbf{N}\|} = \frac{1}{\sqrt{30}}(2\mathbf{i} - \mathbf{j} + 5\mathbf{k}) \quad \text{and} \quad -\frac{\mathbf{N}}{\|\mathbf{N}\|} = -\frac{1}{\sqrt{30}}(2\mathbf{i} - \mathbf{j} + 5\mathbf{k})$$

13. Intercept form: $\dfrac{x}{15} + \dfrac{y}{12} - \dfrac{z}{10} = 1$ x-intercept: $(15, 0, 0)$

    y-intercept: $(0, 12, 0)$

    z-intercept: $(0, 0, -10)$

15. $\mathbf{u_{N_1}} = \dfrac{\sqrt{38}}{38}(5\mathbf{i} - 3\mathbf{j} + 2\mathbf{k}), \quad \mathbf{u_{N_2}} = \dfrac{\sqrt{14}}{14}(\mathbf{i} + 3\mathbf{j} + 2\mathbf{k}), \quad \cos\theta = \left| \mathbf{u_{N_1}} \cdot \mathbf{u_{N_2}} \right| = 0.$

    Therefore $\theta = \pi/2$ radians.

17. $\mathbf{u_{N_1}} = \dfrac{\sqrt{3}}{3}(\mathbf{i} - \mathbf{j} + \mathbf{k}), \quad \mathbf{u_{N_2}} = \dfrac{\sqrt{14}}{14}(2\mathbf{i} + \mathbf{j} + 3\mathbf{k}), \quad \cos\theta = \left| \mathbf{u_{N_1}} \cdot \mathbf{u_{N_2}} \right| = \dfrac{2}{21}\sqrt{42} \cong 0.617.$

    Therefore $\theta \cong 0.91$ radians.

19. coplanar since $0(4\mathbf{j} - \mathbf{k}) + 0(3\mathbf{i} + \mathbf{j} + 2\mathbf{k}) + 1(\mathbf{0}) = 0$

21. We need to determine whether there exist scalars s, t, u not all zero such that

$$s(\mathbf{i}+\mathbf{j}+\mathbf{k})+t(2\mathbf{i}-\mathbf{j})+u(3\mathbf{i}-\mathbf{j}-\mathbf{k})=\mathbf{0}$$

$$(s+2t+3u)\,\mathbf{i}+(s-t-u)\,\mathbf{j}+(s-u)\,\mathbf{k}=\mathbf{0}.$$

The only solution of the system

$$s+2t+3u=0, \quad s-t-u=0, \quad s-u=0$$

is $s=t=u=0.$ Thus, the vectors are not coplanar.

23. By (13.6.5), $\quad d(P,p)=\dfrac{|2(2)+4(-1)-(3)+1|}{\sqrt{4+16+1}}=\dfrac{2}{\sqrt{21}}=\dfrac{2}{21}\sqrt{21}.$

25. By (13.6.5), $\quad d(P,p)=\dfrac{|(-3)(1)+0(-3)+4(5)+5|}{\sqrt{9+16}}=\dfrac{22}{5}.$

27. $\overrightarrow{P_1P}=(x-1)\mathbf{i}+y\mathbf{j}+(z-1)\mathbf{k}, \quad \overrightarrow{P_1P_2}=\mathbf{i}+\mathbf{j}-\mathbf{k}, \quad \overrightarrow{P_1P_3}=\mathbf{j}.$
Therefore

$$(\overrightarrow{P_1P_2} \times \overrightarrow{P_1P_3})=(\mathbf{i}+\mathbf{j}-\mathbf{k})\times\mathbf{j}=\mathbf{i}+\mathbf{k}$$

and

$$\overrightarrow{P_1P} \cdot (\overrightarrow{P_1P_2} \times \overrightarrow{P_1P_3})=[(x-1)\mathbf{i}+y\mathbf{j}+(z-1)\mathbf{k}]\cdot[\mathbf{i}+\mathbf{k}]=x-1+z-1.$$

An equation for the plane can be written $x+z=2.$

29. $\overrightarrow{P_1P}=(x-3)\mathbf{i}+(y+4)\mathbf{j}+(z-1)\mathbf{k}, \quad \overrightarrow{P_1P_2}=6\mathbf{j}, \quad \overrightarrow{P_1P_3}=-2\mathbf{i}+5\mathbf{j}-3\mathbf{k}.$
Therefore

$$(\overrightarrow{P_1P_2} \times \overrightarrow{P_1P_3})=6\mathbf{j}\times(-4\mathbf{i}+5\mathbf{j}-3\mathbf{k})=-18\mathbf{i}+12\mathbf{k}$$

and

$$\overrightarrow{P_1P} \cdot (\overrightarrow{P_1P_2} \times \overrightarrow{P_1P_3})=[(x-3)\mathbf{i}+(y+4)\mathbf{j}+(z-1)\mathbf{k}]\cdot[-18\mathbf{i}+12\mathbf{k}]$$

$$=-18(x-3)+12(z-1)$$

An equation for the plane can be written $-18(x-3)+12(z-1)=0$ or $3x-2z-7=0.$

31. The line passes through the point $P_0(x_0,y_0,z_0)$ with direction numbers: $A, B, C.$
Equations for the line written in symmetric form are:

$$\frac{x-x_0}{A}=\frac{y-y_0}{B}=\frac{z-z_0}{C}, \quad \text{provided} \quad A\neq0, \ B\neq0, \ C\neq0.$$

33. $\dfrac{x - x_0}{d_1} = \dfrac{y - y_0}{d_2}, \qquad \dfrac{y - y_0}{d_2} = \dfrac{z - z_0}{d_3}$

35. We set $x = 0$ and find that $P_0(0,0,0)$ lies on the line of intersection. As normals to the plane we use

$$\mathbf{N}_1 = \mathbf{i} + 2\mathbf{j} + 3\mathbf{k} \quad \text{and} \quad \mathbf{N}_2 = -3\mathbf{i} + 4\mathbf{j} + \mathbf{k}.$$

Note that

$$\mathbf{N}_1 \times \mathbf{N}_2 = (\mathbf{i} + 2\mathbf{j} + 3\mathbf{k}) \times (-3\mathbf{i} + 4\mathbf{j} + \mathbf{k}) = -10\mathbf{i} - 10\mathbf{j} + 10\mathbf{k}.$$

We take $-\frac{1}{10}(\mathbf{N}_1 \times \mathbf{N}_2) = \mathbf{i} + \mathbf{j} - \mathbf{k}$ as a direction vector for the line through $P_0(0,0,0)$. Then

$$x(t) = t, \quad y(t) = t, \quad z(t) = -t.$$

37. Straightforward computations give us

$$l: x(t) = 1 - 3t, \quad y(t) = -1 + 4t, \quad z(t) = 2 - t$$

and

$$p: x + 4y - z = 6.$$

Substitution of the scalar parametric equations for l in the equation for p gives

$$(1 - 3t) + 4(-1 + 4t) - (2 - t) = 6 \quad \text{and thus} \quad t = 11/14.$$

Using $t = 11/14$, we get $\quad x = -19/14, \quad y = 15/7, \quad z = 17/14.$

39. Let $\quad \mathbf{N} = A\mathbf{i} + B\mathbf{j} + C\mathbf{k} \quad$ be normal to the plane. Then
$$\mathbf{N} \cdot \mathbf{d} = (\mathbf{i} + B\mathbf{j} + C\mathbf{k}) \cdot (\mathbf{i} + 2\mathbf{j} + 4\mathbf{k}) = 1 + 2B + 4C = 0$$
and
$$\mathbf{N} \cdot \mathbf{D} = (\mathbf{i} + B\mathbf{j} + C\mathbf{k}) \cdot (-\mathbf{i} - \mathbf{j} + 3\mathbf{k}) = -1 - B + 3C = 0.$$
This gives $\quad B = -7/10 \quad$ and $\quad C = 1/10.$ The equation for the plane can be written
$$1(x - 0) - \tfrac{7}{10}(y - 0) + \tfrac{1}{10}(z - 0) = 0, \quad \text{which simplifies to} \quad 10x - 7y + z = 0.$$

41. $\mathbf{N} + \overrightarrow{PQ}$ and $\mathbf{N} - \overrightarrow{PQ}$ are the diagonals of a rectangle with sides $\mathbf{N}$ and $\overrightarrow{PQ}$. Since the diagonals are perpendicular, the rectangle is a square; that is $\|\mathbf{N}\| = \| \overrightarrow{PQ} \|$. Thus, the points Q form a circle centered at P with radius $\|\mathbf{N}\|$.

43. If $\alpha > 0$, then P_0 lies on the same side of the plane as the tip of $\mathbf{N}$; if $\alpha < 0$, then P_0 and the tip of $\mathbf{N}$ lie on opposite sides of the plane.

To see this, suppose that $\mathbf{N}$ emanates from the point $P_1(x_1, y_1, z_1)$ on the plane. Then

$$\mathbf{N} \cdot \overrightarrow{P_1P_0} = A(x_0 - x_1) + B(y_0 - y_1) + C(z_0 - z_1) = Ax_0 + By_0 + Cz_0 + D = \alpha.$$

If $\alpha > 0$, $0 \leq \sphericalangle \left(\mathbf{N}, \overrightarrow{P_0P_1}\right) < \pi/2$; if $\alpha < 0$, then $\pi/2 < \sphericalangle \left(\mathbf{N}, \overrightarrow{P_0P_1}\right) < \pi$. Since $\mathbf{N}$ is perpendicular to the plane, the result follows.

45. (a) intercepts:

$(4, 0, 0), \ (0, 5, 0), \ (0, 0, 2)$

(b) traces:

in the x, y-plane: $5x + 4y = 20$

in the x, z-plane: $x + 2z = 4$

in the y, z-plane: $2y + 5z = 10$

(c) unit normals: $\pm \dfrac{1}{\sqrt{141}} (5\mathbf{i} + 4\mathbf{j} + 10\mathbf{k})$

(d)

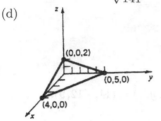

47. (a) intercepts:

$(4, 0, 0), \ \text{no } y\text{-intercept}, \ (0, 0, 6)$

(b) traces:

in the x, y-plane: $x = 4$

in the x, z-plane: $3x + 2z = 12$

in the y, z-plane: $z = 6$

(c) unit normals: $\pm \dfrac{1}{\sqrt{13}} (3\mathbf{i} + 2\mathbf{k})$

(d)

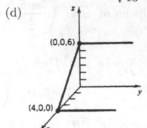

49. The normal vectors to the planes are: $\mathbf{N}_1 = 2\mathbf{i} + \mathbf{j} + 3\mathbf{k}, \ \mathbf{N}_2 = \mathbf{i} + 5\mathbf{j} - 2\mathbf{k}$.

The cosine of the angle θ between the planes is: $\cos \theta = \dfrac{|\mathbf{N}_1 \cdot \mathbf{N}_2|}{\|\mathbf{N}_1\| \, \|\mathbf{N}_2\|} = \dfrac{1}{2\sqrt{105}}$;

$\theta \cong 1.52$ radians $\cong 87.2°$.

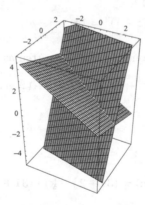

51. An equation of the plane that passes through $(2, 7, -3)$ with normal vector $\mathbf{N} = 3\mathbf{i} + \mathbf{j} + 4\mathbf{k}$ is:

$3(x - 2) + (y - 7) + 4(z + 3) = 0$ or $3x + y + 4z = 1$.

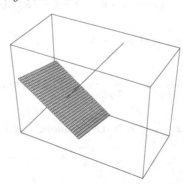

53. $\dfrac{x}{2} + \dfrac{y}{5} + \dfrac{z}{4} = 1$

$10x + 4y + 5z = 20$

55. $\dfrac{x}{3} + \dfrac{y}{5} = 1$

$5x + 3y = 15$

REVIEW EXERCISES

1. (a) $\overline{PQ} = \sqrt{(7-3)^2 + (-5-2)^2 + (-4-[-1])^2} = \sqrt{16 + 49 + 25} = 3\sqrt{10}$

(b) Midpoint of $\overline{QR}$: $\left(\dfrac{7+5}{2}, \dfrac{-5+6}{2}, \dfrac{4-3}{2}\right) = \left(6, \dfrac{1}{2}, \dfrac{1}{2}\right)$

(c) Let $X = (x, y, z)$. Then

$$(7, -5, 4) = \left(\dfrac{3+x}{2}, \dfrac{2+y}{2}, \dfrac{-1+z}{2}\right) \implies (x, y, z) = (11, -12, 9).$$

(d) Midpoint of $\overline{PR}$: $(4, 4, -2)$; radius of the sphere:

$$r = \dfrac{1}{2}|\overline{PR}| = \dfrac{1}{2}\sqrt{(5-3)^2 + (6-2)^2 + (-3+1)^2} = \sqrt{6}.$$

The equation of the sphere is: $(x-4)^2 + (y-4)^2 + (z+2)^2 = 6$.

3. radius: $\sqrt{2^2 + (-3)^2 + 1^2} = \sqrt{14}$

equation: $(x-2)^2 + (y+3)^2 + (z-1)^2 = 14$

5. By completing the square, the equation can be written as

$$(x+1)^2 + (y+2)^2 + (z-4)^2 = 4 = 2^2.$$

center: $(-1, -2, 4)$. radius: 2

7. $\frac{3}{2}\mathbf{i} + \mathbf{j} - \frac{1}{2}\mathbf{k}$

9. $\mathbf{b} + \mathbf{c} = 3\mathbf{i} + 7\mathbf{j} + \mathbf{k}$, $\mathbf{a} \cdot (\mathbf{b} + \mathbf{c}) = (3\mathbf{i} + 2\mathbf{j} - \mathbf{k}) \cdot (3\mathbf{i} + 7\mathbf{j} + \mathbf{k}) = 22$

9. $\|\mathbf{c}\|^2 = (-2)^2 + 4^2 + 1^2 = 21$

13. $2\mathbf{a} - \mathbf{b} = \mathbf{i} + \mathbf{j} - 2\mathbf{k}$; $(2\mathbf{a} - \mathbf{b}) \cdot \mathbf{c} = (\mathbf{i} + \mathbf{j} - 2\mathbf{k}) \cdot (-2\mathbf{i} + 4\mathbf{j} + \mathbf{k}) = 0$

15. $\|\mathbf{a}\| = \sqrt{14}$; $\mathbf{u_a} = \dfrac{1}{\sqrt{14}}(3\mathbf{i} + 2\mathbf{j} - \mathbf{k})$

17. $\cos\theta = \dfrac{\mathbf{a} \cdot \mathbf{c}}{\|\mathbf{a}\|\|\mathbf{c}\|} = \dfrac{1}{\sqrt{14}\sqrt{21}} = \dfrac{1}{7\sqrt{6}}$; $\theta \cong 1.51$ radians

19. $\|\mathbf{a}\| = \sqrt{14}$; $\cos\alpha = \dfrac{3}{\sqrt{14}}$, $\alpha \cong 0.64$ radians, $\cos\beta = \dfrac{2}{\sqrt{14}}$, $\beta \cong 1.01$, $\cos\gamma = \dfrac{-1}{\sqrt{14}}$, $\gamma \approx 1.84$

21. $\mathbf{b} \times \mathbf{c} = \begin{vmatrix} \mathbf{i} & \mathbf{j} & \mathbf{k} \\ 5 & 3 & 0 \\ -2 & 4 & 1 \end{vmatrix} = 3\mathbf{i} - 5\mathbf{j} + 26\mathbf{k};$

$\text{comp}_{\mathbf{a}}(\mathbf{b} \times \mathbf{c}) = (\mathbf{b} \times \mathbf{c}) \cdot \mathbf{u_a} = (3\mathbf{i} - 5\mathbf{j} + 26\mathbf{k}) \cdot \dfrac{1}{\sqrt{14}}(3\mathbf{i} + 2\mathbf{j} - \mathbf{k}) = -\dfrac{27}{\sqrt{14}}$

23. $V = |(\mathbf{a} \times \mathbf{b}) \cdot \mathbf{c}|; \quad (\mathbf{a} \times \mathbf{b}) \cdot \mathbf{c} = \begin{vmatrix} 3 & 2 & -1 \\ 5 & 3 & 0 \\ -2 & 4 & 1 \end{vmatrix} = -27; \quad V = |-27| = 27$

25. (a) Direction vector: $\overrightarrow{QR} = (6, -3, 3);$ scalar parametric equations for the line:

$$x = 1 + 6t, \quad y = 1 - 3t, \quad z = 1 + 3t.$$

(b) Normal vector: $\overrightarrow{PR} = (3, -3, 2);$ equation of the plane:

$$3(x - 1) + (-3)(y - 1) + 2(z - 1) = 0.$$

(c) A normal vector for the plane is: $\overrightarrow{QR} \times \overrightarrow{PR} = (3, -3, -9)$ or $\mathbf{N} = (1, -1, -3);$

an equation for the plane: $(x - 1) - (y - 1) - 3(z - 1) = 0$

27. Solve, if possible, the system of equations: $t = 1 - u, \quad -t = 1 + 3u, \quad -6 + 2t = 2u.$ In this case, the solution is $t = 2, u = -1.$ The lines intersect at the point $(2, -2, -2).$

29. The lines l_1 and l_2 written in scalar parametric form are:

$$l_1 : x = 1 + 2t, \quad y = -2 - t, \quad z = 3 + 4t; \quad l_2 : x = -2 + u, \quad y = 3 + 3u, \quad z = u.$$

Solve, if possible, the system of equations: $1 + 2t = -2 + u, \quad -2 - t = 3 + 3u, \quad 3 + 4t = u.$ In this case there is no solution; the lines are skew.

31. (a) No. $\overrightarrow{PQ} = (4, -7, 5), \quad \overrightarrow{PR} = (2, -3, 2);$ the vectors are not parallel; the points are not collinear.

(b) $\overrightarrow{PQ} = (4, -7, 5), \quad \overrightarrow{PR} = (2, -3, 2), \quad \overrightarrow{PS} = (-2, 0, 1)$

$$(\overrightarrow{PQ} \times \overrightarrow{PR}) \cdot \overrightarrow{PS} = \begin{vmatrix} 4 & -7 & 5 \\ 2 & -3 & 2 \\ -2 & 0 & 1 \end{vmatrix} = 0.$$

The points are coplanar.

33. $\overrightarrow{PQ} \times \overrightarrow{PR} = -10\mathbf{i} + 5\mathbf{k}$ is a normal vector for the plane; so is $\mathbf{N} = 2\mathbf{i} - \mathbf{k}.$

An equation for the plane is: $2(x - 1) - (z - 1) = 0$ or $2x - z = 1$

35. $\mathbf{N} = 3\mathbf{i} + 2\mathbf{j} - 1\mathbf{k}$ is a normal vector for the plane. An equation for the plane is:

$$3(x - 1) + 2(y + 2) - (z + 1) = 0 \quad \text{or} \quad 3x + 2y - z = 0.$$

37. Let P be the plane that satisfies the conditions. A direction vector for the given line is $\mathbf{d} = (3, 2, 4)$; a normal vector for the given plane is $\mathbf{N} = (2, 1, -3)$. The cross product $\mathbf{d} \times \mathbf{N}$ is a normal vector for P.

$$\mathbf{d} \times \mathbf{N} = \begin{vmatrix} \mathbf{i} & \mathbf{j} & \mathbf{k} \\ 3 & 2 & 4 \\ 2 & 1 & -3 \end{vmatrix} = -10\,\mathbf{i} + 17\,\mathbf{j} - \mathbf{k}.$$

The point $Q(-1, 1, 2)$ is on the plane. An equation for P is:

$$-10(x + 1) + 17(y - 1) - (z - 2) = 0 \quad \text{or} \quad 10x - 17y + z + 25 = 0.$$

39. The line l which passes through Q and R has direction vector $\mathbf{d} = \overrightarrow{QR} = (2, 1, -2)$. By (13.5.6), the distance from P to l is given by

$$d(P, l) = \frac{\|\overrightarrow{QP} \times \mathbf{d}\|}{\|\mathbf{d}\|} = \frac{\|(2, 4, -5) \times (2, 1, -2)\|}{3} = \frac{9}{3} = 3.$$

41. The normals are: $\mathbf{N}_1 = (2, 1, 1)$, $\mathbf{N}_2 = (2, 2, -1)$. The cosine of the angle between the planes is:

$$\cos\theta = \frac{|\mathbf{N}_1 \cdot \mathbf{N}_2|}{\|\mathbf{N}_1\|\|\mathbf{N}_2\|} = \frac{5}{\sqrt{54}} \quad \text{and} \quad \theta \cong 0.822 \text{ radians.}$$

43. The normal vectors to the two planes are: $\mathbf{N}_1 = 3\,\mathbf{i} + 5\,\mathbf{j} + 2\,\mathbf{k}$, $\mathbf{N}_2 = \mathbf{i} + 2\,\mathbf{j} - \mathbf{k}$. A direction vector for the line of intersection is:

$$\mathbf{N}_1 \times \mathbf{N}_2 = \begin{vmatrix} \mathbf{i} & \mathbf{j} & \mathbf{k} \\ 3 & 5 & 2 \\ 1 & 2 & -1 \end{vmatrix} = -9\,\mathbf{i} + 5\,\mathbf{j} + \mathbf{k}.$$

A solution of the pair of equations $3x + 5y + 2z - 4 = 0$ $\quad x + 2y - z - 2 = 0$ is $x = -2$, $y = 2$, $z = 0$ (set $z = 0$ and solve for x and y). Scalar parametric equations for the line of intersection are:

$$x = -2 - 9t, \quad y = 2 + 5t, \quad z = t.$$

45. $\mathbf{a} \times \mathbf{b} = -5\,\mathbf{i} + 11\,\mathbf{j} + 7\,\mathbf{k}$ is perpendicular to both $\mathbf{a}$ and $\mathbf{b}$; $\quad \|\mathbf{a} \times \mathbf{b}\| = \sqrt{195}$. The vectors are:

$$\pm \frac{4}{\sqrt{195}}(-5\,\mathbf{i} + 11\,\mathbf{j} + 7\,\mathbf{k}).$$

47. $(\|\mathbf{b}\|\mathbf{a} - \|\mathbf{a}\|\mathbf{b}) \cdot (\|\mathbf{b}\|\mathbf{a} + \|\mathbf{a}\|\mathbf{b}) = \|\mathbf{a}\|^2\|\mathbf{b}\|^2 + \|\mathbf{a}\|\|\mathbf{b}\|\mathbf{a} \cdot \mathbf{b} - \|\mathbf{a}\|\|\mathbf{b}\|\mathbf{a} \cdot \mathbf{b} - \|\mathbf{a}\|^2\|\mathbf{b}\|^2 = 0.$
Therefore, $\quad (\|\mathbf{b}\|\mathbf{a} - \|\mathbf{a}\|\mathbf{b}) \perp (\|\mathbf{b}\|\mathbf{a} + \|\mathbf{a}\|\mathbf{b})$

49. Let $\mathbf{a}$ and $\mathbf{b}$ be adjacent sides of a parallelogram. Then the diagonals of the parallelogram are $\mathbf{a} + \mathbf{b}$ and $\mathbf{a} - \mathbf{b}$. By Exercise 48, the diagonals have equal length iff $\mathbf{a} \perp \mathbf{b}$, which means that the parallelogram is a rectangle.

51. Let A, B, C be the vertices of a triangle. Without loss of generality, assume that $A(0,0)$, $B(x_1, y_1)$, $C(x_2, 0)$. Let D and E be the midpoints of $\overline{AB}$ and $\overline{BC}$, respectively. Then $D\left(\dfrac{x_1}{2}, \dfrac{y_1}{2}\right)$ and $E\left(\dfrac{x_1 + x_2}{2}, \dfrac{y_1}{2}\right)$. Now

$$\overrightarrow{DE} = \left(\frac{x_2}{2}, 0\right), \quad \text{and} \quad \overrightarrow{AC} = (x_2, 0).$$

Therefore $\overrightarrow{DE} \| \overrightarrow{AC}$ and $\| \overrightarrow{DE} \| = \frac{1}{2} \| \overrightarrow{AC} \|$.

CHAPTER 14

SECTION 14.1

1. $\mathbf{f}'(t) = 2\mathbf{i} - \mathbf{j} + 3\mathbf{k}$

3. $\mathbf{f}'(t) = -\dfrac{1}{2\sqrt{1-t}}\,\mathbf{i}\,\dfrac{1}{2\sqrt{1+t}}\,\mathbf{j} + \dfrac{1}{(1-t)^2}\,\mathbf{k}$

5. $\mathbf{f}'(t) = \cos t\,\mathbf{i} - \sin t\,\mathbf{j} + \sec^2 t\,\mathbf{k}$

7. $\mathbf{f}'(t) = \dfrac{-1}{1-t}\,\mathbf{i} - \sin t\,\mathbf{j} + 2t\,\mathbf{k}$

9. $\mathbf{f}'(t) = 4\mathbf{i} + 6t^2\mathbf{j} + (2t+2)\,\mathbf{k};$ $\mathbf{f}''(t) = 12t\mathbf{j} + 2\mathbf{k}$

11. $\mathbf{f}'(t) = -2\sin 2t\,\mathbf{i} + 2\cos 2t\,\mathbf{j} + 4t\,\mathbf{k};$ $\mathbf{f}''(t) = -4\cos 2t\,\mathbf{i} - 4\sin 2t\,\mathbf{j}$

13. (a) $\mathbf{r}'(t) = -2te^{-t^2}\,\mathbf{i} - e^{-t}\,\mathbf{j};$ $\mathbf{r}'(0) = -\mathbf{j}$

    (b) $\mathbf{r}'(t) = \cot t\,\mathbf{i} - \tan t\,\mathbf{j} + (2\cos t + 3\sin t)\,\mathbf{k};$ $\mathbf{r}'(\pi/4) = \mathbf{i} - \mathbf{j} + \dfrac{5}{\sqrt{2}}\,\mathbf{k}$

15. $\displaystyle\int_1^2 (\mathbf{i} + 2t\,\mathbf{j})\,dt = \big[t\mathbf{i} + t^2\mathbf{j}\big]_1^2 = \mathbf{i} + 3\mathbf{j}$

17. $\displaystyle\int_0^1 \left(e^t\,\mathbf{i} + e^{-t}\,\mathbf{k}\right)dt = \big[e^t\,\mathbf{i} - e^{-t}\,\mathbf{k}\big]_0^1 = (e-1)\mathbf{i} + \left(1 - \dfrac{1}{e}\right)\mathbf{k}$

19. $\displaystyle\int_0^1 \left(\dfrac{1}{1+t^2}\,\mathbf{i} + \sec^2 t\,\mathbf{j}\right)dt = \big[\tan^{-1} t\,\mathbf{i} + \tan t\,\mathbf{j}\big]_0^1 = \dfrac{\pi}{4}\,\mathbf{i} + \tan(1)\,\mathbf{j}$

21. $\displaystyle\lim_{t\to 0} \mathbf{f}(t) = \left(\lim_{t\to 0} \dfrac{\sin t}{2t}\right)\mathbf{i} + \left(\lim_{t\to 0} e^{2t}\right)\mathbf{j} + \left(\lim_{t\to 0} \dfrac{t^2}{e^t}\right)\mathbf{k} = \dfrac{1}{2}\mathbf{i} + \mathbf{j}$

23. $\displaystyle\lim_{t\to 0} \mathbf{f}(t) = \left(\lim_{t\to 0} t^2\right)\mathbf{i} + \left(\lim_{t\to 0} \dfrac{1-\cos t}{3t}\right)\mathbf{j} + \left(\lim_{t\to 0} \dfrac{t}{t+1}\right)\mathbf{k} = 0\mathbf{i} + \dfrac{1}{3}\left(\lim_{t\to 0} \dfrac{1-\cos t}{t}\right)\mathbf{j} + 0\mathbf{k} = \mathbf{0}$

25. (a) $\displaystyle\int_0^1 \left(te^t\,\mathbf{i} + te^{t^2}\,\mathbf{j}\right)dt = \mathbf{i} + \dfrac{e-1}{2}\,\mathbf{j}$

    (b) $\displaystyle\int_3^8 \left(\dfrac{t}{t+1}\,\mathbf{i} + \dfrac{t}{(t+1)^2}\,\mathbf{j} + \dfrac{t}{(t+1)^3}\,\mathbf{k}\right)dt = \left[5 + \ln\left(\tfrac{4}{9}\right)\right]\mathbf{i} + \left[-\tfrac{5}{36} + \ln\left(\tfrac{4}{9}\right)\right]\mathbf{j} + \tfrac{295}{2592}\,\mathbf{k}$

27.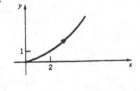

29.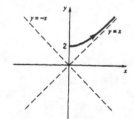

31.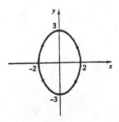

33. 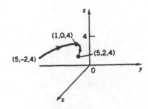   **35.**    **37.**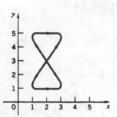

39. (a) $\mathbf{f}(t) = 3\cos t\,\mathbf{i} + 2\sin t\,\mathbf{j}$ (b) $\mathbf{f}(t) = 3\cos t\,\mathbf{i} - 2\sin t\,\mathbf{j}$

41. (a) $\mathbf{f}(t) = t\,\mathbf{i} + t^2\,\mathbf{j}$ (b) $\mathbf{f}(t) = -t\,\mathbf{i} + t^2\,\mathbf{j}$

43. $\mathbf{f}(t) = (1 + 2t)\,\mathbf{i} + (4 + 5t)\,\mathbf{j} + (-2 + 8t)\,\mathbf{k}, \quad 0 \le t \le 1$

45. $\mathbf{f}'(t_0) = \mathbf{i} + m\,\mathbf{j}$,

$$\int_a^b \mathbf{f}(t)\,dt = \left[\frac{1}{2}t^2\,\mathbf{i}\right]_a^b + \left[\int_a^b f(t)\,dt\right]\mathbf{j} = \frac{1}{2}\left(b^2 - a^2\right)\mathbf{i} + A\,\mathbf{j},$$

$$\int_a^b \mathbf{f}'(t)\,dt = [t\,\mathbf{i} + f(t)\,\mathbf{j}]_a^b = (b - a)\,\mathbf{i} + (d - c)\,\mathbf{j}$$

47.
$$\mathbf{f}'(t) = \mathbf{i} + t^2\,\mathbf{j}$$
$$\mathbf{f}(t) = (t + C_1)\,\mathbf{i} + \left(\tfrac{1}{3}t^3 + C_2\right)\mathbf{j} + C_3\,\mathbf{k}$$
$$\mathbf{f}(0) = \mathbf{j} - \mathbf{k} \implies C_1 = 0, \quad C_2 = 1, \quad C_3 = -1$$
$$\mathbf{f}(t) = t\,\mathbf{i} + \left(\tfrac{1}{3}t^3 + 1\right)\mathbf{j} - \mathbf{k}$$

49. $\mathbf{f}'(t) = \alpha\,\mathbf{f}(t) \implies \mathbf{f}(t) = e^{\alpha t}\,\mathbf{f}(0) = e^{\alpha t}\,\mathbf{c}$

51. (a) If $\mathbf{f}'(t) = \mathbf{0}$ on an interval, then the derivative of each component is 0 on that interval, each component is constant on that interval, and therefore $\mathbf{f}$ itself is constant on that interval.

(b) Set $\mathbf{h}(t) = \mathbf{f}(t) - \mathbf{g}(t)$ and apply part (a).

53. If $\mathbf{f}$ is differentiable at t, then each component is differentiable at t, each component is continuous at t, and therefore $\mathbf{f}$ is continuous at t.

55. no; as a counter-example, set $\mathbf{f}(t) = \mathbf{i} = \mathbf{g}(t)$.

57. Suppose $\mathbf{f}(t) = f_1(t)\,\mathbf{i} + f_2(t)\,\mathbf{j} + f_3(t)\,\mathbf{k}$. Then $\|\mathbf{f}(t)\| = \sqrt{f_1^2(t) + f_2^2(t) + f_3^2(t)}$ and

$$\frac{d}{dt}\left(\|\mathbf{f}\|\right) = \frac{1}{2}\left[f_1^2 + f_2^2 + f_3^2\right]^{-1/2}\left(2f_1 \cdot f_1' + 2f_2 \cdot f_2' + 2f_3 \cdot f_3'\right) = \frac{\mathbf{f}(t) \cdot \mathbf{f}'(t)}{\|\mathbf{f}(t)\|}$$

The Answer Section of the text gives an alternative approach.

SECTION 14.2

1. $\mathbf{f}'t = \mathbf{b}, \quad \mathbf{f}''(t) = \mathbf{0}$

3. $\mathbf{f}'(t) = 2e^{2t}\mathbf{i} - \cos t\mathbf{j}, \quad \mathbf{f}''(t) = 4e^{2t}\mathbf{i} + \sin t\mathbf{j}$

5. $\mathbf{f}'t = [(t^2\mathbf{i} - 2t\mathbf{j}) \cdot (\mathbf{i} + 3t^2\mathbf{j}) + (2t\mathbf{i} - 2\mathbf{j}) \cdot (t\mathbf{i} + t^3\mathbf{j})]\mathbf{j} = [3t^2 - 8t^3]\mathbf{j}$
   $\mathbf{f}''(t) = (6t - 24t^2)\mathbf{j}$

7.
$$\mathbf{f}'(t) = \left[(e^t\mathbf{i} + t\mathbf{k}) \times \frac{d}{dt}(t\mathbf{j} + e^{-t}\mathbf{k}) \right] + \left[\frac{d}{dt}(e^t\mathbf{i} + t\mathbf{k}) \times (t\mathbf{j} + e^{-t}\mathbf{k}) \right]$$
$$= [(e^t\mathbf{i} + t\mathbf{k}) \times (\mathbf{j} - e^{-t}\mathbf{k})] + [(e^t\mathbf{i} + \mathbf{k}) \times (t\mathbf{j} + e^{-t}\mathbf{k})]$$
$$= (-t\mathbf{i} + \mathbf{j} + e^t\mathbf{k}) + (-t\mathbf{i} - \mathbf{j} + te^t\mathbf{k})$$
$$= -2t\mathbf{i} + e^t(t+1)\mathbf{k}$$
$$\mathbf{f}''(t) = -2\mathbf{i} + e^t(t+2)\mathbf{k}$$

9. $\mathbf{f}'(t) = (\mathbf{a} \times t\mathbf{b}) \times 2t\mathbf{b} + (\mathbf{a} \times \mathbf{b}) \times (\mathbf{a} + t^2\mathbf{b}), \quad \mathbf{f}''(t) = (\mathbf{a} \times t\mathbf{b}) \times 2\mathbf{b} + 4t(\mathbf{a} \times \mathbf{b}) \times \mathbf{b}$

11. $\mathbf{f}'(t) = \frac{1}{2}\sqrt{t}\,\mathbf{g}'\left(\sqrt{t}\right) + \mathbf{g}\left(\sqrt{t}\right), \quad \mathbf{f}''(t) = \frac{1}{4}\mathbf{g}''\left(\sqrt{t}\right) + \frac{3}{4\sqrt{t}}\mathbf{g}'\left(\sqrt{t}\right)$

13. $-(\sin t)\,e^{\cos t}\,\mathbf{i} + (\cos t)\,e^{\sin t}\,\mathbf{j}$

15. $(e^t\mathbf{i} + e^{-t}\mathbf{j}) \cdot (e^t\mathbf{i} - e^{-t}\mathbf{j}) = e^{2t} - e^{-2t}; \quad$ therefore
$$\frac{d^2}{dt^2}\left[(e^t\mathbf{i} + e^{-t}\mathbf{j}) \cdot (e^t\mathbf{i} - e^{-t}\mathbf{j}) \right] = \frac{d^2}{dt^2}\left[e^{2t} - e^{-2t} \right] = \frac{d}{dt}\left[2e^{2t} + 2e^{-2t} \right] = 4e^{2t} - 4e^{-2t}$$

17. $\frac{d}{dt}\left[(a + t\mathbf{b}) \times (c + t\mathbf{d}) \right] = [(a + t\mathbf{b}) \times \mathbf{d}] + [\mathbf{b} \times (c + t\mathbf{d})] = (a \times \mathbf{d}) + (\mathbf{b} \times c) + 2t(\mathbf{b} \times \mathbf{d})$

19. $\frac{d}{dt}\left[(a + t\mathbf{b}) \cdot (c + t\mathbf{d}) \right] = [(a + t\mathbf{b}) \cdot \mathbf{d}] + [\mathbf{b} \cdot (c + t\mathbf{d})] = (a \cdot \mathbf{d}) + (\mathbf{b} \cdot c) + 2t(\mathbf{b} \cdot \mathbf{d})$

21. $\mathbf{r}(t) = \mathbf{a} + t\mathbf{b}$

23. $\mathbf{r}(t) = \frac{1}{2}t^2\mathbf{a} + \frac{1}{6}t^3\mathbf{b} + t\mathbf{c} + \mathbf{d}$ 24. $\mathbf{r}(t) = \left(1 + 2t - \frac{1}{4}\cos 2t\right)\mathbf{i} + \left(1 - \frac{1}{4}\sin 2t\right)\mathbf{j}$

25. $\mathbf{r}(t) = \sin t\,\mathbf{i} + \cos t\,\mathbf{j}, \quad \mathbf{r}'(t) = \cos t\,\mathbf{i} - \sin t\,\mathbf{j}, \quad \mathbf{r}''(t) = -\sin t\,\mathbf{i} - \cos t\,\mathbf{j} = -\mathbf{r}(t)$.

   Thus $\mathbf{r}(t)$ and $\mathbf{r}''(t)$ are parallel, and they always point in opposite directions.

27.
$$\mathbf{r}(t) \cdot \mathbf{r}'(t) = (\cos t\,\mathbf{i} + \sin t\,\mathbf{j}) \cdot (-\sin t\,\mathbf{i} + \cos t\,\mathbf{j}) = 0$$
$$\mathbf{r}(t) \times \mathbf{r}'(t) = (\cos t\,\mathbf{i} + \sin t\,\mathbf{j}) \times (-\sin t\,\mathbf{i} + \cos t\,\mathbf{j})$$
$$= \cos^2 t\,\mathbf{k} + \sin^2 t\,\mathbf{k} = (\cos^2 t + \sin^2 t)\,\mathbf{k} = \mathbf{k}$$

29. $\dfrac{d}{dt}\left[\mathbf{f}\left(t\right)\times\mathbf{f}'\left(t\right)\right]=\left[\mathbf{f}\left(t\right)\times\mathbf{f}''\left(t\right)\right]+\underbrace{\left[\mathbf{f}'\left(t\right)\times\mathbf{f}'\left(t\right)\right]}_{0}=\mathbf{f}\left(t\right)\times\mathbf{f}''\left(t\right).$

31. $\left[\mathbf{f}\cdot\mathbf{g}\times\mathbf{h}\right]'=\mathbf{f}'\cdot\left(\mathbf{g}\times\mathbf{h}\right)+\mathbf{f}\cdot\left(\mathbf{g}\times\mathbf{h}\right)'=\mathbf{f}'\cdot\left(\mathbf{g}\times\mathbf{h}\right)+\mathbf{f}\cdot\left[\mathbf{g}'\times\mathbf{h}+\mathbf{g}\times\mathbf{h}'\right]$
and the result follows.

33. $\qquad\|\mathbf{r}\left(t\right)\|$ is constant $\quad\Longleftrightarrow\quad\|\mathbf{r}\left(t\right)\|^2=\mathbf{r}\left(t\right)\cdot\mathbf{r}\left(t\right)$ is constant

$\qquad\qquad\qquad\Longleftrightarrow\quad\dfrac{d}{dt}\left[\mathbf{r}\left(t\right)\cdot\mathbf{r}\left(t\right)\right]=2\left[\mathbf{r}\left(t\right)\cdot\mathbf{r}'\left(t\right)\right]=0$ identically

$\qquad\qquad\qquad\Longleftrightarrow\quad\mathbf{r}\left(t\right)\cdot\mathbf{r}'\left(t\right)=0$ identically

35. Write

$$\frac{\left[\mathbf{f}\left(t+h\right)\times\mathbf{g}\left(t+h\right)\right]-\left[\mathbf{f}\left(t\right)\times\mathbf{g}\left(t\right)\right]}{h}$$

as

$$\left(\mathbf{f}\left(t+h\right)\times\left[\frac{\mathbf{g}\left(t+h\right)-\mathbf{g}\left(t\right)}{h}\right]\right)+\left(\left[\frac{\mathbf{f}\left(t+h\right)-\mathbf{f}\left(t\right)}{h}\right]\times\mathbf{g}\left(t\right)\right)$$

and take the limit as $h\to 0$. (Appeal to Theorem 13.1.3.)

SECTION 14.3

1. $\mathbf{r}'\left(t\right)=-\pi\sin\pi t\,\mathbf{i}+\pi\cos\pi t\,\mathbf{j}+\mathbf{k},\quad\mathbf{r}'(2)=\pi\,\mathbf{j}+\mathbf{k}$
$\mathbf{R}\left(u\right)=\left(\mathbf{i}+2\,\mathbf{k}\right)+u(\pi\,\mathbf{j}+\mathbf{k})$

3. $\mathbf{r}'(t)=\mathbf{b}+2t\,\mathbf{c},\quad\mathbf{r}'(-1)=\mathbf{b}-2\,\mathbf{c},\quad\mathbf{R}\left(u\right)=\left(\mathbf{a}-\mathbf{b}+\mathbf{c}\right)+u(\mathbf{b}-2\,\mathbf{c})$

5. $\mathbf{r}'(t)=4t\,\mathbf{i}-\mathbf{j}+4t\,\mathbf{k},\quad P$ is tip of $\mathbf{r}\left(1\right),\quad\mathbf{r}'(1)=4\,\mathbf{i}-\mathbf{j}+4\,\mathbf{k}$
$\mathbf{R}\left(u\right)=\left(2\,\mathbf{i}+5\,\mathbf{k}\right)+u\left(4\,\mathbf{i}-\mathbf{j}+4\,\mathbf{k}\right)$

7. $\mathbf{r}'(t)=-2\,\sin t\,\mathbf{i}+3\,\cos t\,\mathbf{j}+\mathbf{k},\quad\mathbf{r}'(\pi/4)=-\sqrt{2}\,\mathbf{i}+\frac{3}{2}\,\sqrt{2}\,\mathbf{j}+\mathbf{k}$
$\mathbf{R}\left(u\right)=\left(\sqrt{2}\,\mathbf{i}+\frac{3}{2}\,\sqrt{2}\,\mathbf{j}+\frac{\pi}{4}\,\mathbf{k}\right)+u\left(-\sqrt{2}\,\mathbf{i}+\frac{3}{2}\,\sqrt{2}\,\mathbf{j}+\mathbf{k}\right)$

9. The scalar components $x(t)=at$ and $y(t)=bt^2$ satisfy the equation
$$a^2y(t)=a^2(bt^2)=b\left(a^2t^2\right)=b\left[x(t)\right]^2$$
and generate the parabola $\quad a^2y=bx^2$.

11. $\mathbf{r}\left(t\right)=t\,\mathbf{i}+\left(1+t^2\right)\mathbf{j},\qquad\mathbf{r}'(t)=\mathbf{i}+2t\,\mathbf{j}$

(a) $\mathbf{r}\left(t\right)\perp\mathbf{r}'(t)\quad\Longrightarrow\quad\mathbf{r}\left(t\right)\cdot\mathbf{r}'(t)=\left[t\,\mathbf{i}+\left(1+t^2\right)\mathbf{j}\right]\cdot\left(\mathbf{i}+2t\,\mathbf{j}\right)$

$\qquad\qquad\qquad\qquad\qquad\qquad=t(2t^2+3)=0\quad\Longrightarrow\quad t=0$

$\mathbf{r}\left(t\right)$ and $\mathbf{r}'(t)$ are perpendicular at $(0,1)$.

(b) and (c) $\mathbf{r}(t) = \alpha \mathbf{r}'(t)$ with $\alpha \neq 0$ $\implies$ $t = \alpha$ and $1 + t^2 = 2t\alpha$ $\implies$ $t = \pm 1$.

If $\alpha > 0$, then $t = 1$. $\mathbf{r}(t)$ and $\mathbf{r}'(t)$ have the same direction at $(1, 2)$.

If $\alpha < 0$, then $t = -1$. $\mathbf{r}(t)$ and $\mathbf{r}'(t)$ have opposite directions at $(-1, 2)$.

13. The tangent line at $t = t_0$ has the form $\mathbf{R}(u) = \mathbf{r}(t_0) + u\mathbf{r}'(t_0)$. If $\mathbf{r}'(t_0) = \alpha \mathbf{r}(t_0)$, then

$$\mathbf{R}(u) = \mathbf{r}(t_0) + u\,\alpha\,\mathbf{r}(t_0) = (1 + u\alpha)\,\mathbf{r}(t_0).$$

The tangent line passes through the origin at $u = -1/\alpha$.

15. $\mathbf{r}_1(t)$ passes through $P(0,0,0)$ at $t = 0$; $\quad \mathbf{r}_2(u)$ passes through $P(0,0,0)$ at $u = -1$.

$$\mathbf{r}_1'(t) = e^t\,\mathbf{i} + 2\cos t\,\mathbf{j} + \frac{1}{t+1}\,\mathbf{k}; \quad \mathbf{r}_1'(0) = \mathbf{i} + 2\,\mathbf{j} + \mathbf{k}$$

$$\mathbf{r}_2'(u) = \mathbf{i} + 2u\,\mathbf{j} + 3u^2\,\mathbf{k}; \quad \mathbf{r}_2'(-1) = \mathbf{i} - 2\,\mathbf{j} + 3\,\mathbf{k}$$

$$\cos\theta = \frac{\mathbf{r}_1'(0) \cdot \mathbf{r}_2'(1)}{\|\mathbf{r}_1'(0)\|\,\|\mathbf{r}_2'(1)\|} = 0; \quad \theta = \frac{\pi}{2} \cong 1.57, \text{ or } 90°.$$

17. $\mathbf{r}_1(t) = \mathbf{r}_2(u)$ implies

$$\left.\begin{array}{r} e^t = u \\ 2\sin\left(t + \tfrac{1}{2}\pi\right) = 2 \\ t^2 - 2 = u^2 - 3 \end{array}\right\} \quad \text{so that} \quad t = 0, \quad u = 1.$$

The point of intersection is $(1, 2, -2)$.

$$\mathbf{r}_1'(t) = e^t\,\mathbf{i} + 2\cos\left(t + \frac{\pi}{2}\right)\mathbf{j} + 2t\,\mathbf{k}, \quad \mathbf{r}_1'(0) = \mathbf{i}$$

$$\mathbf{r}_2'(u) = \mathbf{i} + 2u\,\mathbf{k}, \quad \mathbf{r}_2'(1) = \mathbf{i} + 2\,\mathbf{k}$$

$$\cos\theta = \frac{\mathbf{r}_1'(0) \cdot \mathbf{r}_2'(1)}{\|\mathbf{r}_1'(0)\|\,\|\mathbf{r}_2'(1)\|} = \frac{1}{5}\sqrt{5} \cong 0.447, \quad \theta \cong 1.11 \text{ radians}$$

19. (a) $\mathbf{r}(t) = a\cos t\,\mathbf{i} + b\sin t\,\mathbf{j}$ $\qquad\qquad$ (b) $\mathbf{r}(t) = a\cos t\,\mathbf{i} - b\sin t\,\mathbf{j}$

(c) $\mathbf{r}(t) = a\cos 2t\,\mathbf{i} + b\sin 2t\,\mathbf{j}$ $\qquad\qquad$ (d) $\mathbf{r}(t) = a\cos 3t\,\mathbf{i} - b\sin 3t\,\mathbf{j}$

21. $\mathbf{r}'(t) = t^3\,\mathbf{i} + 2t\,\mathbf{j}$ $\qquad$ **23.** $\mathbf{r}'(t) = 2e^{2t}\,\mathbf{i} - 4e^{-4t}\,\mathbf{j}$ $\qquad$ **25.** $\mathbf{r}'(t) = -2\sin t\,\mathbf{i} + 3\cos t\,\mathbf{j}$

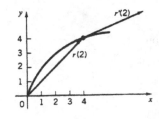 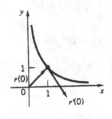 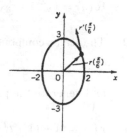

27. $\mathbf{r}(t) = (t^2 + 1)\,\mathbf{i} + t\,\mathbf{j}, \quad t \geq 1;$ or, $\mathbf{r}(t) = \sec^2 t\,\mathbf{i} + \tan t\,\mathbf{j}, \quad t \in \left[\tfrac{1}{4}\pi, \tfrac{1}{2}\pi\right)$

29. $\mathbf{r}(t) = \cos t \sin 3t\,\mathbf{i} + \sin t \sin 3t\,\mathbf{j}, \quad t \in [0, \pi]$

31. $y^3 = x^2$

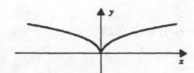

There is no tangent vector at the origin.

33. We substitute $x = t$, $y = t^2$, $z = t^3$ in the plane equation to obtain

$$4t + 2t^2 + t^3 = 24, \quad (t-2)\left(t^2 + 4t + 12\right) = 0, \quad t = 2.$$

The twisted cubic intersects the plane at the tip of $\mathbf{r}(2)$, the point $(2, 4, 8)$.

The angle between the curve and the normal line at the point of intersection is the angle between the tangent vector $\mathbf{r}'(2) = \mathbf{i} + 4\mathbf{j} + 12\mathbf{k}$ and the normal $\mathbf{N} = 4\mathbf{i} + 2\mathbf{j} + \mathbf{k}$:

$$\cos\theta = \frac{(\mathbf{i} + 4\mathbf{j} + 12\mathbf{k}) \cdot (4\mathbf{i} + 2\mathbf{j} + \mathbf{k})}{\|\mathbf{i} + 4\mathbf{j} + 12\mathbf{k}\| \, \|4\mathbf{i} + 2\mathbf{j} + \mathbf{k}\|} = \frac{24}{\sqrt{161}\,\sqrt{21}} \cong 0.412, \quad \theta \cong 1.15 \text{ radians.}$$

35. $\mathbf{r}'(t) = 2\mathbf{j} + 2t\mathbf{k}, \quad \|\mathbf{r}'(t)\| = 2\sqrt{1 + t^2}$

$$\mathbf{T}(t) = \frac{\mathbf{r}'(t)}{\|\mathbf{r}'(t)\|} = \frac{1}{\sqrt{1 + t^2}}\,(\mathbf{j} + t\mathbf{k}),$$

$$\mathbf{T}'(t) = \frac{1}{(1 + t^2)^{3/2}}\,[-t\mathbf{j} + \mathbf{k}]$$

at $t = 1$: tip of $\mathbf{r} = (1, 2, 1)$, $\mathbf{T} = \mathbf{T}(1) = \dfrac{1}{\sqrt{2}}\mathbf{j} + \dfrac{1}{\sqrt{2}}\mathbf{k}$;

$$\mathbf{T}'(1) = -\frac{1}{2\sqrt{2}}\mathbf{j} + \frac{1}{2\sqrt{2}}\mathbf{k}; \quad \|\mathbf{T}'(1)\| = \frac{1}{2}; \quad \mathbf{N} = \mathbf{N}(1) = \frac{\mathbf{T}'(1)}{\|\mathbf{T}'(1)\|} = -\frac{1}{\sqrt{2}}\mathbf{j} + \frac{1}{\sqrt{2}}\mathbf{k}$$

normal for osculating plane:

$$\mathbf{T} \times \mathbf{N} = \left(\frac{1}{\sqrt{2}}\mathbf{j} + \frac{1}{\sqrt{2}}\mathbf{k}\right) \times \left(-\frac{1}{\sqrt{2}}\mathbf{j} + \frac{1}{\sqrt{2}}\mathbf{k}\right) = \frac{1}{2}\mathbf{i}$$

equation for osculating plane:

$$\frac{1}{2}(x - 1) + 0(y - 2) + 0(z - 1) = 0, \quad \text{which gives} \quad x - 1 = 0$$

37.
$$\mathbf{r}'(t) = -2\sin 2t\,\mathbf{i} + 2\cos 2t\,\mathbf{j} + \mathbf{k}, \quad \|\mathbf{r}'(t)\| = \sqrt{5}$$

$$\mathbf{T}(t) = \frac{\mathbf{r}'(t)}{\|\mathbf{r}'(t)\|} = \frac{1}{5}\sqrt{5}\,(-2\sin 2t\,\mathbf{i} + 2\cos 2t\,\mathbf{j} + \mathbf{k})$$

$$\mathbf{T}'(t) = -\frac{4}{5}\sqrt{5}\,(\cos 2t\,\mathbf{i} + \sin 2t\,\mathbf{j}), \quad \|\mathbf{T}'(t)\| = \frac{4}{5}\sqrt{5}$$

$$\mathbf{N}(t) = \frac{\mathbf{T}'(t)}{\|\mathbf{T}'(t)\|} = -(\cos 2t\,\mathbf{i} + \sin 2t\,\mathbf{j})$$

at $t = \pi/4$:   tip of $\mathbf{r} = (0, 1, \pi/4)$, $\mathbf{T} = \dfrac{1}{5}\sqrt{5}\,(-2\,\mathbf{i} + \mathbf{k})$, $\mathbf{N} = -\mathbf{j}$

normal for osculating plane:
$$\mathbf{T} \times \mathbf{N} = \frac{1}{5}\sqrt{5}\,(-2\,\mathbf{i} + \mathbf{k}) \times (-\mathbf{j}) = \frac{1}{5}\sqrt{5}\,\mathbf{i} + \frac{2}{5}\sqrt{5}\,\mathbf{k}$$

equation for osculating plane:
$$\frac{1}{5}\sqrt{5}\,(x - 0) + \frac{2}{5}\sqrt{5}\left(z - \frac{\pi}{4}\right) = 0, \quad \text{which gives} \quad x + 2z = \frac{\pi}{2}$$

39.
$$\mathbf{r}'(t) = \cosh t\,\mathbf{i} + \sinh t\,\mathbf{j} + \mathbf{k}, \quad \|\mathbf{r}'(t)\| = \sqrt{\cosh^2 t + \sinh^2 t + 1} = \sqrt{2}\,\cosh t$$

$$\mathbf{T}(t) = \frac{\mathbf{r}'(t)}{\|\mathbf{r}'(t)\|} = \frac{1}{\sqrt{2}}\,(\mathbf{i} + \tanh t\,\mathbf{j} + \operatorname{sech} t\,\mathbf{k}),$$

$$\mathbf{T}'(t) = \frac{1}{\sqrt{2}}\left(\operatorname{sech}^2 t\,\mathbf{j} - \operatorname{sech} t\,\tanh t\,\mathbf{k}\right)$$

at $t = 0$:   tip of $\mathbf{r} = (0, 1, 0)$, $\mathbf{T} = \dfrac{1}{\sqrt{2}}\,(\mathbf{i} + \mathbf{k})$, $\mathbf{T}'(0) = \dfrac{1}{\sqrt{2}}\,\mathbf{j}$; $\mathbf{N} = \mathbf{j}$

normal for osculating plane:
$$\mathbf{T} \times \mathbf{N} = \left(\frac{1}{\sqrt{2}}\,(-\mathbf{i} + \mathbf{k})\right) \times \mathbf{j} = \frac{1}{\sqrt{2}}\,(-\mathbf{i} + \mathbf{k}) \quad \text{or} \quad \mathbf{i} - \mathbf{k}$$

equation for osculating plane: $x - z = 0$

41.
$$\mathbf{r}'(t) = e^t\,[(\sin t + \cos t)\,\mathbf{i} + (\cos t - \sin t)\,\mathbf{j} + \mathbf{k}], \quad \|\mathbf{r}'(t)\| = e^t\sqrt{3}$$

$$\mathbf{T}(t) = \frac{\mathbf{r}'(t)}{\|\mathbf{r}'(t)\|} = \frac{1}{\sqrt{3}}\,[(\sin t + \cos t)\,\mathbf{i} + (\cos t - \sin t)\,\mathbf{j} + \mathbf{k}],$$

$$\mathbf{T}'(t) = \frac{1}{\sqrt{3}}\,[(\cos t - \sin t)\,\mathbf{i} - (\sin t + \cos t)\,\mathbf{j}]$$

at $t = 0$:   tip of $\mathbf{r} = (0, 1, 1)$, $\mathbf{T} = \mathbf{T}(0) = \dfrac{1}{\sqrt{3}}\,(\mathbf{i} + \mathbf{j} + \mathbf{k})$;

$\mathbf{T}'(0) = \dfrac{1}{\sqrt{3}}\,(\mathbf{i} - \mathbf{j})$; $\|\mathbf{T}'(0)\| = \dfrac{\sqrt{2}}{\sqrt{3}}$; $\mathbf{N} = \mathbf{N}(0) = \dfrac{\mathbf{T}'(0)}{\|\mathbf{T}'(0)\|} = \dfrac{1}{\sqrt{2}}\,(\mathbf{i} - \mathbf{j})$

normal for osculating plane:
$$\mathbf{T} \times \mathbf{N} = \frac{1}{\sqrt{3}}(\mathbf{i} + \mathbf{j} + \mathbf{k}) \times \frac{1}{\sqrt{2}}(\mathbf{i} - \mathbf{j}) = \frac{1}{\sqrt{6}}(\mathbf{i} + \mathbf{j} - 2\,\mathbf{k})$$

equation for osculating plane:
$$\frac{1}{\sqrt{6}}(x - 0) + \frac{1}{\sqrt{6}}(y - 1) - \frac{2}{\sqrt{6}}(z - 1) = 0, \quad \text{or} \quad x + y - 2z + 1 = 0$$

43. $\mathbf{T}_1 = \dfrac{\mathbf{R}'(u)}{\|\mathbf{R}'(u)\|} = -\dfrac{\mathbf{r}'(a + b - u)}{\|\mathbf{r}'(a + b - u)\|} = -\mathbf{T}.$

Therefore $\mathbf{T}_1'(u) = \mathbf{T}'(a + b - u)$ and $\mathbf{N}_1 = \mathbf{N}.$

45. (a) $\mathbf{r}'(t) = -\sqrt{2}\sin t\,\mathbf{i} + \sqrt{2}\cos t\,\mathbf{j} + 5\cos t\,\mathbf{k}$

tangent line $(t = \pi/4)$: $x = 1 - t$, $y = 1 + t$, $z = -\dfrac{\sqrt{2}}{2} - \dfrac{5\sqrt{2}}{2}t$

(c) The tangent line is parallel to the x, y-plane at the points where $t = \dfrac{(2n+1)\pi}{10}$, $n = 0, 1, 2, \ldots, 9$.

SECTION 14.4

1. $\mathbf{r}'(t) = \mathbf{i} + t^{1/2}\,\mathbf{j}$, $\|\mathbf{r}'(t)\| = \sqrt{1+t}$

$$L = \int_0^8 \sqrt{1+t}\, dt = \left[\frac{2}{3}(1+t)^{3/2}\right]_0^8 = \frac{52}{3} \qquad L = \int_0^2 (t^2 + 1)\, dt = \frac{14}{3}$$

3. $\mathbf{r}'(t) = -a\sin t\,\mathbf{i} + a\cos t\,\mathbf{j} + b\,\mathbf{k}$, $\|\mathbf{r}'(t)\| = \sqrt{a^2 + b^2}$

$$L = \int_0^{2\pi} \sqrt{a^2 + b^2}\, dt = 2\pi\sqrt{a^2 + b^2}$$

5. $\mathbf{r}'(t) = \mathbf{i} + \tan t\,\mathbf{j}$, $\|\mathbf{r}'(t)\| = \sqrt{1 + \tan^2 t} = |\sec t|$

$$L = \int_0^{\pi/4} |\sec t|\, dt = \int_0^{\pi/4} \sec t\, dt = [\ln|\sec t + \tan t|]_0^{\pi/4} = \ln(1 + \sqrt{2})$$

7. $\mathbf{r}'(t) = 3t^2\,\mathbf{i} + 2t\,\mathbf{j}$, $\|\mathbf{r}'(t)\| = \sqrt{9t^4 + 4t^2} = |t|\sqrt{4 + 9t^2}$

$$L = \int_0^1 \left| t\sqrt{4 + 9t^2} \right|\, dt = \int_0^1 t\sqrt{4 + 9t^2}\, dt = \left[\frac{1}{27}\left(4 + 9t^2\right)^{3/2}\right]_0^1 = \frac{1}{27}\left(13\sqrt{13} - 8\right)$$

9. $\mathbf{r}'(t) = (\cos t - \sin t)e^t\,\mathbf{i} + (\sin t + \cos t)e^t\,\mathbf{j}$, $\|\mathbf{r}'(t)\| = \sqrt{2}\,e^t$

$$L = \int_0^{\pi} \sqrt{2}\,e^t\, dt = \sqrt{2}\,(e^{\pi} - 1)$$

11. $\mathbf{r}'(t) = \dfrac{1}{t}\,\mathbf{i} + 2\,\mathbf{j} + 2t\,\mathbf{k}$, $\|\mathbf{r}'(t)\| = \sqrt{\dfrac{1}{t^2} + 4 + 4t^2}$

$$L = \int_1^e \sqrt{\frac{1}{t^2} + 4 + 4t^2}\, dt = \int_1^e \left(\frac{1}{t} + 2t\right) dt = \left[\ln|t| + t^2\right]_1^e = e^2$$

13. $\mathbf{r}'(t) = t\cos t\,\mathbf{i} + t\sin t\,\mathbf{j} + \sqrt{3}\,t\,\mathbf{k}$, $\|\mathbf{r}'(t)\| = \sqrt{t^2\cos^2 t + t^2\sin^2 t + 3t^2} = \sqrt{4t^2} = 2t$

$$L = \int_0^{2\pi} 2t\, dt = \left[t^2\right]_0^{2\pi} = 4\pi^2$$

15. $\mathbf{r}'(t) = 2\,\mathbf{i} + 2t\,\mathbf{j} - 2t\,\mathbf{k}$, $\|\mathbf{r}'(t)\| = 2\sqrt{1 + 2t^2}$

$$L = \int_0^2 2\sqrt{1 + 2t^2}\, dt = \sqrt{2}\int_0^{\tan^{-1}(2\sqrt{2})} \sec^3 u\, du$$

$$\underset{(t\sqrt{2} = \tan u)}{\Big\uparrow}$$

$$= \tfrac{1}{2}\sqrt{2}\left[\sec u\tan u + \ln|\sec u + \tan u|\right]_0^{\tan^{-1}(2\sqrt{2})} = 6 + \tfrac{1}{2}\sqrt{2}\,\ln(3 + 2\sqrt{2})$$

17.
$$s = s(t) = \int_a^t \|\mathbf{r}'(u)\| \, du$$
$$s'(t) = \|\mathbf{r}'(t)\| = \|x'(t)\,\mathbf{i} + y'(t)\,\mathbf{j} + z'(t)\,\mathbf{k}\|$$
$$= \sqrt{[x'(t)]^2 + [y'(t)]^2 + [z'(t)]^2}.$$

In the Leibniz notation this translates to
$$\frac{ds}{dt} = \sqrt{\left(\frac{dx}{dt}\right)^2 + \left(\frac{dy}{dt}\right)^2 + \left(\frac{dz}{dt}\right)^2}.$$

19.
$$s = s(x) = \int_a^x \sqrt{1 + [f'(t)]^2} \, dt$$
$$s'(x) = \sqrt{1 + [f'(x)]^2}.$$

In the Leibniz notation this translates to
$$\frac{ds}{dx} = \sqrt{1 + \left(\frac{dy}{dx}\right)^2}.$$

21. $\mathbf{r}'(t) = -\sin t\,\mathbf{i} + \cos t\,\mathbf{j}$. Since $\|\mathbf{r}'\| \equiv 1$, the parametrization is by arc length.

23. $\mathbf{r}'(t) = t \sin t\,\mathbf{i} + t \cos t\,\mathbf{j} + t\,\mathbf{k}; \quad \|\mathbf{r}'\| = t\sqrt{2}$.
$$s = \int_0^t u\sqrt{2}\,du = \frac{\sqrt{2}}{2}t^2; \quad t = 2^{1/4}\sqrt{s}.$$
$$\mathbf{R}(s) = \left(\sin 2^{1/4}\sqrt{s} - 2^{1/4}\sqrt{s}\cos 2^{1/4}\sqrt{s}\right)\mathbf{i} + \left(\cos 2^{1/4}\sqrt{s} + 2^{1/4}\sqrt{s}\sin 2^{1/4}\sqrt{s}\right)\mathbf{j} + \frac{1}{\sqrt{2}}s\,\mathbf{k}.$$

25. $\mathbf{r}'(t) = t^{3/2}\,\mathbf{j} + \mathbf{k}, \quad \|\mathbf{r}'(t)\| = \sqrt{\left(t^{3/2}\right)^2 + 1} = \sqrt{t^3 + 1}$
$$s = \int_0^{1/2} \sqrt{t^3 + 1}\,dt \cong 0.5077$$

27. $\mathbf{r}'(t) = -3 \sin t\,\mathbf{i} + 4 \cos t\,\mathbf{j}, \quad \|\mathbf{r}'(t)\| = \sqrt{9 \sin^2 t + 16 \cos^2 t}\,dt$
$$s = \int_0^{2\pi} \sqrt{9 \sin^2 t + 16 \cos^2 t}\,dt \cong 22.0939$$

29. (a)   (b) $s = \int_0^{2\pi} \sqrt{1 + 16 \cos^2 4t}\,dt \cong 17.6286$

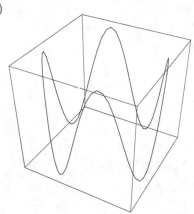

PROJECT 14.4

1. Given the differentiable curve $\mathbf{r} = \mathbf{r}(t)$, $t \in I$. Let $t = \phi(u)$ be a continuously differentiable one-to-one function that maps the interval J onto the interval I, and let $\mathbf{R}(u) = \mathbf{r}(\phi(u))$, $u \in J$. Suppose that $\phi'(u) > 0$ on J. Then, as u increases across J, $t = \phi(u)$ increases across I. As a result, $\mathbf{R}(u)$ takes on exactly the same values in exactly the same order as $\mathbf{r}$. If $\phi'(u) < 0$ on J, then $\mathbf{R}(u)$ takes on exactly the same values as $\mathbf{r}$ but in the reverse order.

3. Suppose that $\phi'(u) < 0$ on J. Then

$$\mathbf{R}'(u) = [\mathbf{r}(\phi(u))]' = \mathbf{r}'(\phi(u))\phi'(u); \quad \frac{\mathbf{R}'(u)}{\mathbf{R}'(u)} = \frac{\mathbf{r}'(\phi(u))\phi'(u)}{\|\mathbf{r}'(\phi(u))\phi'(u)\|} = -\frac{\mathbf{r}'(\phi(u))\phi'(u)}{\phi'(u)\|\mathbf{r}'(\phi(u))\|} = -\frac{\mathbf{r}'(t)}{\|\mathbf{r}'(t)\|}.$$

$$\text{since } \phi'(u) < 0 \nearrow$$

Thus, the unit tangent is reversed by a sense-reversing change of parameter. That is, if $T_{\mathbf{R}}$ and $T_{\mathbf{r}}$ are the respective unit tangents, then $T_{\mathbf{R}} = -T_{\mathbf{r}}$.

Now consider the principal normals:

$$\frac{T'_{\mathbf{R}}}{\|T'_{\mathbf{R}}\|} = -\frac{T'_{\mathbf{r}}(\phi(u)\phi'(u)}{\|T'_{\mathbf{r}}(\phi(u)\phi'(u)\|} = \frac{T'_{\mathbf{r}}(\phi(u)\phi'(u)}{\phi'(u)\|T'_{\mathbf{r}}(\phi(u)\phi'(u)\|} = \frac{T'_{\mathbf{r}}(t)}{\|T'_{\mathbf{r}}(t)\|}.$$

$$\text{since } \phi'(u) < 0 \nearrow$$

Thus the principal normal is unchanged by a sense-reversing change of parameter. The osculating plane is also unchanged since $\mathbf{T} \times \mathbf{N}$ and $-\mathbf{T} \times \mathbf{N}$ are each normal to the osculating plane.

5. Let L be the length as computed from $\mathbf{r}$ and L^* the length as computed from $\mathbf{R}$.   Then

$$L^* = \int_c^d \|\mathbf{R}'(u)\| \, du = \int_c^d \|\mathbf{r}'(\phi(u))\| \, \phi'(u) \, du = \int_a^b \|\mathbf{r}'(t)\| \, dt = L.$$

$$t = \phi(u) \nearrow$$

SECTION 14.5

1. $\mathbf{r}(t) = a[\cos\theta(t)\,\mathbf{i} + \sin\theta(t)\,\mathbf{j}], \quad \mathbf{r}'(t) = a[-\sin\theta(t)\,\mathbf{i} + \cos\theta(t)\,\mathbf{j}]\theta'(t)$

   $\|\mathbf{r}'(t)\| = v \implies a|\theta'(t)| = v \implies |\theta'(t)| = v/a$

   $\mathbf{r}''(t) = a[-\cos\theta(t)\,\mathbf{i} - \sin\theta(t)\,\mathbf{j}][\theta'(t)]^2, \quad \|\mathbf{r}''(t)\| = a[\theta'(t)]^2 = v^2/a$

3. $\mathbf{r}(t) = at\,\mathbf{i} + b\sin at\,\mathbf{j}, \quad = a\,\mathbf{i} + ab\cos at\,\mathbf{j}$

   $\mathbf{r}'(t) = -a^2 b\sin at\,\mathbf{j}, \quad \|\mathbf{r}'(t)\| = a^2|b\sin at| = a^2|y(t)|$

5. $y = \cos \pi x, \quad 0 \le x \le 2$ **7.** $x = \sqrt{1 + y^2}, \quad y \ge -1$

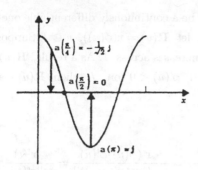 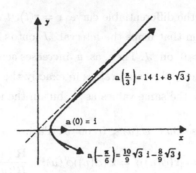

9.   (a) initial position is tip of $\mathbf{r}(0) = x_0\, \mathbf{i} + y_0\, \mathbf{j} + z_0\, \mathbf{k}$

     (b) $\mathbf{r}'(t) = (\alpha \cos \theta)\, \mathbf{j} + (\alpha \sin \theta - 32t)\, \mathbf{k}, \quad \mathbf{r}'(0) = (\alpha \cos \theta)\, \mathbf{j} + (\alpha \sin \theta)\, \mathbf{k}$

     (c) $|\mathbf{r}'(0)| = |\alpha|$ (d) $\mathbf{r}''(t) = -32\, \mathbf{k}$

     (e) a parabolic arc from the parabola

$$z = z_0 + (\tan \theta)\,(y - y_0) - 16\,\frac{(y - y_0)^2}{\alpha^2 \cos^2 \theta}$$

     in the plane $x = x_0$

11. $\|\mathbf{r}(t)\| = C$ iff $\|\mathbf{r}(t)\|^2 = \mathbf{r}(t) \cdot \mathbf{r}(t) = C$

                       iff $\dfrac{d}{dt}\|\mathbf{r}(t)\| = 2\mathbf{r}(t) \cdot \mathbf{r}'(t) = 0$

                       iff $\mathbf{r}(t) \perp \mathbf{r}'(t)$

13. $\kappa = \dfrac{e^{-x}}{(1 + e^{-2x})^{3/2}}$

15. $y' = \dfrac{1}{2x^{1/2}}; \quad y'' = \dfrac{-1}{4x^{3/2}} \quad \kappa = \dfrac{\left|-1/4x^{3/2}\right|}{\left[1 + \left(1/2x^{1/2}\right)^2\right]^{3/2}} = \dfrac{2}{(1 + 4x)^{3/2}}$

17. $\kappa = \dfrac{\sec^2 x}{(1 + \tan^2 x)^{3/2}} = |\cos x|$

19. $\kappa = \dfrac{|\sin x|}{(1 + \cos^2 x)^{3/2}}$

21. $\kappa = \dfrac{|x|}{(1 + x^4/4)^{3/2}}; \quad$ at $\left(2, \dfrac{4}{3}\right), \quad \kappa = \dfrac{2}{5\sqrt{5}}$

23. $\kappa = \dfrac{\left|-1/y^3\right|}{(1 + 1/y^2)^{3/2}} = \dfrac{1}{(1 + y^2)^{3/2}}; \quad$ at $(2, 2), \quad \kappa = \dfrac{1}{5\sqrt{5}}$

25. $y'(x) = \dfrac{1}{x+1}$, $\quad y'(2) = \dfrac{1}{3}$;$\qquad y''(x) = \dfrac{-1}{(x+1)^2}$, $\quad y''(2) = -\dfrac{1}{9}$.

At $\ x = 2$, $\ \kappa = \dfrac{\left|-\dfrac{1}{9}\right|}{\left[1+\left(\dfrac{1}{3}\right)^2\right]^{3/2}} = \dfrac{3}{10\sqrt{10}}$

27. $\kappa(x) = \dfrac{\left|-1/x^2\right|}{(1+1/x^2)^{3/2}} = \dfrac{x}{(x^2+1)^{3/2}}$, $\quad x > 0$

$\kappa'(x) = \dfrac{(1-2x^2)}{(x^2+1)^{5/2}}$, $\qquad \kappa'(x) = 0 \implies x = \dfrac{1}{2}\sqrt{2}$

Since κ increases on $\ \left(0, \tfrac{1}{2}\sqrt{2}\right]$ and decreases on $\ \left[\tfrac{1}{2}\sqrt{2}, \infty\right)$, κ is maximal at $\ \left(\tfrac{1}{2}\sqrt{2}, \tfrac{1}{2}\ln\tfrac{1}{2}\right)$.

29. $x(t) = t$, $\quad x'(t) = 1$, $\quad x''(t) = 0$;$\qquad y(t) = \tfrac{1}{2}t^2$, $\quad y'(t) = t$, $\quad y''(t) = 1$ $\quad \kappa = \dfrac{1}{(1+t^2)^{3/2}}$

31. $x(t) = 2t$, $\quad x'(t) = 2$, $\quad x''(t) = 0$;$\qquad y(t) = t^3$, $\quad y'(t) = 3t^2$, $\quad y''(t) = 6t$; $\quad \kappa = \dfrac{12|t|}{(4+9t^4)^{3/2}}$

33. $x(t) = e^t\cos t$, $\quad x'(t) = e^t(\cos t - \sin t)$, $\quad x''(t) = -2\,e^t\sin t$

$y(t) = e^t\sin t$, $\quad y'(t) = e^t(\sin t + \cos t)$, $\quad y''(t) = 2\,e^t\cos t$

$\kappa = \dfrac{\left|2e^{2t}\cos t\,(\cos t - \sin t) + 2e^{2t}\sin t\,(\cos t + \sin t)\right|}{\left[e^{2t}(\cos t - \sin t)^2 + e^{2t}(\cos t + \sin t)^2\right]^{3/2}} = \dfrac{2e^{2t}}{(2e^{2t})^{3/2}} = \dfrac{1}{2}\sqrt{2}\,e^{-t}$

35. $x(t) = t\cos t$, $\quad x'(t) = \cos t - t\sin t$, $\quad x''(t) = -2\sin t - t\cos t$

$y(t) = t\sin t$, $\quad y'(t) = \sin t + t\cos t$, $\quad y''(t) = 2\cos t - t\sin t$

$\kappa = \dfrac{\left|(\cos t - t\sin t)(2\cos t - t\sin t) - (\sin t + t\cos t)(-2\sin t - t\cos t)\right|}{\left[(\cos t - t\sin t)^2 + (\sin t + t\cos t)^2\right]^{3/2}} = \dfrac{2+t^2}{[1+t^2]^{3/2}}$

37. $\kappa = \dfrac{\left|2/x^3\right|}{[1+1/x^4]^{3/2}} = \dfrac{2\left|x^3\right|}{(x^4+1)^{3/2}}$;$\quad$ at $x = \pm1$, $\quad \kappa = \dfrac{\sqrt{2}}{2}$

39. We use (14.5.3) and the hint to obtain

$$\kappa = \dfrac{\left|ab\sinh^2 t - ab\cosh^2 t\right|}{\left[a^2\sinh^2 t + b^2\cosh^2 t\right]^{3/2}} = \dfrac{\left|\dfrac{a}{b}y^2 - \dfrac{b}{a}x^2\right|}{\left[\left(\dfrac{ay}{b}\right)^2 + \left(\dfrac{bx}{a}\right)^2\right]^{3/2}}$$

$$= \dfrac{a^3b^3\left|\dfrac{a}{b}y^2 - \dfrac{b}{a}x^2\right|}{\left[a^4y^2 + b^4x^2\right]^{3/2}} = \dfrac{a^4b^4}{\left[a^4y^2 + b^4x^2\right]^{3/2}}.$$

41. $\mathbf{r}'(t) = e^t(\cos t - \sin t)\,\mathbf{i} + e^t(\sin t + \cos t)\,\mathbf{j} + e^t\,\mathbf{k}$

$$\frac{ds}{dt} = \|\mathbf{r}'(t)\| = \sqrt{3}\,e^t, \quad \frac{d^2s}{dt^2} = \sqrt{3}\,e^t$$

$$\mathbf{T}(t) = \frac{\mathbf{r}'(t)}{\|\mathbf{r}'(t)\|} = \frac{1}{\sqrt{3}}\left[(\cos t - \sin t)\,\mathbf{i} + (\sin t + \cos t)\,\mathbf{j} + \mathbf{k}\right]$$

$$\mathbf{T}'(t) = \frac{1}{\sqrt{3}}\left[(-\sin t - \cos t)\,\mathbf{i} + (\cos t - \sin t)\,\mathbf{j}\right]; \quad \|\mathbf{T}'(t)\| = \sqrt{2/3}$$

$$\kappa = \frac{\|\mathbf{T}'(t)\|}{ds/dt} = \frac{\sqrt{2/3}}{\sqrt{3}\,e^t} = \frac{1}{3}\sqrt{2}\,e^{-t}; \quad \mathbf{a_T} = \frac{d^2s}{dt^2} = \sqrt{3}\,e^t, \quad \mathbf{a_N} = \kappa\left(\frac{ds}{dt}\right)^2 = \sqrt{2}\,e^t$$

43. $\mathbf{r}'(t) = -2\sin 2t\,\mathbf{i} + 2\cos 2t\,\mathbf{j}; \quad \dfrac{ds}{dt} = \|\mathbf{r}'(t)\| = 2, \quad \dfrac{d^2s}{dt^2} = 0$

$$\mathbf{T}(t) = \frac{\mathbf{r}'(t)}{\|\mathbf{r}'(t)\|} = -\sin 2t\,\mathbf{i} + \cos 2t\,\mathbf{j}$$

$$\mathbf{T}'(t) = -2\,(\cos 2t\,\mathbf{i} + \sin 2t\,\mathbf{j}); \quad \|\mathbf{T}'(t)\| = 2$$

$$\kappa = \frac{\|\mathbf{T}'(t)\|}{ds/dt} = \frac{2}{2} = 1; \quad \mathbf{a_T} = \frac{d^2s}{dt^2} = 0, \quad \mathbf{a_N} = \kappa\left(\frac{ds}{dt}\right)^2 = 1 \cdot 4 = 4.$$

45. $\mathbf{r}'(t) = -3\sin 3t\,\mathbf{i} + 4\,\mathbf{j} - 3\cos 3t\,\mathbf{k}; \quad \dfrac{ds}{dt} = \|\mathbf{r}'\| = 5; \quad \dfrac{d^2s}{dt^2} = 0.$

$$\mathbf{T}(t) = \frac{\mathbf{r}'(t)}{\|\mathbf{r}'(t)\|} = -\frac{3}{5}\sin 3t\,\mathbf{i} + \frac{4}{5}\,\mathbf{j} - \frac{3}{5}\cos 3t\,\mathbf{k}$$

$$\mathbf{T}'(t) = -\frac{9}{5}\cos 3t\,\mathbf{i} + \frac{9}{5}\sin 3t\,\mathbf{k}; \quad \|\mathbf{T}'(t)\| = \frac{9}{5}$$

$$\kappa = \frac{\|\mathbf{T}'(t)\|}{ds/dt} = \frac{9/5}{5} = \frac{9}{25}; \quad \mathbf{a_T} = \frac{d^2s}{dt^2} = 0, \quad \mathbf{a_N} = \kappa\left(\frac{ds}{dt}\right)^2 = \frac{9}{25} \cdot 25 = 9.$$

47. $\mathbf{r}'(t) = \sqrt{1+t}\,\mathbf{i} - \sqrt{1-t}\,\mathbf{j} + \sqrt{2}\,\mathbf{k}, \quad \dfrac{ds}{dt} = \|\mathbf{r}'(t)\| = \sqrt{(1+t)+(1-t)+2} = 2, \quad \dfrac{d^2s}{dt^2} = 0$

$$\mathbf{T}(t) = \frac{\mathbf{r}'(t)}{\|\mathbf{r}'(t)\|} = \frac{\sqrt{1+t}}{2}\,\mathbf{i} - \frac{\sqrt{1-t}}{2}\,\mathbf{j} + \frac{\sqrt{2}}{2}\,\mathbf{k}$$

$$\mathbf{T}'(t) = \frac{1}{4\sqrt{1+t}}\,\mathbf{i} + \frac{1}{4\sqrt{1-t}}\,\mathbf{j}.$$

$$\|\mathbf{T}'(t)\| = \sqrt{\frac{1}{16(1+t)} + \frac{1}{16(1-t)}} = \frac{1}{4}\sqrt{\frac{2}{1-t^2}}$$

Then, $\quad \kappa = \dfrac{\|\mathbf{T}'(t)\|}{ds/dt} = \dfrac{1}{8}\sqrt{\dfrac{2}{1-t^2}}$

$$\mathbf{a_T} = \frac{d^2s}{dt^2} = 0, \quad \mathbf{a_N} = \kappa\left(\frac{ds}{dt}\right)^2 = \frac{1}{2}\sqrt{\frac{2}{1-t^2}}.$$

49. tangential component: $\quad \mathbf{a_T} = \dfrac{6t + 12t^3}{\sqrt{1+t^2+t^4}};$ $\quad$ normal component: $\quad \mathbf{a_N} = 6\sqrt{\dfrac{1+4t^2+t^4}{1+t^2+t^4}}$

51. By Exercise 50

$$\kappa = \frac{\left| \left(e^{a\theta}\right)^2 + 2\left(ae^{a\theta}\right)^2 - \left(e^{a\theta}\right)\left(a^2 e^{a\theta}\right) \right|}{\left[\left(e^{a\theta}\right)^2 + \left(ae^{a\theta}\right)^2\right]^{3/2}} = \frac{e^{-a\theta}}{\sqrt{1+a^2}}.$$

53. By Exercise 50,

$$\kappa = \frac{\left| a^2(1-\cos\theta)^2 + 2a^2\sin^2\theta - a^2(1-\cos\theta)(\cos\theta)\right|}{\left[a^2(1-\cos\theta)^2 + a^2\sin^2\theta\right]^{3/2}} = \frac{3a^2(1-\cos\theta)}{\left[2a^2(1-\cos\theta)\right]^{3/2}} = \frac{3ar}{\left[2ar\right]^{3/2}} = \frac{3}{2\sqrt{2ar}}.$$

PROJECT 14.5A

1. The system of equations generated by the specified conditions is:

$a + b + c + d = 3$ $27a + 9b + 3c + d = 7$

$6a + 2b = 0$ $27\alpha + 9\beta + 3\gamma + \delta = 7$

$729\alpha + 81\beta + 9\gamma + \delta = -2$ $54\alpha + 2\beta = 0$

$27a + 6b + c = 27\alpha + 6\beta + \gamma$ $18a + 2b = 18\alpha + 2\beta$

$a \cong -0.1094$ $b \cong 0.3281$ $c \cong 2.1094$ $d \cong 0.6719$

$\alpha \cong 0.0365$ $\beta \cong -0.9844$ $\gamma \cong 6.0469$ $\delta \cong -3.2656$

3. (a), (b) The system of equations generated by the specified conditions (and the derivative conditions of Problem 1) is:

$27\alpha + 9b + 3c + d = 10$ $64\alpha + 16b + 4c + d = 15$

$18a + 2b = 0$ $64\alpha + 64\beta + 4\gamma + \delta = 15$

$216\alpha + 36\beta + 6\gamma + \delta = 35$ $36\alpha + 2\beta = 0$

$48\alpha + 8b + c = 48\alpha + 8\beta + \gamma$ $24\alpha + 2b = 24\alpha + 2\beta$

(c) $a = b = 0,\ c = 5,\ d = -5;\ \alpha = 1.25,\ \beta = -15,\ \gamma = 65,\ \delta = -85$

PROJECT 14.5B

1. $\dfrac{d\mathbf{T}}{dt} = \dfrac{d\mathbf{T}}{ds}\dfrac{ds}{dt} \implies \dfrac{d\mathbf{T}}{ds} = \dfrac{d\mathbf{T}/dt}{ds/dt} = \dfrac{\mathbf{T}'(t)}{\|\mathbf{T}'(t)\|}\dfrac{\|\mathbf{T}/(t)\|}{ds/dt} = \kappa\,\mathbf{N}.$

3. $\dfrac{d\mathbf{N}}{ds} = \dfrac{d}{ds}(\mathbf{B} \times \mathbf{T}) = \left(\mathbf{B} \times \dfrac{d\mathbf{T}}{ds}\right) + \left(\dfrac{d\mathbf{B}}{ds} \times \mathbf{T}\right) = (\mathbf{B} \times \kappa\mathbf{N}) + \tau(\mathbf{N} \times \mathbf{T})$

$$= -\kappa(\mathbf{N} \times \mathbf{B}) - \tau(\mathbf{T} \times \mathbf{N}) = -\kappa\mathbf{T} - \tau\mathbf{B}$$

SECTION 14.6

1. (a) $\mathbf{r}'(t) = \dfrac{a\omega}{2}\left(e^{\omega t} - e^{-\omega t}\right)\mathbf{i} + \dfrac{b\omega}{2}\left(e^{\omega t} + e^{-\omega t}\right)\mathbf{j},\quad \mathbf{r}'(0) = b\omega\mathbf{j}$

(b) $\mathbf{r}''(t) = \dfrac{a\omega^2}{2}\left(e^{\omega t} + e^{-\omega t}\right)\mathbf{i} + \dfrac{b\omega^2}{2}\left(e^{\omega t} - e^{-\omega t}\right)\mathbf{j} = \omega^2\mathbf{r}(t)$

(c) The torque τ is $\mathbf{0}$: $\tau(t) = \mathbf{r}(t) \times m\mathbf{a}(t) = \mathbf{r}(t) \times m\omega^2\mathbf{r}(t) = \mathbf{0}.$

The angular momentum $\mathbf{L}(t)$ is constant since $\mathbf{L}'(t) = \tau(t) = \mathbf{0}.$

3. We begin with the force equation $\mathbf{F}(t) = \alpha\mathbf{k}$. In general, $\mathbf{F}(t) = m\,\mathbf{a}(t)$, so that here

$$\mathbf{a}(t) = \frac{\alpha}{m}\,\mathbf{k}.$$

Integration gives

$$\mathbf{v}(t) = C_1\mathbf{i} + C_2\,\mathbf{j} + \left(\frac{\alpha}{m}t + C_3\right)\mathbf{k}.$$

Since $\mathbf{v}(0) = 2\,\mathbf{j}$, we can conclude that $C_1 = 0$, $C_2 = 2$, $C_3 = 0$. Thus

$$\mathbf{v}(t) = 2\,\mathbf{j} + \frac{\alpha}{m}\,t\mathbf{k}.$$

Another integration gives

$$\mathbf{r}(t) = D_1\mathbf{i} + (2t + D_2)\,\mathbf{j} + \left(\frac{\alpha}{2m}t^2 + D_3\right)\mathbf{k}.$$

Since $\mathbf{r}(0) = y_0\,\mathbf{j} + z_0\,\mathbf{k}$, we have $D_1 = 0$, $D_2 = y_0$, $D_3 = z_0$, and therefore

$$\mathbf{r}(t) = (2t + y_0)\,\mathbf{j} + \left(\frac{\alpha}{2m}t^2 + z_0\right)\mathbf{k}.$$

The conditions of the problem require that t be restricted to nonnegative values.

To obtain an equation for the path in Cartesian coordinates, we write out the components

$$x(t) = 0, \quad y(t) = 2t + y_0, \quad z(t) = \frac{\alpha}{2m}t^2 + z_0. \qquad (t \geq 0)$$

From the second equation we have

$$t = \tfrac{1}{2}\,[y(t) - y_0]. \qquad (y(t) \geq y_0)$$

Substituting this into the third equation, we get

$$z(t) = \frac{\alpha}{8m}\,[y(t) - y_0]^2 + z_0. \qquad (y(t) \geq y_0)$$

Eliminating t altogether, we have

$$z = \frac{\alpha}{8m}\,(y - y_0)^2 + z_0. \qquad (y \geq y_0)$$

Since $x = 0$, the path of the object is a parabolic arc in the yz-plane.

Answers to (a) through (d):

(a) velocity: $\mathbf{v}(t) = 2\,\mathbf{j} + \dfrac{\alpha}{m}\,t\,\mathbf{k}.$ (b) speed: $v(t) = \dfrac{1}{m}\,\sqrt{4m^2 + \alpha^2 t^2}.$

(c) momentum: $\mathbf{p}(t) = 2m\,\mathbf{j} + \alpha\,t\,\mathbf{k}.$

(d) path in vector form: $\mathbf{r}(t) = (2t + y_0)\,\mathbf{j} + \left(\dfrac{\alpha}{2m}t^2 + z_0\right)\mathbf{k}, \quad t \geq 0.$

   path in Cartesian coordinates: $z = \dfrac{\alpha}{8m}\,(y - y_0)^2 + z_0, \quad y \geq y_0, \quad x = 0.$

5. $\mathbf{F}(t) = m\,\mathbf{a}(t) = m\,\mathbf{r}''(t) = 2m\mathbf{k}$

7. From $\mathbf{F}(t) = m\,\mathbf{a}(t)$ we obtain

$$\mathbf{a}(t) = \pi^2[a\cos\pi t\,\mathbf{i} + b\sin\pi t\,\mathbf{j}].$$

By direct calculation using $\mathbf{v}(0) = -\pi b\mathbf{j} + \mathbf{k}$ and $\mathbf{r}(0) = b\mathbf{j}$ we obtain

$$\mathbf{v}(t) = a\pi \sin \pi t\, \mathbf{i} - b\pi \cos \pi t\, \mathbf{j} + \mathbf{k}$$

$$\mathbf{r}(t) = a(1 - \cos \pi t)\,\mathbf{i} + b(1 - \sin \pi t)\,\mathbf{j} + t\mathbf{k}.$$

(a) $\mathbf{v}(1) = b\pi\mathbf{j} + \mathbf{k}$

(b) $\|\mathbf{v}(1)\| = \sqrt{\pi^2 b^2 + 1}$

(c) $\mathbf{a}(1) = -\pi^2 a\mathbf{i}$

(d) $m\,\mathbf{v}(1) = m\,(\pi b\mathbf{j} + \mathbf{k})$

(e) $\mathbf{L}(1) = \mathbf{r}(1) \times m\,\mathbf{v}(1) = [2a\mathbf{i} + b\mathbf{j} + \mathbf{k}] \times [m\,(b\pi\mathbf{j} + \mathbf{k})]$

$\qquad = m\,[b(1 - \pi)\mathbf{i} - 2a\mathbf{j} + 2ab\pi\mathbf{k}]$

(f) $\boldsymbol{\tau}(1) = \mathbf{r}(1) \times \mathbf{F}(1) = [2a\mathbf{i} + b\mathbf{j} + \mathbf{k}] \times [-m\pi^2 a\mathbf{i}] = -m\pi^2 a\,[\mathbf{j} - b\mathbf{k}]$

9. We have $\qquad\qquad m\mathbf{v} = m\mathbf{v}_1 + m\mathbf{v}_2$ and $\quad \frac{1}{2}mv^2 = \frac{1}{2}mv_1{}^2 + \frac{1}{2}mv_2{}^2$.

Therefore $\qquad\qquad \mathbf{v} = \mathbf{v}_1 + \mathbf{v}_2$ and $\quad v^2 = v_1{}^2 + v_2{}^2$.

Since $\qquad\qquad v^2 = \mathbf{v}\cdot\mathbf{v} = (\mathbf{v}_1 + \mathbf{v}_2)\cdot(\mathbf{v}_1 + \mathbf{v}_2) = v_1{}^2 + v_2{}^2 + 2(\mathbf{v}_1 \cdot \mathbf{v}_2)$,

we have $\qquad\qquad \mathbf{v}_1 \cdot \mathbf{v}_2 = 0$ and $\quad \mathbf{v}_1 \perp \mathbf{v}_2$.

11. $\mathbf{r}''(t) = \mathbf{a}$, $\quad \mathbf{r}'(t) = \mathbf{v}(0) + t\mathbf{a}$, $\quad \mathbf{r}(t) = \mathbf{r}(0) + t\mathbf{v}(0) + \frac{1}{2}t^2\,\mathbf{a}$.

If neither $\mathbf{v}(0)$ nor $\mathbf{a}$ is zero, the displacement $\mathbf{r}(t) - \mathbf{r}(0)$ is a linear combination of $\mathbf{v}(0)$ and $\mathbf{a}$ and thus remains on the plane determined by these vectors. The equation of this plane can be written

$$[\mathbf{a} \times \mathbf{v}(0)] \cdot [\mathbf{r} - \mathbf{r}(0)] = 0.$$

(If either $\mathbf{v}(0)$ or $\mathbf{a}$ is zero, the motion is restricted to a straight line; if both of these vectors are zero, the particle remains at its initial position $\mathbf{r}(0)$.)

13. $\mathbf{r}(t) = \mathbf{i} + t\mathbf{j} + \left(\dfrac{qE_0}{2m}\right)t^2\mathbf{k}$
$\qquad\qquad\qquad$ 15. $\mathbf{r}(t) = \left(1 + \dfrac{t^3}{6m}\right)\mathbf{i} + \dfrac{t^4}{12m}\,\mathbf{j} + t\,\mathbf{k}$

17.
$$\frac{d}{dt}\left(\frac{1}{2}mv^2\right) = mv\frac{dv}{dt} = m\left(\mathbf{v}\cdot\frac{d\mathbf{v}}{dt}\right) = m\frac{d\mathbf{v}}{dt}\cdot\mathbf{v} = \mathbf{F}\cdot\frac{d\mathbf{r}}{dt}$$

$$= 4r^2\left(\mathbf{r}\cdot\frac{d\mathbf{r}}{dt}\right) = 4r^2\left(r\frac{dr}{dt}\right) = 4r^3\frac{dr}{dt} = \frac{d}{dt}\left(r^4\right).$$

Therefore $d/dt\left(\frac{1}{2}mv^2 - r^4\right) = 0$ and $\quad \frac{1}{2}mv^2 - r^4$ is a constant E. Evaluating E from $t = 0$, we find that $E = 2m$.

Thus $\frac{1}{2}mv^2 - r^4 = 2m$ and $\quad v = \sqrt{4 + (2/m)\,r^4}$.

SECTION 14.7

1. On Earth: year of length T, average distance from sun d.

On Venus: year of length αT, average distance from sun $0.72d$.

Therefore

$$\frac{(\alpha T)^2}{T^2} = \frac{(0.72d)^3}{d^3}.$$

This gives $\quad \alpha^2 = (0.72)^3 \cong 0.372$ and $\quad \alpha \cong 0.615$. Answer: about 61.5% of an Earth year.

3.

$$\left(\frac{dx}{dt}\right)^2 + \left(\frac{dy}{dt}\right)^2 = \left[\frac{d}{dt}(r\cos\theta)\right]^2 + \left[\frac{d}{dt}(r\sin\theta)\right]^2$$

$$= \left[r(-\sin\theta)\frac{d\theta}{dt} + \frac{dr}{dt}\cos\theta\right]^2 + \left[r\cos\theta\frac{d\theta}{dt} + \frac{dr}{dt}\sin\theta\right]^2$$

$$= r^2\sin^2\theta\left(\frac{d\theta}{dt}\right)^2 + \left(\frac{dr}{dt}\right)^2\cos^2\theta + r^2\cos^2\theta\left(\frac{d\theta}{dt}\right)^2 + \left(\frac{dr}{dt}\right)^2\sin^2\theta$$

$$= \left(\frac{dr}{dt}\right)^2 + r^2\left(\frac{d\theta}{dt}\right)^2$$

5. Substitute

$$r = \frac{a}{1+e\cos\theta}, \quad \left(\frac{dr}{d\theta}\right)^2 = \left[\frac{-a}{(1+e\cos\theta)^2}\cdot(-e\sin\theta)\right]^2 = \frac{(ae\sin\theta)^2}{(1+e\cos\theta)^4}$$

into the right side of the equation and you will see that, with a and e^2 as given, the expression reduces to E.

REVIEW EXERCISES

1. $\mathbf{f}'(t) = 6t\,\mathbf{i} - 15t^2\,\mathbf{j}, \qquad \mathbf{f}''(t) = 6\,\mathbf{i} - 30t\,\mathbf{j}$

3. $\mathbf{f}'(t) = (e^t\cos t - e^t\sin t)\,\mathbf{i} + 2\sin 2t\,\mathbf{j}, \qquad \mathbf{f}''(t) = -2e^t\sin t\,\mathbf{i} + 4\cos 2t\,\mathbf{j}$

5. $\displaystyle\int_0^2 \left[2t\,\mathbf{i} + (t^2 - 1)\,\mathbf{j}\right]\,dt = \left[t^2\,\mathbf{i} + \left(\frac{1}{3}t^3 - t\right)\mathbf{j}\right]_0^2 = 4\,\mathbf{i} + \frac{2}{3}\,\mathbf{j}$

7.

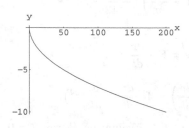

9.

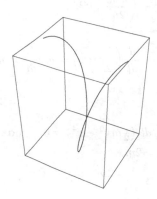

11. (a) $\mathbf{r}(t) = 2\cos\left(t + \frac{\pi}{2}\right)\mathbf{i} + 4\sin\left(t + \frac{\pi}{2}\right)\mathbf{j}$ (b) $\mathbf{r}(t) = -2\cos 2t\,\mathbf{i} + 4\sin 2t\,\mathbf{j}$

13. $\mathbf{f}(t) = \frac{1}{3}t^3\,\mathbf{i} + \left(\frac{1}{2}e^{2t} + t\right)\mathbf{j} + \frac{1}{3}(2t + 1)^{3/2}\,\mathbf{k} + \mathbf{C}.$

    $\mathbf{f}(0) = \mathbf{i} - 3\,\mathbf{j} + 3\,\mathbf{k} \Longrightarrow \mathbf{C} = \mathbf{i} - \frac{7}{2}\,\mathbf{j} + \frac{8}{3}\,\mathbf{k}; \quad \mathbf{f}(t) = \left(\frac{1}{3}t^3 + 1\right)\mathbf{i} + \left(\frac{1}{2}e^{2t} + t - \frac{7}{2}\right)\mathbf{j} + \left(\frac{1}{3}(2t+1)^{3/2} + \frac{8}{3}\right)\mathbf{k}$

15. $\mathbf{f}'(t) = (6\,\mathbf{i} + 12t^3\,\mathbf{j}) + (8t\,\mathbf{i} - 12\,\mathbf{k}) = (6 + 8t)\,\mathbf{i} + 12t^3\,\mathbf{j} - 12\,\mathbf{k}$

17. $\mathbf{f}(t) = (t^2 + 2t^3)\,\mathbf{i} - \left(2t^2 + \dfrac{1}{t^2}\right)\mathbf{j} + (t^4 - t)\,\mathbf{k}, \quad \mathbf{f}'(t) = (2t + 6t^2)\,\mathbf{i} - \left(4t - \dfrac{2}{t^3}\right)\mathbf{j} + (4t^3 - 1)\,\mathbf{k}$

19. $\mathbf{r}'(t) = 2\mathbf{r}(t) \Longrightarrow \mathbf{r}(t) = \mathbf{r}_0 e^{2t}$

$\mathbf{r}(0) = (1, 2, 1) \Longrightarrow \mathbf{r}_0 = (1, 2, 1) \quad \text{and} \quad \mathbf{r}(t) = (e^{2t}, 2e^{2t}, e^{2t})$

21. The tip of $\mathbf{r}(t)$ is $P(1, 1, 1)$ when $t = 0$.

$\mathbf{r}'(t) = (2t + 2)\,\mathbf{i} + 3\,\mathbf{j} + (3t^2 + 1)\,\mathbf{k}, \quad \mathbf{r}'(0) = 2\,\mathbf{i} + 3\,\mathbf{j} + \mathbf{k}$

Scalar parametric equations for the tangent line are: $x = 1 + 2t, \;\; y = 1 + 3t, \;\; z = 1 + t$.

23. $\mathbf{r}_1(t) = (2, 1, 1)$ at $t = 1$; $\quad \mathbf{r}_2(u) = (2, 1, 1)$ at $u = -1$. Therefore the curves intersect at the point $(2, 1, 1)$.

$\mathbf{r}_1'(t) = (2\,\mathbf{i} + 2t\,\mathbf{j} + \mathbf{k}, \quad \mathbf{r}_1'(1) = 2\,\mathbf{i} + 2\,\mathbf{j} + \mathbf{k}; \quad \mathbf{r}_2'(u) = -\mathbf{i} - 2u\,\mathbf{j} + 2u\,\mathbf{k}, \quad \mathbf{r}_2'(-1) = -\mathbf{i} + 2\,\mathbf{j} - 2\,\mathbf{k}$.

Since $\mathbf{r}_1'(1) \cdot \mathbf{r}_2'(-1) = 0$, the angle of intersection is $\pi/2$ radians

25. $\mathbf{r}(t) = t\,\mathbf{i} + e^{2t}\,\mathbf{j}, \quad \mathbf{r}(0) = \mathbf{j}$;

$\mathbf{r}'(t) = \mathbf{i} + 2e^{2t}\,\mathbf{j}, \quad \mathbf{r}'(0) = \mathbf{i} + 2\,\mathbf{j}$

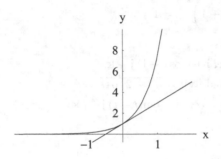

27. $\mathbf{r}'(t) = t\cos t\,\mathbf{i} + t\sin t\,\mathbf{j} + \sqrt{3}t\mathbf{k}; \quad \|\mathbf{r}'\| = \dfrac{ds}{dt} = 2t$

unit tangent vector: $\mathbf{T} = \dfrac{\mathbf{r}'(t)}{\|\mathbf{r}'(t)\|} = \dfrac{1}{2}(\cos t\,\mathbf{i} + \sin t\,\mathbf{j} + \sqrt{3}\,\mathbf{k})$.

$\mathbf{T}'(t) = -\dfrac{1}{2}\sin t\,\mathbf{i} + \dfrac{1}{2}\cos t\,\mathbf{j}; \quad \|\mathbf{T}'(t)\| = \dfrac{1}{2}$.

principal normal vector: $\mathbf{N} = \dfrac{\mathbf{T}'(t)}{\|\mathbf{T}'(t)\|} = -\sin t\,\mathbf{i} + \cos t\,\mathbf{j}$

29. $\mathbf{r}'(t) = 2\,\mathbf{i} + \dfrac{1}{t}\,\mathbf{j} - 2t\mathbf{k}; \quad \|\mathbf{r}'(t)\| = \dfrac{1 + 2t^2}{t}$.

$\mathbf{T}(t) = \dfrac{\mathbf{r}'(t)}{\|\mathbf{r}'(t)\|} = \dfrac{2t}{2t^2 + 1}\,\mathbf{i} + \dfrac{1}{2t^2 + 1}\,\mathbf{j} - \dfrac{2t^2}{2t^2 + 1}\,\mathbf{k}; \quad \mathbf{T}(1) = \dfrac{2}{3}\,\mathbf{i} + \dfrac{1}{3}\,\mathbf{j} - \dfrac{2}{3}\,\mathbf{k}$

$\mathbf{T}'(t) = \dfrac{2 - 4t^2}{(2t^2 + 1)^2}\,\mathbf{i} - \dfrac{4t}{(2t^2 + 1)^2}\,\mathbf{j} - \dfrac{4t}{(2t^2 + 1)^2}\,\mathbf{k}; \quad \mathbf{T}'(1) = -\frac{2}{9}\,\mathbf{i} - \frac{4}{9}\,\mathbf{j} - \frac{4}{9}\,\mathbf{k}, \quad \|\mathbf{T}'(1)\| = \frac{2}{3}$

$\mathbf{N}(1) = \frac{\mathbf{T}'(1)}{\|\mathbf{T}'(1)\|} = -\frac{1}{3}\,\mathbf{i} - \frac{2}{3}\,\mathbf{j} - \frac{2}{3}\,\mathbf{k}$

A normal vector for the osculating plane is: $(2\,\mathbf{i} + \mathbf{j} - 2\,\mathbf{k}) \times (\mathbf{i} + 2\,\mathbf{j} + 2\,\mathbf{k}) = 6\,\mathbf{i} - 6\,\mathbf{j} + 3\,\mathbf{k}$.

Since $\mathbf{r}(1) = 2\mathbf{i} - \mathbf{k}$, an equation for the osculating plane is

$$6(x-2) - 6y + 3(z+1) = 0 \quad \text{or} \quad 2x - 2y + z = 3.$$

31. $\mathbf{r}'(t) = 2\mathbf{i} + t^{1/2}\mathbf{j}; \quad L = \int_0^5 ||\mathbf{r}'(t)||dt = \int_0^5 \sqrt{4+t}\,dt = \dfrac{38}{3}$

33. $\mathbf{r}'(t) = \cosh t\,\mathbf{i} + \sinh t\,\mathbf{j} + \mathbf{k}; \quad ||\mathbf{r}'(t)|| = \sqrt{\cosh^2 t + \sinh^2 t + 1} = \sqrt{2}\cosh t;$

$L = \int_0^1 \sqrt{2}\cosh t\,dt = \left[\sqrt{2}\sinh t\right]_0^1 = \sqrt{2}\sinh 1.$

35. (a) $\mathbf{r}'(t) = -\sin t\,\mathbf{i} + \cos t\,\mathbf{j} + t^{1/2}\mathbf{k}; \quad ||\mathbf{r}'(t)|| = \sqrt{1+t}.$

$s = \int_0^t ||\mathbf{r}'(u)||\,du = \int_0^t \sqrt{1+t}\,dt = \left[\dfrac{2}{3}(1+u)^{3/2}\right]_0^t = \dfrac{2}{3}(1+t)^{3/2} - \dfrac{2}{3}$

(b) $t = \left(\dfrac{3}{2}s+1\right)^{2/3} - 1 = \phi(s); \quad \mathbf{R}(s) = \cos\phi(s)\,\mathbf{i} + \sin\phi(s)\,\mathbf{j} + \dfrac{2}{3}[\phi(s)]^{3/2}\mathbf{k}$

(c) $\mathbf{R}'(s) = \left[-\sin\phi(s)\,\mathbf{i} + \cos\phi(s)\,\mathbf{j} + \phi(s)^{1/2}\mathbf{k}\right]\phi'(s)$

$||\mathbf{R}'(s)|| = \phi'(s)\sqrt{1+\phi(s)} = \dfrac{2}{3}\left[\dfrac{3}{2}s+1\right]^{-1/3}\left(\dfrac{3}{2}\right)\sqrt{\left(\dfrac{3}{2}s+1\right)^{2/3}} = 1$

37. $\mathbf{r}''(t) = -\cos t\,\mathbf{i} - \sin t\,\mathbf{j}$ and $\mathbf{r}'(0) = \mathbf{k} \Longrightarrow \mathbf{r}'(t) = -\sin t\,\mathbf{i} + (\cos t - 1)\mathbf{j} + \mathbf{k}.$

Thus: velocity $\mathbf{v} = -\sin t\,\mathbf{i} + (\cos t - 1)\mathbf{j} + \mathbf{k}$ and speed $||\mathbf{v}|| = \sqrt{3 - 2\cos t}.$

$\mathbf{r}'(t) = -\sin t\,\mathbf{i} + (\cos t - 1)\mathbf{j} + \mathbf{k}$ and $\mathbf{r}(0) = \mathbf{i} \Longrightarrow \mathbf{r}(t) = \cos t\,\mathbf{i} + (\sin t - t)\mathbf{j} + t\mathbf{k}.$

39. $y' = \dfrac{3}{2}x^{1/2}, \quad y'' = \dfrac{3}{4}x^{-1/2};$

$\kappa = \dfrac{|y''|}{[1+(y')^2]^{3/2}} = \dfrac{\frac{3}{4}x^{-1/2}}{[1+\frac{9}{4}x]^{3/2}} = \dfrac{6}{\sqrt{x}(4+9x)^{3/2}}$

41. $x(t) = 2e^{-t}, \ y(t) = e^{-2t} \Longrightarrow x'(t) = -2e^{-t}, \ y'(t) = -2e^{-2t} \Longrightarrow x''(t) = 2e^{-t}, \ y''(t) = 4e^{-2t}$

$\kappa = \dfrac{|(-2e^{-t})(4e^{-2t}) - (-2e^{-2t})(2e^{-t})|}{[4e^{-2t} + 4e^{-4t}]^{3/2}} = \dfrac{1}{2(1+e^{-2t})^{3/2}} = \dfrac{e^{3t}}{2(e^{2t}+1)^{3/2}}$

43. $\mathbf{r}'(t) = -3\sin 3t\,\mathbf{i} - 4\mathbf{j} + 3\cos 3t\,\mathbf{k}, \quad \dfrac{ds}{dt} = |\mathbf{r}'(t)| = 5$

$\mathbf{T}(t) = -\dfrac{3}{5}\sin 3t\,\mathbf{i} - \dfrac{4}{5}\mathbf{j} + \dfrac{3}{5}\cos 3t\,\mathbf{k}, \quad \mathbf{T}'(t) = -\dfrac{9}{5}\cos 3t\,\mathbf{i} - \dfrac{9}{5}\sin 3t\,\mathbf{k}; \quad ||\mathbf{T}'(t)|| = 9/5$

$\kappa = \dfrac{||\mathbf{T}'(t)||}{ds/dt} = \dfrac{9}{25}$

45. $y' = \sinh(x/a), \quad y'' = \dfrac{1}{a}\cosh(x/a); \quad \kappa = \dfrac{|y''|}{[1+(y')^2]^{\frac{3}{2}}} = \dfrac{1}{a\cosh^2(x/a)} = \dfrac{a}{y^2}$

47. $\mathbf{r}'(t) = -\dfrac{4}{5}\sin t\,\mathbf{i} + \dfrac{3}{5}\sin t\,\mathbf{j} + \cos t\,\mathbf{k}; \quad \dfrac{ds}{dt} = \|\mathbf{r}'(t)\| = 1$

$\mathbf{T} = \mathbf{r}'(t) = -\dfrac{4}{5}\sin t\,\mathbf{i} + \dfrac{3}{5}\sin t\,\mathbf{j} + \cos t\,\mathbf{k}, \quad \mathbf{T}'(t) = -\dfrac{4}{5}\cos t\,\mathbf{i} + \dfrac{3}{5}\cos t\,\mathbf{j} - \sin t\,\mathbf{k}; \quad \|\mathbf{T}'(t)\| = 1$

$\kappa = 1; \qquad \mathbf{a_T} = \dfrac{d^2 s}{dt^2} = 0, \quad \mathbf{a_N} = \kappa \left(\dfrac{ds}{dt}\right)^2 = 1$

CHAPTER 15

SECTION 15.1

1. dom (f) = the first and third quadrants, including the axes;   range $(f) = [0, \infty)$

3. dom (f) = the set of all points (x, y) except those on the line $y = -x$;   range $(f) = (-\infty, 0) \cup (0, \infty)$

5. dom (f) = the entire plane;   range $(f) = (-1, 1)$ since

$$\frac{e^x - e^y}{e^x + e^y} = \frac{e^x + e^y - 2e^y}{e^x + e^y} = 1 - \frac{2}{e^{x-y} + 1}$$

and the last quotient takes on all values between 0 and 2.

7. dom (f) = the first and third quadrants, excluding the axes;   range $(f) = (-\infty, \infty)$

9. dom (f) = the set of all points (x, y) with $x^2 < y$ —in other words, the set of all points of the plane above the parabola $y = x^2$;   range $(f) = (0, \infty)$

11. dom (f) = the set of all points (x, y) with $-3 \leq x \leq 3$, $-2 \leq y \leq 2$ (a rectangle);
range $(f) = [-2, 3]$

13. dom (f) = the set of all points (x, y, z) not on the plane $x + y + z = 0$;   range $(f) = \{-1, 1\}$

15. dom (f) = the set of all points (x, y, z) with $|y| < |x|$;   range $(f) = (-\infty, 0]$

17. dom (f) = the set of all points (x, y) with $x^2 + y^2 < 9$ —in other words, the set of all points of the plane inside the circle $x^2 + y^2 = 9$;   range $(f) = [2/3, \infty)$

19. dom (f) = the set of all points (x, y, z) with $x + 2y + 3z > 0$ — in other words, the set of all points in space that lie on the same side of the plane $x + 2y + 3z = 0$ as the point $(1, 1, 1)$;   range $(f) = (-\infty, \infty)$

21. dom (f) = all of space;   range $(f) = (0, 1]$

23. dom $(f) = \{x : x \geq 0\}$;   range $(f) = [0, \infty)$
dom $(g) = \{(x, y) : x \geq 0,\ y \text{ real}\}$;   range $(g) = [0, \infty)$
dom $(h) = \{(x, y, z) : x \geq 0,\ y, z \text{ real}\}$;   range $(h) = [0, \infty)$

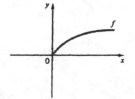
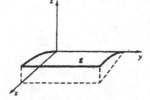

25. $\lim\limits_{h\to 0}\dfrac{f(x+h,y)-f(x,y)}{h}=\lim\limits_{h\to 0}\dfrac{2(x+h)^2-y-(2x^2-y)}{h}=\lim\limits_{h\to 0}\dfrac{4xh+2h^2}{h}=4x$

$\lim\limits_{h\to 0}\dfrac{f(x,y+h)-f(x,y)}{h}=\lim\limits_{h\to 0}\dfrac{2x^2-(y+h)-(2x^2-y)}{h}=-1$

27. $\lim\limits_{h\to 0}\dfrac{f(x+h,y)-f(x,y)}{h}=\lim\limits_{h\to 0}\dfrac{3(x+h)-(x+h)y+2y^2-(3x-xy+2y^2)}{h}=\lim\limits_{h\to 0}\dfrac{3h-hy}{h}=3-y$

$\lim\limits_{h\to 0}\dfrac{f(x,y+h)-f(x,y)}{h}=\lim\limits_{h\to 0}\dfrac{3x-x(y+h)+2(y+h)^2-(3x-xy+2y^2)}{h}$

$\qquad\qquad\qquad\qquad\qquad =\lim\limits_{h\to 0}\dfrac{-xh+4yh+2h^2}{h}=-x+4y$

29. $\lim\limits_{h\to 0}\dfrac{f(x+h,y)-f(x,y)}{h}=\lim\limits_{h\to 0}\dfrac{\cos[(x+h)y]-\cos[xy]}{h}$

$\qquad\qquad\qquad\qquad\qquad =\lim\limits_{h\to 0}\dfrac{\cos[xy]\cos[hy]-\sin[xy]\sin[hy]-\cos[xy]}{h}$

$\qquad\qquad\qquad\qquad\qquad =\cos[xy]\left(\lim\limits_{h\to 0}\dfrac{\cos[hy]-1}{h}\right)-\sin[xy]\lim\limits_{h\to 0}\dfrac{\sin hy}{h}$

$\qquad\qquad\qquad\qquad\qquad =y\cos[xy]\left(\lim\limits_{h\to 0}\dfrac{\cos[hy]-1}{hy}\right)-y\sin[xy]\lim\limits_{h\to 0}\dfrac{\sin hy}{hy}$

$\qquad\qquad\qquad\qquad\qquad =-y\sin[xy]$

and

$\qquad\lim\limits_{h\to 0}\dfrac{f(x,y+h)-f(x,y)}{h}=\lim\limits_{h\to 0}\dfrac{\cos[x(y+h)]-\cos[xy]}{h}$

$\qquad\qquad\qquad\qquad\qquad =\lim\limits_{h\to 0}\dfrac{\cos[xy]\cos[hx]-\sin[xy]\sin[hx]-\cos[xy]}{h}$

$\qquad\qquad\qquad\qquad\qquad =\cos[xy]\left(\lim\limits_{h\to 0}\dfrac{\cos[hx]-1}{h}\right)-\sin[xy]\lim\limits_{h\to 0}\dfrac{\sin hx}{h}$

$\qquad\qquad\qquad\qquad\qquad =x\cos[xy]\left(\lim\limits_{h\to 0}\dfrac{\cos[hx]-1}{hx}\right)-x\sin[xy]\lim\limits_{h\to 0}\dfrac{\sin hx}{hx}$

$\qquad\qquad\qquad\qquad\qquad =-x\sin[xy]$

31. (a) $f(x,y)=Ay$ $\qquad\qquad$ (b) $f(x,y)=\pi x^2 y$ $\qquad\qquad$ (b) $f(x,y)=|2\,\mathbf{i}\times(x\,\mathbf{i}+y\,\mathbf{j})|=2|y|$

33. Surface area: $S=2lw+2lh+2hw=20\implies w=\dfrac{20-2lh}{2l+2h}=\dfrac{10-lh}{l+h}$

Volume: $V=lwh=\dfrac{lh(10-lh)}{l+h}$

35. $V=\pi r^2 h+\dfrac{4}{3}\pi r^3$

SECTION 15.2

1. an elliptic cone

3. a parabolic cylinder

5. a hyperboloid of one sheet

7. a sphere

9. an elliptic paraboloid

11. a hyperbolic paraboloid

13.

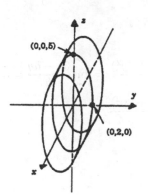

15.

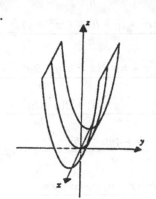

17.

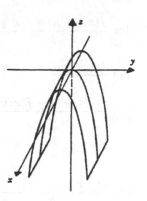

19.

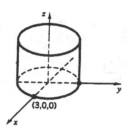

21.

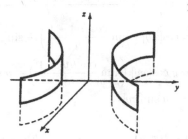

23.

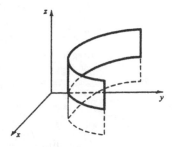

25. elliptic paraboloid
xy-trace: the origin
xz-trace: the parabola $x^2 = 4z$
yz-trace: the parabola $y^2 = 9z$
surface has the form of Figure 15.2.5

27. elliptic cone
xy-trace: the origin
xz-trace: the lines $x = \pm 2z$
yz-trace: the lines $y = \pm 3z$
surface has the form of Figure 15.2.4

29. hyperboloid of two sheets
xy-trace: none
xz-trace: the hyperbola $4z^2 - x^2 = 4$
yz-trace: the hyperbola $9z^2 - y^2 = 9$
surface has the form of Figure 15.2.3

31. hyperboloid of two sheets
xy-trace: the hyperbola $9x^2 - 4y^2 = 36$
xz-trace: the hyperbola $x^2 - 4z^2 = 4$
yz-trace: none
see Figure 15.2.3

33. elliptic paraboloid
xy-trace: the parabola $x^2 = 9y$
xz-trace: the origin
yz-trace: the parabola $z^2 = 4y$
surface has the form of Figure 15.2.5

35. hyperboloid of two sheets
xy-trace: the hyperbola $9y^2 - 4x^2 = 36$
xz-trace: none
yz-trace: the hyperbola $y^2 - 4z^2 = 4$
see Figure 15.2.3

37. paraboloid of revolution
xy-trace: the origin
xz-trace: the parabola $x^2 = 4z$
yz-trace: the parabola $y^2 = 4z$
surface has the form of Figure 15.2.5

39. (a) an elliptic paraboloid (vertex down if A and B are both positive, vertex up if A and B are both negative)

(b) a hyperbolic paraboloid

(c) the xy-plane if A and B are both zero; otherwise a parabolic cylinder

41. $x^2 + y^2 - 4z = 0$ (paraboloid of revolution)

43. (a) a circle

(b) (i) $\sqrt{x^2 + y^2} = -3z$ (ii) $\sqrt{x^2 + z^2} = \frac{1}{3}y$

45. $x + 2y + 3\left(\dfrac{x + y - 6}{2}\right) = 6$ or $5x + 7y = 30$,   a line

47. $\left.\begin{array}{r} x^2 + y^2 + (z-1)^2 = \frac{3}{2} \\ x^2 + y^2 - z^2 = 1 \end{array}\right\}$ $(z^2 + 1) + (z-1)^2 = \dfrac{3}{2};$ $(2z-1)^2 = 0,$ $z = \dfrac{1}{2}$ so that $x^2 + y^2 = \dfrac{5}{4}$

49. $x^2 + y^2 + \left(x^2 + 3y^2\right) = 4$ or $x^2 + 2y^2 = 2$,   an ellipse

51. $x^2 + y^2 = (2 - y)^2$ or $x^2 = -4(y-1)$,   a parabola

53. (a) Set $x = a\cos u \cos v,\ y = b\cos u \sin v,\ z = c\sin u.$ Then: $\dfrac{x^2}{a^2} + \dfrac{y^2}{b^2} + \dfrac{z^2}{c^2} = 1.$

(b)

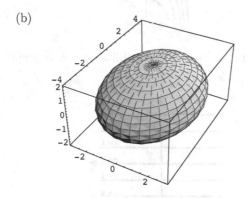

55. (a) Set $x = av\cos u,$

$y = bv\sin u,$

$z = cv$

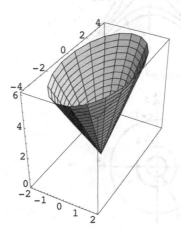

SECTION 15.3

1.   lines of slope 1: $y = x - c$

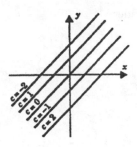

3.   parabolas: $y = x^2 - c$

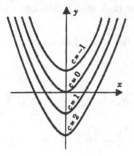

5.   the y-axis and the lines $y = \left(\dfrac{1-c}{c}\right) x$ with the origin omitted throughout

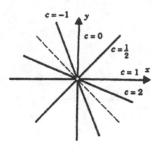

7.   the cubics $y = x^3 - c$

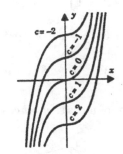

9.   the lines $y = \pm x$ and the hyperbolas $x^2 - y^2 = c$

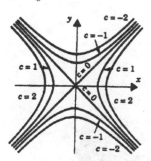

11.   pairs of horizontal lines $y = \pm\sqrt{c}$ and the x-axis

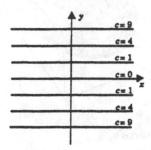

13.   the circles $x^2 + y^2 = e^c$, c real

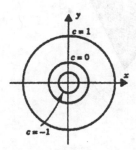

15.   the curves $y = e^{cx^2}$ with the point $(0,1)$ omitted

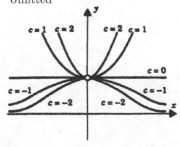

17. the coordinate axes and pairs of lines

$$y = \pm \frac{\sqrt{1-c}}{\sqrt{c}}\, x, \quad \text{with the origin omitted}$$

throughout

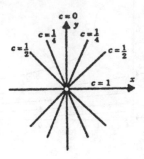

19. $x + 2y + 3z = 0$, plane through the origin

21. $z = \sqrt{x^2 + y^2}$, the upper nappe of the circular cone $z^2 = x^2 + y^2$ (Figure 15.2.4)

23. the elliptic paraboloid $\dfrac{x^2}{18} + \dfrac{y^2}{8} = z$ (Figure 15.2.5)

25. (i) hyperboloid of two sheets   (Figure 15.2.3)

(ii) circular cone   (Figure 15.2.4)

(iii) hyperboloid of one sheet   (Figure 15.2.2)

27. The level curves of f are: $1 - 4x^2 - y^2 = c$. Substituting $P(0,1)$ into this equation, we have

$$1 - 4(0)^2 - (1)^2 = c \quad \Longrightarrow \quad c = 0$$

The level curve that contains P is: $1 - 4x^2 - y^2 = 0$, or $4x^2 + y^2 = 1$.

29. The level curves of f are: $y^2 \arctan x = c$. Substituting $P(1,2)$ into this equation, we have

$$4 \arctan 1 = c \quad \Longrightarrow \quad c = \pi$$

The level curve that contains P is: $y^2 \tan^{-1} x = \pi$.

31. The level surfaces of f are: $x^2 + 2y^2 - 2xyz = c$. Substituting $P(-1, 2, 1)$ into this equation, we have

$$(-1)^2 + 2(2)^2 - 2(-1)(2)(1) = c \quad \Longrightarrow \quad c = 13$$

The level surface that contains P is: $x^2 + 2y^2 - 2xyz = 13$.

33. (a)

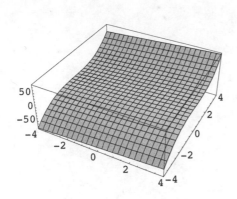

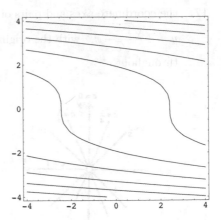

(b)

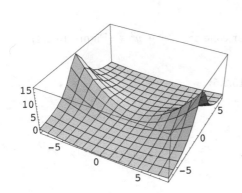

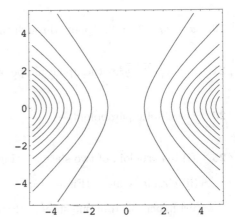

35. (a) $\dfrac{3x + 2y + 1}{4x^2 + 9} = \dfrac{3}{5}$　　　　　　(b) $x^2 + 2y^2 - z^2 = 21$

37. $\dfrac{GmM}{x^2 + y^2 + z^2} = c \implies x^2 + y^2 + z^2 = \dfrac{GmM}{c}$; the surfaces of constant gravitational force are concentric spheres.

39. (a) $T(x,y,z) = \dfrac{k}{\sqrt{x^2 + y^2 + z^2}}$, where k is a constant.

(b) $\dfrac{k}{\sqrt{x^2 + y^2 + z^2}} = c \implies x^2 + y^2 + z^2 = \dfrac{k^2}{c^2}$; the level surfaces are concentric spheres.

(c) $T(1,2,1) = \dfrac{k}{\sqrt{1^2 + 2^2 + 1^2}} = 50 \implies k = 50\sqrt{6} \implies T(x,y,z) = \dfrac{50\sqrt{6}}{\sqrt{x^2 + y^2 + z^2}}$

41. $f(x,y) = y^2 - y^3$; F　　　　**43.** $f(x,y) = \cos\sqrt{x^2 + y^2}$; A　　**45.** $f(x,y) = xye^{-(x^2+y^2)/2}$; E

PROJECT 15.3

1.

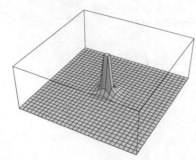

3.

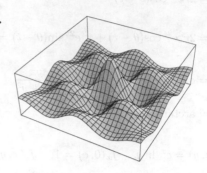

SECTION 15.4

1. $\dfrac{\partial f}{\partial x} = 6x - y, \quad \dfrac{\partial f}{\partial y} = 1 - x$

3. $\dfrac{\partial \rho}{\partial \phi} = \cos\phi\cos\theta, \quad \dfrac{\partial \rho}{\partial \theta} = -\sin\phi\sin\theta$

5. $\dfrac{\partial f}{\partial x} = e^{x-y} + e^{y-x}, \quad \dfrac{\partial f}{\partial y} = -e^{x-y} - e^{y-x}$

7. $\dfrac{\partial g}{\partial x} = \dfrac{(AD - BC)y}{(Cx + Dy)^2}, \quad \dfrac{\partial g}{\partial y} = \dfrac{(BC - AD)x}{(Cx + Dy)^2}$

9. $\dfrac{\partial u}{\partial x} = y + z, \quad \dfrac{\partial u}{\partial y} = x + z, \quad \dfrac{\partial u}{\partial z} = x + y$

11. $\dfrac{\partial f}{\partial x} = z\cos(x - y), \quad \dfrac{\partial f}{\partial y} = -z\cos(x - y), \quad \dfrac{\partial f}{\partial z} = \sin(x - y)$

13. $\dfrac{\partial \rho}{\partial \theta} = e^{\theta+\phi}\left[\cos(\theta - \phi) - \sin(\theta - \phi)\right], \quad \dfrac{\partial \rho}{\partial \phi} = e^{\theta+\phi}\left[\cos(\theta - \phi) + \sin(\theta - \phi)\right]$

15. $\dfrac{\partial f}{\partial x} = 2xy\sec xy + x^2 y(\sec xy)(\tan xy)y = 2xy\sec xy + x^2 y^2 \sec xy\tan xy$

$\dfrac{\partial f}{\partial y} = x^2 \sec xy + x^2 y(\sec xy)(\tan xy)x = x^2 \sec xy + x^3 y\sec xy\tan xy$

17. $\dfrac{\partial h}{\partial x} = \dfrac{x^2 + y^2 - x(2x)}{(x^2 + y^2)^2} = \dfrac{y^2 - x^2}{(x^2 + y^2)^2} \qquad \dfrac{\partial h}{\partial y} = \dfrac{-2xy}{(x^2 + y^2)^2}$

19. $\dfrac{\partial f}{\partial x} = \dfrac{(y\cos x)\sin y - (x\sin y)(-y\sin x)}{(y\cos x)^2} = \dfrac{\sin y(\cos x + x\sin x)}{y\cos^2 x}$

$\dfrac{\partial f}{\partial y} = \dfrac{(y\cos x)(x\cos y) - (x\sin y)\cos x}{(y\cos x)^2} = \dfrac{x(y\cos y - \sin y)}{y^2\cos x}$

21. $\dfrac{\partial h}{\partial x} = 2f(x)f'(x)g(y), \quad \dfrac{\partial h}{\partial y} = [f(x)]^2 g'(y)$

23. $\dfrac{\partial f}{\partial x} = (y^2 \ln z)z^{xy^2}, \quad \dfrac{\partial f}{\partial y} = (2xy\ln z)z^{xy^2}, \quad \dfrac{\partial f}{\partial z} = xy^2 z^{xy^2-1}$

25. $\dfrac{\partial h}{\partial r} = 2re^{2t}\cos(\theta - t)$ $\dfrac{\partial h}{\partial \theta} = -r^2 e^{2t}\sin(\theta - t)$

$\dfrac{\partial h}{\partial t} = 2r^2 e^{2t}\cos(\theta - t) + r^2 e^{2t}\sin(\theta - t) = r^2 e^{2t}[2\cos(\theta - t) + \sin(\theta - t)]$

27. $\dfrac{\partial f}{\partial x} = z\,\dfrac{1}{1+(y/x)^2}\left(\dfrac{-y}{x^2}\right) = -\dfrac{yz}{x^2+y^2}$ $\dfrac{\partial f}{\partial y} = z\,\dfrac{1}{1+(y/x)^2}\left(\dfrac{1}{x}\right) = \dfrac{xz}{x^2+y^2}$

$\dfrac{\partial f}{\partial x} = \arctan(y/x)$

29. $f_x(x,y) = e^x\ln y,\quad f_x(0,e) = 1;\quad f_y(x,y) = \dfrac{1}{y}e^x,\quad f_y(0,e) = e^{-1}$

31. $f_x(x,y) = \dfrac{y}{(x+y)^2},\quad f_x(1,2) = \dfrac{2}{9};\quad f_y(x,y) = \dfrac{-x}{(x+y)^2},\quad f_y(1,2) = -\dfrac{1}{9}$

33. $f_x(x,y) = \lim\limits_{h\to 0}\dfrac{(x+h)^2 y - x^2 y}{h} = \lim\limits_{h\to 0} y\left(\dfrac{2xh + h^2}{h}\right) = y\lim\limits_{h\to 0}(2x+h) = 2xy$

$f_x(x,y) = \lim\limits_{h\to 0}\dfrac{x^2(y+h) - x^2 y}{h} = \lim\limits_{h\to 0}\dfrac{x^2 h}{h} = \lim\limits_{h\to 0} x^2 = x^2$

35. $f_x(x,y) = \lim\limits_{h\to 0}\dfrac{\ln\left(y(x+h)^2\right) - \ln x^2 y}{h} = \lim\limits_{h\to 0}\dfrac{\ln y + 2\ln(x+h) - 2\ln x - \ln y}{h}$

$= 2\lim\limits_{h\to 0}\dfrac{\ln(x+h) - \ln x}{h} = 2\dfrac{d}{dx}(\ln x) = \dfrac{2}{x}$

$f_y(x,y) = \lim\limits_{h\to 0}\dfrac{\ln\left(x^2(y+h)\right) - \ln x^2 y}{h} = \lim\limits_{h\to 0}\dfrac{\ln x^2 + \ln(y+h) - \ln x^2 - \ln y}{h}$

$= \lim\limits_{h\to 0}\dfrac{\ln(y+h) - \ln y}{h} = \dfrac{d}{dy}(\ln y) = \dfrac{1}{y}$

37. $f_y(x,y) = \lim\limits_{h\to 0}\dfrac{1}{h}\left\{\dfrac{1}{(x+h)-y} - \dfrac{1}{x-y}\right\} = \lim\limits_{h\to 0}\dfrac{1}{h}\left\{\dfrac{-h}{(x+h-y)(x-y)}\right\}$

$= \lim\limits_{h\to 0}\dfrac{-1}{(x+h-y)(x-y)} = \dfrac{-1}{(x-y)^2}$

$f_y(x,y) = \lim\limits_{h\to 0}\dfrac{1}{h}\left\{\dfrac{1}{x-(y+h)} - \dfrac{1}{x-y}\right\} = \lim\limits_{h\to 0}\dfrac{1}{h}\left\{\dfrac{h}{(x-y-h)(x-y)}\right\}$

$= \lim\limits_{h\to 0}\dfrac{1}{(x-y-h)(x-y)} = \dfrac{1}{(x-y)^2}$

39. $f_x(x,y,z) = \lim\limits_{h\to 0}\dfrac{(x+h)y^2 z - xy^2 z}{h} = \lim\limits_{h\to 0} y^2 z = y^2 z$

$f_y(x,y,z) = \lim\limits_{h\to 0}\dfrac{x(y+h)^2 z - xy^2 z}{h} = \lim\limits_{h\to 0}\dfrac{xz(2yh + h^2)}{h}$

$= \lim\limits_{h\to 0} xz(2y + h) = 2xyz$

$f_z(x,y,z) = \lim\limits_{h\to 0}\dfrac{xy^2(z+h) - xy^2 z}{h} = \lim\limits_{h\to 0} xy^2 = xy^2$

41. (b) The slope of the tangent line to C at the point $P(x_0, y_0, f(x_0, y_0))$ is $f_y(x_0, y_0)$
Thus, equations for the tangent line are:

$$x = x_0, \quad z - z_0 = f_y(x_0, y_0)(y - y_0)$$

43. Let $z = f(x, y) = x^2 + y^2$. Then $f(2, 1) = 5$, $\quad f_y(x, y) = 2y$ and $\quad f_y(2, 1) = 2$;
equations for the tangent line are: $\quad x = 2, \quad z - 5 = 2(y - 1)$

45. Let $z = f(x, y) = \dfrac{x^2}{y^2 - 3}$. Then $f(3, 2) = 9$, $\quad f_x(x, y) = \dfrac{2x}{y^2 - 3}$ and $\quad f_x(3, 2) = 6$;
equations for the tangent line are: $\quad y = 2, \quad z - 9 = 6(x - 3)$

47. (a) $m_x = -6$; $\qquad$ tangent line: $\quad y = 2, \quad z = -6x + 13$

$\quad\;$ (b) $m_y = 18$; $\qquad$ tangent line: $\quad x = 1, \quad z = 18y - 29$

49. $u_x(x, y) = 2x = v_y(x, y)$; $\qquad u_y(x, y) = -2y = -v_x(x, y)$

51. $u_x(x, y) = \dfrac{1}{2} \dfrac{1}{x^2 + y^2} 2x = \dfrac{x}{x^2 + y^2}$; $\quad v_y(x, y) = \dfrac{1}{1 + (y/x)^2}\left(\dfrac{1}{x}\right) = \dfrac{x}{x^2 + y^2}$

Thus, $u_x(x, y) = v_y(x, y)$.

$u_y(x, y) = \dfrac{1}{2} \dfrac{1}{x^2 + y^2} 2y = \dfrac{y}{x^2 + y^2}$; $\quad v_x(x, y) = \dfrac{1}{1 + (y/x)^2}\left(\dfrac{-y}{x^2}\right) = \dfrac{-y}{x^2 + y^2}$

Thus, $u_y(x, y) = -v_x(x, y)$.

53. (a) f depends only on y. $\qquad\qquad\qquad$ (b) f depends only on x.

55. (a) $\dfrac{75\sqrt{3}}{2}$ in.^{2}

$\quad\;$ (b) $\dfrac{\partial A}{\partial b} = \dfrac{1}{2} c \sin\theta$; $\quad$ at time t_0, $\quad \dfrac{\partial A}{\partial b} = \dfrac{15\sqrt{3}}{4}$

$\quad\;$ (c) $\dfrac{\partial A}{\partial \theta} = \dfrac{1}{2} bc \cos\theta$; $\quad$ at time t_0, $\quad \dfrac{\partial A}{\partial \theta} = \dfrac{75}{2}$

$\quad\;$ (d) with $\quad h = \dfrac{\pi}{180}$, $\quad A(b, c, \theta + h) - A(b, c, \theta) \cong h\dfrac{\partial A}{\partial \theta} = \dfrac{\pi}{180}\dfrac{75}{2} = \dfrac{5\pi}{24}$ in.^{2}

$\quad\;$ (e) $0 = \dfrac{1}{2} \sin\theta\left(b\dfrac{\partial c}{\partial b} + c\right)$; $\quad$ at time t_0, $\quad \dfrac{\partial c}{\partial b} = \dfrac{-c}{b} = -\dfrac{3}{2}$

57. (a) y_0-section: $\mathbf{r}(x) = x\mathbf{i} + y_0\mathbf{j} + f(x, y_0)\mathbf{k}$

$\qquad$ tangent line: $\mathbf{R}(t) = [x_0\mathbf{i} + y_0\mathbf{j} + f(x_0, y_0)\mathbf{k}] + t\left[\mathbf{i} + \dfrac{\partial f}{\partial x}(x_0, y_0)\mathbf{k}\right]$

$\quad\;$ (b) x_0-section: $\mathbf{r}(y) = x_0\mathbf{i} + y\mathbf{j} + f(x_0, y)\mathbf{k}$

$\qquad$ tangent line: $\mathbf{R}(t) = [x_0\mathbf{i} + y_0\mathbf{j} + f(x_0, y_0)\mathbf{k}] + t\left[\mathbf{j} + \dfrac{\partial f}{\partial y}(x_0, y_0)\mathbf{k}\right]$

(c) For (x, y, z) in the plane

$$[(x - x_0)\mathbf{i} + (y - y_0)\mathbf{j} + (z - f(x_0, y_0))\mathbf{k}] \cdot \left[\left(\mathbf{i} + \frac{\partial f}{\partial x}(x_0, y_0)\mathbf{k}\right) \times \left(\mathbf{j} + \frac{\partial f}{\partial y}(x_0, y_0)\mathbf{k}\right)\right] = 0.$$

From this it follows that

$$z - f(x_0, y_0) = (x - x_0)\frac{\partial f}{\partial x}(x_0, y_0) + (y - y_0)\frac{\partial f}{\partial y}(x_0, y_0).$$

59. (a) Set $u = ax + by$. Then

$$b\frac{\partial w}{\partial x} - a\frac{\partial w}{\partial y} = b(a\,g'(u)) - a(b\,g'(u)) = 0.$$

(b) Set $u = x^m y^n$. Then

$$nx\frac{\partial w}{\partial x} - my\frac{\partial w}{\partial y} = nx\left[mx^{m-1}y^n g'(u)\right] - my\left[nx^m y^{n-1}g'(u)\right] = 0.$$

61. $\quad V\dfrac{\partial P}{\partial V} = V\left(-\dfrac{kT}{V^2}\right) = -k\dfrac{T}{V} = -P; \qquad V\dfrac{\partial P}{\partial V} + T\dfrac{\partial P}{\partial T} = -k\dfrac{T}{V} + T\left(\dfrac{k}{V}\right) = 0$

SECTION 15.5

1. interior $= \{(x, y) : 2 < x < 4, \quad 1 < y < 3\}$ (the inside of the rectangle), boundary $=$ the union of the four boundary line segments; set is closed.

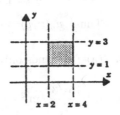

3. interior $=$ the entire set (region between the two concentric circles), boundary $=$ the two circles, one of radius 1, the other of radius 2; set is open.

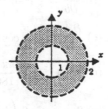

5. interior $= \{(x, y) : 1 < x^2 < 4\} =$ $\{(x, y) : -2 < x < -1\} \cup \{(x, y) : 1 < x < 2\}$ (two vertical strips without the boundary lines), boundary $= \{(x, y) : x = -2,\ x = -1,\ x = 1,$ or $x = 2\}$ (four vertical lines); set is neither open nor closed.

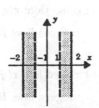

7. interior = region below the parabola $y = x^2$,
    boundary = the parabola $y = x^2$;   the set is closed.

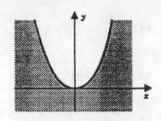

9. interior = $\{ (x,y,z) : x^2 + y^2 < 1, 0 < z \le 4 \}$
    (the inside of the cylinder), boundary = the total
    surface of the cylinder (the curved part, the top, and
    the bottom);   the set is closed.

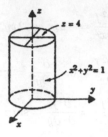

11. (a) ϕ (b) S (c) closed

13. interior = $\{x : 1 < x < 3\}$, boundary = $\{1, 3\}$;   set is closed.

15. interior = the entire set, boundary = $\{1\}$;   set is open.

17. interior = $\{x : |x| > 1\}$, boundary = $\{1, -1\}$;   set is neither open nor closed.

19. interior = ϕ, boundary = {the entire set} $\cup$ $\{0\}$;   the set is neither open nor closed.

SECTION 15.6

1. $\dfrac{\partial^2 f}{\partial x^2} = 2A, \quad \dfrac{\partial^2 f}{\partial y^2} = 2C, \quad \dfrac{\partial^2 f}{\partial y \partial x} = \dfrac{\partial^2 f}{\partial x \partial y} = 2B$

3. $\dfrac{\partial^2 f}{\partial x^2} = Cy^2 e^{xy}, \quad \dfrac{\partial^2 f}{\partial y^2} = Cx^2 e^{xy}, \quad \dfrac{\partial^2 f}{\partial y \partial x} = \dfrac{\partial^2 f}{\partial x \partial y} = Ce^{xy}(xy + 1)$

5. $\dfrac{\partial^2 f}{\partial x^2} = 2, \quad \dfrac{\partial^2 f}{\partial y^2} = 4(x + 3y^2 + z^3), \quad \dfrac{\partial^2 f}{\partial z^2} = 6z(2x + 2y^2 + 5z^3)$

    $\dfrac{\partial^2 f}{\partial x \partial y} = \dfrac{\partial^2 f}{\partial y \partial x} = 4y, \quad \dfrac{\partial^2 f}{\partial z \partial x} = \dfrac{\partial^2 f}{\partial x \partial z} = 6z^2, \quad \dfrac{\partial^2 f}{\partial z \partial y} = \dfrac{\partial^2 f}{\partial y \partial z} = 12yz^2$

7. $\dfrac{\partial^2 f}{\partial x^2} = \dfrac{1}{(x + y)^2} - \dfrac{1}{x^2}, \quad \dfrac{\partial^2 f}{\partial y^2} = \dfrac{1}{(x + y)^2}, \quad \dfrac{\partial^2 f}{\partial y \partial x} = \dfrac{\partial^2 f}{\partial x \partial y} = \dfrac{1}{(x + y)^2}$

9. $\dfrac{\partial^2 f}{\partial x^2} = 2(y + z), \quad \dfrac{\partial^2 f}{\partial y^2} = 2(x + z), \quad \dfrac{\partial^2 f}{\partial z^2} = 2(x + y)$

    all the second mixed partials are $2(x + y + z)$

11. $\dfrac{\partial^2 f}{\partial x^2} = y(y-1)x^{y-2}$, $\dfrac{\partial^2 f}{\partial y^2} = (\ln x)^2 x^y$, $\dfrac{\partial^2 f}{\partial y \partial x} = \dfrac{\partial^2 f}{\partial x \partial y} = x^{y-1}(1 + y\ln x)$

13. $\dfrac{\partial^2 f}{\partial x^2} = ye^x$, $\dfrac{\partial^2 f}{\partial y^2} = xe^y$, $\dfrac{\partial^2 f}{\partial y \partial x} = \dfrac{\partial^2 f}{\partial x \partial y} = e^y + e^x$

15. $\dfrac{\partial^2 f}{\partial x^2} = \dfrac{y^2 - x^2}{(x^2 + y^2)^2}$, $\dfrac{\partial^2 f}{\partial y^2} = \dfrac{x^2 - y^2}{(x^2 + y^2)^2}$, $\dfrac{\partial^2 f}{\partial y \partial x} = \dfrac{\partial^2 f}{\partial x \partial y} = -\dfrac{2xy}{(x^2 + y^2)^2}$

17. $\dfrac{\partial^2 f}{\partial x^2} = -2\,y^2 \cos 2xy$, $\dfrac{\partial^2 f}{\partial y^2} = -2\,x^2 \cos 2xy$, $\dfrac{\partial^2 f}{\partial y \partial x} = \dfrac{\partial^2 f}{\partial x \partial y} = -[\sin 2xy + 2xy \cos 2xy]$

19. $\dfrac{\partial^2 f}{\partial x^2} = 0$, $\dfrac{\partial^2 f}{\partial y^2} = xz \sin y$, $\dfrac{\partial^2 f}{\partial z^2} = -xy \sin z$,

$\dfrac{\partial^2 f}{\partial y \partial x} = \dfrac{\partial^2 f}{\partial x \partial y} = \sin z - z \cos y$, $\dfrac{\partial^2 f}{\partial x \partial z} = \dfrac{\partial^2 f}{\partial z \partial x} = y \cos z - \sin y$, $\dfrac{\partial^2 f}{\partial y \partial z} = \dfrac{\partial^2 f}{\partial z \partial y} = x \cos z - x \cos y$

21. $x^2 \dfrac{\partial^2 u}{\partial x^2} + 2xy \dfrac{\partial^2 u}{\partial x \partial y} + y^2 \dfrac{\partial^2 u}{\partial y^2} = x^2 \left(\dfrac{-2y^2}{(x+y)^3} \right) + 2xy \left(\dfrac{2xy}{(x+y)^3} \right) + y^2 \left(\dfrac{-2x^2}{(x+y)^3} \right) = 0$

23. (a) no, since $\dfrac{\partial^2 f}{\partial y \partial x} \neq \dfrac{\partial^2 f}{\partial x \partial y}$ (b)   no, since $\dfrac{\partial^2 f}{\partial y \partial x} \neq \dfrac{\partial^2 f}{\partial x \partial y}$ for $x \neq y$

25. $\dfrac{\partial^3 f}{\partial x^2 \partial y} = \dfrac{\partial}{\partial x}\left(\dfrac{\partial^2 f}{\partial x \partial y} \right) = \dfrac{\partial}{\partial x}\left(\dfrac{\partial^2 f}{\partial y \partial x} \right) = \dfrac{\partial^2}{\partial x \partial y}\left(\dfrac{\partial f}{\partial x} \right) = \dfrac{\partial^2}{\partial y \partial x}\left(\dfrac{\partial f}{\partial x} \right) = \dfrac{\partial}{\partial y}\left(\dfrac{\partial^2 f}{\partial x^2} \right) = \dfrac{\partial^3 f}{\partial y \partial x^2}$

     ↑ by def.   ↑ (15.6.5)   ↑ by def.   ↑ (15.6.5)   ↑ by def.   ↑ by def.

27. (a) $\displaystyle\lim_{x \to 0} \dfrac{(x)(0)}{x^2 + 0} = \lim_{x \to 0} 0 = 0$ (b) $\displaystyle\lim_{y \to 0} \dfrac{(0)(y)}{0 + y^2} = \lim_{y \to 0} 0 = 0$

(c) $\displaystyle\lim_{x \to 0} \dfrac{(x)(mx)}{x^2 + (mx)^2} = \lim_{x \to 0} \dfrac{m}{1 + m^2} = \dfrac{m}{1 + m^2}$

(d) $\displaystyle\lim_{\theta \to 0^+} \dfrac{(\theta \cos \theta)(\theta \sin \theta)}{(\theta \cos \theta)^2 + (\theta \sin \theta)^2} = \lim_{\theta \to 0^+} \cos \theta \sin \theta = 0$

(e) By L'Hospital's rule $\displaystyle\lim_{x \to 0} \dfrac{f(x)}{x} = \lim_{x \to 0} f'(x) = f'(0)$.   Thus

$$\lim_{x \to 0} \dfrac{xf(x)}{x^2 + [\,f(x)\,]^2} = \lim_{x \to 0} \dfrac{f(x)/x}{1 + [\,f(x)/x\,]^2} = \dfrac{f'(0)}{1 + [\,f'(0)\,]^2}.$$

(f) $\displaystyle\lim_{\theta \to (\pi/3)^-} = \dfrac{(\cos \theta \sin 3\theta)(\sin \theta \sin 3\theta)}{(\cos \theta \sin 3\theta)^2 + (\sin \theta \sin 3\theta)^2} = \lim_{\theta \to (\pi/3)^-} \cos \theta \sin \theta = \dfrac{1}{4}\sqrt{3}$

(g) $\displaystyle\lim_{t \to \infty} \dfrac{(1/t)(\sin t)/t}{1/t^2 + \left(\sin^2 t\right)/t^2} = \lim_{t \to \infty} \dfrac{\sin t}{1 + \sin^2 t};$ does not exist

29. (a) $\dfrac{\partial g}{\partial x}(0,0) = \lim\limits_{h \to 0} \dfrac{g(h,0) - g(0,0)}{h} = \lim\limits_{h \to 0} 0 = 0,$

$\dfrac{\partial g}{\partial y}(0,0) = \lim\limits_{h \to 0} \dfrac{g(0,h) - g(0,0)}{h} = \lim\limits_{h \to 0} 0 = 0$

(b) as (x,y) tends to $(0,0)$ along the x-axis, $\;g(x,y) = g(x,0) = 0\;$ tends to 0;

as (x,y) tends to $(0,0)$ along the line $y = x$, $\;g(x,y) = g(x,x) = \frac{1}{2}\;$ tends to $\frac{1}{2}$

31. For $y \neq 0$, $\qquad \dfrac{\partial f}{\partial x}(0,y) = \lim\limits_{h \to 0} \dfrac{f(h,y) - f(0,y)}{h} = \lim\limits_{h \to 0} \dfrac{y(y^2 - h^2)}{h^2 + y^2} = y.$

Since $\qquad \dfrac{\partial f}{\partial x}(0,0) = \lim\limits_{h \to 0} \dfrac{f(h,0) - f(0,0)}{h} = \lim\limits_{h \to 0} 0 = 0,$

we have $\qquad \dfrac{\partial f}{\partial x}(0,y) = y$ for all y.

For $x \neq 0$, $\qquad \dfrac{\partial f}{\partial y}(x,0) = \lim\limits_{h \to 0} \dfrac{f(x,h) - f(x,0)}{h} = \lim\limits_{h \to 0} \dfrac{x(h^2 - x^2)}{x^2 + h^2} = -x.$

Since $\qquad \dfrac{\partial f}{\partial y}(0,0) = \lim\limits_{h \to 0} \dfrac{f(0,h) - f(0,0)}{h} = \lim\limits_{h \to 0} 0 = 0,$

we have $\qquad \dfrac{\partial f}{\partial y}(x,0) = -x\quad$ for all x.

Therefore $\qquad \dfrac{\partial^2 f}{\partial y \partial x}(0,y) = 1$ for all $y\quad$ and $\quad \dfrac{\partial^2 f}{\partial x \partial y}(x,0) = -1$ for all x.

In particular $\qquad \dfrac{\partial^2 f}{\partial y \partial x}(0,0) = 1\quad$ while $\quad \dfrac{\partial^2 f}{\partial x \partial y}(0,0) = -1.$

33. Since $f_{xy}(x,y) = 0$, $\;f_x(x,y)\;$ must be a function of x alone, and $f_y(x,y)\;$ must be a function of y alone. Then f must be of the form

$f(x,y) = g(x) + h(y).$

35.

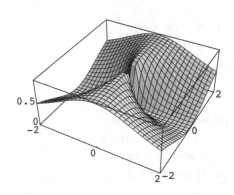

PROJECT 15.6

1. (a) $\dfrac{\partial u}{\partial x} = \dfrac{x^2 y^2 + 2xy^3}{(x+y)^2}, \quad \dfrac{\partial u}{\partial y} = \dfrac{x^2 y^2 + 2x^3 y}{(x+y)^2}$

$$x\dfrac{\partial u}{\partial x} + y\dfrac{\partial u}{\partial y} = \dfrac{3x^2 y^2 (x+y)}{(x+y)^2} = 3u$$

(b) $\dfrac{\partial u}{\partial x} = 2xy + z^2, \quad \dfrac{\partial u}{\partial y} = 2yz + x^2, \quad \dfrac{\partial u}{\partial z} = 2xz + y^2$

$$\dfrac{\partial u}{\partial x} + \dfrac{\partial u}{\partial y} + \dfrac{\partial u}{\partial z} = 2xy + z^2 + 2yz + x^2 + 2xz + y^2 = (x+y+z)^2$$

3. (i) $\dfrac{\partial^2 f}{\partial t^2} = \dfrac{\partial^2 f}{\partial x^2} = 0 \implies \dfrac{\partial^2 f}{\partial t^2} - c^2 \dfrac{\partial^2 f}{\partial x^2} = 0$

(ii) $\dfrac{\partial^2 f}{\partial t^2} = -5c^2 \sin(x+ct)\cos(2x+2ct) - 4c^2 \cos(x+ct)\sin(2x+2ct)$

$$\dfrac{\partial^2 f}{\partial x^2} = -5\sin(x+ct)\cos(2x+2ct) - 4\cos(x+ct)\sin(2x+2ct)$$

It now follows that $\dfrac{\partial^2 f}{\partial t^2} - c^2\dfrac{\partial^2 f}{\partial x^2} = 0$

(iii) $\dfrac{\partial^2 f}{\partial t^2} = -\dfrac{c^2}{(x+ct)^2}, \quad \dfrac{\partial^2 f}{\partial x^2} = -\dfrac{1}{(x+ct)^2} \implies \dfrac{\partial^2 f}{\partial t^2} - c^2 \dfrac{\partial^2 f}{\partial x^2} = 0$

(iv) $\dfrac{\partial^2 f}{\partial t^2} = c^2 k^2 \left(Ae^{kx} + Be^{-kx}\right)\left(Ce^{ckt} + De^{-ckt}\right), \quad \dfrac{\partial^2 f}{\partial x^2} = k^2\left(Ae^{kx} + Be^{-kx}\right)\left(Ce^{ckt} + De^{-ckt}\right)$

It now follows that $\dfrac{\partial^2 f}{\partial t^2} - c^2 \dfrac{\partial^2 f}{\partial x^2} = 0$

REVIEW EXERCISES

1. domain $\{(x,y) : y > x^2\}$,   range $(0, \infty)$

3. domain $\{(x,y,x) : z \geq x^2 + y^2\}$,   dange $[0, +\infty)$

5. (a) $f(x,y) = \dfrac{1}{3}\pi x^2 y$;

(b) $f(x,y) = \dfrac{1}{2}yx^2$;

(c) $\theta = \arccos \dfrac{x + 2y}{\sqrt{5}\sqrt{x^2 + y^2}}$

7. ellipsoid
xy−trace: ellipse $4x^2 + 9y^2 = 36$
xz−trace: ellipse $4x^2 + 36z^2 = 36$
yz−trace: ellipse $9y^2 + 36z^2 = 36$

9. hyperbolic paraboloid
xy−trace: lines $x = \pm y$
xz−trace: parabola $z = -x^2$
yz−trace: parabola $z = y^2$

11.   cone

     xy−trace: lines $x = \pm y$

     xz−trace: lines $x = \pm z$

     yz−trace: $(0,0)$

13.

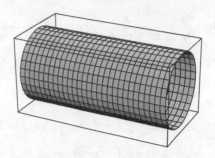

15.

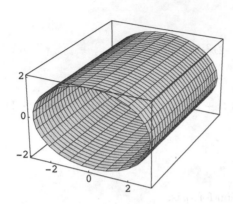

17. $c = 0, \Longrightarrow 0 = 2x^2 + 3y^2 \Longrightarrow (0,0)$

     $c = 6, \Longrightarrow 6 = 2x^2 + 3y^2$, ellipse

     $c = 12, \Longrightarrow 6 = 2x^2 + 3y^2$, ellipse

19. $c = -4, \Longrightarrow x = -4y^2$, parabola

     $c = -1, \Longrightarrow x = -y^2$, parabola

     $c = 1, \Longrightarrow x = y^2$, parabola

     $c = 4, \Longrightarrow x = 4y^2$, parabola

     the origin is omitted

21. $c = 6$, $2x + y + 3z = 6$, plane

23.   (a) $f(0,0) = 1$, level curve: $f(x,y) = 1$

     (b) $f(\ln 2, 1) = 4$, level curve: $f(x,y) = 4$

     (c) $f(1,-1) = 2e$, level curve: $f(x,y) = 2e$

25.
$$f_x = \lim_{h \to 0} \frac{f(x+h, y) - f(x,y)}{h} = \lim_{h \to 0} \frac{(x+h)^2 + 2(x+h)y - x^2 - 2xy}{h}$$
$$= \lim_{h \to 0}(2x + h + 2y) = 2x + 2y$$
$$f_y = \lim_{h \to 0} \frac{f(x, y+h) - f(x,y)}{h} = \lim_{h \to 0} \frac{x^2 + 2x(y+h) - x^2 - 2xy}{h} = 2x$$

27. $f_x = 2xy - 2y^3$; $f_y = x^2 - 6xy^2$

29. $\dfrac{\partial z}{\partial x} = 2x \sin(xy^2) + x^2 y^2 \cos(xy^2)$; $\dfrac{\partial z}{\partial y} = 2x^3 y \cos(xy^2)$.

31. $h_x = -e^{-x} \cos(2x - y) - 2e^{-x} \sin(2x - y)$ $h_y = e^{-x} \sin(2x - y)$

33. $f_x = \dfrac{2y^2 + 2yz}{(x + y + z)^2}$; $f_y = \dfrac{2x^2 + 2xz}{(x + y + z)^2}$; $f_z = \dfrac{-2xy}{(x + y + z)^2}$

35. $\dfrac{\partial g}{\partial x} = \dfrac{x}{x^2 + y^2 + z^2};$ $\dfrac{\partial g}{\partial y} = \dfrac{y}{x^2 + y^2 + z^2};$ $\dfrac{\partial g}{\partial z} = \dfrac{z}{x^2 + y^2 + z^2}.$

37. $f_x = 3x^2 y^2 - 4y^3 + 2,$ $f_y = 2x^3 y - 12xy^2 - 1;$

$f_{xx} = 6xy^2,$ $f_{yy} = 2x^3 - 24xy,$ $f_{yx} = f_{xy} = 6x^2 y - 12y^2$

39. $g_x = y \sin xy + xy^2 \cos xy,$ $g_{xx} = 2y^2 \cos xy - xy^3 \sin xy;$

$g_y = x \sin xy + x^2 y \cos xy,$ $g_{yy} = 2x^2 \cos xy - yx^3 \sin xy,$

$g_{xy} = g_{yx} = \sin xy + 3xy \cos xy - x^2 y^2 \sin xy$

41. $f_x = 2xe^{2y} \cos(2z+1),$ $f_y = 2x^2 e^{2y} \cos(2z+1),$ $f_z = -2x^2 e^{2y} \sin(2z+1);$

$f_{xx} = 2e^{2y} \cos(2z+1),$ $f_{yy} = 4x^2 e^{2y} \cos(2z+1),$ $f_{zz} = -4x^2 e^{2y} \cos(2z+1);$

$f_{xy} = f_{yx} = 4xe^{2y} \cos(2z+1),$ $f_{xz} = f_{zx} = -4xe^{2y} \sin(2z+1),$ $f_{yz} = f_{zy} = -4x^2 e^{2y} \sin(2z+1)$

43. $\dfrac{\partial z}{\partial x} = 4x + 6|_{(1,2,8)} = 10$

$x = 1 + t,$ $y = 2,$ $z = 8 + 10t$

45. (a) $z_y(2,1) = \dfrac{-6y}{2\sqrt{20 - 2x^2 - 3y^2}}(2,1) = -1;$ the equation for l_1 is:

$x = 2;$ $y = 1 - t;$ $z = 3 + t$

   (b) $z_x(2,1) = \dfrac{-4x}{2\sqrt{20 - 2x^2 - 3y^2}}(2,1) = -\dfrac{4}{3};$ the equation for l_2 is:

$x = 2 - \dfrac{3}{4}t;$ $y = 1;$ $z = 3 + t$

   (c) The normal vector for this plane is: $-\mathbf{i} - \dfrac{3}{4}\mathbf{j} - \dfrac{3}{4}\mathbf{k}$ or $4\mathbf{i} + 3\mathbf{j} + 3\mathbf{k};$

   an equation for the plane is: $4(x - 2) + 3(y - 1) + 3(z - 3) = 0.$

47. Open.

interior: $\{(x,y) : 0 < x^2 + y^2 < 4\}$

boundary: $\{(0,0)\} \cup \{(x,y) : x^2 + y^2 = 4\}$

49. Closed.

interior: $\{(x,y,z) : 0 < x < 2,\ 0 < y,\ 0 < z,\ y^2 + z^2 < 4\}$

boundary: the quarter disks $x = 0,\ y^2 + z^2 \leq 4;$ $x = 2,\ y^2 + z^2 \leq 4;$

   the squares $z = 0,\ 0 \leq x, y \leq 2;$ $y = 0, 0 \leq x,\ z \leq 2;$ and

   the cylindrical surface $y^2 + z^2 = 4,\ 0 \leq x \leq 2,\ y,\ z \geq 0$

51. (a) $f_x = yg'(xy),$ $f_y = xg'(xy);$ $xf_x - yf_y = xyg' - xyg' = 0$

   (b) $f_{xx} = y^2 g''(xy),$ $f_{yy} = x^2 g''(xy);$ $x^2 f_{xx} - y^2 f_{yy} = x^2 y^2 g'' - x^2 y^2 g'' = 0$

53. No. $\dfrac{\partial^2 f}{\partial y \partial x} = x^2 e^{xy} \neq y^2 e^{xy} = \dfrac{\partial^2 f}{\partial x \partial y}$

CHAPTER 16

SECTION 16.1

1. $\nabla f = (6x - y)\,\mathbf{i} + (1 - x)\,\mathbf{j}$

3. $\nabla f = e^{xy}[\,(xy + 1)\,\mathbf{i} + x^2\mathbf{j}]$

5. $\nabla f = \left[2y^2 \sin(x^2 + 1) + 4x^2y^2 \cos(x^2 + 1)\right]\mathbf{i} + 4xy \sin(x^2 + 1)\,\mathbf{j}$

7. $\nabla f = (e^{x-y} + e^{y-x})\,\mathbf{i} + (-e^{x-y} - e^{y-x})\,\mathbf{j} = (e^{x-y} + e^{y-x})(\mathbf{i} - \mathbf{j})$

9. $\nabla f = (z^2 + 2xy)\,\mathbf{i} + (x^2 + 2yz)\,\mathbf{j} + (y^2 + 2zx)\,\mathbf{k}$

11. $\nabla f = e^{-z}(2xy\,\mathbf{i} + x^2\,\mathbf{j} - x^2y\,\mathbf{k})$

13. $\nabla f = e^{x+2y} \cos\left(z^2 + 1\right)\mathbf{i} + 2e^{x+2y} \cos\left(z^2 + 1\right)\mathbf{j} - 2ze^{x+2y} \sin\left(z^2 + 1\right)\mathbf{k}$

15. $\nabla f = \left[2y \cos(2xy) + \dfrac{2}{x}\right]\mathbf{i} + 2x \cos(2xy)\,\mathbf{j} + \dfrac{1}{z}\,\mathbf{k}$

17. $\nabla f = (4x - 3y)\,\mathbf{i} + (8y - 3x)\,\mathbf{j};$ at $(2, 3)$, $\nabla f = -\mathbf{i} + 18\mathbf{j}$

19. $\nabla f = \dfrac{2x}{x^2 + y^2}\,\mathbf{i} + \dfrac{2y}{x^2 + y^2}\,\mathbf{j};$ at $(2, 1)$, $\nabla f = \dfrac{4}{5}\mathbf{i} + \dfrac{2}{5}\mathbf{j}$

21. $\nabla f = (\sin xy + xy \cos xy)\,\mathbf{i} + x^2 \cos xy\,\mathbf{j};$ at $(1, \pi/2)$, $\nabla f = \mathbf{i}$

23. $\nabla f = -e^{-x} \sin(z + 2y)\,\mathbf{i} + 2e^{-x} \cos(z + 2y)\,\mathbf{j} + e^{-x} \cos(z + 2y)\,\mathbf{k};$
    at $(0, \pi/4, \pi/4)$, $\nabla f = -\tfrac{1}{2}\sqrt{2}\,(\mathbf{i} + 2\mathbf{j} + \mathbf{k})$

25. $\nabla f = \mathbf{i} - \dfrac{y}{\sqrt{y^2 + z^2}}\,\mathbf{j} - \dfrac{z}{\sqrt{y^2 + z^2}}\,\mathbf{k};$ at $(2, -3, 4)$, $\nabla f = \mathbf{i} + \dfrac{3}{5}\mathbf{j} - \dfrac{4}{5}\mathbf{k}$

27. (a) $\nabla f(0, 2) = 4\mathbf{i}$ (b) $\nabla f\left(\tfrac{1}{4}\pi, \tfrac{1}{6}\pi\right) = \left(-1 - \dfrac{-1 + \sqrt{3}}{2\sqrt{2}}\right)\mathbf{i} + \left(-\dfrac{1}{2} - \dfrac{-1 + \sqrt{3}}{\sqrt{2}}\right)\mathbf{j}$
    (c) $\nabla f(1, e) = (1 - 2e)\,\mathbf{i} - 2\mathbf{j}$

29. For the function $f(x, y) = 3x^2 - xy + y$, we have

$$f(\mathbf{x} + \mathbf{h}) - f(\mathbf{x}) = f(x + h_1, y + h_2) - f(x, y)$$
$$= 3(x + h_1)^2 - (x + h_1)(y + h_2) + (y + h_2) - [3x^2 - xy + y]$$
$$= [(6x - y)\,\mathbf{i} + (1 - x)\,\mathbf{j}] \cdot (h_1\,\mathbf{i} + h_2\,\mathbf{j}) + 3h_1^2 - h_1h_2$$
$$= [(6x - y)\,\mathbf{i} + (1 - x)\,\mathbf{j}] \cdot \mathbf{h} + 3h_1^2 - h_1h_2$$

The remainder $g(\mathbf{h}) = 3h_1^2 - h_1 h_2 = (3h_1\,\mathbf{i} - h_1\,\mathbf{j}) \cdot (h_1\,\mathbf{i} + h_2\,\mathbf{j})$,   and

$$\frac{|g(\mathbf{h})|}{\|\mathbf{h}\|} = \frac{\|3h_1\,\mathbf{i} - h_1\,\mathbf{j}\| \cdot \|\mathbf{h}\| \cdot \cos\theta}{\|\mathbf{h}\|} \leq \|3h_1\,\mathbf{i} - h_1\,\mathbf{j}\|$$

Since $\|3h_1\,\mathbf{i} - h_1\,\mathbf{j}\| \to 0$ as $\mathbf{h} \to \mathbf{0}$ it follows that

$$\nabla f = (6x - y)\,\mathbf{i} + (1 - x)\,\mathbf{j}$$

31. For the function $f(x, y, z) = x^2 y + y^2 z + z^2 x$, we have

$$f(\mathbf{x} + \mathbf{h}) - f(\mathbf{x}) = f(x + h_1, y + h_2, z + h_3) - f(x, y, z)$$

$$= (x + h_1)^2 (y + h_2) + (y + h_2)^2 (z + h_3) + (z + h_3)^2 (x + h_1) - \left(x^2 y + y^2 z + z^2 x\right)$$

$$= \left(2xy + z^2\right) h_1 + \left(2yz + x^2\right) h_2 + \left(2xz + y^2\right) h_3 + (2xh_2 + yh_1 + h_1 h_2) h_1 +$$

$$\quad (2yh_3 + zh_2 + h_2 h_3) h_2 + (2zh_1 + xh_3 + h_1 h_3) h_3$$

$$= \left[\left(2xy + z^2\right)\mathbf{i} + \left(2yz + x^2\right)\mathbf{j} + \left(2xz + y^2\right)\mathbf{k}\right] \cdot \mathbf{h} + g(\mathbf{h}) \cdot \mathbf{h},$$

where $g(\mathbf{h}) = (2xh_2 + yh_1 + h_1 h_2)\,\mathbf{i} + (2yh_3 + zh_2 + h_2 h_3)\,\mathbf{j} + (2zh_1 + xh_3 + h_1 h_3)\,\mathbf{k}$

Since $\dfrac{|g(\mathbf{h})|}{\|\mathbf{h}\|} \to 0$ as $\mathbf{h} \to \mathbf{0}$ it follows that

$$\nabla f = \left(2xy + z^2\right)\mathbf{i} + \left(2yz + x^2\right)\mathbf{j} + \left(2xz + y^2\right)\mathbf{k}$$

33. $\nabla f = \mathbf{F}(x, y) = 2xy\,\mathbf{i} + \left(1 + x^2\right)\mathbf{j} \;\Rightarrow\; \dfrac{\partial f}{\partial x} = 2xy \;\Rightarrow\; f(x, y) = x^2 y + g(y)$ for some function g.

Now, $\dfrac{\partial f}{\partial y} = x^2 + g'(y) = 1 + x^2 \;\Rightarrow\; g'(y) = 1 \;\Rightarrow\; g(y) = y + C$, C a constant.

Thus, $f(x, y) = x^2 y + y + C$

35. $\nabla f = \mathbf{F}(x, y) = (x + \sin y)\,\mathbf{i} + (x \cos y - 2y)\,\mathbf{j} \;\Rightarrow\; \dfrac{\partial f}{\partial x} = x + \sin y \;\Rightarrow\; f(x, y) = \tfrac{1}{2} x^2 + x \sin y + g(y)$

for some function g.

Now, $\dfrac{\partial f}{\partial y} = x \cos y + g'(y) = x \cos y - 2y \;\Rightarrow\; g'(y) = -2y \;\Rightarrow\; g(y) = -y^2 + C$, C a constant.

Thus, $f(x, y) = \tfrac{1}{2} x^2 + x \sin y - y^2 + C$.

37. With $r = (x^2 + y^2 + z^2)^{1/2}$ we have

$$\frac{\partial r}{\partial x} = \frac{x}{r}, \quad \frac{\partial r}{\partial y} = \frac{y}{r}, \quad \frac{\partial r}{\partial z} = \frac{z}{r}.$$

(a)

$$\nabla(\ln r) = \frac{\partial}{\partial x}(\ln r)\,\mathbf{i} + \frac{\partial}{\partial y}(\ln r)\,\mathbf{j} + \frac{\partial}{\partial z}(\ln r)\mathbf{k}$$

$$= \frac{1}{r}\frac{\partial r}{\partial x}\,\mathbf{i} + \frac{1}{r}\frac{\partial r}{\partial y}\,\mathbf{j} + \frac{1}{r}\frac{\partial r}{\partial z}\,\mathbf{k}$$

$$= \frac{x}{r^2}\,\mathbf{i} + \frac{y}{r^2}\,\mathbf{j} + \frac{z}{r^2}\,\mathbf{k} = \frac{\mathbf{r}}{r^2}$$

(b)
$$\nabla(\sin r) = \frac{\partial}{\partial x}(\sin r)\,\mathbf{i} + \frac{\partial}{\partial y}(\sin r)\,\mathbf{j} + \frac{\partial}{\partial z}(\sin r)\mathbf{k}$$

$$= \cos r\,\frac{\partial r}{\partial x}\,\mathbf{i} + \cos r\,\frac{\partial r}{\partial y}\,\mathbf{j} + \cos r\,\frac{\partial r}{\partial z}\,\mathbf{k}$$

$$= (\cos r)\frac{x}{r}\,\mathbf{i} + (\cos r)\frac{y}{r}\,\mathbf{j} + (\cos r)\frac{z}{r}\,\mathbf{k}$$

$$= \left(\frac{\cos r}{r}\right)\mathbf{r}$$

(c) $\nabla e^r = \left(\dfrac{e^r}{r}\right)\mathbf{r}$ [same method as in (a) and (b)]

39. (a) $\nabla f = 2x\,\mathbf{i} + 2y\,\mathbf{j} = \mathbf{0} \implies x = y = 0;\quad \nabla f = \mathbf{0}$ at $(0,0)$.

(b)

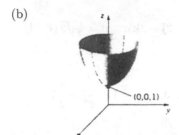

(c) f has an absolute minimum at $(0,0)$

41. (a) Let $\mathbf{c} = c_1\mathbf{i} + c_2\mathbf{j} + c_3\mathbf{k}$. First, we take $\mathbf{h} = h\mathbf{i}$. Since $\mathbf{c} \cdot \mathbf{h}$ is $o(\mathbf{h})$,
$$0 = \lim_{\mathbf{h}\to\mathbf{0}} \frac{\mathbf{c} \cdot \mathbf{h}}{\|\mathbf{h}\|} = \lim_{h\to 0} \frac{c_1 h}{h} = c_1.$$

Similarly, $c_2 = 0$ and $c_3 = 0$.

(b) $(\mathbf{y} - \mathbf{z}) \cdot \mathbf{h} = [f(\mathbf{x}+\mathbf{h}) - f(\mathbf{x}) - \mathbf{z}\cdot\mathbf{h}] + [\mathbf{y}\cdot\mathbf{h} - f(\mathbf{x}+\mathbf{h}) + f(\mathbf{x})] = o(\mathbf{h}) + o(\mathbf{h}) = o(\mathbf{h})$,
so that, by part (a), $\mathbf{y} - \mathbf{z} = \mathbf{0}$.

43. (a) In Section 15.6 we showed that f was not continuous at $(0,0)$. It is therefore not differentiable at $(0,0)$.

(b) For $(x,y) \neq (0,0)$, $\quad \dfrac{\partial f}{\partial x} = \dfrac{2y(y^2 - x^2)}{(x^2 + y^2)^2}$. As (x,y) tends to $(0,0)$ along the positive y-axis,
$\dfrac{\partial f}{\partial x} = \dfrac{2y^3}{y^4} = \dfrac{2}{y}$ tends to ∞.

SECTION 16.2

1. $\nabla f = 2x\mathbf{i} + 6y\mathbf{j},\quad \nabla f(1,1) = 2\mathbf{i} + 6\mathbf{j},\quad \mathbf{u} = \tfrac{1}{2}\sqrt{2}\,(\mathbf{i} - \mathbf{j}),\qquad f'_{\mathbf{u}}(1,1) = \nabla f(1,1) \cdot \mathbf{u} = -2\sqrt{2}$

3. $\nabla f = (e^y - ye^x)\,\mathbf{i} + (xe^y - e^x)\,\mathbf{j},\quad \nabla f(1,0) = \mathbf{i} + (1-e)\mathbf{j},\quad \mathbf{u} = \dfrac{1}{5}(3\mathbf{i} + 4\mathbf{j}),$

$f'_{\mathbf{u}}(1,0) = \nabla f(1,0) \cdot \mathbf{u} = \dfrac{1}{5}(7 - 4e)$

5. $\nabla f = \dfrac{(a-b)y}{(x+y)^2}\mathbf{i} + \dfrac{(b-a)x}{(x+y)^2}\mathbf{j}, \quad \nabla f(1,1) = \dfrac{a-b}{4}(\mathbf{i}-\mathbf{j}), \quad \mathbf{u} = \dfrac{1}{2}\sqrt{2}\,(\mathbf{i}-\mathbf{j}),$

$f'_{\mathbf{u}}(1,1) = \nabla f(1,1) \cdot \mathbf{u} = \dfrac{1}{4}\sqrt{2}\,(a-b)$

7. $\nabla f = \dfrac{2x}{x^2+y^2}\mathbf{i} + \dfrac{2y}{x^2+y^2}\mathbf{j}, \quad \nabla f(0,1) = 2\mathbf{j}, \quad \mathbf{u} = \dfrac{1}{\sqrt{65}}(8\mathbf{i}+\mathbf{j}),$

$f'_{\mathbf{u}}(0,1) = \nabla f(0,1) \cdot \mathbf{u} = \dfrac{2}{\sqrt{65}}$

9. $\nabla f = (y+z)\mathbf{i} + (x+z)\mathbf{j} + (y+x)\mathbf{k}, \quad \nabla f(1,-1,1) = 2\mathbf{j}, \quad \mathbf{u} = \frac{1}{6}\sqrt{6}\,(\mathbf{i}+2\mathbf{j}+\mathbf{k}),$

$f'_{\mathbf{u}}(1,-1,1) = \nabla f(1,-1,1) \cdot \mathbf{u} = \dfrac{2}{3}\sqrt{6}$

11. $\nabla f = 2\left(x+y^2+z^3\right)\left(\mathbf{i}+2y\mathbf{j}+3z^2\mathbf{k}\right), \quad \nabla f(1,-1,1) = 6(\mathbf{i}-2\mathbf{j}+3\mathbf{k}), \quad \mathbf{u} = \frac{1}{2}\sqrt{2}\,(\mathbf{i}+\mathbf{j}),$

$f'_{\mathbf{u}}(1,-1,1) = \nabla f(1,-1,1) \cdot \mathbf{u} = -3\sqrt{2}$

13. $\nabla f = \tan^{-1}(y+z)\mathbf{i} + \dfrac{x}{1+(y+z)^2}\mathbf{j} + \dfrac{x}{1+(y+z)^2}\mathbf{k}, \quad \nabla f(1,0,1) = \dfrac{\pi}{4}\mathbf{i} + \dfrac{1}{2}\mathbf{j} + \dfrac{1}{2}\mathbf{k},$

$\mathbf{u} = \dfrac{1}{\sqrt{3}}(\mathbf{i}+\mathbf{j}-\mathbf{k}), \qquad f'_{\mathbf{u}}(1,0,1) = \nabla f(1,0,1) \cdot \mathbf{u} = \dfrac{\pi}{4\sqrt{3}} = \dfrac{\sqrt{3}}{12}\pi$

15. $\nabla f = \dfrac{x}{x^2+y^2}\mathbf{i} + \dfrac{y}{x^2+y^2}\mathbf{j}, \quad \mathbf{u} = \dfrac{1}{\sqrt{x^2+y^2}}\,(-x\mathbf{i}-y\mathbf{j}), \quad f'_{\mathbf{u}}(x,y) = \nabla f \cdot \mathbf{u} = -\dfrac{1}{\sqrt{x^2+y^2}}$

17. $\nabla f = (2Ax+2By)\,\mathbf{i} + (2Bx+2Cy)\,\mathbf{j}, \quad \nabla f(a,b) = (2aA+2bB)\mathbf{i} + (2aB+2bC)\,\mathbf{j}$

(a) $\mathbf{u} = \frac{1}{2}\sqrt{2}\,(-\mathbf{i}+\mathbf{j}), \quad f'_{\mathbf{u}}(a,b) = \nabla f(a,b) \cdot \mathbf{u} = \sqrt{2}\,[a(B-A)+b(C-B)]$

(b) $\mathbf{u} = \frac{1}{2}\sqrt{2}\,(\mathbf{i}-\mathbf{j}), \quad f'_{\mathbf{u}}(a,b) = \nabla f(a,b) \cdot \mathbf{u} = \sqrt{2}\,[a(A-B)+b(B-C)]$

19. $\nabla f = e^{y^2-z^2}(\mathbf{i}+2xy\mathbf{j}-2xz\mathbf{k}), \quad \nabla f(1,2,-2) = \mathbf{i}+4\mathbf{j}+4\mathbf{k}, \quad \mathbf{r}'(t) = \mathbf{i} - 2\sin(t-1)\mathbf{j} - 2e^{t-1}\mathbf{k},$

at $(1,2,-2) \quad t=1, \quad \mathbf{r}'(1) = \mathbf{i}-2\mathbf{k}, \quad \mathbf{u} = \frac{1}{5}\sqrt{5}\,(\mathbf{i}-2\mathbf{k}), \quad f'_{\mathbf{u}}(1,2,-2) = \nabla f(1,2,-2) \cdot \mathbf{u} = -\dfrac{7}{5}\sqrt{5}$

21. $\nabla f = (2x+2yz)\,\mathbf{i} + \left(2xz-z^2\right)\mathbf{j} + (2xy-2yz)\,\mathbf{k}, \quad \nabla f(1,1,2) = 6\,\mathbf{i} - 2\,\mathbf{k}$

The vectors $\mathbf{v} = \pm(2\mathbf{i}+\mathbf{j}-3\mathbf{k})$ are direction vectors for the given line; $\mathbf{u} = \pm\left(\dfrac{1}{\sqrt{14}}[2\mathbf{i}+\mathbf{j}-3\mathbf{k}]\right)$

are corresponding unit vectors; $\quad f'_{\mathbf{u}}(1,1,2) = \nabla f(1,1,2) \cdot (\pm\mathbf{u}) = \pm\dfrac{18}{\sqrt{14}}$

23. $\nabla f = 2y^2 e^{2x}\,\mathbf{i} + 2y e^{2x}\,\mathbf{j}, \quad \nabla f(0,1) = 2\,\mathbf{i} + 2\,\mathbf{j}, \quad \|\nabla f\| = 2\sqrt{2}, \quad \dfrac{\nabla f}{\|\nabla f\|} = \dfrac{1}{\sqrt{2}}(\mathbf{i}+\mathbf{j})$

f increases most rapidly in the direction $\mathbf{u} = \dfrac{1}{\sqrt{2}}(\mathbf{i}+\mathbf{j})$; the rate of change is $2\sqrt{2}$.

f decreases most rapidly in the direction $\mathbf{v} = -\dfrac{1}{\sqrt{2}}(\mathbf{i}+\mathbf{j})$; the rate of change is $-2\sqrt{2}$.

25. $\nabla f = \dfrac{x}{\sqrt{x^2+y^2+z^2}}\mathbf{i} + \dfrac{y}{\sqrt{x^2+y^2+z^2}}\mathbf{j} + \dfrac{z}{\sqrt{x^2+y^2+z^2}}\mathbf{k}$,

$\nabla f(1,-2,1) = \dfrac{1}{\sqrt{6}}(\mathbf{i}-2\mathbf{j}+\mathbf{k}), \quad \|\nabla f\| = 1$

f increases most rapidly in the direction $\mathbf{u} = \dfrac{1}{\sqrt{6}}(\mathbf{i}-2\mathbf{j}+\mathbf{k})$; the rate of change is 1.

f decreases most rapidly in the direction $\mathbf{v} = -\dfrac{1}{\sqrt{6}}(\mathbf{i}-2\mathbf{j}+\mathbf{k})$; the rate of change is -1.

27. $\nabla f = f'(x_0)\,\mathbf{i}$. If $f'(x_0) \neq 0$, the gradient points in the direction in which f increases: to the right if $f'(x_0) > 0$, to the left if $f'(x_0) < 0$.

29. (a) $\displaystyle\lim_{h\to 0}\dfrac{f(h,0)-f(0,0)}{h} = \lim_{h\to 0}\dfrac{\sqrt{h^2}}{h} = \lim_{h\to 0}\dfrac{|h|}{h}$ does not exist

(b) no; by Theorem 16.2.5 f cannot be differentiable at $(0,0)$

31. $\nabla\lambda(x,y) = -\frac{8}{3}x\mathbf{i} - 6y\mathbf{j}$

(a) $\nabla\lambda(1,-1) = -\frac{8}{3}\mathbf{i} = 6\mathbf{j}, \quad \mathbf{u} = \dfrac{-\nabla\lambda(1,-1)}{\|\nabla\lambda(1,-1)\|} = \dfrac{\frac{8}{3}\mathbf{i}-6\mathbf{j}}{\frac{2}{3}\sqrt{97}}, \quad \lambda'_{\mathbf{u}}(1,-1) = \nabla\lambda(1,-1)\cdot\mathbf{u} = -\frac{2}{3}\sqrt{97}$

(b) $\mathbf{u} = \mathbf{i}, \quad \lambda'_{\mathbf{u}}(1,2) = \nabla\lambda(1,2)\cdot\mathbf{u} = \left(-\frac{8}{3}\mathbf{i}-12\mathbf{j}\right)\cdot\mathbf{i} = -\frac{8}{3}$

(c) $\mathbf{u} = \frac{1}{2}\sqrt{2}\,(\mathbf{i}+\mathbf{j}), \quad \lambda'_{\mathbf{u}}(2,2) = \nabla\lambda(2,2)\cdot\mathbf{u} = \left(-\frac{16}{3}\mathbf{i}-12\mathbf{j}\right)\cdot\left[\frac{1}{2}\sqrt{2}\,(\mathbf{i}+\mathbf{j})\right] = -\frac{26}{3}\sqrt{2}$

33. (a) The projection of the path onto the xy-plane is the curve

$$C: \ \mathbf{r}(t) = x(t)\mathbf{i} + y(t)\mathbf{j}$$

which begins at $(1,1)$ and at each point has its tangent vector in the direction of $-\nabla f$. Since

$$\nabla f = 2x\mathbf{i} + 6y\mathbf{j},$$

we have the initial-value problems

$$x'(t) = -2x(t), \quad x(0) = 1 \quad \text{and} \quad y'(t) = -6y(t), \quad y(0) = 1.$$

From Theorem 7.6.1 we find that

$$x(t) = e^{-2t} \quad \text{and} \quad y(t) = e^{-6t}.$$

Eliminating the parameter t, we find that C is the curve $y = x^3$ from $(1,1)$ to $(0,0)$.

(b) Here

$$x'(t) = -2x(t), \quad x(0) = 1 \quad \text{and} \quad y'(t) = -6y(t), \quad y(0) = -2$$

so that

$$x(t) = e^{-2t} \quad \text{and} \quad y(t) = -2e^{-6t}.$$

Eliminating the parameter t, we find that the projection of the path onto the xy-plane is the curve $y = -2x^3$ from $(1,-2)$ to $(0,0)$.

35. The projection of the path onto the xy-plane is the curve

$$C: \ \mathbf{r}(t) = x(t)\mathbf{i} + y(t)\mathbf{j}$$

which begins at (a, b) and at each point has its tangent vector in the direction of $-\nabla f = -\left(2a^2x\mathbf{i} + 2b^2y\mathbf{j}\right)$. We can satisfy these conditions by setting

$$x'(t) = -2a^2x(t), \quad x(0) = a^2 \quad \text{and} \quad y'(t) = -2b^2y(t), \quad y(0) = b$$

so that

$$x(t) = ae^{-2a^2t} \quad \text{and} \quad y(t) = be^{-2b^2t}.$$

Since

$$\left[\frac{x}{a}\right]^{b^2} = \left(e^{-2a^2t}\right)^{b^2} = \left[\frac{y}{b}\right]^{a^2},$$

C is the curve $(b)^{a^2}x^{b^2} = (a)^{b^2}y^{a^2}$ from (a, b) to $(0, 0)$.

37. We want the curve

$$C: \ \mathbf{r}(t) = x(t)\mathbf{i} + y(t)\mathbf{j}$$

which begins at $(\pi/4, 0)$ and at each point has its tangent vector in the direction of

$$\nabla T = -\sqrt{2}\,e^{-y}\sin x\,\mathbf{i} - \sqrt{2}\,e^{-y}\cos x\,\mathbf{j}.$$

From

$$x'(t) = -\sqrt{2}\,e^{-y}\sin x \quad \text{and} \quad y'(t) = -\sqrt{2}\,e^{-y}\cos x$$

we obtain

$$\frac{dy}{dx} = \frac{y'(t)}{x'(t)} = \cot x$$

so that

$$y = \ln|\sin x| + C.$$

Since $y = 0$ when $x = \pi/4$, we get $C = \ln\sqrt{2}$ and $y = \ln|\sqrt{2}\sin x|$. As $\nabla T(\pi/4, 0) = -\mathbf{i} - \mathbf{j}$, the curve $y = \ln|\sqrt{2}\sin x|$ is followed in the direction of decreasing x.

39. (a)
$$\lim_{h\to 0}\frac{f\left(2+h, (2+h)^2\right) - f(2, 4)}{h} = \lim_{h\to 0}\frac{3(2+h)^2 + (2+h)^2 - 16}{h}$$

$$= \lim_{h\to 0} 4\left[\frac{4h + h^2}{h}\right] = \lim_{h\to 0} 4(4 + h) = 16$$

(b)
$$\lim_{h\to 0}\frac{f\left(\dfrac{h+8}{4}, 4+h\right) - f(2, 4)}{h} = \lim_{h\to 0}\frac{3\left(\dfrac{h+8}{4}\right)^2 + (4+h) - 16}{h}$$

$$= \lim_{h\to 0}\frac{\frac{3}{16}h^2 + 3h + 12 + 4 + h - 16}{h}$$

$$= \lim_{h\to 0}\left(\tfrac{3}{16}h + 4\right) = 4$$

(c) The limits computed in (a) and (b) are not directional derivatives. In (a) and (b) we have, in essence, computed $\nabla f(2,4) \cdot \mathbf{r}_0$ taking $\mathbf{r}_0 = \mathbf{i} + 4\mathbf{j}$ in (a) and $\mathbf{r}_0 = \frac{1}{4}\mathbf{i} + \mathbf{j}$ in (b). In neither case is $\mathbf{r}_0$ a unit vector.

41. (a) $\mathbf{u} = \cos\theta\,\mathbf{i} + \sin\theta\,\mathbf{j}, \qquad \nabla f(x,y) = \dfrac{\partial f}{\partial x}\mathbf{i} + \dfrac{\partial f}{\partial y}\mathbf{j};$

$$f_{\mathbf{u}}'(x,y) = \nabla f \cdot \mathbf{u} = \left(\dfrac{\partial f}{\partial x}\mathbf{i} + \dfrac{\partial f}{\partial y}\mathbf{j}\right)\cdot(\cos\theta\,\mathbf{i} + \sin\theta\,\mathbf{j}) = \dfrac{\partial f}{\partial x}\cos\theta + \dfrac{\partial f}{\partial y}\sin\theta$$

(b) $\nabla f = (3x^2 + 2y - y^2)\,\mathbf{i} + (2x - 2xy)\,\mathbf{j}, \qquad \nabla f(-1,2) = 3\,\mathbf{i} + 2\,\mathbf{j}$

$$f_{\mathbf{u}}'(-1,2) = 3\cos(2\pi/3) + 2\sin(2\pi/3) = \dfrac{2\sqrt{3}-3}{2}$$

43. $\nabla(fg) = \dfrac{\partial(fg)}{\partial x}\mathbf{i} + \dfrac{\partial(fg)}{\partial y}\mathbf{j} + \dfrac{\partial(fg)}{\partial z}\mathbf{k} = \left(f\dfrac{\partial g}{\partial x} + g\dfrac{\partial f}{\partial x}\right)\mathbf{i} + \left(f\dfrac{\partial g}{\partial y} + g\dfrac{\partial f}{\partial y}\right)\mathbf{j} + \left(f\dfrac{\partial g}{\partial z} + g\dfrac{\partial f}{\partial z}\right)\mathbf{k}$

$$= f\left(\dfrac{\partial g}{\partial x}\mathbf{i} + \dfrac{\partial g}{\partial y}\mathbf{j} + \dfrac{\partial g}{\partial z}\mathbf{k}\right) + g\left(\dfrac{\partial f}{\partial x}\mathbf{i} + \dfrac{\partial f}{\partial y}\mathbf{j} + \dfrac{\partial f}{\partial z}\mathbf{k}\right)$$

$$= f\,\nabla g + g\,\nabla f$$

45. $\nabla f^n = \dfrac{\partial f^n}{\partial x}\mathbf{i} + \dfrac{\partial f^n}{\partial y}\mathbf{j} + \dfrac{\partial f^n}{\partial z}\mathbf{k} = nf^{n-1}\dfrac{\partial f}{\partial x}\mathbf{i} + nf^{n-1}\dfrac{\partial f}{\partial y}\mathbf{j} + nf^{n-1}\dfrac{\partial f}{\partial z}\mathbf{k} = nf^{n-1}\,\nabla f$

SECTION 16.3

1. $f(\mathbf{b}) = f(1,3) = -2; \quad f(\mathbf{a}) = f(0,1) = 0; \quad f(\mathbf{b}) - f(\mathbf{a}) = -2$

$\nabla f = (3x^2 - y)\,\mathbf{i} - x\,\mathbf{j}; \quad \mathbf{b} - \mathbf{a} = \mathbf{i} + 2\mathbf{j} \quad$ and $\quad \nabla f \cdot (\mathbf{b} - \mathbf{a}) = 3x^2 - y - 2x$

The line segment joining $\mathbf{a}$ and $\mathbf{b}$ is parametrized by

$$x = t, \quad y = 1 + 2t, \quad 0 \le t \le 1$$

Thus, we need to solve the equation

$$3t^2 - (1+2t) - 2t = -2, \quad \text{which is the same as} \quad 3t^2 - 4t + 1 = 0, \quad 0 \le t \le 1$$

The solutions are: $t = \frac{1}{3}, t = 1$. Thus, $\mathbf{c} = (\frac{1}{3}, \frac{5}{3})$ satisfies the equation.
Note that the endpoint $\mathbf{b}$ also satisfies the equation.

3. (a) $f(x,y,z) = a_1 x + a_2 y + a_3 z + C$ (b) $f(x,y,z) = g(x,y,z) + a_1 x + a_2 y + a_3 z + C$

5. (a) U is not connected
    (b) (i) $g(\mathbf{x}) = f(\mathbf{x}) - 1$ (ii) $g(\mathbf{x}) = -f(\mathbf{x})$

7. $\nabla f = 2xy\mathbf{i} + x^2\mathbf{j};$

$\nabla f(\mathbf{r}(t)) \cdot \mathbf{r}'(t) = (2\mathbf{i} + e^{2t}\mathbf{j}) \cdot (e^t\mathbf{i} - e^{-t}\mathbf{j}) = e^t$

9. $\nabla f = \dfrac{-2x}{1+(y^2-x^2)^2}\,\mathbf{i} + \dfrac{2y}{1+(y^2-x^2)^2}\,\mathbf{j},\quad \nabla f(\mathbf{r}(t)) = \dfrac{-2\sin t}{1+\cos^2 2t}\,\mathbf{i} + \dfrac{2\cos t}{1+\cos^2 2t}\,\mathbf{j}$

$\nabla f(\mathbf{r}(t))\cdot\mathbf{r}'(t) = \left(\dfrac{-2\sin t}{1+\cos^2 2t}\,\mathbf{i} + \dfrac{2\cos t}{1+\cos^2 2t}\,\mathbf{j}\right)\cdot(\cos t\,\mathbf{i} - \sin t\,\mathbf{j}) = \dfrac{-4\sin t\cos t}{1+\cos^2 2t} = \dfrac{-2\sin 2t}{1+\cos^2 2t}$

11. $\nabla f = (e^y - ye^{-x})\,\mathbf{i} + (xe^y + e^{-x})\,\mathbf{j};\quad \nabla f(\mathbf{r}(t)) = (t^t - \ln t)\,\mathbf{i} + \left(t^t\ln t + \dfrac{1}{t}\right)\mathbf{j}$

$\nabla f(\mathbf{r}(t))\cdot\mathbf{r}'(t) = \left((t^t - \ln t)\,\mathbf{i} + \left(t^t\ln t + \dfrac{1}{t}\right)\mathbf{j}\right)\cdot\left(\dfrac{1}{t}\mathbf{i} + [1+\ln t]\mathbf{j}\right) = t^t\left(\dfrac{1}{t} + \ln t + [\ln t]^2\right) + \dfrac{1}{t}$

13. $\nabla f = y\mathbf{i} + (x-z)\mathbf{j} - y\mathbf{k};$

$\nabla f(\mathbf{r}(t))\cdot\mathbf{r}'(t) = \left(t^2\mathbf{i} + (t - t^3)\,\mathbf{j} - t^2\mathbf{k}\right)\cdot\left(\mathbf{i} + 2t\mathbf{j} + 3t^2\mathbf{k}\right) = 3t^2 - 5t^4$

15. $\nabla f = 2x\mathbf{i} + 2y\mathbf{j} + \mathbf{k};$

$\nabla f(\mathbf{r}(t))\cdot\mathbf{r}'(t) = (2a\cos\omega t\,\mathbf{i} + 2b\sin\omega t\,\mathbf{j} + \mathbf{k})\cdot(-a\omega\sin\omega t\,\mathbf{i} + b\omega\cos\omega t\,\mathbf{j} + b\omega\mathbf{k})$

$= 2\omega\left(b^2 - a^2\right)\sin\omega t\cos\omega t + b\omega$

17. $\dfrac{du}{dt} = \dfrac{\partial u}{\partial x}\dfrac{dx}{dt} + \dfrac{\partial u}{\partial y}\dfrac{dy}{dt} = (2x - 3y)(-\sin t) + (4y - 3x)(\cos t)$

$= 2\cos t\sin t + 3\sin^2 t - 3\cos^2 t = \sin 2t - 3\cos 2t$

19. $\dfrac{du}{dt} = \dfrac{\partial u}{\partial x}\dfrac{dx}{dt} + \dfrac{\partial u}{\partial y}\dfrac{dy}{dt}$

$= (e^x\sin y + e^y\cos x)\left(\tfrac{1}{2}\right) + (e^x\cos y + e^y\sin x)\,(2)$

$= e^{t/2}\left(\tfrac{1}{2}\sin 2t + 2\cos 2t\right) + e^{2t}\left(\tfrac{1}{2}\cos\tfrac{1}{2}t + 2\sin\tfrac{1}{2}t\right)$

21. $\dfrac{du}{dt} = \dfrac{\partial u}{\partial x}\dfrac{dx}{dt} + \dfrac{\partial u}{\partial y}\dfrac{dy}{dt} = (e^x\sin y)\,(2t) + (e^x\cos y)\,(\pi)$

$= e^{t^2}[2t\,\sin(\pi t) + \pi\,\cos(\pi t)]$

23. $\dfrac{du}{dt} = \dfrac{\partial u}{\partial x}\dfrac{dx}{dt} + \dfrac{\partial u}{\partial y}\dfrac{dy}{dt} + \dfrac{\partial u}{\partial z}\dfrac{dz}{dt}$

$= (y + z)(2t) + (x + z)(1 - 2t) + (y + x)(2t - 2)$

$= (1 - t)(2t) + (2t^2 - 2t + 1)(1 - 2t) + t(2t - 2)$

$= 1 - 4t + 6t^2 - 4t^3$

25. $V = \dfrac{1}{3}\pi r^2 h,\quad \dfrac{dV}{dt} = \dfrac{\partial V}{\partial r}\dfrac{dr}{dt} + \dfrac{\partial V}{\partial h}\dfrac{dh}{dt} = \left(\dfrac{2}{3}\pi rh\right)\dfrac{dr}{dt} + \left(\dfrac{1}{3}\pi r^2\right)\dfrac{dh}{dt}.$

At the given instant,

$$\dfrac{dV}{dt} = \dfrac{2}{3}\pi(280)(3) + \dfrac{1}{3}\pi(196)(-2) = \dfrac{1288}{3}\pi.$$

The volume is increasing at the rate of $\dfrac{1288}{3}\pi$ in.^{3}/ sec .

27. $A = \frac{1}{2} xy \sin\theta$; $\dfrac{dA}{dt} = \dfrac{\partial A}{\partial x}\dfrac{dx}{dt} + \dfrac{\partial A}{\partial y}\dfrac{dy}{dt} + \dfrac{\partial A}{\partial \theta}\dfrac{d\theta}{dt} = \frac{1}{2}\left[(y\sin\theta)\dfrac{dx}{dt} + (x\sin\theta)\dfrac{dy}{dt} + (xy\cos\theta)\dfrac{d\theta}{dt} \right].$

At the given instant

$$\dfrac{dA}{dt} = \dfrac{1}{2}\left[(2\sin 1)(0.25) + (1.5\sin 1)(0.25) - (2(1.5)\cos 1)(0.1) \right] \cong 0.2871\,\text{ft}^2/s \cong 41.34\,\text{in}^2/s$$

29. $\dfrac{\partial u}{\partial s} = \dfrac{\partial u}{\partial x}\dfrac{\partial x}{\partial s} + \dfrac{\partial u}{\partial y}\dfrac{\partial y}{\partial s} = (2x - y)(\cos t) + (-x)(t\cos s)$

$\qquad = 2s\cos^2 t - t\sin s\cos t - st\cos s\cos t$

$\dfrac{\partial u}{\partial t} = \dfrac{\partial u}{\partial x}\dfrac{\partial x}{\partial t} + \dfrac{\partial u}{\partial y}\dfrac{\partial y}{\partial t} = (2x - y)(-s\sin t) + (-x)(\sin s)$

$\qquad = -2s^2\cos t\sin t + st\sin s\sin t - s\cos t\sin s$

31. $\dfrac{\partial u}{\partial s} = \dfrac{\partial u}{\partial x}\dfrac{\partial x}{\partial s} + \dfrac{\partial u}{\partial y}\dfrac{\partial y}{\partial s} = (2x\tan y)(2st) + \left(x^2\sec^2 y\right)(1)$

$\qquad = 4s^3t^2\tan\left(s + t^2\right) + s^4t^2\sec^2\left(s + t^2\right)$

$\dfrac{\partial u}{\partial t} = \dfrac{\partial u}{\partial x}\dfrac{\partial x}{\partial t} + \dfrac{\partial u}{\partial y}\dfrac{\partial y}{\partial t} = (2x\tan y)\left(s^2\right) + \left(x^2\sec^2 y\right)(2t)$

$\qquad = 2s^4t\tan\left(s + t^2\right) + 2s^4t^3\sec^2\left(s + t^2\right)$

33. $\dfrac{\partial u}{\partial s} = \dfrac{\partial u}{\partial x}\dfrac{\partial x}{\partial s} + \dfrac{\partial u}{\partial y}\dfrac{\partial y}{\partial s} + \dfrac{\partial u}{\partial z}\dfrac{\partial z}{\partial s}$

$\qquad = (2x - y)(\cos t) + (-x)(-\cos(t - s)) + 2z(t\cos s)$

$\qquad = 2s\cos^2 t - \sin(t - s)\cos t + s\cos t\cos(t - s) + 2t^2\sin s\cos s$

$\dfrac{\partial u}{\partial t} = \dfrac{\partial u}{\partial x}\dfrac{\partial x}{\partial t} + \dfrac{\partial u}{\partial y}\dfrac{\partial y}{\partial t} + \dfrac{\partial u}{\partial z}\dfrac{\partial z}{\partial t}$

$\qquad = (2x - y)(-s\sin t) + (-x)(\cos(t - s)) + 2z(\sin s)$

$\qquad = -2s^2\cos t\sin t + s\sin(t - s)\sin t - s\cos t\cos(t - s) + 2t\sin^2 s$

35. $\dfrac{d}{dt}\left[f(\mathbf{r}(t)) \right] = \left[\boldsymbol{\nabla} f(\mathbf{r}(t)) \cdot \dfrac{\mathbf{r}'(t)}{\|\mathbf{r}'(t)\|} \right] \|\mathbf{r}'(t)\|$

$\qquad = f'_{\mathbf{u}(t)}(\mathbf{r}(t))\,\|\mathbf{r}'(t)\| \quad \text{where} \quad \mathbf{u}(t) = \dfrac{\mathbf{r}'(t)}{\|\mathbf{r}'(t)\|}$

37. (a) $(\cos r)\dfrac{\mathbf{r}}{r}$ (b) $(r\cos r + \sin r)\dfrac{\mathbf{r}}{r}$

39. (a) $(r\cos r - \sin r)\dfrac{\mathbf{r}}{r^3}$ (b) $\left(\dfrac{\sin r - r\cos r}{\sin^2 r} \right)\dfrac{\mathbf{r}}{r}$

41. (a)

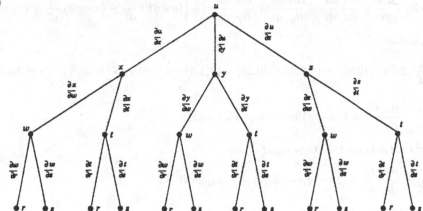

(b) $\dfrac{\partial u}{\partial r} = \dfrac{\partial u}{\partial x}\left(\dfrac{\partial x}{\partial w}\dfrac{\partial w}{\partial r} + \dfrac{\partial x}{\partial t}\dfrac{\partial t}{\partial r}\right) + \dfrac{\partial u}{\partial y}\left(\dfrac{\partial y}{\partial w}\dfrac{\partial w}{\partial r} + \dfrac{\partial y}{\partial t}\dfrac{\partial t}{\partial r}\right) + \dfrac{\partial u}{\partial z}\left(\dfrac{\partial z}{\partial w}\dfrac{\partial w}{\partial r} + \dfrac{\partial z}{\partial t}\dfrac{\partial t}{\partial r}\right).$

To obtain $\partial u/\partial s$, replace each r by s.

43. $\dfrac{du}{dt} = \dfrac{\partial u}{\partial x}\dfrac{dx}{dt} + \dfrac{\partial u}{\partial y}\dfrac{dy}{dt}$

$\dfrac{d^2 u}{dt^2} = \dfrac{\partial u}{\partial x}\dfrac{d^2 x}{dt^2} + \dfrac{dx}{dt}\left[\dfrac{\partial^2 u}{\partial x^2}\dfrac{dx}{dt} + \dfrac{\partial^2 u}{\partial y \partial x}\dfrac{dy}{dt}\right] + \dfrac{\partial u}{\partial y}\dfrac{d^2 y}{dt^2} + \dfrac{dy}{dt}\left[\dfrac{\partial^2 u}{\partial x \partial y}\dfrac{dx}{dt} + \dfrac{\partial^2 u}{\partial y^2}\dfrac{dy}{dt}\right]$

and the result follows.

45. (a) $\dfrac{\partial u}{\partial r} = \dfrac{\partial u}{\partial x}\dfrac{\partial x}{\partial r} + \dfrac{\partial u}{\partial y}\dfrac{\partial y}{\partial r} = \dfrac{\partial u}{\partial x}\cos\theta + \dfrac{\partial u}{\partial y}\sin\theta$

$\dfrac{\partial u}{\partial \theta} = \dfrac{\partial u}{\partial x}\dfrac{\partial x}{\partial \theta} + \dfrac{\partial u}{\partial y}\dfrac{\partial y}{\partial \theta} = \dfrac{\partial u}{\partial x}(-r\sin\theta) + \dfrac{\partial u}{\partial y}(r\cos\theta)$

(b) $\left(\dfrac{\partial u}{\partial r}\right)^2 = \left(\dfrac{\partial u}{\partial x}\right)^2 \cos^2\theta + 2\dfrac{\partial u}{\partial x}\dfrac{\partial u}{\partial y}\cos\theta\sin\theta + \left(\dfrac{\partial u}{\partial y}\right)^2 \sin^2\theta,$

$\dfrac{1}{r^2}\left(\dfrac{\partial u}{\partial \theta}\right)^2 = \left(\dfrac{\partial u}{\partial x}\right)^2 \sin^2\theta - 2\dfrac{\partial u}{\partial x}\dfrac{\partial u}{\partial y}\cos\theta\sin\theta + \left(\dfrac{\partial u}{\partial y}\right)^2 \cos^2\theta,$

$\left(\dfrac{\partial u}{\partial r}\right)^2 + \dfrac{1}{r^2}\left(\dfrac{\partial u}{\partial \theta}\right)^2 = \left(\dfrac{\partial u}{\partial x}\right)^2 (\cos^2\theta + \sin^2\theta) + \left(\dfrac{\partial u}{\partial y}\right)^2 (\sin^2\theta + \cos^2\theta) = \left(\dfrac{\partial u}{\partial x}\right)^2 + \left(\dfrac{\partial u}{\partial y}\right)^2$

47. Solve the equations in Exercise 45 (a) for $\dfrac{\partial u}{\partial x}$ and $\dfrac{\partial u}{\partial y}$:

$$\dfrac{\partial u}{\partial x} = \dfrac{\partial u}{\partial r}\cos\theta - \dfrac{1}{r}\dfrac{\partial u}{\partial \theta}\sin\theta, \quad \dfrac{\partial u}{\partial y} = \dfrac{\partial u}{\partial r}\sin\theta + \dfrac{1}{r}\dfrac{\partial u}{\partial \theta}\cos\theta$$

Then $\quad \nabla u = \dfrac{\partial u}{\partial x}\mathbf{i} + \dfrac{\partial u}{\partial y}\mathbf{j} = \dfrac{\partial u}{\partial r}(\cos\theta\,\mathbf{i} + \sin\theta\,\mathbf{j}) + \dfrac{1}{r}\dfrac{\partial u}{\partial \theta}(-\sin\theta\,\mathbf{i} + \cos\theta\,\mathbf{j})$

49. $u(x,y) = x^2 - xy + y^2 = r^2 - r^2 \cos\theta \sin\theta = r^2\left(1 - \frac{1}{2}\sin 2\theta\right)$

$$\frac{\partial u}{\partial r} = r(2 - \sin 2\theta), \quad \frac{\partial u}{\partial \theta} = -r^2 \cos 2\theta$$

$$\nabla u = \frac{\partial u}{\partial r}\,\mathbf{e_r} + \frac{1}{r}\frac{\partial u}{\partial \theta}\,\mathbf{e_\theta} = r(2 - \sin 2\theta)\mathbf{e_r} - r\cos 2\theta\,\mathbf{e_\theta}$$

51. From Exercise 45 (a),

$$\frac{\partial^2 u}{\partial r^2} = \frac{\partial^2 u}{\partial x^2}\cos^2\theta + 2\frac{\partial^2 u}{\partial y\,\partial x}\sin\theta\cos\theta + \frac{\partial^2 u}{\partial y^2}\sin^2\theta$$

$$\frac{\partial^2 u}{\partial \theta^2} = \frac{\partial^2 u}{\partial x^2}r^2\sin^2\theta - 2\frac{\partial^2 u}{\partial y\,\partial x}r^2\sin\theta\cos\theta + \frac{\partial^2 u}{\partial y^2}r^2\cos^2\theta - r\left(\frac{\partial u}{\partial x}\cos\theta + \frac{\partial u}{\partial y}\sin\theta\right).$$

The term in parentheses is $\dfrac{\partial u}{\partial r}$. Now divide the second equation by r^2 and add the two equations. The result follows.

53. Set $u = xe^y + ye^x - 2x^2 y.$ Then

$$\frac{\partial u}{\partial x} = e^y + ye^x - 4xy, \qquad \frac{\partial u}{\partial y} = xe^y + e^x - 2x^2$$

$$\frac{dy}{dx} = -\frac{\partial u/\partial x}{\partial u/\partial y} = -\frac{e^y + ye^x - 4xy}{xe^y + e^x - 2x^2}.$$

55. Set $u = x\cos xy + y\cos x - 2.$ Then

$$\frac{\partial u}{\partial x} = \cos xy - xy\sin xy - y\sin x, \qquad \frac{\partial u}{\partial y} = -x^2\sin xy + \cos x$$

$$\frac{dy}{dx} = -\frac{\partial u/\partial x}{\partial u/\partial y} = \frac{\cos xy - xy\sin xy - y\sin x}{x^2\sin xy - \cos x}.$$

57. Set $u = \cos xyz + \ln\left(x^2 + y^2 + z^2\right).$ Then

$$\frac{\partial u}{\partial x} = -yz\sin xyz + \frac{2x}{x^2+y^2+z^2}, \quad \frac{\partial u}{\partial y} = -xz\sin xyz + \frac{2y}{x^2+y^2+z^2}, \quad \text{and}$$

$$\frac{\partial u}{\partial z} = -xy\sin xyz + \frac{2z}{x^2+y^2+z^2}.$$

$$\frac{\partial z}{\partial x} = -\frac{\partial u/\partial x}{\partial u/\partial z} = -\frac{2x - yz\left(x^2+y^2+z^2\right)\sin xyz}{2z - xy\left(x^2+y^2+z^2\right)\sin xyz},$$

$$\frac{\partial z}{\partial y} = -\frac{\partial u/\partial y}{\partial u/\partial z} = -\frac{2y - xz\left(x^2+y^2+z^2\right)\sin xyz}{2z - xy\left(x^2+y^2+z^2\right)\sin xyz}.$$

59. $\dfrac{\partial \mathbf{u}}{\partial s} = \dfrac{\partial \mathbf{u}}{\partial x}\dfrac{\partial x}{\partial s} + \dfrac{\partial \mathbf{u}}{\partial y}\dfrac{\partial y}{\partial s}, \quad \dfrac{\partial \mathbf{u}}{\partial t} = \dfrac{\partial \mathbf{u}}{\partial x}\dfrac{\partial x}{\partial t} + \dfrac{\partial \mathbf{u}}{\partial y}\dfrac{\partial y}{\partial t}$

SECTION 16.4

1. Set $f(x, y) = x^2 + xy + y^2$.   Then,

$$\nabla f = (2x + y)\mathbf{i} + (x + 2y)\mathbf{j}, \quad \nabla f(-1, -1) = -3\mathbf{i} - 3\mathbf{j}.$$

normal vector $\mathbf{i} + \mathbf{j}$; tangent vector $\mathbf{i} - \mathbf{j}$
tangent line $x + y + 2 = 0$; normal line $x - y = 0$

3. Set $f(x, y) = \left(x^2 + y^2\right)^2 - 9\left(x^2 - y^2\right)$.   Then,

$$\nabla f = [4x(x^2 + y^2) - 18x]\mathbf{i} + \left[4y\left(x^2 + y^2\right) + 18y\right]\mathbf{j}, \quad \nabla f\left(\sqrt{2}, 1\right) = -6\sqrt{2}\,\mathbf{i} + 30\mathbf{j}.$$

normal vector $\sqrt{2}\,\mathbf{i} - 5\mathbf{j}$; tangent vector $5\mathbf{i} + \sqrt{2}\,\mathbf{j}$
tangent line $\sqrt{2}x - 5y + 3 = 0$; normal line $5x + \sqrt{2}\,y - 6\sqrt{2} = 0$

5. Set $f(x, y) = xy^2 - 2x^2 + y + 5x$.   Then,

$$\nabla f = \left(y^2 - 4x + 5\right)\mathbf{i} + (2xy + 1)\mathbf{j}, \quad \nabla f(4, 2) = -7\mathbf{i} + 17\mathbf{j}.$$

normal vector $7\mathbf{i} - 17\mathbf{j}$; tangent vector $17\mathbf{i} + 7\mathbf{j}$
tangent line $7x - 17y + 6 = 0$; normal line $17x + 7y - 82 = 0$

7. Set $f(x, y) = 2x^3 - x^2y^2 - 3x + y$.   Then,

$$\nabla f = (6x^2 - 2xy^2 - 3)\mathbf{i} + (-2x^2y + 1)\mathbf{j}, \quad \nabla f(1, -2) = -5\mathbf{i} + 5\mathbf{j}.$$

normal vector $\mathbf{i} - \mathbf{j}$; tangent vector $\mathbf{i} + \mathbf{j}$
tangent line $x - y - 3 = 0$; normal line $x + y + 1 = 0$

9. Set $f(x, y) = x^2y + a^2y$.   By (15.4.4)

$$m = -\frac{\partial f / \partial x}{\partial f / \partial y} = -\frac{2xy}{x^2 + a^2}.$$

At $(0, a)$ the slope is 0.

11. Set $f(x, y, z) = x^3 + y^3 - 3xyz$.   Then,

$$\nabla f = (3x^2 - 3yz)\mathbf{i} + (3y^2 - 3xz)\mathbf{j} - 3xy\mathbf{k}, \quad \nabla f\left(1, 2, \tfrac{3}{2}\right) = -6\mathbf{i} + \tfrac{15}{2}\mathbf{j} - 6\mathbf{k};$$

tangent plane at $\left(1, 2, \tfrac{3}{2}\right)$: $-6(x - 1) + \tfrac{15}{2}(y - 2) - 6\left(z - \tfrac{3}{2}\right) = 0$, which reduces to $4x - 5y + 4z = 0$.
Normal: $x = 1 + 4t$, $y = 2 - 5t$, $z = \tfrac{3}{2} + 4t$

13. Set $z = g(x, y) = axy$.   Then, $\nabla g = ay\mathbf{i} + ax\mathbf{j}$, $\nabla g\left(1, \dfrac{1}{a}\right) = \mathbf{i} + a\mathbf{j}$.

tangent plane at $\left(1, \dfrac{1}{a}, 1\right)$: $z - 1 = 1(x - 1) + a\left(y - \dfrac{1}{a}\right)$, which reduces to $x + ay - z - 1 = 0$
Normal: $x = 1 + t$, $y = \tfrac{1}{a} + at$, $z = 1 - t$

15. Set $z = g(x,y) = \sin x + \sin y + \sin(x+y)$.   Then,

$$\nabla g = [\cos x + \cos(x+y)]\,\mathbf{i} + [\cos y + \cos(x+y)]\,\mathbf{j}, \quad \nabla g(0,0) = 2\mathbf{i} + 2\mathbf{j};$$

tangent plane at $(0,0,0)$: $z - 0 = 2(x-0) + 2(y-0), \quad 2x + 2y - z = 0.$
Normal: $x = 2t, \quad y = 2t, \quad z = -t$

17. Set $f(x,y,z) = b^2c^2x^2 - a^2c^2y^2 - a^2b^2z^2$.   Then,

$$\nabla f(x_0, y_0, z_0) = 2b^2c^2x_0\mathbf{i} - 2a^2c^2y_0\mathbf{j} - 2a^2b^2z_0\mathbf{k};$$

tangent plane at (x_0, y_0, z_0):

$$2b^2c^2x_0(x - x_0) - 2a^2c^2y_0(y - y_0) - 2a^2b^2z_0(z - z_0) = 0,$$

which can be rewritten as follows:

$$b^2c^2x_0x - a^2c^2y_0y - a^2b^2z_0z = b^2c^2x_0^2 - a^2c^2y_0^2 - a^2b^2z_0^2$$

$$= f(x_0, y_0, z_0) = a^2b^2c^2.$$

Normal: $x = x_0 + 2b^2c^2x_0t, \quad y = y_0 - 2a^2c^2y_0t, \quad z = z_0 - 2a^2b^2z_0t$

19. Set $z = g(x,y) = xy + a^3x^{-1} + b^3y^{-1}$.

$$\nabla g = (y - a^3x^{-2})\mathbf{i} + (x - b^3y^{-2})\mathbf{j}, \quad \nabla g = \mathbf{0} \implies y = a^3x^{-2} \text{ and } x = b^3y^{-2}.$$

Thus,

$$y = a^3b^{-6}y^4, \quad y^3 = b^6a^{-3}, \quad y = b^2/a, \quad x = b^3y^{-2} = a^2/b \quad \text{and} \quad g(a^2/b, b^2/a) = 3ab.$$

The tangent plane is horizontal at $\left(a^2/b, b^2/a, 3ab\right)$.

21. Set $z = g(x,y) = xy$.   Then, $\nabla g = y\mathbf{i} + x\mathbf{j}$.

$$\nabla g = \mathbf{0} \implies x = y = 0.$$

The tangent plane is horizontal at $(0,0,0)$.

23. Set $z = g(x,y) = 2x^2 + 2xy - y^2 - 5x + 3y - 2$.   Then,

$$\nabla g = (4x + 2y - 5)\,\mathbf{i} + (2x - 2y + 3)\,\mathbf{j}.$$

$$\nabla g = \mathbf{0} \implies 4x + 2y - 5 = 0 = 2x - 2y + 3 \implies x = \tfrac{1}{3}, \quad y = \tfrac{11}{6}.$$

The tangent plane is horizontal at $\left(\tfrac{1}{3}, \tfrac{11}{6}, -\tfrac{1}{12}\right)$.

25. $\dfrac{x - x_0}{(\partial f/\partial x)(x_0, y_0, z_0)} = \dfrac{y - y_0}{(\partial f/\partial y)(x_0, y_0, z_0)} = \dfrac{z - z_0}{(\partial f/\partial z)(x_0, y_0, z_0)}$

27. Since the tangent planes meet at right angles, the normals ∇F and ∇G meet at right angles:

$$\frac{\partial F}{\partial x}\frac{\partial G}{\partial x} + \frac{\partial F}{\partial y}\frac{\partial G}{\partial y} + \frac{\partial F}{\partial z}\frac{\partial G}{\partial z} = 0.$$

29. The tangent plane at an arbitrary point (x_0, y_0, z_0) has equation

$$y_0 z_0 (x - x_0) + x_0 z_0 (y - y_0) + x_0 y_0 (z - z_0) = 0,$$

which simplifies to

$$y_0 z_0 x + x_0 z_0 y + x_0 y_0 z = 3 x_0 y_0 z_0 \quad \text{and thus to} \quad \frac{x}{3x_0} + \frac{y}{3y_0} + \frac{z}{3z_0} = 1.$$

The volume of the pyramid is

$$V = \frac{1}{3} Bh = \frac{1}{3} \left[\frac{(3x_0)(3y_0)}{2} \right] (3z_0) = \frac{9}{2} x_0 y_0 z_0 = \frac{9}{2} a^3.$$

31. The point $(2, 3, -2)$ is the tip of $\mathbf{r}(1)$.

Since $\quad \mathbf{r}'(t) = 2\mathbf{i} - \dfrac{3}{t^2}\mathbf{j} - 4t\mathbf{k}, \quad$ we have $\quad \mathbf{r}'(1) = 2\mathbf{i} - 3\mathbf{j} - 4\mathbf{k}.$

Now set $\quad f(x, y, z) = x^2 + y^2 + 3z^2 - 25.$ The function has gradient $2x\mathbf{i} + 2y\mathbf{j} + 6z\mathbf{k}.$

At the point $(2, 3, -2)$,

$$\nabla f = 2(2\mathbf{i} + 3\mathbf{j} - 6\mathbf{k}).$$

The angle θ between $\mathbf{r}'(1)$ and the gradient gives

$$\cos \theta = \frac{(2\mathbf{i} - 3\mathbf{j} - 4\mathbf{k})}{\sqrt{29}} \cdot \frac{(2\mathbf{i} + 3\mathbf{j} - 6\mathbf{k})}{7} = \frac{19}{7\sqrt{29}} \cong 0.504.$$

Therefore $\theta \cong 1.043$ radians. The angle between the curve and the plane is

$$\frac{\pi}{2} - \theta \cong 1.571 - 1.043 \cong 0.528 \text{ radians.}$$

33. Set $\quad f(x, y, z) = x^2 y^2 + 2x + z^3.$ Then,

$$\nabla f = \left(2xy^2 + 2 \right) \mathbf{i} + 2x^2 y\mathbf{j} + 3z^2\mathbf{k}, \quad \nabla f(2, 1, 2) = 6\mathbf{i} + 8\mathbf{j} + 12\mathbf{k}.$$

The plane tangent to $f(x, y, z) = 16$ at $(2, 1, 2)$ has equation

$$6(x - 2) + 8(y - 1) + 12(z - 2) = 0, \quad \text{or} \quad 3x + 4y + 6z = 22.$$

Next, set $\quad g(x, y, z) = 3x^2 + y^2 - 2z.$ Then,

$$\nabla g = 6x\mathbf{i} + 2y\mathbf{j} - 2\mathbf{k}, \quad \nabla g(2, 1, 2) = 12\mathbf{i} + 2\mathbf{j} - 2\mathbf{k}.$$

The plane tangent to $g(x, y, z) = 9$ at $(2, 1, 2)$ is

$$12(x - 2) + 2(y - 1) - 2(z - 2) = 0, \quad \text{or} \quad 6x + y - z = 11.$$

35. A normal vector to the sphere at $(1, 1, 2)$ is

$$2x\mathbf{i} + (2y - 4)\mathbf{j} + (2z - 2)\mathbf{k} = 2\mathbf{i} - 2\mathbf{j} + 2\mathbf{k}.$$

A normal vector to the paraboloid at $(1, 1, 2)$ is

$$6x\mathbf{i} + 4y\mathbf{j} - 2\mathbf{k} = 6\mathbf{i} + 4\mathbf{j} - 2\mathbf{k}.$$

Since

$$(2\mathbf{i} - 2\mathbf{j} + 2\mathbf{k}) \cdot (6\mathbf{i} + 4\mathbf{j} - 2\mathbf{k}) = 0,$$

the surfaces intersect at right angles.

37. (a) $3x + 4y + 6 = 0$ since plane p is vertical.

    (b) $y = -\frac{1}{4}(3x + 6) = -\frac{1}{4}[3(4t - 2) + 6] = -3t$

       $z = x^2 + 3y^2 + 2 = (4t - 2)^2 + 3(-3t)^2 + 2 = 43t^2 - 16t + 6$

       $\mathbf{r}(t) = (4t - 2)\mathbf{i} - 3t\mathbf{j} + (43t^2 - 16t + 6)\mathbf{k}$

    (c) From part (b) the tip of $\mathbf{r}(1)$ is $(2, -3, 33)$.   We take

       $\mathbf{r}'(1) = 4\mathbf{i} - 3\mathbf{j} + 70\mathbf{j}$ as $\mathbf{d}$ to write

$$\mathbf{R}(s) = (2\mathbf{i} - 3\mathbf{j} + 33\mathbf{k}) + s(4\mathbf{i} - 3\mathbf{j} + 70\mathbf{k}).$$

    (d) Set $g(x, y) = x^2 + 3y^2 + 2$.   Then,

$$\nabla g = 2x\mathbf{i} + 6y\mathbf{j} \quad \text{and} \quad \nabla g(2, -3) = 4\mathbf{i} - 18\mathbf{j}.$$

       An equation for the plane tangent to $z = g(x, y)$ at $(2, -3, 33)$ is

$$z - 33 = 4(x - 2) - 18(y + 3) \quad \text{which reduces to} \quad 4x - 18y - z = 29.$$

    (e) Substituting t for x in the equations for p and p_1, we obtain

$$3t + 4y + 6 = 0 \quad \text{and} \quad 4t - 18y - z = 29.$$

    From the first equation

$$y = -\frac{3}{4}(t + 2)$$

    and then from the second equation

$$z = 4t - 18\left[-\frac{3}{4}(t + 2)\right] - 29 = \frac{35}{2}t - 2.$$

    Thus,

    $(*)$ $\mathbf{r}(t) = t\mathbf{i} - \left(\frac{3}{4}t + \frac{3}{2}\right)\mathbf{j} + \left(\frac{35}{2}t - 2\right)\mathbf{k}.$

    Lines l and l' are the same. To see this, consider how l and l' are formed; to assure yourself, replace t in $(*)$ by $4s + 2$ to obtain $\mathbf{R}(s)$ found in part (c).

39. (a) normal vector: $2\mathbf{i} + 2\mathbf{j} + 4\mathbf{k}$;   normal line: $x = 1 + 2t, \; y = 2 + 2t, \; z = 2 + 4t$

    (b) tangent plane: $2(x - 1) + 2(y - 2) + 4(z - 2) = 0$ or $x + y + 2z - 7 = 0$

    (c)

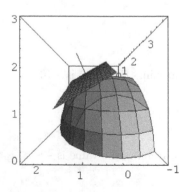

41. (a)

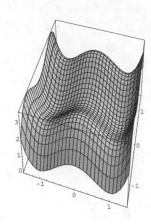

(b)

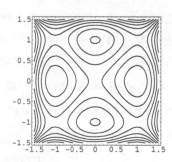

(c) $\nabla f = \left(4x^3 - 4x\right)\mathbf{i} - \left(4y^3 - 4y\right)\mathbf{j}$;

$\nabla f = 0:$ $4x^3 - 4x = 0 \implies x = 0, \pm 1;$ $4y^3 - 4y = 0 \implies y = 0, \pm 1$

$\nabla f = 0$ at $(0,0),\ (\pm 1, 0),\ (0, \pm 1),\ (\pm 1, \pm 1)$

SECTION 16.5

1. $\nabla f = (2 - 2x)\mathbf{i} - 2y\,\mathbf{j} = \mathbf{0}$ only at $(1, 0)$.

The difference

$$f(1 + h, k) - f(1, 0) = \left[2(1 + h) - (1 + h)^2 - k^2\right] - 1 = -h^2 - k^2 \leq 0$$

for all small h and k; there is a local maximum of 1 at $(1, 0)$.

3. $\nabla f = (2x + y + 3)\mathbf{i} + (x + 2y)\mathbf{j} = \mathbf{0}$ only at $(-2, 1)$.

The difference

$f(-2 + h,\ 1 + k) - f(-2, 1)$

$\quad = [(-2 + h)^2 + (-2 + h)(1 + k) + (1 + k)^2 + 3(-2 + h) + 1] - (-2) = h^2 + hk + k^2$

is nonnegative for all small h and k. To see this, note that

$$h^2 + hk + k^2 \geq h^2 - 2|h||k| + k^2 = (|h| - |k|)^2 \geq 0;$$

there is a local minimum of -2 at $(-2, 1)$.

5. $\nabla f = (2x + y - 6)\mathbf{i} + (x + 2y)\mathbf{j} = \mathbf{0}$ only at $(4, -2)$.

$f_{xx} = 2, \quad f_{xy} = 1, \quad f_{yy} = 2.$

At $(4, -2)$, $D = 3 > 0$ and $A = 2 > 0$ so we have a local min; the value is -10.

7. $\nabla f = \left(3x^2 - 6y\right)\mathbf{i} + \left(3y^2 - 6x\right)\mathbf{j} = \mathbf{0}$ at $(2, 2)$ and $(0, 0)$.

$f_{xx} = 6x, \quad f_{xy} = -6, \quad f_{yy} = 6y, \quad D = 36xy - 36.$

At $(2, 2)$, $D = 108 > 0$ and $A = 12 > 0$ so we have a local min; the value is -8.

At $(0, 0)$, $D = -36 < 0$ so we have a saddle point.

9. $\nabla f = (3x^2 - 6y + 6)\,\mathbf{i} + (2y - 6x + 3)\,\mathbf{j} = \mathbf{0}$ at $(5, \frac{27}{2})$ and $(1, \frac{3}{2})$.

    $f_{xx} = 6x$, $\quad f_{xy} = -6$, $\quad f_{yy} = 2$, $\quad D = 12x - 36$.

    At $(5, \frac{27}{2})$, $D = 24 > 0$ and $A = 30 > 0$ so we have a local min; the value is $-\frac{117}{4}$.

    At $(1, \frac{3}{2})$, $D = -24 < 0$ so we have a saddle point.

11. $\nabla f = \sin y\,\mathbf{i} + x\cos y\,\mathbf{j} = \mathbf{0}$ $\quad$ at $\quad$ $(0, n\pi)$ for all integral n.

    $f_{xx} = 0$, $\quad f_{xy} = \cos y$, $\quad f_{yy} = -x\sin y$.

    Since $D = -\cos^2 n\pi = -1 < 0$, each stationary point is a saddle point.

13. $\nabla f = (2xy + 1 + y^2)\,\mathbf{i} + (x^2 + 2xy + 1)\,\mathbf{j} = \mathbf{0}$ $\quad$ at $\quad$ $(1, -1)$ and $(-1, 1)$.

    $f_{xx} = 2y$, $\quad f_{xy} = 2x + 2y$, $\quad f_{yy} = 2x$, $\quad D = 4xy - 4(x+y)^2$.

    At both $(1, -1)$ and $(-1, 1)$ we have saddle points since $D = -4 < 0$.

15. $\nabla f = (y - x^{-2})\,\mathbf{i} + (x - 8y^{-2})\,\mathbf{j} = \mathbf{0}$ $\quad$ only at $\quad$ $(\frac{1}{2}, 4)$.

    $f_{xx} = 2x^{-3}$, $\quad f_{xy} = 1$, $\quad f_{yy} = 16y^{-3}$, $\quad D = 32x^{-3}y^{-3} - 1$.

    At $(\frac{1}{2}, 4)$, $D = 3 > 0$ and $A = 16 > 0$ so we have a local min; the value is 6.

17. $\nabla f = (y - x^{-2})\,\mathbf{i} + (x - y^{-2})\,\mathbf{j} = \mathbf{0}$ $\quad$ only at $\quad$ $(1, 1)$.

    $f_{xx} = 2x^{-3}$, $\quad f_{xy} = 1$, $\quad f_{yy} = 2y^{-3}$, $\quad D = 4x^{-3}y^{-3} - 1$.

    At $(1, 1)$, $D = 3 > 0$ and $A = 2 > 0$ so we have a local min; the value is 3.

19. $\nabla f = \dfrac{2(x^2 - y^2 - 1)}{(x^2 + y^2 + 1)^2}\,\mathbf{i} + \dfrac{4xy}{(x^2 + y^2 + 1)^2}\,\mathbf{j} = \mathbf{0}$ $\quad$ at $\quad$ $(1, 0)$ and $(-1, 0)$.

    $f_{xx} = \dfrac{-4x^3 + 12xy^2 + 12x}{(x^2 + y^2 + 1)^3}$, $\quad f_{xy} = \dfrac{4y^3 + 4y - 12x^2y}{(x^2 + y^2 + 1)^3}$, $\quad f_{yy} = \dfrac{4x^3 + 4x - 12xy^2}{(x^2 + y^2 + 1)^3}$.

point	A	B	C	D	result
$(1, 0)$	1	0	1	1	loc. min.
$(-1, 0)$	-1	0	-1	1	loc. max.

    $f(1, 0) = -1$; $\quad f(-1, 0) = 1$

21. $\nabla f = (4x^3 - 4x)\,\mathbf{i} + 2y\,\mathbf{j} = \mathbf{0}$ $\quad$ at $\quad$ $(0, 0)$, $(1, 0)$, and $(-1, 0)$.

    $f_{xx} = 12x^2 - 4$, $\quad f_{xy} = 0$, $\quad f_{yy} = 2$.

point	A	B	C	D	result
$(0, 0)$	-4	0	2	-8	saddle
$(1, 0)$	8	0	2	16	loc. min.
$(-1, 0)$	8	0	2	16	loc. min.

    $f(\pm 1, 0) = -3$.

23. $\nabla f = \cos x \, \sin y \, \mathbf{i} + \sin x \, \cos y \, \mathbf{j} = \mathbf{0}$ at $\left(\frac{1}{2}\pi, \frac{1}{2}\pi\right)$, $\left(\frac{1}{2}\pi, \frac{3}{2}\pi\right)$, (π, π), $\left(\frac{3}{2}\pi, \frac{1}{2}\pi\right)$, $\left(\frac{3}{2}\pi, \frac{3}{2}\pi\right)$.

$f_{xx} = -\sin x \, \sin y, \quad f_{xy} = \cos x \, \cos y, \quad f_{yy} = -\sin x \, \sin y$

point	A	B	C	D	result
$\left(\frac{1}{2}\pi, \frac{1}{2}\pi\right)$	-1	0	-1	1	loc. max.
$\left(\frac{1}{2}\pi, \frac{3}{2}\pi\right)$	1	0	1	1	loc. min.
(π, π)	0	1	0	-1	saddle
$\left(\frac{3}{2}\pi, \frac{1}{2}\pi\right)$	1	0	1	1	loc. min.
$\left(\frac{3}{2}\pi, \frac{3}{2}\pi\right)$	-1	0	-1	1	loc. max.

$f\left(\frac{1}{2}\pi, \frac{1}{2}\pi\right) = f\left(\frac{3}{2}\pi, \frac{3}{2}\pi\right) = 1; \quad f\left(\frac{1}{2}\pi, \frac{3}{2}\pi\right) = f\left(\frac{3}{2}\pi, \frac{1}{2}\pi\right) = -1$

25. (a) $\nabla f = (2x + ky)\,\mathbf{i} + (2y + kx)\,\mathbf{j}$ and $\nabla f(0,0) = \mathbf{0}$ independent of the value of k.

(b) $f_{xx} = 2, \quad f_{xy} = k, \quad f_{yy} = 2, \quad D = 4 - k^2$. Thus, $D < 0$ for $|k| > 2$ and $(0,0)$ is a saddle point

(c) $D = 4 - k^2 > 0$ for $|k| < 2$. Since $A = f_{xx} = 2 > 0$, $(0,0)$ is a local minimum.

(d) The test is inconclusive when $D = 4 - k^2 = 0$ i.e., for $k = \pm 2$. (If $k = \pm 2$, $f(x,y) = (x \pm y)^2$ and $(0, 0)$ is a minimum.)

27. Let $P(x,y,z)$ be a point in the plane. We want to find the minimum of $f(x,y,z) = \sqrt{x^2 + y^2 + z^2}$. However, it is sufficient to minimize the square of the distance: $F(x,y,z) = x^2 + y^2 + z^2$. It is clear that F has a minimum value, but no maximum value. Since P lies in the plane, $2x - y + 2z = 16$ which implies $y = 2x + 2z - 16 = 2(x + z - 8)$. Thus, we want to find the minimum value of

$$F(x,z) = x^2 + 4(x + z - 8)^2 + z^2$$

Now,

$$\nabla F = [2x + 8(x + z - 8)]\,\mathbf{i} + [8(x + z - 8) + 2z]\,\mathbf{k}$$

The gradient is $\mathbf{0}$ when

$$2x + 8(x + z - 8) = 0 \quad \text{and} \quad 8(x + z - 8) + 2z = 0$$

The only solution to this pair of equations is: $x = z = \dfrac{32}{9}$, from which it follows that $y = -\dfrac{16}{9}$.

The point in the plane that is closest to the origin is $P\left(\frac{32}{9}, -\frac{16}{9}, \frac{32}{9}\right)$.

The distance from the origin to the plane is: $F(P) = \frac{16}{3}$.

Check using (13.6.5): $d(P, 0) = \dfrac{|2 \cdot 0 - 0 + 2 \cdot 0 - 16|}{\sqrt{2^2 + (-1)^2 + 2^2}} = \dfrac{16}{3}$.

29. $f(x,y) = (x-1)^2 + (y-2)^2 + z^2 = (x-1)^2 + (y-2)^2 + x^2 + 2y^2 \quad \left[\text{since } z = \sqrt{x^2 + 2y^2}\right]$

$\nabla f = [2(x-1) + 2x]\,\mathbf{i} + [2(y-2) + 4y]\,\mathbf{j} = \mathbf{0} \implies x = \dfrac{1}{2}, \ y = \dfrac{2}{3}$.

$f_{xx} = 4 > 0, \quad f_{xy} = 0, \quad f_{yy} = 6, \quad D = 24 > 0$. Thus, f has a local minimum at $(1/2, 2/3)$.

The shortest distance from $(1, 2, 0)$ to the cone is $\sqrt{f\left(\frac{1}{2}, \frac{2}{3}\right)} = \frac{1}{6}\sqrt{114}$

31. (a)

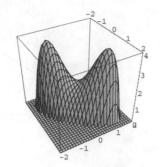

(b)

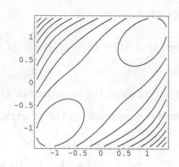

(c) $\nabla f = (4y - 4x^3)\,\mathbf{i} + (4x - 4y^3)\,\mathbf{j} = \mathbf{0}$ at $(0,0)$, $(1,1)$, $(-1,-1)$.

$f_{xx} = -12x^2$, $f_{xy} = 4$, $f_{yy} = -12y^2$, $D = 144x^2y^2 - 16$

point	A	B	C	D	result
$(0,0)$	0	4	0	-16	saddle
$(1,1)$	-12	4	-12	128	loc. max.
$(-1,-1)$	-12	4	-12	128	loc. max.

$f(1,1) = f(-1,-1) = 3$

33. (a)

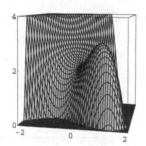

(b)

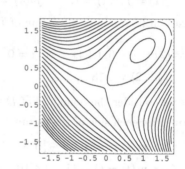

$f(1,1) = 3$ is a local max.; f has a saddle at $(0,0)$.

35. (a)

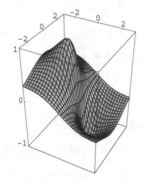

(b)

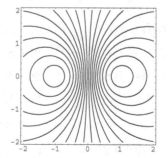

$f(1,0) = -1$ is a local min.; $f(-1,0) = 1$ is a loc. max.

SECTION 16.6

1. $\nabla f = (4x - 4)\,\mathbf{i} + (2y - 2)\,\mathbf{j} = \mathbf{0}$ at $(1,1)$ in D;

$f(1,1) = -1$

Next we consider the boundary of D. We

parametrize each side of the triangle:

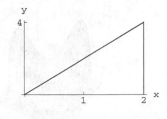

$$C_1 : \mathbf{r}_1(t) = t\,\mathbf{i}, \quad t \in [0,2],$$
$$C_2 : \mathbf{r}_2(t) = 2\,\mathbf{i} + t\,\mathbf{j}, \quad t \in [0,4],$$
$$C_3 : \mathbf{r}_3(t) = t\,\mathbf{i} + 2t\,\mathbf{j}, \quad t \in [0,2],$$

Now,

$$f_1(t) = f(\mathbf{r}_1(t)) = 2(t-1)^2, \quad t \in [0,2]; \quad \text{critical number: } t = 1,$$
$$f_2(t) = f(\mathbf{r}_2(t)) = (t-1)^2 + 1, \quad t \in [0,4]; \quad \text{critical number: } t = 1,$$
$$f_3(t) = f(\mathbf{r}_3(t)) = 6t^2 - 8t + 2, \quad t \in [0,2]; \quad \text{critical number: } t = \tfrac{2}{3}.$$

Evaluating these functions at the endpoints of their domains and at the critical numbers, we find that:

$$f_1(0) = f_3(0) = f(0,0) = 2; \qquad f_1(1) = f(1,0) = 0; \qquad f_1(2) = f_2(0) = f(2,0) = 2;$$
$$f_2(1) = f(2,1) = 1; \qquad f_2(4) = f_3(2) = f(2,4) = 10; \qquad f_3(2/3) = f(2/3, 4/3) = -\tfrac{2}{3}.$$

f takes on its absolute maximum of 10 at $(2,4)$ and its absolute minimum of -1 at $(1,1)$.

3. $\nabla f = (2x + y - 6)\,\mathbf{i} + (x + 2y)\,\mathbf{j} = \mathbf{0}$ at $(4,-2)$ in

D; $\quad f(4,-2) = -13$

Next we consider the boundary of D. We

parametrize each side of the rectangle:

$$C_1 : \mathbf{r}_1(t) = -t\,\mathbf{j}, \quad t \in [0,3]$$
$$C_2 : \mathbf{r}_2(t) = t\,\mathbf{i} - 3\,\mathbf{j}, \quad t \in [0,5]$$
$$C_3 : \mathbf{r}_3(t) = 5\,\mathbf{i} - t\,\mathbf{j}, \quad t \in [0,3]$$
$$C_4 : \mathbf{r}_4(t) = t\,\mathbf{i}, \quad t \in [0,5]$$

Now,

$$f_1(t) = f(\mathbf{r}_1(t)) = t^2 - 1, \quad t \in [0,3]; \quad \text{no critical numbers}$$
$$f_2(t) = f(\mathbf{r}_2(t)) = t^2 - 9t + 8, \quad t \in [0,5]; \quad \text{critical number: } t = \tfrac{9}{2}$$
$$f_3(t) = f(\mathbf{r}_3(t)) = t^2 - 5t - 6, \quad t \in [0,3]; \quad \text{critical number: } t = \tfrac{5}{2}$$
$$f_4(t) = f(\mathbf{r}_4(t)) = t^2 - 6t - 1, \quad t \in [0,5]; \quad \text{critical number: } t = 3$$

Evaluating these functions at the endpoints of their domains and at the critical numbers, we find that:

$$f_1(0) = f_4(0) = f(0,0) = -1; \qquad f_1(-3) = f_2(0) = f(0,-3) = 8; \qquad f_2(9/2) = f(9/2,-3) = -\tfrac{49}{4};$$
$$f_2(5) = f_3(3) = f(5,-3) = -12; \qquad f_3(5/2) = f(5,-5/2) = -\tfrac{49}{4}; \qquad f_3(0) = f_4(5) = f(5,0) = -6.$$
$$f_4(3) = f(3,0) = -10$$

f takes on its absolute maximum of 8 at $(0,-3)$ and its absolute minimum of -13 at $(4,-2)$.

5. $\nabla f = (2x + 3y)\,\mathbf{i} + (2y + 3x)\,\mathbf{j} = \mathbf{0}$ at $(0,0)$ in $D;$ $f(0,0) = 2$

Next we consider the boundary of $D.$ We parametrize the circle by:

$$C: \ \mathbf{r}(t) = 2\cos t\,\mathbf{i} + 2\sin t\,\mathbf{j}, \quad t \in [\,0, 2\pi\,]$$

The values of f on the boundary are given by the function

$$F(t) = f(\mathbf{r}(t)) = 6 + 12\sin t\,\cos t, \quad t \in [\,0, 2\pi\,]$$

$$F'(t) = 12\cos^2 t - 12\sin^2 t: \quad F'(t) = 0 \implies \cos t = \pm\sin t \implies t = \tfrac{1}{4}\pi, \ \tfrac{3}{4}\pi, \ \tfrac{5}{4}\pi, \ \tfrac{7}{4}\pi$$

Evaluating F at the endpoints and critical numbers, we have:

$$F(0) = F(2\pi) = f(2,0) = 6; \quad F\left(\tfrac{1}{4}\pi\right) = F\left(\tfrac{5}{4}\pi\right) = f\left(\sqrt{2},\sqrt{2}\right) = f\left(\left(-\sqrt{2},-\sqrt{2}\right)\right) = 12;$$

$$F\left(\tfrac{3}{4}\pi\right) = f\left(-\sqrt{2},\sqrt{2}\right) = F\left(\tfrac{7}{4}\pi\right) = f\left(\sqrt{2},-\sqrt{2}\right) = 0$$

f takes on its absolute maximum of 12 at $\left(\sqrt{2},\sqrt{2}\right)$ and at $\left(-\sqrt{2},-\sqrt{2}\right);$ f takes on its absolute minimum of 0 at $\left(-\sqrt{2},\sqrt{2}\right)$ and at $\left(\sqrt{2},-\sqrt{2}\right).$

7. $\nabla f = 2(x-1)\mathbf{i} + 2(y-1)\,\mathbf{j} = \mathbf{0}$ only at $(1,1)$ in $D.$ As the sum of two squares, $f(x,y) \geq 0.$ Thus, $f(1,1) = 0$ is a minimum. To examine the behavior of f on the boundary of $D,$ we note that f represents the square of the distance between (x,y) and $(1,1).$ Thus, f is maximal at the point of the boundary furthest from $(1,1).$ This is the point $\left(-\sqrt{2},\ -\sqrt{2}\right);$ the maximum value of f is $f\left(-\sqrt{2},\ -\sqrt{2}\right) = 6 + 4\sqrt{2}.$

9. $\nabla f = \dfrac{2x^2 - 2y^2 - 2}{(x^2 + y^2 + 1)^2}\,\mathbf{i} + \dfrac{4xy}{(x^2 + y^2 + 1)^2}\,\mathbf{j} = \mathbf{0}$ at $(1,0)$ and $(-1,0)$ in $D;$ $f(1,0) = -1,$ $f(-1,0) = 1.$

Next we consider the boundary of $D.$ We parametrize each side of the square:

$$C_1: \ \mathbf{r}_1(t) = -2\,\mathbf{i} + t\,\mathbf{j}, \quad t \in [-2,2]$$
$$C_2: \ \mathbf{r}_2(t) = t\,\mathbf{i} + 2\,\mathbf{j}, \quad t \in [-2,2]$$
$$C_3: \ \mathbf{r}_3(t) = 2\,\mathbf{i} + t\,\mathbf{j}, \quad t \in [-2,2]$$
$$C_4: \ \mathbf{r}_4(t) = t\,\mathbf{i}, \quad t \in [-2,2]$$

Now,

$$f_1(t) = f(\mathbf{r}_1(t)) = \frac{4}{t^2 + 5}, \quad t \in [-2,2]; \quad \text{critical number: } t = 0$$

$$f_2(t) = f(\mathbf{r}_2(t)) = \frac{-2t}{t^2 + 5}, \quad t \in [-2,2]; \quad \text{no critical numbers}$$

$$f_3(t) = f(\mathbf{r}_3(t)) = \frac{-4}{t^2 + 5}, \quad t \in [-2,2]; \quad \text{critical number: } t = 0$$

$$f_4(t) = f(\mathbf{r}_4(t)) = \frac{-2t}{t^2 + 5}, \quad t \in [-2,2]; \quad \text{no critical numbers}$$

Evaluating these functions at the endpoints of their domains and at the critical numbers, we find that:

$$f_1(-2) = f_4(-2) = f(-2,-2) = \tfrac{4}{9}; \quad f_1(0) = f(-2,0) = \tfrac{4}{5}; \quad f_1(2) = f_2(-2) = f(-2,2) = \tfrac{4}{9};$$

$$f_4(2) = f_3(-2) = f(2,-2) = -\tfrac{4}{9}; \quad f_3(0) = f(2,0) = -\tfrac{4}{5}; \quad f_2(2) = f_3(2) = f(2,2) = -\tfrac{4}{9}.$$

f takes on its absolute maximum of 1 at $(-1,0)$ and its absolute minimum of -1 at $(1,0).$

11. $\nabla f = (4 - 4x)\cos y\,\mathbf{i} - (4x - 2x^2)\sin y\,\mathbf{j} = \mathbf{0}$ at $(1,0)$ in D: $f(1,0) = 2$

Next we consider the boundary of D. We parametrize each side of the rectangle:

$$C_1: \mathbf{r}_1(t) = t\,\mathbf{j}, \quad t \in \left[-\tfrac{1}{4}\pi, \tfrac{1}{4}\pi\right]$$

$$C_2: \mathbf{r}_2(t) = t\,\mathbf{i} - \tfrac{1}{4}\pi\,\mathbf{j}, \quad t \in [0,2]$$

$$C_3: \mathbf{r}_3(t) = 2\,\mathbf{i} + t\,\mathbf{j}, \quad t \in \left[-\tfrac{1}{4}\pi, \tfrac{1}{4}\pi\right]$$

$$C_4: \mathbf{r}_4(t) = t\,\mathbf{i} + \tfrac{1}{4}\pi\,\mathbf{j}, \quad t \in [0,2]$$

Now,

$$f_1(t) = f(\mathbf{r}_1(t)) = 0;$$

$$f_2(t) = f(\mathbf{r}_2(t)) = \frac{\sqrt{2}}{2}(4t - 2t^2), \quad t \in [0,2]; \quad \text{critical number: } t = 1;$$

$$f_3(t) = f(\mathbf{r}_3(t)) = 0;$$

$$f_4(t) = f(\mathbf{r}_4(t)) = \frac{\sqrt{2}}{2}(4t - 2t^2), \quad t \in [0,2]; \quad \text{critical number: } t = 1;$$

f at the vertices of the rectangle has the value 0; $\quad f_2(1) = f_4(1) = f\left(1, -\tfrac{1}{4}\pi\right) = f\left(1, \tfrac{1}{4}\pi\right) = \sqrt{2}$.

f takes on its absolute maximum of 2 at $(1,0)$ and its absolute minimum of 0 along the lines $x = 0$ and $x = 2$.

13. $\nabla f = (3x^2 - 3y)\,\mathbf{i} + (-3x - 3y^2)\,\mathbf{j} = \mathbf{0}$ at $(-1,1)$ in D;

$f(-1,1) = 1$

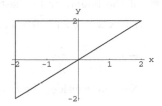

Next we consider the boundary of D. We parametrize each side of the triangle:

$$C_1: \mathbf{r}_1(t) = -2\,\mathbf{i} + t\,\mathbf{j}, \quad t \in [-2,2],$$

$$C_2: \mathbf{r}_2(t) = t\,\mathbf{i} + t\,\mathbf{j}, \quad t \in [-2,2],$$

$$C_3: \mathbf{r}_3(t) = t\,\mathbf{i} + 2\,\mathbf{j}, \quad t \in [-2,2],$$

and evaluate f:

$$f_1(t) = f(\mathbf{r}_1(t)) = -8 + 6t - t^3, \quad t \in [-2,2]; \quad \text{critical numbers: } t = \pm\sqrt{2},$$

$$f_2(t) = f(\mathbf{r}_2(t)) = -3t^2, \quad t \in [-2,2]; \quad \text{critical number: } t = 0,$$

$$f_3(t) = f(\mathbf{r}_3(t)) = t^3 - 6t - 8, \quad t \in [-2,2]; \quad \text{critical numbers: } t = \pm\sqrt{2}.$$

Evaluating these functions at the endpoints of their domains and at the critical numbers, we find that:

$$f_1(-2) = f_2(-2) = f(-2,-2) = -12; \qquad f_1(-\sqrt{2}) = f(-2,-\sqrt{2}) = -8 - 4\sqrt{2} \cong -13.66;$$

$$f_1(\sqrt{2}) = f(-2,\sqrt{2}) = -8 + 4\sqrt{2} \cong -2.34; \qquad f_1(2) = f_3(-2) = f(-2,2) = -4;$$

$$f_2(0) = f(0,0) = 0; \qquad f_2(2) = f_3(2) = f(2,2) = -12;$$

$$f_3(-\sqrt{2}) = f(-\sqrt{2},2) = -8 + 4\sqrt{2}; \qquad f_3(\sqrt{2}) = f(\sqrt{2},2) = -8 - 4\sqrt{2}$$

f takes on its absolute maximum of 1 at $(-1,1)$ and its absolute minimum of $-8 - 4\sqrt{2}$ at $(\sqrt{2},2)$ and $(-2,-\sqrt{2})$.

15. $\nabla f = \dfrac{4xy}{(x^2 + y^2 + 1)^2}\,\mathbf{i} + \dfrac{2y^2 - 2x^2 - 2}{(x^2 + y^2 + 1)^2}\,\mathbf{j} = \mathbf{0}$ at $(0,1)$ and $(0,-1)$ in D;

$f(0,1) = -1, \quad f(0,-1) = 1$

Next we consider the boundary of D.   We parametrize the circle by:

$$C: \ \mathbf{r}(t) = 2\cos t\,\mathbf{i} + 2\sin t\,\mathbf{j}, \quad t \in [0, 2\pi]$$

The values of f on the boundary are given by the function

$$F(t) = f(\mathbf{r}(t)) = -\tfrac{4}{5}\sin t, \quad t \in [0, 2\pi]$$

$$F'(t) = -\tfrac{4}{5}\cos t: \quad F'(t) = 0 \implies \cos t = 0 \implies t = \tfrac{1}{2}\pi, \ \tfrac{3}{2}\pi.$$

Evaluating F at the endpoints and critical numbers, we have:

$$F(0) = F(2\pi) = f(2,0) = 0; \quad F\left(\tfrac{1}{2}\pi\right) = f(0,2) = -\tfrac{4}{5}; \quad F\left(\tfrac{3}{2}\pi\right) = f(0,-2) = \tfrac{4}{5}$$

f takes on its absolute maximum of 1 at $(0,-1)$ and its absolute minimum of -1 at $(0,1)$.

17. $\nabla f = 2(x - y)\mathbf{i} - 2(x - y)\mathbf{j} = \mathbf{0}$ at each point of the
line segment $y = x$ from $(0,0)$ to $(4,4)$. Since
$f(x,x) = 0$ and $f(x,y) \geq 0$, f takes on its minimum
of 0 at each of these points.

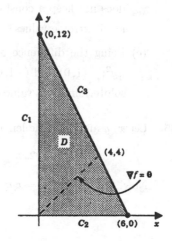

Next we consider the boundary of D. We
parametrize each side of the triangle:

$$C_1: \ \mathbf{r}_1(t) = t\mathbf{j}, \quad t \in [0,12]$$
$$C_2: \ \mathbf{r}_2(t) = t\mathbf{i}, \quad t \in [0,6]$$
$$C_3: \ \mathbf{r}_3(t) = t\mathbf{i} + (12 - 2t)\mathbf{j}, \quad t \in [0,6]$$

and observe from

$$f(\mathbf{r}_1(t)) = t^2, \quad t \in [0,12]$$
$$f(\mathbf{r}_2(t)) = t^2, \quad t \in [0,6]$$
$$f(\mathbf{r}_3(t)) = (3t - 12)^2, \quad t \in [0,6]$$

that f takes on its maximum of 144 at the point $(0,12)$.

19. Using the hint, we want to find the maximum value of $f(x,y) = 18xy - x^2 y - xy^2$ in the triangular
region. The gradient of f is:

$$\nabla D = \left(18y - 2xy - y^2\right)\mathbf{i} + \left(18x - x^2 - 2xy\right)\mathbf{j}$$

The gradient is $\mathbf{0}$ when

$$18y - 2xy - y^2 = 0 \quad \text{and} \quad 18x - x^2 - 2xy = 0$$

The solution set of this pair of equations is: $(0,0), \ (18,0), \ (0,18), \ (6,6)$.

It is easy to verify that f is a maximum when $x = y = 6$. The three numbers that satisfy $x + y + z = 18$
and maximize the product xyz are: $x = 6, \ y = 6, \ z = 6$.

21. $f(x, y) = xy(1 - x - y), \quad 0 \le x \le 1, \quad 0 \le y \le 1 - x.$

[dom (f) is the triangle with vertices $(0, 0)$, $(1, 0)$, $(0, 1)$.]

$\nabla f = (y - 2xy - y^2)\mathbf{i} + (x - 2xy - x^2)\mathbf{j} = \mathbf{0} \implies x = y = 0, \ x = 1, \ y = 0, \ x = 0, \ y = 1, \ x = y = \frac{1}{3}.$

(Note that $[0, 0]$ is not an interior point of the domain of f.)

$f_{xx} = -2y, \quad f_{xy} = 1 - 2x - 2y, \quad f_{yy} = -2x.$

At $\left(\frac{1}{3}, \frac{1}{3}\right)$, $D = \frac{1}{3} > 0$ and $A < 0$ so we have a local max; the value is $1/27$.

Since $f(x, y) = 0$ at each point on the boundary of the domain, the local max of $1/27$ is also the absolute max.

23. (a) $\nabla f = \frac{1}{2}x\,\mathbf{i} - \frac{2}{9}y\,\mathbf{j} = \mathbf{0}$ only at $(0, 0)$.

(b) The difference

$$f(h, k) - f(0, 0) = \tfrac{1}{4}h^2 - \tfrac{1}{9}k^2$$

does not keep a constant sign for all small h and k; $(0, 0)$ is a saddle point. The function has no local extreme values.

(c) Being the difference of two squares, f can be maximized by maximizing $\frac{1}{4}x^2$ and minimizing $\frac{1}{9}y^2$; $(1, 0)$ and $(-1, 0)$ give absolute maximum value $\frac{1}{4}$. Similarly, $(0, 1)$ and $(0, -1)$ give absolute minimum value $-\frac{1}{9}$.

25. Let x, y and z be the length, width and height of the box. The surface area is given by

$$S = 2xy + 2xz + 2yz, \quad \text{so} \quad z = \frac{S - 2xy}{2(x + y)}, \quad \text{where } S \text{ is a constant, and } x, y, z > 0.$$

Now, the volume $V = xyz$ is given by:

$$V(x, y) = xy \left[\frac{S - 2xy}{2(x + y)} \right]$$

and

$$\nabla V = \left\{ y \left[\frac{S - 2xy}{2(x + y)} \right] + xy \, \frac{2(x + y)(-2y) - (S - 2xy)(2)}{4(x + y)^2} \right\} \mathbf{i}$$

$$+ \left\{ x \left[\frac{S - 2xy}{2(x + y)} \right] + xy \, \frac{2(x + y)(-2x) - (S - 2xy)(2)}{4(x + y)^2} \right\} \mathbf{j}$$

Setting $\dfrac{\partial V}{\partial x} = \dfrac{\partial V}{\partial y} = 0$ and simplifying, we get the pair of equations

$$2S - 4x^2 - 8xy = 0$$

$$2S - 4y^2 - 8xy = 0$$

from which it follows that $x = y = \sqrt{S/6}$. From practical considerations, we conclude that V has a maximum value at $(\sqrt{S/6}, \sqrt{S/6})$. Substituting these values into the equation for z, we get $z = \sqrt{S/6}$ and so the box of maximum volume is a cube.

27.
$$f(x,y) = \sum_{i=1}^{3}\left[(x-x_i)^2 + (y-y_i)^2\right]$$

$$\nabla f(x,y) = 2\left[(3x-x_1-x_2-x_3)\,\mathbf{i} + (3y-y_1-y_2-y_3)\,\mathbf{j}\right]$$

$$\nabla f = \mathbf{0} \quad \text{only at} \quad \left(\frac{x_1+x_2+x_3}{3}, \frac{y_1+y_2+y_3}{3}\right) = (x_0, y_0).$$

The difference $f(x_0 + h, y_0 + k) - f(x_0, y_0)$

$$= \sum_{i=1}^{3}\left[(x_0+h-x_i)^2 + (y_0+k-y_i)^2 - (x_0-x_i)^2 - (y_0-y_i)^2\right]$$

$$= \sum_{i=1}^{3}\left[2h(x_0-x_i) + h^2 + 2k(y_0-y_i) + k^2\right]$$

$$= 2h(3x_0-x_1-x_2-x_3) + 2k(3y_0-y_1-y_2-y_3) + 3h^2 + 3k^2$$

$$= 3h^2 + 3k^2$$

is nonnegative for all h and k. Thus, f has its absolute minimum at (x_0, y_0).

29.

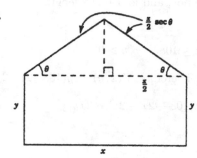

$$A = xy + \frac{1}{2}x\left(\frac{x}{2}\tan\theta\right),$$

$$P = x + 2y + 2\left(\frac{x}{2}\sec\theta\right), \quad y = \tfrac{1}{2}(P - x - x\sec\theta)$$

$$0 < \theta < \frac{1}{2}\pi, \quad 0 < x < \frac{P}{1+\sec\theta}.$$

$$A(x,\theta) = \tfrac{1}{2}x(P - x - x\sec\theta) + \tfrac{1}{4}x^2\tan\theta,$$

$$\nabla A = \left(\frac{P}{2} - x - x\sec\theta + \frac{x}{2}\tan\theta\right)\mathbf{i} + \left(\frac{x^2}{4}\sec^2\theta - \frac{x^2}{2}\sec\theta\tan\theta\right)\mathbf{j},$$

(Here $\mathbf{j}$ is the unit vector in the direction of increasing θ.)

$$\nabla A = \frac{1}{2}\left[P + x(\tan\theta - 2\sec\theta - 2)\right]\mathbf{i} + \frac{x^2}{4}\sec\theta\,(\sec\theta - 2\tan\theta)\,\mathbf{j}.$$

From $\dfrac{\partial A}{\partial \theta} = 0$ we get $\theta = \tfrac{1}{6}\pi$ and then from $\dfrac{\partial A}{\partial x} = 0$ we get

$$P + x\left(\tfrac{1}{3}\sqrt{3} - \tfrac{4}{3}\sqrt{3} - 2\right) = 0 \quad \text{so that} \quad x = (2-\sqrt{3})P.$$

Next,

$$A_{xx} = \tfrac{1}{2}(\tan\theta - 2\sec\theta - 2),$$

$$A_{x\theta} = \frac{x}{2}\sec\theta\,(\sec\theta - 2\tan\theta),$$

$$A_{\theta\theta} = \frac{x^2}{2}\sec\theta\,\left(\sec\theta\tan\theta - \sec^2\theta - \tan^2\theta\right).$$

Apply the second-partials test:

$$A = -\tfrac{1}{2}(2+\sqrt{3}), \quad B = 0, \quad C = -\tfrac{1}{3}P^2\sqrt{3}\,(2-\sqrt{3})^2, \quad D < 0.$$

Since, $D > 0$ and $A < 0$, the area is a maximum when $\theta = \tfrac{1}{6}\pi$, $x = (2-\sqrt{3})P$ and $y = \tfrac{1}{6}(3-\sqrt{3})P$.

31. From
$$x = \tfrac{1}{2}y = \tfrac{1}{3}z = t \quad \text{and} \quad x = y - 2 = z = s$$
we take
$$(t, 2t, 3t) \quad \text{and} \quad (s, 2 + s, s)$$

as arbitrary points on the lines. It suffices to minimize the square of the distance between these points:

$$f(t, s) = (t - s)^2 + (2t - 2 - s)^2 + (3t - s)^2$$
$$= 14t^2 - 12ts + 3s^2 - 8t + 4s + 4, \qquad t, s \text{ real.}$$

Let $\mathbf{i}$ and $\mathbf{j}$ be the unit vectors in the direction of increasing t and s, respectively.

$$\nabla f = (28t - 12s - 8)\mathbf{i} + (-12t + 6s + 4)\mathbf{j}; \qquad \nabla f = \mathbf{0} \implies t = 0, \ s = -2/3.$$

$$f_{tt} = 28, \quad f_{ts} = -12, \quad f_{ss} = 6, \quad D = 6(28) - (-12)^2 = -24 < 0.$$

By the second-partials test, the distance is a minimum when $t = 0$, $s = -2/3$; the nature of the problem tells us the minimum is absolute. The distance is $\sqrt{f(0, -2/3)} = \tfrac{2}{3}\sqrt{6}$.

33. (a) Let x and y be the cross-sectional measurements of the box, and let l be its length. Then

$$V = xyl, \quad \text{where} \quad 2x + 2y + l \leq 108, \quad x, y > 0$$

To maximize V we will obviously take $2x + 2y + l = 108$. Therefore, $V(x, y) = xy(108 - 2x - 2y)$ and

$$\nabla V = [y(108 - 2x - 2y) - 2xy]\mathbf{i} + [x(108 - 2x - 2y) - 2xy]\mathbf{j}$$

Setting $\dfrac{\partial V}{\partial x} = \dfrac{\partial V}{\partial y} = 0$, we get the pair of equations

$$\frac{\partial V}{\partial x} = 108y - 4xy - 2y^2 = 0$$

$$\frac{\partial V}{\partial y} = 108x - 4xy - 2x^2 = 0$$

from which it follows that $x = y = 18 \implies l = 36$.

Now, at $(18, 18)$, we have

$$A = V_{xx} = -4y = -72 < 0, \quad B = V_{xy} = 108 - 4x - 4y = -36,$$

$$C = V_{yy} = -4x = -72, \quad \text{and} \quad D = (36)^2 - (72)^2 < 0.$$

Thus, V is a maximum when $x = y = 18$ inches and $l = 36$ inches.

(b) Let r be the radius of the tube and let l be its length. Then

$$V = \pi r^2 l, \quad \text{where} \quad 2\pi r + l \leq 108, \quad r > 0$$

To maximize V we take $2\pi r + l = 108$. Then $V(r) = \pi r^2(108 - 2\pi r) = 108\pi r^2 - 2\pi^2 r^3$. Now

$$\frac{dV}{dr} = 216\pi r - 6\pi^2 r^2$$

Setting $\dfrac{dV}{dr} = 0$, we get

$$216\pi\, r - 6\pi^2 r^2 = 0 \quad\Longrightarrow\quad r = \frac{36}{\pi} \quad\Longrightarrow\quad l = 36$$

Now, at $r = 36/\pi$, we have

$$\frac{d^2V}{dr^2} = 216\pi - 12\pi^2 \frac{36}{\pi} = -216\pi < 0$$

Thus, V is a maximum when $r = 36/\pi$ inches and $l = 36$ inches.

35. Let S denote the cross-sectional area. Then

$$S = \frac{1}{2}\,(12 - 2x + 12 - 2x + 2x\cos\theta)\,x\,\sin\theta = 12x\sin\theta - 2x^2\sin\theta + \frac{1}{2}x^2\sin 2\theta,$$

where $0 < x < 6, \;\; 0 < \theta < \pi/2$

Now, with $\mathbf{j}$ in the direction of increasing θ,

$$\nabla S = (12\sin\theta - 4x\sin\theta + x\sin 2\theta)\,\mathbf{i} + (12x\cos\theta - 2x^2\cos\theta + x^2\cos 2\theta)\,\mathbf{j}$$

Setting $\dfrac{\partial S}{\partial x} = \dfrac{\partial S}{\partial \theta} = 0$, we get the pair of equations

$$12\sin\theta - 4x\sin\theta + x\sin 2\theta = 0$$

$$12x\cos\theta - 2x^2\cos\theta + x^2\cos 2\theta = 0$$

from which it follows that $x = 4, \theta = \pi/3$.

Now, at $(4, \pi/3)$, we have

$$A = S_{xx} = -4\sin\theta + \sin 2\theta = -\frac{3}{2}\sqrt{3}, \quad B = S_{x\theta} = 12\cos\theta - 4x\cos\theta + 2x\cos 2\theta = -6,$$

$$C = S_{\theta\theta} = -12x\sin\theta + 2x^2\sin\theta - 2x^2\sin 2\theta = -24\sqrt{3} \quad \text{and} \quad D = 108 - 36 > 0.$$

Thus, S is a maximum when $x = 4$ inches and $\theta = \pi/3$.

37. (a) $f(m, b) = [2 - b]^2 + [-5 - (m + b)]^2 + [4 - (2m + b)]^2.$

$f_m = 10m + 6b - 6, \quad f_b = 6m + 6b - 2; \qquad f_m = f_b = 0 \qquad\Longrightarrow\qquad m = 1, \;\; b = -\frac{2}{3}.$

$f_{mm} = 10, \quad f_{mb} = 6, \quad f_{bb} = 6, \quad D = 24 > 0 \implies$ a minimum.

Answer: the line $y = x - \frac{2}{3}$.

(b) $f(\alpha, \beta) = [2 - \beta]^2 + [-5 - (\alpha + \beta)]^2 + [4 - (4\alpha + \beta)]^2.$

$f_\alpha = 34\alpha + 10\beta - 22, \quad f_\beta = 10\alpha + 6\beta - 2; \qquad f_\alpha = f_\beta = 0 \qquad\Longrightarrow\qquad \left[\begin{array}{l} \alpha = \dfrac{14}{13} \\[2mm] \beta = -\dfrac{19}{13} \end{array}\right].$

$f_{\alpha\alpha} = 34, \quad f_{\alpha\beta} = 10, \quad f_{\beta\beta} = 6, \quad D = 104 > 0 \implies$ a minimun.

Answer: the parabola $y = \frac{1}{13}\left(14x^2 - 19\right)$.

1.
$$f(x,y) = x^2 + y^2, \qquad g(x,y) = xy - 1$$
$$\nabla f = 2x\mathbf{i} + 2y\mathbf{j}, \qquad \nabla g = y\mathbf{i} + x\mathbf{j}.$$
$$\nabla f = \lambda \nabla g \implies 2x = \lambda y \text{ and } 2y = \lambda x.$$

Multiplying the first equation by x and the second equation by y, we get
$$2x^2 = \lambda xy = 2y^2.$$

Thus, $x = \pm y$. From $g(x,y) = 0$ we conclude that $x = y = \pm 1$. The points $(1, 1)$ and $(-1, -1)$ clearly give a minimum, since f represents the square of the distance of a point on the hyperbola from the origin. The minimum is 2.

3.
$$f(x,y) = xy, \qquad g(x,y) = b^2 x^2 + a^2 y^2 - a^2 b^2$$
$$\nabla f = y\mathbf{i} + x\mathbf{j}, \qquad \nabla g = 2b^2 x\mathbf{i} + 2a^2 y\mathbf{j}.$$
$$\nabla f = \lambda \nabla g \implies y = 2\lambda b^2 x \text{ and } x = 2\lambda a^2 y.$$

Multiplying the first equation by $a^2 y$ and the second equation by $b^2 x$, we get
$$a^2 y^2 = 2\lambda a^2 b^2 xy = b^2 x^2.$$

Thus, $ay = \pm bx$. From $g(x,y) = 0$ we conclude that $x = \pm \frac{1}{2} a\sqrt{2}$ and $y = \pm \frac{1}{2} b\sqrt{2}$.

Since f is continuous and the ellipse is closed and bounded, the minimum exists. It occurs at $\left(\frac{1}{2} a\sqrt{2}, -\frac{1}{2} b\sqrt{2} \right)$ and $\left(-\frac{1}{2} a\sqrt{2}, \frac{1}{2} b\sqrt{2} \right)$; the minimum is $-\frac{1}{2} ab$.

5. Since f is continuous and the ellipse is closed and bounded, the maximum exists.
$$f(x,y) = xy^2, \qquad g(x,y) = b^2 x^2 + a^2 y^2 - a^2 b^2$$
$$\nabla f = y^2 \mathbf{i} + 2xy\mathbf{j}, \qquad \nabla g = 2b^2 x\mathbf{i} + 2a^2 y\mathbf{j}.$$
$$\nabla f = \lambda \nabla g \implies y^2 = 2\lambda b^2 x \text{ and } 2xy = 2\lambda a^2 y.$$

Multiplying the first equation by $a^2 y$ and the second equation by $b^2 x$, we get
$$a^2 y^3 = 2\lambda a^2 b^2 xy = 2b^2 x^2 y.$$

We can exclude $y = 0$; it clearly cannot produce the maximum. Thus,
$$a^2 y^2 = 2b^2 x^2 \text{ and, from } g(x,y) = 0, \ 3b^2 x^2 = a^2 b^2.$$

This gives us $x = \pm \frac{1}{3}\sqrt{3}\, a$ and $y = \pm \frac{1}{3}\sqrt{6}\, b$. The maximum occurs at $x = \frac{1}{3}\sqrt{3}\, a$, $y = \pm \frac{1}{3}\sqrt{6}\, b$; the value there is $\frac{2}{9}\sqrt{3}\, ab^2$.

7. The given curve is closed and bounded. Since $x^2 + y^2$ represents the square of the distance from points on this curve to the origin, the maximum exists.

$$f(x, y) = x^2 + y^2, \qquad g(x, y) = x^4 + 7x^2y^2 + y^4 - 1$$

$$\nabla f = 2x\mathbf{i} + 2y\mathbf{j}, \qquad \nabla g = \left(4x^3 + 14xy^2\right)\mathbf{i} + \left(4y^3 + 14x^2y\right)\mathbf{j}.$$

We use the cross-product equation (16.7.4):

$$2x(4y^3 + 14x^2y) - 2y(4x^3 + 14xy^2) = 0,$$

$$20x^3y - 20xy^3 = 0,$$

$$xy(x^2 - y^2) = 0.$$

Thus, $x = 0$, $y = 0$, or $x = \pm y$. From $g(x, y) = 0$ we conclude that the points to examine are

$$(0, \pm 1), \quad (\pm 1, 0), \quad \left(\pm\tfrac{1}{3}\sqrt{3}, \pm\tfrac{1}{3}\sqrt{3}\right).$$

The value of f at each of the first four points is 1; the value at the last four points is $2/3$. The maximum is 1.

9. The maximum exists since xyz is continuous and the ellipsoid is closed and bounded.

$$f(x, y, z) = xyz, \qquad g(x, y, z) = \frac{x^2}{a^2} + \frac{y^2}{b^2} + \frac{z^2}{c^2} - 1$$

$$\nabla f = yz\mathbf{i} + xz\mathbf{j} + xy\mathbf{k}, \qquad \nabla g = \frac{2x}{a^2}\mathbf{i} + \frac{2y}{b^2}\mathbf{j} + \frac{2z}{c^2}\mathbf{k}.$$

$$\nabla f = \lambda \nabla g \quad \Longrightarrow \quad yz = \frac{2x}{a^2}\lambda, \quad xz = \frac{2y}{b^2}\lambda, \quad xy = \frac{2z}{c^2}\lambda.$$

We can assume x, y, z are non-zero, for otherwise $f(x, y, z) = 0$, which is clearly not a maximum. Then from the first two equations

$$\frac{yza^2}{x} = 2\lambda = \frac{xzb^2}{y} \quad \text{so that} \quad a^2y^2 = b^2x^2 \quad \text{or} \quad \frac{x^2}{a^2} = \frac{y^2}{b^2}.$$

Similarly from the second and third equations we get

$$b^2z^2 = c^2y^2 \quad \text{or} \quad \frac{y^2}{b^2} = \frac{z^2}{c^2}.$$

From $g(x, y, z) = 0$, we get $\dfrac{3x^2}{a^2} = 1 \quad \Longrightarrow \quad x \pm \dfrac{a}{\sqrt{3}}$, from which it follows that $y = \pm\dfrac{b}{\sqrt{3}}$, $z = \pm\dfrac{c}{\sqrt{3}}$. The maximum value is $\frac{1}{9}\sqrt{3}\,abc$.

11. Since the sphere is closed and bounded and $2x + 3y + 5z$ is continuous, the maximum exists.

$$f(x, y, z) = 2x + 3y + 5z, \qquad g(x, y, z) = x^2 + y^2 + z^2 - 19$$

$$\nabla f = 2\mathbf{i} + 3\mathbf{j} + 5\mathbf{k}, \qquad \nabla g = 2x\mathbf{i} + 2y\mathbf{j} + 2z\mathbf{k}.$$

$$\nabla f = \lambda \nabla g \quad \Longrightarrow \quad 2 = 2\lambda x, \quad 3 = 2\lambda y, \quad 5 = 2\lambda z.$$

Since $\lambda \neq 0$ here, we solve the equations for x, y and z:

$$x = \frac{1}{\lambda}, \quad y = \frac{3}{2\lambda}, \quad z = \frac{5}{2\lambda},$$

and substitute these results in $g(x, y, z) = 0$ to obtain

$$\frac{1}{\lambda^2} + \frac{9}{4\lambda^2} + \frac{25}{4\lambda^2} - 19 = 0, \quad \frac{38}{4\lambda^2} - 19 = 0, \quad \lambda = \pm\frac{1}{2}\sqrt{2}.$$

The positive value of λ will produce positive values for x, y, z and thus the maximum for f. We get $x = \sqrt{2}$, $y = \frac{3}{2}\sqrt{2}$, $z = \frac{5}{2}\sqrt{2}$, and $2x + 3y + 5z = 19\sqrt{2}$.

13.
$$f(x, y, z) = xyz, \qquad g(x, y, z) = \frac{x}{a} + \frac{y}{b} + \frac{z}{c} - 1$$

$$\nabla f = yz\mathbf{i} + xz\mathbf{j} + xy\mathbf{k}, \qquad \nabla g = \frac{1}{a}\mathbf{i} + \frac{1}{b}\mathbf{j} + \frac{1}{c}\mathbf{k}.$$

$$\nabla f = \lambda\nabla g \implies yz = \frac{\lambda}{a}, \quad xz = \frac{\lambda}{b}, \quad xy = \frac{\lambda}{c}.$$

Multiplying these equations by x, y, z respectively, we obtain

$$xyz = \frac{\lambda x}{a}, \quad xyz = \frac{\lambda y}{b}, \quad xyz = \frac{\lambda z}{c}.$$

Adding these equations and using the fact that $g(x, y, z) = 0$, we have

$$3xyz = \lambda\left(\frac{x}{a} + \frac{y}{b} + \frac{z}{c}\right) = \lambda.$$

Since x, y, z are non-zero,

$$yz = \frac{\lambda}{a} = \frac{3xyz}{a}, \quad 1 = \frac{3x}{a}, \quad x = \frac{a}{3}.$$

Similarly, $y = \frac{b}{3}$ and $z = \frac{c}{3}$. The maximum is $\frac{1}{27}\,abc$.

15. It suffices to minimize the square of the distance from $(0, 1)$ to a point on the parabola. Clearly, the minimum exists.

$$f(x, y) = x^2 + (y - 1)^2, \qquad g(x, y) = x^2 - 4y$$

$$\nabla f = 2x\mathbf{i} + 2(y - 1)\mathbf{j}, \qquad \nabla g = 2x\mathbf{i} - 4\mathbf{j}.$$

We use the cross-product equation (16.7.4):

$$2x(-4) - 2x(2y - 2) = 0, \quad 4x + 4xy = 0, \quad x(y + 1) = 0.$$

Since $y \geq 0$, we have $x = 0$ and thus $y = 0$. The minimum is 1.

17. It suffices to maximize and minimize the square of the distance from $(2, 1, 2)$ to a point on the sphere. Clearly, these extreme values exist.

$$f(x, y, z) = (x - 2)^2 + (y - 1)^2 + (z - 2)^2, \qquad g(x, y, z) = x^2 + y^2 + z^2 - 1$$

$$\nabla f = 2(x - 2)\,\mathbf{i} + 2(y - 1)\,\mathbf{j} + 2(z - 2)\,\mathbf{k}, \qquad \nabla g = 2x\,\mathbf{i} + 2y\,\mathbf{j} + 2z\,\mathbf{k}.$$

$$\nabla f = \lambda\nabla g \implies 2(x - 2) = 2x\lambda, \quad 2(y - 1) = 2y\lambda, \quad 2(z - 2) = 2z\lambda$$

Thus,

$$x = \frac{2}{1-\lambda}, \quad y = \frac{1}{1-\lambda}, \quad z = \frac{2}{1-\lambda}.$$

Using the fact that $x^2 + y^2 + z^2 = 1$, we have

$$\left(\frac{2}{1-\lambda}\right)^2 + \left(\frac{1}{1-\lambda}\right)^2 + \left(\frac{2}{1-\lambda}\right)^2 = 1 \implies \lambda = -2, 4$$

At $\lambda = -2$, $(x, y, z) = (2/3, 1/3, 2/3)$ and $f(2/3, 1/3, 2/3) = 4$

At $\lambda = 4$, $(x, y, z) = (-2/3, -1/3, -2/3)$ and $f(-2/3, -1/3, -2/3) = 16$

Thus, $(2/3, 1/3, 2/3)$ is the closest point and $(-2/3, -1/3, -2/3)$ is the furthest point.

19.
$$f(x, y, z) = 3x - 2y + z, \qquad g(x, y, z) = x^2 + y^2 + z^2 - 14$$

$$\nabla f = 3\mathbf{i} - 2\mathbf{j} + \mathbf{k}, \qquad \nabla g = 2x\,\mathbf{i} + 2y\,\mathbf{j} + 2z\,\mathbf{k}.$$

$$\nabla f = \lambda \nabla g \implies 3 = 2x\lambda, \quad -2 = 2y\lambda, \quad 1 = 2z\lambda.$$

Thus,

$$x = \frac{3}{2\lambda}, \quad y = -\frac{1}{\lambda}, \quad z = \frac{1}{2\lambda}.$$

Using the fact that $x^2 + y^2 + z^2 = 14$, we have

$$\left(\frac{3}{2\lambda}\right)^2 + \left(-\frac{1}{\lambda}\right)^2 + \left(\frac{1}{2\lambda}\right)^2 = 14 \implies \lambda = \pm\frac{1}{2}.$$

At $\lambda = \frac{1}{2}$, $(x, y, z) = (3, -2, 1)$ and $f(3, -2, 1) = 14$

At $\lambda = -\frac{1}{2}$, $(x, y, z) = (-3, 2, -1)$ and $f(-3, 2, -1) = -14$

Thus, the maximum value of f on the sphere is 14.

21. It's easier to work with the square of the distance; the minimum certainly exists.

$$f(x, y, z) = x^2 + y^2 + z^2, \qquad g(x, y, z) = Ax + By + Cz + D$$

$$\nabla f = 2x\mathbf{i} + 2y\mathbf{j} + 2z\mathbf{k}, \qquad \nabla g = A\mathbf{i} + B\mathbf{j} + C\mathbf{k}.$$

$$\nabla f = \lambda \nabla g \implies 2x = A\lambda, \quad 2y = B\lambda, \quad 2z = C\lambda.$$

Substituting these equations in $g(x, y, z) = 0$, we have

$$\frac{1}{2}\lambda\left(A^2 + B^2 + C^2\right) + D = 0, \quad \lambda = \frac{-2D}{A^2 + B^2 + C^2}.$$

Thus, in turn,

$$x = \frac{-DA}{A^2 + B^2 + C^2}, \quad y = \frac{-DB}{A^2 + B^2 + C^2}, \quad z = \frac{-DC}{A^2 + B^2 + C^2}$$

so the minimum value of $\sqrt{x^2 + y^2 + z^2}$ is $|D|\left(A^2 + B^2 + C^2\right)^{-1/2}$.

23.

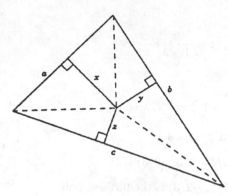

area $A = \frac{1}{2}ax + \frac{1}{2}by + \frac{1}{2}cz$.

The geometry suggests that

$$x^2 + y^2 + z^2$$

has a minimum.

$$f(x,y,z) = x^2 + y^2 + z^2, \qquad g(x,y,z) = ax + by + cz - 2A$$

$$\nabla f = 2x\mathbf{i} + 2y\mathbf{j} + 2z\mathbf{k}, \qquad \nabla g = a\mathbf{i} + b\mathbf{j} + c\mathbf{k}.$$

$$\nabla f = \lambda \nabla g \implies 2x = a\lambda, \ 2y = b\lambda, \ 2z = c\lambda.$$

Solving these equations for x, y, z and substituting the results in $g(x,y,z) = 0$, we have

$$\frac{a^2\lambda}{2} + \frac{b^2\lambda}{2} + \frac{c^2\lambda}{2} - 2A = 0, \quad \lambda = \frac{4A}{a^2 + b^2 + c^2}$$

and thus

$$x = \frac{2aA}{a^2 + b^2 + c^2}, \quad y = \frac{2bA}{a^2 + b^2 + c^2}, \quad z = \frac{2cA}{a^2 + b^2 + c^2}.$$

The minimum is $4A^2(a^2 + b^2 + c^2)^{-1}$.

25. Since the curve is asymptotic to the line $y = x$ as $x \to -\infty$ and as $x \to \infty$, the maximum exists. The distance between the point (x,y) and the line $y - x = 0$ is given by

$$\frac{|y - x|}{\sqrt{1+1}} = \frac{1}{2}\sqrt{2}\,|y - x|. \qquad \text{(see Section 1.4)}$$

Since the points on the curve are below the line $y = x$, we can replace $|y - x|$ by $x - y$. To simplify the work we drop the constant factor $\frac{1}{2}\sqrt{2}$.

$$f(x,y) = x - y, \qquad g(x,y) = x^3 - y^3 - 1$$

$$\nabla f = \mathbf{i} - \mathbf{j}, \qquad \nabla g = 3x^2\mathbf{i} - 3y^2\mathbf{j}.$$

We use the cross-product equation (16.7.4):

$$1\left(-3y^2\right) - \left(3x^2\right)(-1) = 0, \quad 3x^2 - 3y^2 = 0, \quad x = -y \ \ (x \neq y).$$

Now $g(x,y) = 0$ gives us

$$x^3 - (-x)^3 - 1 = 0, \quad 2x^3 = 1, \quad x = 2^{-1/3}.$$

The point is $\left(2^{-1/3}, -2^{-1/3}\right)$.

27. It suffices to show that the square of the area is a maximum when $a = b = c$.

$$f(a,b,c) = s(s-a)(s-b)(s-c), \quad g(a,b,c) = a+b+c-2s$$

$$\nabla f = -s(s-b)(s-c)\mathbf{i} - s(s-a)(s-c)\mathbf{j} - s(s-a)(s-b)\mathbf{k}, \quad \nabla g = \mathbf{i}+\mathbf{j}+\mathbf{k}.$$

(Here $\mathbf{i}$, $\mathbf{j}$, $\mathbf{k}$ are the unit vectors in the directions of increasing a, b, c.)

$$\nabla f = \lambda \nabla g \implies -s(s-b)(s-c) = -s(s-a)(s-c) = -s(s-a)(s-b) = \lambda.$$

Thus, $s-b = s-a = s-c$ so that $a = b = c$. This gives us the maximum, as no minimum exists. [The area can be made arbitrarily small by taking a close to s.]

29. (a) $f(x,y) = (xy)^{1/2}, \qquad g(x,y) = x+y-k, (x,y \geq 0,$ k a nonnegative constant)

$$\nabla f = \frac{y^{1/2}}{2x^{1/2}}\mathbf{i} + \frac{x^{1/2}}{2y^{1/2}}\mathbf{j}, \qquad \nabla g = \mathbf{i}+\mathbf{j}.$$

$$\nabla f = \lambda \nabla g \implies \frac{y^{1/2}}{2x^{1/2}} = \lambda = \frac{x^{1/2}}{2y^{1/2}} \implies x = y = \frac{k}{2}.$$

Thus, the maximum value of f is: $f(k/2, k/2) = \dfrac{k}{2}$.

(b) For all x, y $(x, y \geq 0)$ we have

$$(xy)^{1/2} = f(x,y) \leq f(k/2, k/2) = \frac{k}{2} = \frac{x+y}{2}.$$

31. Simply extend the arguments used in Exercises 29 and 30.

33. $$S(r,h) = 2\pi r^2 + 2\pi rh, \qquad\qquad g(r,h) = \pi r^2 h - V, \quad (V \text{ constant})$$

$$\nabla S = (4\pi r + 2\pi h)\mathbf{i} + 2\pi r\mathbf{j}, \qquad\qquad \nabla g = 2\pi rh\mathbf{i} + \pi r^2 \mathbf{j}.$$

$$\nabla S = \lambda \nabla g \implies 4\pi r + 2\pi h = 2\pi rh\lambda, \quad 2\pi r = \pi r^2 \lambda \implies r = \frac{2}{\lambda}, \quad h = \frac{4}{\lambda}.$$

Now $\pi r^2 h = V, \implies \lambda = \sqrt[3]{\dfrac{16\pi}{V}} \implies r = \sqrt[3]{\dfrac{V}{2\pi}}, \quad h = \sqrt[3]{\dfrac{4V}{\pi}}.$

To minimize the surface area, take $r = \sqrt[3]{\dfrac{V}{2\pi}}, \quad$ and $\quad h = \sqrt[3]{\dfrac{4V}{\pi}}.$

35. Same as Exercise 13.

37. Let x, y, z denote the length, width and height of the box. We want to maximize the volume V of the box given that the surface area S is constant. That is:

maximize $V(x,y,z) = xyz \qquad$ subject to $\qquad S(x,y,z) = 2xy + 2xz + 2yz = S$ constant

Let $g(x,y,z) = 2xy + 2xz + 2yz - S$. Then

$$\nabla V = yz\,\mathbf{i} + xz\,\mathbf{j} + xy\,\mathbf{k}, \qquad \nabla g = (2y+2z)\,\mathbf{i} + (2x+2z)\,\mathbf{j} + (2x+2y)\,\mathbf{k}$$

$\nabla V = \lambda \nabla g$ and the side condition yield the system of equations:

$$yz = \lambda\,(2y + 2z)$$

$$xz = \lambda\,(2x + 2z)$$

$$xy = \lambda\,(2x + 2y)$$

$$xy + 2xz + 2yz = S.$$

Multiply the first equation by x, the second by y and subtract. This gives

$$0 = 2\lambda\,z(x - y) \implies x = y \quad \text{since } z = 0 \implies V = 0.$$

Multiply the second equation by y, the third by z and subtract. This gives

$$0 = 2\lambda\,x(y - z) \implies y = z \quad \text{since } x = 0 \implies V = 0.$$

Thus the closed rectangular box of maximum volume is a cube. The cube has side length $x = \sqrt{S/6}$.

---

39. $S(r, h) = 4\pi r^2 + 2\pi rh, \quad g(r, h) = \dfrac{4}{3}\pi r^3 + \pi r^2 h - 10{,}000$

$\nabla S = (8\pi r + 2\pi h)\mathbf{i} + 2\pi r\mathbf{j}, \quad \nabla g = (4\pi r^2 + 2\pi rh)\mathbf{i} + \pi r^2\mathbf{j}$

(Here $\mathbf{i}$, $\mathbf{j}$ are the unit vectors in the directions of increasing r and h.)

$\nabla S = \lambda\nabla g \implies 2\pi(4r + h) = 2\pi r\lambda(2r + h), \quad 2\pi r = \lambda\pi r^2 \implies h = 0$

Maximum volume for sphere of radius $r = \sqrt[3]{7500/\pi}$ meters.

---

41. $f(x, y, z) = 8xyz, \quad g(x, y, z) = 4x^2 + 9y^2 + 36z^2 - 36.$

$\nabla f(x, y, z) = 8yz\mathbf{i} + 8xz\mathbf{j} + 8xy\mathbf{k}, \quad \nabla g(x, y, z) = 8x\mathbf{i} + 18y\mathbf{j} + 72z\mathbf{k}.$

$\nabla f = \lambda\nabla g \quad$ gives

$$yz = \lambda x, \quad 4xz = 9\lambda y, \quad xy = 9\lambda z.$$

$$4\frac{xyz}{\lambda} = 4x^2, \quad 4\frac{xyz}{\lambda} = 9y^2, \quad 4\frac{xyz}{\lambda} = 36z^2.$$

Also notice

$$4x^2 + 9y^2 + 36z^2 - 36 = 0$$

We have

$$12\frac{xyz}{\lambda} = 36 \implies x = \sqrt{3}, \quad y = \frac{2}{\sqrt{3}}, \quad z = \frac{1}{\sqrt{3}}.$$

Thus,

$$V = 8xyz = 8 \cdot \sqrt{3} \cdot \frac{2}{\sqrt{3}} \cdot \frac{1}{\sqrt{3}} = \frac{16}{\sqrt{3}}.$$

43. To simplify notation we set $x = Q_1, \quad y = Q_2, \quad z = Q_3.$

$$f(x,y,z) = 2x + 8y + 24z, \qquad g(x,y,z) = x^2 + 2y^2 + 4z^2 - 4,500,000,000$$

$$\nabla f = 2\mathbf{i} + 8\mathbf{j} + 24\mathbf{k}, \qquad \nabla g = 2x\mathbf{i} + 4y\mathbf{j} + 8z\mathbf{k}.$$

$$\nabla f = \lambda \nabla g \implies 2 = 2\lambda x, \quad 8 = 4\lambda y, \quad 24 = 8\lambda z.$$

Since $\lambda \neq 0$ here, we solve the equations for x, y, z:

$$x = \frac{1}{\lambda}, \quad y = \frac{2}{\lambda}, \quad z = \frac{3}{\lambda},$$

and substitute these results in $g(x,y,z) = 0$ to obtain

$$\frac{1}{\lambda^2} + 2\left(\frac{4}{\lambda^2}\right) + 4\left(\frac{9}{\lambda^2}\right) - 45 \times 10^8 = 0, \quad \frac{45}{\lambda^2} = 45 \times 10^8, \quad \lambda = \pm 10^{-4}.$$

Since x, y, z are non-negative, $\lambda = 10^{-4}$ and

$$x = 10^4 = Q_1, \quad y = 2 \times 10^4 = Q_2, \quad z = 3 \times 10^4 = Q_3.$$

PROJECT 16.7

1. $f(x,y,z) = xy + z^2, \quad g(x,y,z) = x^2 + y^2 + z^2 - 4, \quad h(x,y,z) = y - x$

$\nabla f = y\mathbf{i} + x\mathbf{j} + 2z\mathbf{k}, \quad \nabla g = 2x\mathbf{i} + 2y\mathbf{j} + 2z\mathbf{k}, \quad \nabla h = -\mathbf{i} + \mathbf{j}.$

$\nabla f = \lambda \nabla g + \mu \nabla h \implies y = 2\lambda x - \mu, \quad x = 2\lambda y - \mu, \quad 2z = 2\lambda z$

$2z = 2\lambda z \implies \lambda = 0 \quad \text{or} \quad z = 1.$

$\lambda = 0 \implies y = -x$ which contradicts $y = x.$

$z = 1 \implies x^2 + y^2 = 3,$ which, with $y = x$ implies $x = \pm\sqrt{3/2}; \quad \left(\pm\sqrt{3/2}, \pm\sqrt{3/2}\right)$

Adding the first two equations gives

$$x + y = 2\lambda(x + y) \implies (x+y)[2\lambda - 1] = 0 \implies \lambda = \frac{1}{c} \quad \text{or} \quad x = y = 0.$$

$x = y = 0 \implies z = \pm 2; \quad (0, 0, \pm 2).$

$\lambda = \dfrac{1}{2} \implies z = 0$ and $y = x \implies 2x^4 = 4; \quad x = \pm\sqrt{2}; \quad (\pm\sqrt{2}, \pm\sqrt{2}, 0).$

$f\left(\pm\sqrt{3/2}, \pm\sqrt{3/2}, 1\right) = \dfrac{5}{2}; \quad f(0, 0, \pm 2) = 4; \quad f(\pm\sqrt{2}, \pm\sqrt{2}, 0) = 2.$

The maximum value of f is 4; the minimum value is 2.

3. $f(x,y,z) = x^2 + y^2 + z^2, \qquad g(x,y,z) = x + y - z + 1, \qquad h(x,y,z) = x^2 + y^2 - z^2$

$\nabla f = 2x\mathbf{i} + 2y\mathbf{j} + 2z\mathbf{k}, \qquad \nabla g = \mathbf{i} + \mathbf{j} - \mathbf{k}, \qquad \nabla h = 2x\mathbf{i} + 2y\mathbf{j} - 2z\mathbf{k}.$

$\nabla f = \lambda \nabla g + \mu \nabla h \implies 2x = \lambda + 2x\mu, \quad 2y = \lambda + 2y\mu, \quad 2z = -\lambda - 2z\mu$

Multiplying the first equation by y, the second equation by x and subtracting, yields

$$\lambda(y - x) = 0.$$

Now $\lambda = 0 \implies \mu = 1 \implies x = y = z = 0$. This is impossible since $x + y - z = -1$.

Therefore, we must have $y = x \implies z = \pm\sqrt{2}\,x$.

Substituting $y = x$, $z = \sqrt{2}\,x$ into the equation $x + y - z + 1 = 0$, we get

$$x = -1 - \frac{\sqrt{2}}{2} \implies y = -1 - \frac{\sqrt{2}}{2},\ z = -1 - \sqrt{2}$$

Substituting $y = x$, $z = -\sqrt{2}\,x$ into the equation $x + y - z + 1 = 0$, we get

$$x = -1 + \frac{\sqrt{2}}{2} \implies y = -1 + \frac{\sqrt{2}}{2},\ z = -1 + \sqrt{2}$$

Since

$$f\left(-1 - \frac{\sqrt{2}}{2}, -1 - \frac{\sqrt{2}}{2}, -1 - \sqrt{2}\right) = 6 + 4\sqrt{2} \text{ and}$$

$$f\left(-1 + \frac{\sqrt{2}}{2}, -1 + \frac{\sqrt{2}}{2}, -1 + \sqrt{2}\right) = 6 - 4\sqrt{2},$$

it follows that $\left(-1 + \frac{\sqrt{2}}{2}, -1 + \frac{\sqrt{2}}{2}, -1 + \sqrt{2}\right)$ is closest to the origin and

$\left(-1 - \frac{\sqrt{2}}{2}, -1 - \frac{\sqrt{2}}{2}, -1 - \sqrt{2}\right)$ is furthest from the origin.

SECTION 16.8

1. $df = \left(3x^2y - 2xy^2\right)\Delta x + \left(x^3 - 2x^2y\right)\Delta y$

3. $df = (\cos y + y\sin x)\,\Delta x - (x\sin y + \cos x)\,\Delta y$

5. $df = \Delta x - (\tan z)\,\Delta y - \left(y\sec^2 z\right)\Delta z$

7. $df = \dfrac{y(y^2 + z^2 - x^2)}{(x^2 + y^2 + z^2)^2}\,\Delta x + \dfrac{x(x^2 + z^2 - y^2)}{(x^2 + y^2 + z^2)^2}\,\Delta y - \dfrac{2xyz}{(x^2 + y^2 + z^2)^2}\,\Delta z$

9. $df = [\cos(x + y) + \cos(x - y)]\,\Delta x + [\cos(x + y) - \cos(x - y)]\,\Delta y$

11. $df = \left(y^2 z e^{xz} + \ln z\right)\Delta x + 2y e^{xz}\,\Delta y + \left(xy^2 e^{xz} + \dfrac{x}{z}\right)\Delta z$

13.
$$\Delta u = [(x + \Delta x)^2 - 3(x + \Delta x)(y + \Delta y) + 2(y + \Delta y)^2] - (x^2 - 3xy + 2y^2)$$
$$= [(1.7)^2 - 3(1.7)(-2.8) + 2(-2.8)^2] - (2^2 - 3(2)(-3) + 2(-3)^2)$$
$$= (2.89 + 14.28 + 15.68) - 40 = -7.15$$
$$du = (2x - 3y)\,\Delta x + (-3x + 4y)\,\Delta y$$
$$= (4 + 9)(-0.3) + (-6 - 12)(0.2) = -7.50$$

15. $\Delta u = [(x + \Delta x)^2(z + \Delta z) - 2(y + \Delta y)(z + \Delta z)^2 + 3(x + \Delta x)(y + \Delta y)(z + \Delta z)]$
$$- (x^2 z - 2yz^2 + 3xyz)$$
$$= [(2.1)^2(2.8) - 2(1.3)(2.8)^2 + 3(2.1)(1.3)(2.8)] - [(2)^2 3 - 2(1)(3)^2 + 3(2)(1)(3)] = 2.896$$
$$du = (2xz + 3yz)\,\Delta x + (-2z^2 + 3xz)\,\Delta y + (x^2 - 4yz + 3xy)\,\Delta z$$
$$= [2(2)(3) + 3(1)(3)](0.1) + [-2(3)^2 + 3(2)(3)](0.3) + [2^2 - 4(1)(3) + 3(2)(1)](-0.2) = 2.5$$

17. $f(x, y) = x^{1/2} y^{1/4}; \quad x = 121, \quad y = 16, \quad \Delta x = 4, \quad \Delta y = 1$
$$f(x + \Delta x, y + \Delta y) \cong f(x, y) + df$$
$$= x^{1/2} y^{1/4} + \tfrac{1}{2} x^{-1/2} y^{1/4}\,\Delta x + \tfrac{1}{4} x^{1/2} y^{-3/4}\,\Delta y$$
$$\sqrt{125}\ \sqrt[4]{17} \cong \sqrt{121}\ \sqrt[4]{16} + \tfrac{1}{2}(121)^{-1/2}(16)^{1/4}(4) + \tfrac{1}{4}(121)^{1/2}(16)^{-3/4}(1)$$
$$= 11(2) + \tfrac{1}{2}\left(\tfrac{1}{11}\right)(2)(4) + \tfrac{1}{4}(11)\left(\tfrac{1}{8}\right)$$
$$= 22 + \tfrac{4}{11} + \tfrac{11}{32} = 22\tfrac{249}{352} \cong 22.71$$

19.
$$f(x, y) = \sin x \cos y; \quad x = \pi, \quad y = \frac{\pi}{4}, \quad \Delta x = -\frac{\pi}{7}, \quad \Delta y = -\frac{\pi}{20}$$
$$df = \cos x \cos y\,\Delta x - \sin x \sin y\,\Delta y$$
$$f(x + \Delta x, y + \Delta y) \cong f(x, y) + df$$
$$\sin \frac{6}{7}\pi \cos \frac{1}{5}\pi \cong \sin \pi \cos \frac{\pi}{4} + \left(\cos \pi \cos \frac{\pi}{4}\right)\left(-\frac{\pi}{7}\right) - \left(\sin \pi \sin \frac{\pi}{4}\right)\left(-\frac{\pi}{20}\right)$$
$$= 0 + \left(\frac{1}{2}\sqrt{2}\right)\left(\frac{\pi}{7}\right) + 0 = \frac{\pi\sqrt{2}}{14} \cong 0.32$$

21. $f(2.9, 0.01) \cong f(3, 0) + df$, where df is to be evaluated at $x = 3$, $y = 0$, $\Delta x = -0.1$, $\Delta y = 0.01$.
$$df = \left(2xe^{xy} + x^2 ye^{xy}\right)\Delta x + x^3 e^{xy}\,\Delta y = [2(3)e^0 + (3)^2(0)e^0](-0.1) + 3^3 e^0(0.01) = -0.33$$
Thus, $f(2.9, .01) \cong 3^2 e^0 - 0.33 = 8.67$.

23. $f(2.94, 1.1, 0.92) \cong f(3, 1, 1) + df$, where df is to be evaluated at $x = 3$, $y = 1$, $z = 1$,
$\Delta x = -0.06$, $\Delta y = 0.1$, $\Delta z = -0.08$
$$df = \tan^{-1} yz\,\Delta x + \frac{xz}{1 + y^2 z^2}\,\Delta y + \frac{xy}{1 + y^2 z^2}\,\Delta z = \frac{\pi}{4}(-0.06) + (1.5)(0.1) + (1.5)(-0.08) \cong -0.0171$$
Thus, $f(2.94, 1.1, 0.92) \cong \tfrac{3}{4}\pi - 0.0171 \cong 2.3391$

25. $df = \dfrac{\partial z}{\partial x}\Delta x + \dfrac{\partial z}{\partial y}\Delta y = \dfrac{2y}{(x+y)^2}\Delta x - \dfrac{2x}{(x+y)^2}\Delta y$

With $x = 4$, $y = 2$, $\Delta x = 0.1$, $\Delta y = 0.1$, we get

$$df = \tfrac{4}{36}(0.1) - \tfrac{8}{36}(0.1) = -\tfrac{1}{90}.$$

The exact change is $\dfrac{4.1 - 2.1}{4.1 + 2.1} - \dfrac{4 - 2}{4 + 2} = \dfrac{2}{6.2} - \dfrac{1}{3} = -\dfrac{1}{93}.$

27. $S = 2\pi r^2 + 2\pi rh;\quad r = 8,\ h = 12,\ \Delta r = -0.3,\ \Delta h = 0.2$

$$dS = \dfrac{\partial S}{\partial r}\Delta r + \dfrac{\partial S}{\partial h}\Delta h = (4\pi r + 2\pi h)\,\Delta r + (2\pi r)\,\Delta h$$

$$= 56\pi(-0.3) + 16\pi(0.2) = -13.6\pi.$$

The area decreases about 13.6π in.^{2}.

29. $S(9.98, 5.88, 4.08) \cong S(10, 6, 4) + dS = 248 + dS,\ $ where

$dS = (2w + 2h)\,\Delta l + (2l + 2h)\,\Delta w + (2l + 2w)\,\Delta h = 20(-0.02) + 28(-0.12) + 32(0.08) = -1.20$

Thus, $S(9.98, 5.88, 4.08) \cong 248 - 1.20 = 246.80.$

31. (a) $dV = yz\,\Delta x + xz\,\Delta y + xy\,\Delta z = (8)(6)(0.02) + (12)(6)(-0.05) + (12)(8)(0.03) = 0.24$

(b) $\Delta V = (12.02)(7.95)(6.03) - (12)(8)(6) = 0.22077$

33. $T(P) - T(Q) \cong dT = (-2x + 2yz)\,\Delta x + (-2y + 2xz)\,\Delta y + (-2z + 2xy)\,\Delta z$

Letting $x = 1$, $y = 3$, $z = 4$, $\Delta x = 0.15$, $\Delta y = -0.10$, $\Delta z = 0.10$, we have

$$dT = (22)(0.15) + (2)(-0.10) + (-2)(0.10) = 2.9$$

35. (a) $\pi r^2 h = \pi(r + \Delta r)^2(h + \Delta h) \implies \Delta h = \dfrac{r^2 h}{(r + \Delta r)^2} - h = -\dfrac{(2r + \Delta r)h}{(r + \Delta r)^2}\Delta r.$

$$df = (2\pi rh)\,\Delta r + \pi r^2\,\Delta h, \qquad df = 0 \implies \Delta h = \dfrac{-2h}{r}\Delta r.$$

(b) $2\pi r^2 + 2\pi rh = 2\pi(r + \Delta r)^2 + 2\pi(r + \Delta r)(h + \Delta h).$

Solving for Δh,

$$\Delta h = \dfrac{r^2 + rh - (r + \Delta r)^2}{r + \Delta r} - h = -\dfrac{2r + h + \Delta r}{r + \Delta r}\Delta r.$$

$$df = (4\pi r + 2\pi h)\,\Delta r + 2\pi r\,\Delta h, \qquad df = 0 \implies \Delta h = -\left(\dfrac{2r + h}{r}\right)\Delta r.$$

37. (a) $A = \dfrac{1}{2}x^2\sin\theta;\quad \Delta A \cong dA = x\sin\theta\,\Delta x + \dfrac{x^2}{2}\cos\theta\,\Delta\theta$

(b) The area is more sensitive to changes in θ if $x > 2\tan\theta$, otherwise it is more sensitive to changes in x.

39. $s = \dfrac{A}{A - W}$; $A = 9$, $W = 5$, $\Delta A = \pm 0.01$, $\Delta W = \pm 0.02$

$$ds = \frac{\partial s}{\partial A}\,\Delta A + \frac{\partial s}{\partial W}\,\Delta W = \frac{-W}{(A-W)^2}\,\Delta A + \frac{A}{(A-W)^2}\,\Delta W$$

$$= -\frac{5}{16}(\pm 0.01) + \frac{9}{16}(\pm 0.02) \cong \pm 0.014$$

The maximum possible error in the value of s is 0.014 lbs; $2.23 \le s + \Delta s \le 2.27$

SECTION 16.9

1. $\dfrac{\partial f}{\partial x} = xy^2$, $\quad f(x,y) = \frac{1}{2}x^2 y^2 + \phi(y)$, $\quad \dfrac{\partial f}{\partial y} = x^2 y + \phi'(y) = x^2 y$.

Thus, $\quad \phi'(y) = 0$, $\phi(y) = C$, $\quad$ and $\quad f(x,y) = \frac{1}{2}x^2 y^2 + C$.

3. $\dfrac{\partial f}{\partial x} = y$, $\quad f(x,y) = xy + \phi(y)$, $\quad \dfrac{\partial f}{\partial y} = x + \phi'(y) = x$.

Thus, $\quad \phi'(y) = 0$, $\phi(y) = C$, $\quad$ and $\quad f(x,y) = xy + C$.

5. No; $\dfrac{\partial}{\partial y}\left(y^3 + x\right) = 3y^2$ $\quad$ whereas $\quad \dfrac{\partial}{\partial x}\left(x^2 + y\right) = 2x$.

7. $\dfrac{\partial f}{\partial x} = \cos x - y \sin x$, $\quad f(x,y) = \sin x + y \cos x + \phi(y)$, $\quad \dfrac{\partial f}{\partial y} = \cos x + \phi'(y) = \cos x$.

Thus, $\quad \phi'(y) = 0$, $\phi(y) = C$, $\quad$ and $\quad f(x,y) = \sin x + y \cos x + C$.

9. $\dfrac{\partial f}{\partial x} = e^x \cos y^2$, $\quad f(x,y) = e^x \cos y^2 + \phi(y)$, $\quad \dfrac{\partial f}{\partial y} = -2y e^x \sin y^2 + \phi'(y) = -2y e^x \sin y^2$.

Thus, $\quad \phi'(y) = 0$, $\phi(y) = C$, $\quad$ and $\quad f(x,y) = e^x \cos y^2 + C$.

11. $\dfrac{\partial f}{\partial y} = xe^x - e^{-y}$, $\quad f(x,y) = xye^x + e^{-y} + \phi(x)$, $\quad \dfrac{\partial f}{\partial x} = ye^x + xye^x + \phi'(x) = ye^x(1 + x)$.

Thus, $\quad \phi'(x) = 0$, $\phi(x) = C$, $\quad$ and $\quad f(x,y) = xye^x + e^{-y} + C$.

13. No; $\dfrac{\partial}{\partial y}\left(xe^{xy} + x^2\right) = x^2 e^{xy}$ $\quad$ whereas $\quad \dfrac{\partial}{\partial x}\left(ye^{xy} - 2y\right) = y^2 e^{xy}$

15. $\dfrac{\partial f}{\partial x} = 1 + y^2 + xy^2$, $f(x,y) = x + xy^2 + \frac{1}{2}x^2 y^2 + \phi(y)$, $\dfrac{\partial f}{\partial y} = 2xy + x^2 y + \phi'(y) = x^2 y + y + 2xy + 1$.

Thus, $\quad \phi'(y) = y + 1$, $\quad \phi(y) = \frac{1}{2}y^2 + y + C$ $\quad$ and $\quad f(x,y) = x + xy^2 + \frac{1}{2}x^2 y^2 + \frac{1}{2}y^2 + y + C$.

17. $\dfrac{\partial f}{\partial x} = \dfrac{x}{\sqrt{x^2 + y^2}}$, $\quad f(x,y) = \sqrt{x^2 + y^2} + \phi(y)$, $\quad \dfrac{\partial f}{\partial y} = \dfrac{y}{\sqrt{x^2 + y^2}} + \phi'(y) = \dfrac{y}{\sqrt{x^2 + y^2}}$.

Thus, $\quad \phi'(y) = 0$, $\phi(y) = C$, $\quad$ and $\quad f(x,y) = \sqrt{x^2 + y^2} + C$.

19. $\dfrac{\partial f}{\partial x} = x^2 \sin^{-1} y$, $f(x,y) = \frac{1}{3}x^3 \sin^{-1} y + \phi(y)$, $\dfrac{\partial f}{\partial y} = \dfrac{x^3}{3\sqrt{1-y^2}} + \phi'(y) = \dfrac{x^3}{3\sqrt{1-y^2}} - \ln y$.

Thus, $\phi'(y) = -\ln y$, $\implies$ $\phi(y) = y - y\ln y + C$, and

$$f(x,y) = \frac{1}{3}x^3 \sin^{-1} y + y - y\ln y + C.$$

21. (a) Yes   (b) Yes   (c) No

23. $\dfrac{\partial f}{\partial x} = f(x,y)$, $\dfrac{\partial f/\partial x}{f(x,y)} = 1$, $\ln f(x,y) = x + \phi(y)$, $\dfrac{\partial f/\partial y}{f(x,y)} = 0 + \phi'(y)$, $\dfrac{\partial f}{\partial y} = f(x,y)$.

Thus, $\phi'(y) = 1$, $\phi(y) = y + K$, and $f(x,y) = e^{x+y+K} = Ce^{x+y}$.

25. (a) $P = 2x$, $Q = z$, $R = y$; $\dfrac{\partial P}{\partial y} = 0 = \dfrac{\partial Q}{\partial x}$, $\dfrac{\partial P}{\partial z} = 0 = \dfrac{\partial R}{\partial x}$, $\dfrac{\partial Q}{\partial z} = 1 = \dfrac{\partial R}{\partial y}$

(b), (c), and (d)

$$\frac{\partial f}{\partial x} = 2x, \quad f(x,y,z) = x^2 + g(y,z).$$

$$\frac{\partial f}{\partial y} = 0 + \frac{\partial g}{\partial y} \quad \text{with} \quad \frac{\partial f}{\partial y} = z \implies \frac{\partial g}{\partial y} = z.$$

Then,

$$g(y,z) = yz + h(z) \implies f(x,y,z) = x^2 + yz + h(z),$$

$$\frac{\partial f}{\partial z} = 0 + y + h'(z) \quad \text{and} \quad \frac{\partial f}{\partial z} = y \implies h'(z) = 0.$$

Thus, $h(z) = C$ and $f(x,y,z) = x^2 + yz + C$.

27. The function is a gradient by the test stated before Exercise 25.

Take $P = 2x + y$, $Q = 2y + x + z$, $R = y - 2z$. Then

$$\frac{\partial P}{\partial y} = 1 = \frac{\partial Q}{\partial x}, \quad \frac{\partial P}{\partial z} = 0 = \frac{\partial R}{\partial x}, \quad \frac{\partial Q}{\partial z} = 1 = \frac{\partial R}{\partial y}.$$

Next, we find f where $\nabla f = P\mathbf{i} + Q\mathbf{j} + R\mathbf{k}$.

$$\frac{\partial f}{\partial x} = 2x + y \implies f(x,y,z) = x^2 + xy + g(y,z).$$

$$\frac{\partial f}{\partial y} = x + \frac{\partial g}{\partial y} \quad \text{with} \quad \frac{\partial f}{\partial y} = 2y + x + z \implies \frac{\partial g}{\partial y} = 2y + z.$$

Then,

$$g(y,z) = y^2 + yz + h(z),$$

$$f(x,y,z) = x^2 + xy + y^2 + yz + h(z).$$

$$\frac{\partial f}{\partial z} = y + h'(z) = y - 2z \implies h'(z) = -2z.$$

Thus, $h(z) = -z^2 + C$ and $f(x,y,z) = x^2 + xy + y^2 + yz - z^2 + C$.

29. The function is a gradient by the test stated before Exercise 25.

Take $\quad P = y^2 z^3 + 1, \quad Q = 2xyz^3 + y, \quad R = 3xy^2z^2 + 1. \quad$ Then

$$\frac{\partial P}{\partial y} = 2yz^3 = \frac{\partial Q}{\partial x}, \quad \frac{\partial P}{\partial z} = 3y^2z^2 = \frac{\partial R}{\partial x}, \quad \frac{\partial Q}{\partial z} = 6xyz^2 = \frac{\partial R}{\partial y}.$$

Next, we find f where $\quad \nabla f = P\mathbf{i} + Q\mathbf{j} + R\mathbf{k}.$

$$\frac{\partial f}{\partial x} = y^2 z^3 + 1,$$

$$f(x, y, z) = xy^2 z^3 + x + g(y, z).$$

$$\frac{\partial f}{\partial y} = 2xyz^3 + \frac{\partial g}{\partial y} \quad \text{with} \quad \frac{\partial f}{\partial y} = 2xyz^3 + y \quad \Longrightarrow \quad \frac{\partial g}{\partial y} = y.$$

Then,

$$g(y, z) = \tfrac{1}{2} y^2 + h(z),$$

$$f(x, y, z) = xy^2 z^3 + x + \tfrac{1}{2} y^2 + h(z).$$

$$\frac{\partial f}{\partial z} = 3xy^2 z^2 + h'(z) = 3xy^2 z^2 + 1 \quad \Longrightarrow \quad h'(z) = 1.$$

Thus, $\quad h(z) = z + C \quad$ and $\quad f(x, y, z) = xy^2 z^3 + x + \tfrac{1}{2} y^2 + z + C.$

31. $\mathbf{F(r)} = \nabla \left(\dfrac{GmM}{r} \right)$

REVIEW EXERCISES

1. $\nabla f(x, y) = (4x - 4y)\mathbf{i} + (3y^2 - 4x)\mathbf{j}$

3. $\nabla f(x, y) = (ye^{xy} \tan 2x + 2e^{xy} \sec^2 2x)\,\mathbf{i} + xe^{xy} \tan 2x\,\mathbf{j}$

5. $\nabla f(x, y) = 2xe^{-yz} \sec z\,\mathbf{i}, -zx^2 e^{-yz} \sec z\,\mathbf{j} - (x^2 ye^{-yz} \sec z - x^2 e^{-yz} \sec z \tan z)\mathbf{k}$

7. $\nabla f(x, y) = (2x - 2y)\,\mathbf{i} - 2x\,\mathbf{j}, \quad \nabla f(1, -2) = 6\,\mathbf{i} - 2\,\mathbf{j}; \qquad \mathbf{u_a} = \dfrac{1}{\sqrt{5}}\mathbf{i} + \dfrac{2}{\sqrt{5}}\mathbf{j};$

$f'_{\mathbf{u_a}}(1, -2) = \nabla f(1, -2) \cdot \mathbf{u_a} = \dfrac{2}{\sqrt{5}}.$

9. $\nabla f(x, y, z) = (y^2 + 6xz)\,\mathbf{i} + (2xy + 2z)\,\mathbf{j} + (2y + 3x^2)\,\mathbf{k}, \quad \nabla f(1, -2, 3) = 22\,\mathbf{i} + 2\,\mathbf{j} - \mathbf{k};$

$\mathbf{u_a} = \dfrac{1}{3}\mathbf{i} - \dfrac{2}{3}\mathbf{j} + \dfrac{2}{3}\mathbf{k}; f'_{\mathbf{u_a}}(1, -2, 3) = \nabla f(1, -2, 3) \cdot \mathbf{u_a} = \dfrac{16}{3}.$

11. $\nabla f(x, y) = (6x - 2y^2)\,\mathbf{i} - 4xy\,\mathbf{j}, \quad \nabla f(3, -2) = 10\,\mathbf{i} + 24\,\mathbf{j};$

$\mathbf{a} = (0, 0) - (3, -2) = (-3, 2) = -3\,\mathbf{i} + 2\,\mathbf{j}, \quad \mathbf{u_a} = \dfrac{-3}{\sqrt{13}}\,\mathbf{i} + \dfrac{2}{\sqrt{13}}\,\mathbf{j};$

$f'_{\mathbf{u_a}}(3, -2) = \nabla f(3, -2) \cdot \mathbf{u_a} = \dfrac{18}{\sqrt{13}}.$

13. $\nabla f(x,y,z) = \dfrac{1}{\sqrt{x^2+y^2+z^2}}(x\mathbf{i}+y\mathbf{j}+z\mathbf{k}), \quad \nabla f(3,-1,4) = \dfrac{1}{\sqrt{26}}(3\mathbf{i}-\mathbf{j}+4\mathbf{k});$

$\mathbf{a} = \pm(4\mathbf{i}-3\mathbf{j}+\mathbf{k}), \quad \mathbf{u_a} = \pm\dfrac{1}{\sqrt{26}}(4\mathbf{i}-3\mathbf{j}+\mathbf{k}); \qquad f'_{\mathbf{u_a}}(3,-1,4) = \nabla f(3,-1,4)\cdot\mathbf{u_a} = \pm\dfrac{19}{26}.$

15. $\nabla f(x,y,z) = \cos xyz\,(yz\,\mathbf{i}+xz\,\mathbf{j}+xy\,\mathbf{k}), \quad \nabla f(\frac{1}{2},\frac{1}{3},\pi) = \frac{\pi\sqrt{3}}{6}\mathbf{i}+\frac{\pi\sqrt{3}}{4}\mathbf{j}+\frac{\sqrt{3}}{12}\mathbf{k};$

minimum directional derivative: $f'_{\mathbf{u}} = -\|\nabla f(\frac{1}{2},\frac{1}{3},\pi)\| = -\dfrac{\sqrt{39\pi^2+3}}{12}$

17. Let $\mathbf{r}(t) = x(t)\mathbf{i}+y(t)\mathbf{j}$ be the path of the particle. $\nabla T = -e^{-x}\cos y\,\mathbf{i} - e^{-x}\sin y\,\mathbf{j}.$ Then

$$x'(t) = -e^{-x(t)}\cos y(t), \quad y'(t) = -e^{-x(t)}\sin y(t) \implies \frac{y'(t)}{x'(t)} = \tan y(t) \implies \frac{dy}{dx} = \tan y$$

The solution is $\sin y = Ce^x$. Since $\mathbf{r}(0) = 0$, $C = 0$ and $y = 0$. The particle moves to the right the x-axis.

19. $\nabla f(x,y) = e^x \arctan y\,\mathbf{i} + e^x\dfrac{1}{1+y^2}\mathbf{j}; \quad \nabla f(0,1) = \dfrac{\pi}{4}\mathbf{i}+\dfrac{1}{2}\mathbf{j}.$

$\mathbf{u} = \dfrac{\nabla f(0,1)}{\|\nabla f(0,1)\|} = \dfrac{1}{\sqrt{4+\pi^2}}(\pi\,\mathbf{i}+2\,\mathbf{j}); \quad$ rate: $\|\nabla f(0,1)\| = \dfrac{\sqrt{\pi^2+4}}{4}$

21. rate: $\dfrac{df}{dt} = \nabla f\cdot\mathbf{r'} = \left(4x\,\mathbf{i}-9y^2\,\mathbf{j}\right)\cdot\left(\dfrac{1}{2}t^{-1/2}\,\mathbf{i}+2e^{2t}\,\mathbf{j}\right) = 2-18e^{6t}$

23. rate: $\dfrac{df}{dt} = \nabla f\cdot\mathbf{r'} = \left[\left(\dfrac{1}{y}+\dfrac{z}{x^2}\right)\mathbf{i}-\dfrac{x}{y^2}\mathbf{j}-\dfrac{1}{x}\mathbf{k}\right]\cdot(\cos t\,\mathbf{i}-\sin t\,\mathbf{j}+\sec^2 t\,\mathbf{k}) = \dfrac{1-\sin t}{\cos^2 t}$

25. $\dfrac{du}{dt} = \nabla u\cdot\mathbf{r'} = \left[(3y^2-2x)\,\mathbf{i}+6xy\,\mathbf{j}\right]\cdot[(2t+2)\,\mathbf{i}+3\,\mathbf{j}] = 104t^3+150t^2-8t$

27. area $A = \frac{1}{2}x(t)y(t)\sin\theta(t)$

$\dfrac{dA}{dt} = 0 = \frac{1}{2}y(t)x'(t)\sin\theta(t) + \frac{1}{2}x(t)y'(t)\sin\theta(t) + \frac{1}{2}\theta'(t)x(t)y(t)\cos\theta(t) = 0$

At $x=4$, $y=5$, $\theta = \pi/3$, $\dfrac{dx}{dt} = \dfrac{dy}{dt} = 2$, we have

$$5\frac{d\theta}{dt} + 2\sqrt{3} + \frac{5\sqrt{3}}{2} = 0 \implies \frac{d\theta}{dt} = -\frac{9\sqrt{3}}{10}.$$

29. $\dfrac{\partial u}{\partial s} = \dfrac{\partial u}{\partial x} + \dfrac{\partial u}{\partial y}; \qquad \dfrac{\partial u}{\partial t} = \dfrac{\partial u}{\partial x} - \dfrac{\partial u}{\partial y}$

$\dfrac{\partial u}{\partial s}\dfrac{\partial u}{\partial t} = \left(\dfrac{\partial u}{\partial x}+\dfrac{\partial u}{\partial y}\right)\left(\dfrac{\partial u}{\partial x}-\dfrac{\partial u}{\partial y}\right) = \left(\dfrac{\partial u}{\partial x}\right)^2 - \left(\dfrac{\partial u}{\partial y}\right)^2$

31. $\nabla f(x,y) = (3x^2-6xy)\,\mathbf{i} + (-3x^2+2y)\,\mathbf{j}; \quad \nabla f(1,-1) = \mathbf{N} = 9\mathbf{i}-5\mathbf{j}$

normal line: $x = 1+9t, \quad y = -1-5t;$ tangent line: $x = 1+5t, \quad y = -1+9t$

33. Set $f(x, y, z) = x^{1/2} + y^{1/2} - z$

$\nabla f(x, y, z) = \dfrac{1}{2\sqrt{x}}\,\mathbf{i} + \dfrac{1}{2\sqrt{y}}\,\mathbf{j} - \mathbf{k}; \quad \nabla f(1, 1, 2) = \frac{1}{2}\mathbf{i} + \frac{1}{2}\mathbf{j} - \mathbf{k}.$ Take $\mathbf{N} = \mathbf{i} + \mathbf{j} - 2\,\mathbf{k}.$

tangent plane: $(x - 1) + (y - 1) - 2(z - 2) = 0;$ normal line: $x = 1 + t, \; y = 1 + t, \; z = 2 - 2t$

35. Set $f(x, y, z) = z^3 + xyz - 2.$

$\nabla f(x, y, z) = yz\,\mathbf{i} + xz\,\mathbf{j} + (3z^2 + xy)\,\mathbf{k}; \quad \nabla f(1, 1, 1) = \mathbf{i} + \mathbf{j} + 4\mathbf{k}.$

tangent plane: $(x - 1) + (y - 1) + 4(z - 1) = 0;$ normal line: $x = 1 + t; \; y = 1 + t; \; z = 1 + 4t$

37. The point $(2, 2, 1)$ is on each hyperboloid. Set $f(x, y, z) = x^2 + 2y^2 - 4z^2, \; g(x, y, z) = 4x^2 - y^2 + 2z^2.$

$\nabla f = 2x\,\mathbf{i} + 4y\,\mathbf{j} - 8z\,\mathbf{k}, \quad \nabla f(2, 2, 1) = (4, 8, -8); \quad \nabla g = 8x\,\mathbf{i} - 2y\,\mathbf{j} + 4z\,\mathbf{k}, \; \nabla g(2, 2, 1) = (16, -4, 4).$

Since $\nabla f(2, 2, 1) \cdot \nabla g(2, 2, 1) = 0,$ the hyperboloids are mutually perpendicular at $(2, 2, 1).$

39. $\nabla f(x, y) = (2xy - 2y)\,\mathbf{i} + (x^2 - 2x + 4y - 15)\,\mathbf{j} = \mathbf{0}$ at $(5, 0), \; (-3, 0), \; (1, 4).$

$f_{xx} = 2y, \quad f_{xy} = 2x - 2, \quad f_{yy} = 4.$

point	A	B	C	D	result
$(5, 0)$	0	8	4	-64	saddle
$(-3, 0)$	0	-8	4	-64	saddle
$(1, 4)$	8	0	4	32	loc. min.

$f(1, 4) = -34$

41. $\nabla f(x, y) = (3x^2 - 18y)\,\mathbf{i} + (3y^2 - 18x)\,\mathbf{j} = \mathbf{0}$ at $(0, 0), \; (6, 6).$

$f_{xx} = 6x, \quad f_{xy} = -18, \quad f_{yy} = 6y.$

point	A	B	C	D	result
$(0, 0)$	0	-18	0	-18^2	saddle
$(6, 6)$	36	-18	36	> 0	loc. min.

$f(6, 6) = -216$

43. $\nabla f(x, y) = (1 - 2xy + y^2)\,\mathbf{i} + (-1 - x^2 + 2xy)\,\mathbf{j} = \mathbf{0}$ at $(1, 1), \; (-1, -1).$

$f_{xx} = -2y, \quad f_{xy} = -2x + 2y, \quad f_{yy} = 2x.$

point	A	B	C	D	result
$(1, 1)$	-2	0	2	-4	saddle
$(-1, -1)$	2	0	-2	-4	saddle

45. $\nabla f = (2x - 2)\,\mathbf{i} + (2y + 2)\,\mathbf{j} = \mathbf{0}$ at $(1, -1)$ in $D; \quad f(1, -1) = 0$

Next we consider the boundary of $D.$ We parametrize the circle by:

$C: \; \mathbf{r}(t) = 2\cos t\,\mathbf{i} + 2\sin t\,\mathbf{j}, \quad t \in [0, 2\pi]$

The values of f on the boundary are given by the function

$$F(t) = f(\mathbf{r}(t)) = 6 - 4\cos t + 4\sin t, \quad t \in [0, 2\pi]$$

$$F'(t) = 4\sin t + 4\cos t: \quad F'(t) = 0 \implies \sin t = -\cos t \implies t = \frac{3}{4}\pi, \frac{7}{4}\pi$$

Evaluating F at the endpoints and critical numbers, we have:

$$F(0) = F(2\pi) = f(2,0) = 2; \quad F\left(\tfrac{3}{4}\pi\right) = f\left(-\sqrt{2}, \sqrt{2}\right) = 6 + 4\sqrt{2};$$

$$F\left(\tfrac{7}{4}\pi\right) = f\left(\sqrt{2}, -\sqrt{2}\right) = 6 - 4\sqrt{2}.$$

f takes on its absolute maximum of $6 + 4\sqrt{2}$ at $\left(-\sqrt{2}, \sqrt{2}\right)$; f takes on its absolute minimum of 0 at $(1, -1)$.

47. $\nabla f(x,y) = (8x - y)\mathbf{i} + (-x + 2y + 1)\mathbf{j} = \mathbf{0}$ at $(-1/15, -8/15)$ in D; $f(-1/15, -8/15) = -4/15$. On the boundary of $D: x = \cos t, \ y = 2\sin t$. Set

$$F(t) = f(\cos t, 2\sin t) = 4 + 2\sin t - 2\sin t \cos t, \quad 0 \le t \le 2\pi.$$

Then

$$F'(t) = 2\cos t - 4\cos^2 t + 2 = -2(2\cos t + 1)(\cos t - 1); \quad F'(t) = 0 \implies t = \frac{2\pi}{3}, \frac{4\pi}{3}.$$

Evaluating F at the endpoints of the interval and at the critical points, we get

$$F(0) = F(2\pi) = f(1,0) = 4, \ F(2\pi/3) = f(-1/2, \sqrt{3}) = 4 + \frac{3\sqrt{3}}{2},$$

$$F(4\pi/3) = f(-1/2, -\sqrt{3}) = 4 - \frac{3\sqrt{3}}{2} > -\frac{4}{15}$$

f takes on its absolute maximum of 2 at $(0,1)$; f takes on its absolute minimum of $-4/15$ at $(-1/15, -8/15)$.

49. Set $f(x,y,z) = D^2 = (x-1)^2 + (y+2)^2 + (z-3)^2, \quad g(x,y) = 3x + 2y - z - 5.$

$\nabla f = 2(x-1)\mathbf{i} + 2(y+2)\mathbf{j} + 2(z-3)\mathbf{k}, \quad \nabla g = 3\mathbf{i} + 2\mathbf{j} - \mathbf{k}.$

Set $\nabla f = \lambda \nabla g$:

$2(x-1) = 3\lambda \implies x = \tfrac{3}{2}\lambda + 1,$

$2(y+2) = 2\lambda \implies y = \lambda - 2,$

$2(z-3) = -\lambda \implies z = -\tfrac{1}{2}\lambda + 3.$

Substituting these values in $3x + 2y - z = 5$ gives $\lambda = \dfrac{9}{7} \implies x = \dfrac{41}{14}, \ y = -\dfrac{5}{7}, \ z = \dfrac{33}{14}.$

The point on the plane that is closest to $(1, -2, 3)$ is $(41/14, -5/7, 33/14)$. The distance from the point to the plane is $\dfrac{9}{\sqrt{14}}$.

51. Set $f(x, y, z) = x + y - z$, $\quad g(x, y, z) = x^2 + y^2 + 4z^2 - 4$,

$\nabla f = \mathbf{i} + \mathbf{j} - \mathbf{k}$, $\quad \nabla g = 2x\,\mathbf{i} + 2y\,\mathbf{j} + 8z\,\mathbf{k}$.

Set $\nabla f = \lambda \nabla g$:

$$1 = 2\lambda x \implies x = 1/2\lambda, \quad 1 = 2\lambda y \implies y = 1/2\lambda, \quad -1 = 8\lambda z \implies z = -1/8\lambda.$$

Substituting these values in $x^2 + y^2 + 4z^2 = 4$ gives $\lambda = \pm\frac{3}{8} \implies x = 4/3$, $y = 4/3$, $z = -1/3$ or $x = -4/3$, $y = -4/3$, $z = 1/3$. Evaluating f : $f(\frac{4}{3}, \frac{4}{3}, -\frac{1}{3}) = 3$, $f(-\frac{4}{3}, -\frac{4}{3}, \frac{1}{3}) = -3$. The maximum value of f is 3, the minimum value is -3.

53. $df = (9x^2 - 10xy^2 + 2)\,dx + (-10x^2 y - 1)\,dy$

55. $df = \dfrac{y^2 z + z^2 y}{(x + y + z)^2}\,dx + \dfrac{xz^2 + zx^2}{(x + y + z)^2}\,dy + \dfrac{x^2 y + y^2 x}{(x + y + z)^2}\,dz$

57. Set $f(x, y, z) = e^x \sqrt{y + z^3}$. Then

$$df = e^x \sqrt{y + z^3}\,\Delta x + \frac{e^x}{2}\frac{1}{\sqrt{y + z^3}}\,\Delta y + \frac{e^x}{2}\frac{3z^2}{\sqrt{y + z^3}}\,\Delta z.$$

With $x = 0$, $y = 15$, $z = 1$, $\Delta x = 0.02$, $\Delta y = 0.2$, $\Delta z = 0.01$, $df = 4\,\Delta x + \frac{1}{8}\,\Delta y + \frac{3}{8}\,\Delta z \cong 0.1088$. Therefore, $e^{0.02}\sqrt{15.2 + (1.01)^3} \cong e^0 \sqrt{15 + 1} + 0.1088 = 4.1088$.

59. $V = \pi r^2 h$; $\quad r = 5\,\text{ft.}$, $\quad h = 22\,\text{ft.}$, $\quad \Delta r = 0.01\,\text{in.} = \frac{1}{1200}\,\text{ft.}$, $\quad \Delta h = 0.01 = \frac{1}{1200}$

$$dV = 2\pi r h\,\Delta r + \pi r^2\,\Delta h$$

Using the values given above,

$$dV = 2\pi(5)(22)\frac{1}{1200} + \pi(25)\frac{1}{1200} \cong 0.6414\,\text{cu. ft.} \cong 1108.35\,\text{cu. in.}; \quad \frac{1108.35}{231} \cong 4.80.$$

Approximately 4.80 gallons will be needed.

61. $\dfrac{\partial P}{\partial y} = 2x - \sin x = \dfrac{\partial Q}{\partial x}$; the vector function is a gradient.

$\dfrac{\partial f}{\partial x} = 2xy + 3 - y\sin x$, $\quad f(x, y) = x^2 y + 3x + y\cos x + \phi(y)$,

$\dfrac{\partial f}{\partial y} = x^2 + \cos x + \phi'(y) = x^2 + 2y + 1 + \cos x$.

Thus, $\quad \phi'(y) = 2y + 1$, $\quad \phi(y) = y^2 + y + C$, $\quad$ and $\quad f(x, y) = x^2 y + 3x + y\cos x + y^2 + y + C$.

63. $\dfrac{\partial P}{\partial y} = e^y \sin z = \dfrac{\partial Q}{\partial x}, \quad \dfrac{\partial P}{\partial z} = e^y \cos z = \dfrac{\partial R}{\partial x}, \quad \dfrac{\partial Q}{\partial z} = xe^y \cos z = \dfrac{\partial R}{\partial y};$

the vector function is a gradient.

$$f(x, y, z) = \int (e^y \sin z + 2x)\, dx = xe^y \sin z + x^2 + \phi(y, z),$$

$$f_y = xe^y \sin z + \frac{\partial \phi}{\partial y} = xe^y \sin z - y^2 \quad \Longrightarrow \quad \frac{\partial \phi}{\partial y} = -y^2 \quad \Longrightarrow \quad \phi = -\frac{1}{3}y^3 + \psi(z),$$

$$f(x, y, z) = xe^y \sin z + x^2 - \frac{1}{3}y^3 + \psi(z), \quad f_z = xe^y \cos z + \psi'(z) = xe^y \cos z \quad \Longrightarrow \quad \psi'(x) = 0 \quad \Longrightarrow$$
$$\psi(x) = C$$

Therefore $f(x, y, z) = xe^y \sin z + x^2 - \frac{1}{3}y^3 + C.$

CHAPTER 17

SECTION 17.1

1. $\displaystyle\sum_{i=1}^{3}\sum_{j=1}^{3}2^{i-1}3^{j+1} = \left(\sum_{i=1}^{3}2^{i-1}\right)\left(\sum_{j=1}^{3}3^{j+1}\right) = (1+2+4)(9+27+81) = 819$

3. $\displaystyle\sum_{i=1}^{4}\sum_{j=1}^{3}(i^2+3i)(j-2) = \left[\sum_{i=1}^{4}(i^2+3i)\right]\left[\sum_{j=1}^{3}(j-2)\right] = (4+10+18+28)(-1+0+1) = 0$

5. $\displaystyle\sum_{i=1}^{m}\Delta x_i = \Delta x_1 + \Delta x_2 + \cdots + \Delta x_m = (x_1-x_0)+(x_2-x_1)+\cdots+(x_m-x_{m-1})$
$$= x_m - x_0 = a_2 - a_1$$

7. $\displaystyle\sum_{i=1}^{m}\sum_{j=1}^{n}\Delta x_i\,\Delta y_j = \left(\sum_{i=1}^{m}\Delta x_i\right)\left(\sum_{j=1}^{n}\Delta y_j\right) = (a_2-a_1)(b_2-b_1)$

9. $\displaystyle\sum_{i=1}^{m}(x_i+x_{i-1})\Delta x_i = \sum_{i=1}^{m}(x_i+x_{i-1})(x_i-x_{i-1}) = \sum_{i=1}^{m}(x_i{}^2 - x_{i-1}^2)$
$$= x_m{}^2 - x_0{}^2 = a_2{}^2 - a_1{}^2$$

11. $\displaystyle\sum_{i=1}^{m}\sum_{j=1}^{n}(x_i+x_{i-1})\Delta x_i\Delta y_j = \left(\sum_{i=1}^{m}(x_i+x_{i-1})\Delta x_i\right)\left(\sum_{j=1}^{n}\Delta y_j\right)$

(Exercise 9) $\uparrow$

$$= \left(a_2{}^2 - a_1{}^2\right)(b_2-b_1)$$

13. $\displaystyle\sum_{i=1}^{m}\sum_{j=1}^{n}(2\Delta x_i - 3\Delta y_j) = 2\left(\sum_{i=1}^{m}\Delta x_i\right)\left(\sum_{j=1}^{n}1\right) - 3\left(\sum_{i=1}^{m}1\right)\left(\sum_{j=1}^{n}\Delta y_j\right)$
$$= 2n(a_2-a_1) - 3m(b_2-b_1)$$

15. $\displaystyle\sum_{i=1}^{m}\sum_{j=1}^{n}\sum_{k=1}^{q}\Delta x_i\,\Delta y_j\,\Delta z_k = \left(\sum_{i=1}^{m}\Delta x_i\right)\left(\sum_{j=1}^{n}\Delta y_j\right)\left(\sum_{k=1}^{q}\Delta z_k\right)$
$$= (a_2-a_1)(b_2-b_1)(c_2-c_1)$$

17. $\displaystyle\sum_{i=1}^{n}\sum_{j=1}^{n}\sum_{k=1}^{n}\delta_{ijk}a_{ijk} = a_{111} + a_{222} + \cdots + a_{nnn} = \sum_{p=1}^{n}a_{ppp}$

SECTION 17.2

1. $L_f(P) = 2\frac{1}{4}, \quad U_f(P) = 5\frac{3}{4}$

3. (a) $\displaystyle L_f(P) = \sum_{i=1}^{m}\sum_{j=1}^{n}(x_{i-1}+2y_{j-1})\,\Delta x_i\,\Delta y_j, \quad U_f(P) = \sum_{i=1}^{m}\sum_{j=1}^{n}(x_i+2y_j)\,\Delta x_i\,\Delta y_j$

(b) $L_f(P) \le \sum\limits_{i=1}^{m} \sum\limits_{j=1}^{n} \left[\dfrac{x_{i-1}+x_i}{2} + 2\left(\dfrac{y_{j-1}+y_j}{2}\right)\right] \Delta x_i \, \Delta y_j \le U_f(P).$

The middle expression can be written

$$\sum_{i=1}^{m}\sum_{j=1}^{n} \frac{1}{2}\left(x_i{}^2 - x_{i-1}^2\right)\Delta y_j + \sum_{i=1}^{m}\sum_{j=1}^{n}\left(y_j{}^2 - y_{j-1}^2\right)\Delta x_i.$$

The first double sum reduces to

$$\sum_{i=1}^{m}\sum_{j=1}^{n}\frac{1}{2}\left(x_i{}^2 - x_{i-1}^2\right)\Delta y_j = \frac{1}{2}\left(\sum_{i=1}^{m}\left(x_i{}^2 - x_{i-1}^2\right)\right)\left(\sum_{j=1}^{n}\Delta y_j\right) = \frac{1}{2}\left(4-0\right)\left(1-0\right) = 2.$$

In like manner the second double sum also reduces to 2. Thus, $I = 4$; the volume of the prism bounded above by the plane $z = x + 2y$ and below by R.

5. $L_f(P) = -7/24, \quad U_f(P) = 7/24$

7. (a) $L_f(P) = \sum\limits_{i=1}^{m}\sum\limits_{j=1}^{n}\left(4x_{i-1}\, y_{j-1}\right)\Delta x_i \, \Delta y_j, \quad U_f(P) = \sum\limits_{i=1}^{m}\sum\limits_{j=1}^{n}\left(4x_i\, y_j\right)\Delta x_i\, \Delta y_j$

(b) $L_f(P) \le \sum\limits_{i=1}^{m}\sum\limits_{j=1}^{n}\left(x_i + x_{i-1}\right)\left(y_j + y_{j-1}\right)\Delta x_1 \, \Delta y_j \le U_f(P).$

The middle expression can be written

$$\sum_{i=1}^{m}\sum_{j=1}^{n}\left(x_i{}^2 - x_{i-1}^2\right)\left(y_j{}^2 - y_{j-1}^2\right) = \left(\sum_{i=1}^{m} x_i{}^2 - x_{i-1}^2\right)\left(\sum_{j=1}^{n} y_j{}^2 - y_{j-1}^2\right)$$

by (17.1.5)

$$= \left(b^2 - 0^2\right)\left(d^2 - 0^2\right) = b^2 d^2.$$

It follows that $I = b^2 d^2$.

9. (a) $L_f(P) = \sum\limits_{i=1}^{m}\sum\limits_{j=1}^{n} 3\left(x_{i-1}^2 - y_j{}^2\right)\Delta x_i \, \Delta y_j, \quad U_f(P) = \sum\limits_{i=1}^{m}\sum\limits_{j=1}^{n} 3\left(x_i{}^2 - y_{j-1}^2\right)\Delta x_i \, \Delta y_j$

(b) $L_f(P) \le \sum\limits_{i=1}^{m}\sum\limits_{j=1}^{n}\left[\left(x_i{}^2 + x_i x_{i-1} + x_{i-1}^2\right) - \left(y_j{}^2 + y_j y_{j-1} + y_{j-1}^2\right)\right]\Delta x_i\, \Delta y_j \le U_f(P).$

Since in general $\left(A^2 + AB + B^2\right)\left(A - B\right) = A^3 - B^3,$ the middle expression can be written

$$\sum_{i=1}^{m}\sum_{j=1}^{n}\left(x_i{}^3 - x_{i-1}^3\right)\Delta y_j - \sum_{i=1}^{m}\sum_{j=1}^{n}\left(y_j{}^3 - y_{j-1}^3\right)\Delta x_i,$$

which reduces to

$$\left(\sum_{i=1}^{m} x_i{}^3 - x_{i-1}^3\right)\left(\sum_{j=1}^{n}\Delta y_j\right) - \left(\sum_{i=1}^{m}\Delta x_i\right)\left(\sum_{j=1}^{n} y_j{}^3 - y_{j-1}^3\right).$$

This can be evaluated as $b^3 d - bd^3 = bd\left(b^2 - d^2\right)$. It follows that $I = bd\left(b^2 - d^2\right)$.

11. $\displaystyle\iint\limits_{\Omega} dx\,dy = \int_a^b \phi(x)\,dx$

13. Suppose $f(x_0, y_0) \neq 0$. Assume $f(x_0, y_0) > 0$. Since f is continuous, there exists a disc Ω_ϵ with radius ϵ centered at (x_0, y_0) such that $f(x, y) > 0$ on Ω_ϵ. Let R be a rectangle contained in Ω_ϵ. Then $\displaystyle\iint\limits_{R} f(x, y)\,dx\,dy > 0$, which contradicts the hypothesis.

15. By Exercise 7, Section 17.2, $\displaystyle\iint\limits_{R} 4xy\,dx\,dy = 2^2 3^2 = 36$. Thus

$$f_{avg} = \frac{1}{\text{area}\,(R)} \iint\limits_{R} 4xy\,dx\,dy = \frac{1}{6}\,(36) = 6$$

17. By Theorem 16.2.10, there exists a point $(x_1, y_1) \in D_r$ such that

$$\iint\limits_{D_r} f(x, y)\,dx\,dy = f(x_1, y_1) \iint\limits_{R} dx\,dy = f(x_1, y_1)\pi r^2 \quad \Longrightarrow \quad f(x_1, y_1) = \frac{1}{\pi r^2} \iint\limits_{D_r} f(x, y)\,dx\,dy$$

As $r \to 0$, $(x_1, y_1) \to (x_0, y_0)$ and $f(x_1, y_1) \to f(x_0, y_0)$ since f is continuous.

The result follows.

19. $z = \sqrt{4 - x^2 - y^2}$ on $\Omega : x^2 + y^2 \leq 4$, $x \geq 0$, $y \geq 0$; $\displaystyle\iint\limits_{\Omega} \sqrt{4 - x^2 - y^2}\,dx\,dy$ is the volume V of one quarter of a hemisphere; $V = \frac{4}{3}\pi$.

21. $z = 6 - 2x - 3y \;\Rightarrow\; \dfrac{x}{3} + \dfrac{y}{2} + \dfrac{z}{6} = 1$;   the solid is the tetrahedron bounded by the coordinate planes and the plane: $\dfrac{x}{3} + \dfrac{y}{2} + \dfrac{z}{6} = 1$; $V = \frac{1}{6}(3)(2)(6) = 6$

SECTION 17.3

1. $\displaystyle\int_0^1 \int_0^3 x^2\,dy\,dx = \int_0^1 3x^2\,dx = 1$

3. $\displaystyle\int_0^1 \int_0^3 xy^2\,dy\,dx = \int_0^1 x\left[\frac{1}{3}y^3\right]_0^3 dx = \int_0^1 9x\,dx = \frac{9}{2}$

5. $\displaystyle\int_0^1 \int_0^x xy^3\,dy\,dx = \int_0^1 x\left[\frac{1}{4}y^4\right]_0^x dx = \int_0^1 \frac{1}{4}x^5\,dx = \frac{1}{24}$

7. $\displaystyle\int_0^{\pi/2} \int_0^{\pi/2} \sin{(x+y)}\,dy\,dx = \int_0^{\pi/2} [-\cos{(x+y)}]_0^{\pi/2}\,dx = \int_0^{\pi/2} \left[\cos x - \cos\left(x + \frac{\pi}{2}\right)\right] dx = 2$

9. $\displaystyle\int_0^{\pi/2} \int_0^{\pi/2} (1 + xy)\,dy\,dx = \int_0^{\pi/2} \left[y + \frac{1}{2}xy^2\right]_0^{\pi/2} dx = \int_0^{\pi/2} \left(\frac{1}{2}\pi + \frac{1}{8}\pi^2 x\right) dx = \frac{1}{4}\pi^2 + \frac{1}{64}\pi^4$

11. $\displaystyle\int_0^1 \int_{y^2}^y \sqrt{xy}\, dx\, dy = \int_0^1 \sqrt{y}\left[\frac{2}{3}x^{3/2}\right]_{y^2}^y dy = \int_0^1 \frac{2}{3}\left(y^2 - y^{7/2}\right) dy = \frac{2}{27}$

13. $\displaystyle\int_{-2}^2 \int_{\frac{1}{2}y^2}^{4-\frac{1}{2}y^2} \left(4 - y^2\right) dx\, dy = \int_{-2}^2 \left(4 - y^2\right)\left[\left(4 - \frac{1}{2}y^2\right) - \left(\frac{1}{2}y^2\right)\right] dy$

$$= 2\int_0^2 \left(16 - 8y^2 + y^4\right) dy = \frac{512}{15}$$

15. 0   by symmetry (integrand odd in y, Ω symmetric about x-axis)

17. $\displaystyle\int_0^2 \int_0^{x/2} e^{x^2}\, dy\, dx = \int_0^2 \frac{1}{2}xe^{x^2}\, dx = \left[\frac{1}{4}e^{x^2}\right]_0^2 = \frac{1}{4}\left(e^4 - 1\right)$

19.

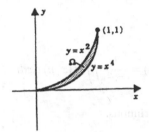

$$\int_0^1 \int_{y^{1/2}}^{y^{1/4}} f(x,y)\, dx\, dy$$

21.

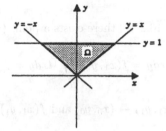

$$\int_{-1}^0 \int_{-x}^1 f(x,y)\, dy\, dx + \int_0^1 \int_x^1 f(x,y)\, dy\, dx$$

23.

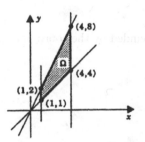

$$\int_1^2 \int_1^y f(x,y)\, dx\, dy + \int_2^4 \int_{y/2}^y f(x,y)\, dx\, dy$$

$$+ \int_4^8 \int_{y/2}^4 f(x,y)\, dx\, dy$$

25. $\displaystyle\int_{-2}^4 \int_{1/4x^2}^{\frac{1}{2}x+2} dy\, dx = \int_{-2}^4 \left[\frac{1}{2}x + 2 - \frac{1}{4}x^2\right] dx = 9$

27. $\displaystyle\int_0^{1/4} \int_{2y^{3/2}}^y dx\, dy = \int_0^{1/4} \left[y - 2y^{3/2}\right] dy = \frac{1}{160}$

29.

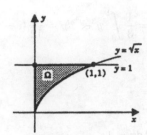

$$\int_0^1 \int_0^{y^2} \sin\left(\frac{y^3+1}{2}\right) dx\, dy = \int_0^1 y^2 \sin\left(\frac{y^3+1}{2}\right) dy$$

$$= \left[-\frac{2}{3}\cos\left(\frac{y^3+1}{2}\right)\right]_0^1$$

$$= \frac{2}{3}\left(\cos\frac{1}{2} - \cos 1\right)$$

31.

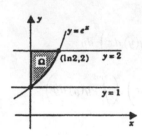

$$\int_0^{\ln 2} \int_{e^x}^2 e^{-x}\, dy\, dx = \int_0^{\ln 2} e^{-x}\left(2 - e^x\right) dx$$

$$= \left[-2e^{-x} - x\right]_0^{\ln 2} = 1 - \ln 2$$

33. $\displaystyle \int_1^2 \int_{y-1}^{2/y} dx\, dy = \int_1^2 \left[\frac{2}{y} - (y-1)\right] dy = \ln 4 - \frac{1}{2}$

35. $\displaystyle \int_0^2 \int_0^{3-\frac{3}{2}x} \left(4 - 2x - \frac{4}{3}y\right) dy\, dx = \int_0^3 \int_0^{2-\frac{2}{3}y} \left(4 - 2x - \frac{4}{3}y\right) dx\, dy = 4$

37. $\displaystyle \int_0^2 \int_0^{1-\frac{1}{2}x} x^3 y\, dy\, dx = \int_0^2 \int_0^{2-2y} x^3 y\, dx\, dy = \frac{2}{15}$

39. $\displaystyle \int_0^2 \int_{-\sqrt{2x-x^2}}^{\sqrt{2x-x^2}} (2x+1)\, dy\, dx = \int_{-1}^1 \int_{1-\sqrt{1-y^2}}^{1+\sqrt{1-y^2}} (2x+1)\, dx\, dy$

$$= \int_{-1}^1 \left[x^2 + x\right]_{1-\sqrt{1-y^2}}^{1+\sqrt{1-y^2}} dy$$

$$= 6\int_{-1}^1 \sqrt{1-y^2}\, dy = 6\left(\frac{\pi}{2}\right) = 3\pi$$

41. $\displaystyle \int_0^1 \int_0^{1-x} (x^2 + y^2)\, dy\, dx = \int_0^1 \left(2x^2 - \frac{4}{3}x^3 - x + \frac{1}{3}\right) dx = \frac{1}{6}$

43. $\displaystyle \int_0^1 \int_{x^2}^x (x^2 + 3y^2)\, dy\, dx = \int_0^1 \left(2x^3 - x^4 - x^6\right) dx = \frac{11}{70}$

45. $\displaystyle \int_0^a \int_0^{\sqrt{a^2-x^2}} \sqrt{a^2 - x^2}\, dy\, dx = \int_0^a (a^2 - x^2)\, dx = \frac{2}{3}a^3$

47. $\displaystyle \int_0^1 \int_y^1 e^{y/x} dx\, dy = \int_0^1 \int_0^x e^{y/x} dy\, dx = \int_0^1 \left[xe^{y/x}\right]_0^x dx = \int_0^1 x(e-1)\, dx = \frac{1}{2}(e-1)$

49. $\displaystyle\int_0^1\int_x^1 x^2 e^{y^4}\,dy\,dx = \int_0^1\int_0^y x^2 e^{y^4}\,dx\,dy = \int_0^1\left[\frac{1}{3}x^3 e^{y^4}\right]_0^y\,dy = \frac{1}{3}\int_0^1 y^3 e^{y^4}\,dy = \frac{1}{12}\left(e-1\right)$

51. $\displaystyle f_{avg} = \frac{1}{8}\int_{-1}^1\int_0^4 x^2 y\,dy\,dx = \frac{1}{8}\int_{-1}^1 8x^2\,dx = \int_{-1}^1 x^2\,dx = \frac{2}{3}$

53. $\displaystyle f_{avg} = \frac{1}{(\ln 2)^2}\int_{\ln 2}^{2\ln 2}\int_{\ln 2}^{2\ln 2}\frac{1}{xy}\,dy\,dx = \frac{1}{(\ln 2)^2}\int_{\ln 2}^{2\ln 2}\frac{1}{x}\ln 2\,dx = 1$

55. $\displaystyle\iint\limits_{R} f(x)g(y)\,dxdy = \int_c^d\int_a^b f(x)g(y)\,dx\,dy = \int_c^d\left(\int_a^b f(x)g(y)\,dx\right)\,dy$

$$= \int_c^d g(y)\left(\int_a^b f(x)\,dx\right)\,dy = \left(\int_a^b f(x)\,dx\right)\left(\int_c^d g(y)\,dy\right)$$

57. Note that $\Omega = \{\,(x,y):\ 0 \le x \le y,\ \ 0 \le y \le 1\,\}.$

Set $\Omega' = \{\,(x,y):\ 0 \le y \le x,\ \ 0 \le x \le 1\,\}.$

$$\iint\limits_{\Omega} f(x)f(y)\,dxdy = \int_0^1\int_0^y f(x)f(y)\,dx\,dy$$

$$= \int_0^1\int_0^x f(y)f(x)\,dy\,dx$$

x and y are dummy variables

$$= \int_0^1\int_0^x f(x)f(y)\,dy\,dx = \iint\limits_{\Omega'} f(x)f(y)\,dxdy.$$

Note that Ω and Ω' don't overlap and their union is the unit square

$$R = \{\,(x,y):\ 0 \le x \le 1,\ \ 0 \le y \le 1\,\}.$$

If $\displaystyle\int_0^1 f(x)\,dx = 0,$ then

$$0 = \left(\int_0^1 f(x)\,dx\right)\left(\int_0^1 f(y)\,dy\right) = \iint\limits_{R} f(x)f(y)\,dxdy$$

by Exercise 55

$$= \iint\limits_{\Omega} f(x)f(y)\,dxdy + \iint\limits_{\Omega'} f(x)f(y)\,dxdy$$

$$= 2\iint\limits_{\Omega} f(x)f(y)\,dxdy$$

and therefore $\displaystyle\iint\limits_{\Omega} f(x)f(y)\,dxdy = 0.$

59. Let M be the maximum value of $|f(x, y)|$ on Ω.

$$\int_{\phi_1(x+h)}^{\phi_2(x+h)} = \int_{\phi_1(x+h)}^{\phi_1(x)} + \int_{\phi_1(x)}^{\phi_2(x)} + \int_{\phi_2(x)}^{\phi_2(x+h)}$$

$$|F(x+h) - F(x)| = \left| \int_{\phi_1(x+h)}^{\phi_2(x+h)} f(x,y)\, dy - \int_{\phi_1(x)}^{\phi_2(x)} f(x,y)\, dy \right|$$

$$= \left| \int_{\phi_1(x+h)}^{\phi_1(x)} f(x,y)\, dy + \int_{\phi_2(x)}^{\phi_2(x+h)} f(x,y)\, dy \right|$$

$$\leq \left| \int_{\phi_1(x+h)}^{\phi_1(x)} f(x,y)\, dy \right| + \left| \int_{\phi_2(x)}^{\phi_2(x+h)} f(x,y)\, dy \right|$$

$$\leq |\phi_1(x) - \phi_1(x+h)|\, M + |\phi_2(x+h) - \phi_2(x)|\, M.$$

The expression on the right tends to 0 as h tends to 0 since ϕ_1 and ϕ_2 are continuous.

61. (a) $\displaystyle \int_1^2 \int_{x^2-2x+2}^{1+\sqrt{x-1}} 1\, dy\, dx = \frac{1}{3}$ (b) $\displaystyle \int_1^2 \int_{y^2-2y+2}^{1+\sqrt{y-1}} 1\, dx\, dy = \frac{1}{3}$

SECTION 17.4

1. $\displaystyle \int_0^{\pi/2} \int_0^{\sin\theta} r \cos\theta\, dr\, d\theta = \int_0^{\pi/2} \frac{1}{2} \sin^2\theta \cos\theta\, d\theta = \left[\frac{1}{6} \sin^3\theta \right]_0^{\pi/2} = \frac{1}{6}$

3. $\displaystyle \int_0^{\pi/2} \int_0^{3\sin\theta} r^2\, dr\, d\theta = \int_0^{\pi/2} 9 \sin^3\theta\, d\theta = 9 \int_0^{\pi/2} (1 - \cos^2\theta) \sin\theta\, d\theta = 9 \left[-\cos\theta + \frac{1}{3} \cos^3\theta \right]_0^{\pi/2} = 6$

5. (a) $\Gamma: \ 0 \leq \theta \leq 2\pi, \ \ 0 \leq r \leq 1$

$$\iint_\Gamma (\cos r^2) r\, dr d\theta = \int_0^{2\pi} \int_0^1 (\cos r^2) r\, dr\, d\theta = 2\pi \int_0^1 r \cos r^2\, dr = \pi \sin 1$$

(b) $\Gamma: \ 0 \leq \theta \leq 2\pi, \ \ 1 \leq r \leq 2$

$$\iint_\Gamma (\cos r^2) r\, dr d\theta = \int_0^{2\pi} \int_1^2 (\cos r^2) r\, dr\, d\theta = 2\pi \int_1^2 r \cos r^2\, dr = \pi(\sin 4 - \sin 1)$$

7. (a) $\Gamma: \ 0 \leq \theta \leq \pi/2, \ \ 0 \leq r \leq 1$

$$\iint_\Gamma (r\cos\theta + r\sin\theta) r\, dr d\theta = \int_0^{\pi/2} \int_0^1 r^2 (\cos\theta + \sin\theta)\, dr\, d\theta$$

$$= \left(\int_0^{\pi/2} (\cos\theta + \sin\theta)\, d\theta \right) \left(\int_0^1 r^2\, dr \right) = 2 \left(\frac{1}{3} \right) = \frac{2}{3}$$

(b) $\Gamma: \ 0 \leq \theta \leq \pi/2, \ \ 1 \leq r \leq 2$

$$\iint_\Gamma (r\cos\theta + r\sin\theta) r\, dr d\theta = \int_0^{\pi/2} \int_1^2 r^2 (\cos\theta + \sin\theta)\, dr\, d\theta$$

$$= \left(\int_0^{\pi/2} (\cos\theta + \sin\theta)\, d\theta \right) \left(\int_1^2 r^2\, dr \right) = 2 \left(\frac{7}{3} \right) = \frac{14}{3}$$

9. $\displaystyle\int_{-\pi/2}^{\pi/2}\int_0^1 r^2\, dr\, d\theta = \frac{1}{3}\,\pi$

11. $\displaystyle\int_{1/2}^1\int_0^{\sqrt{1-x^2}} dy\, dx = \int_0^{\pi/3}\int_{\frac{1}{2}\sec\theta}^1 r\, dr\, d\theta = \int_0^{\pi/3}\left(\frac{1}{2}-\frac{1}{8}\sec^2\theta\right) d\theta = \frac{1}{6}\pi - \frac{\sqrt{3}}{8}$

13. $\displaystyle\int_0^1\int_0^{\sqrt{1-x^2}} \sin\sqrt{x^2+y^2}\, dy\, dx = \int_0^{\pi/2}\int_0^1 \sin(r)\, r\, dr\, d\theta = \int_0^{\pi/2}(\sin 1 - \cos 1)\, d\theta = \frac{\pi}{2}(\sin 1 - \cos 1)$

15. $\displaystyle\int_0^2\int_0^{\sqrt{2x-x^2}} x\, dy\, dx = \int_0^{\pi/2}\int_0^{2\cos\theta} r\cos\theta\, r\, dr\, d\theta = \frac{8}{3}\int_0^{\pi/2}\cos^4\theta\, d\theta = \frac{8}{3}\cdot\frac{3}{4}\cdot\frac{1}{2}\cdot\frac{\pi}{2} = \frac{\pi}{2}$

(See Exercise 62, Section 8.3)⤴

17. $\displaystyle A = \int_0^{\pi/3}\int_0^{3\sin 3\theta} r\, dr\, d\theta = \frac{9}{2}\int_0^{\pi/3}\sin^2 3\theta\, d\theta = \frac{9}{4}\int_0^{\pi/3}(1-6\cos\theta)\, d\theta = \frac{3\pi}{4}$

19. First we find the points of intersection:

$r = 4\cos\theta = 2 \quad\Longrightarrow\quad \cos\theta = \frac{1}{2}$

$\Longrightarrow\quad \theta = \pm\frac{\pi}{3}.$

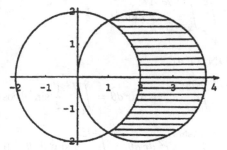

$\displaystyle A = \int_{-\pi/3}^{\pi/3}\int_2^{4\cos\theta} r\, dr\, d\theta = \int_{-\pi/3}^{\pi/3}(8\cos^2\theta - 2)\, d\theta = \int_{-\pi/3}^{\pi/3}(2 + 4\cos 2\theta)\, d\theta = \frac{4\pi}{3} + 2\sqrt{3}$

21. $\displaystyle A = 4\int_0^{\pi/4}\int_0^{2\sqrt{\cos 2\theta}} r\, dr\, d\theta = 8\int_0^{\pi/4}\cos 2\theta\, d\theta = 4$

23. $\displaystyle\int_0^{2\pi}\int_0^b \left(r^2\sin\theta + br\right) dr\, d\theta = \int_0^{2\pi}\left[\frac{1}{3}r^3\sin\theta + \frac{b}{2}r^2\right]_0^b d\theta$

$\displaystyle = b^3\int_0^{2\pi}\left(\frac{1}{3}\sin\theta + \frac{1}{2}\right) d\theta = b^3\pi$

25. $\displaystyle 8\int_0^{\pi/2}\int_0^2 \frac{r}{2}\sqrt{12-3r^2}\, dr\, d\theta = 8\int_0^{\pi/2}\left[-\frac{1}{18}\left(12-3r^2\right)^{3/2}\right]_0^2 d\theta$

$\displaystyle = 8\int_0^{\pi/2}\frac{4}{3}\sqrt{3}\, d\theta = \frac{16}{3}\sqrt{3}\,\pi$

27. $\displaystyle\int_0^{2\pi}\int_0^1 r\sqrt{4-r^2}\, dr\, d\theta = \int_0^{2\pi}\left[-\frac{1}{3}\left(4-r^2\right)^{3/2}\right]_0^1 d\theta$

$\displaystyle = \int_0^{2\pi}\left(\frac{8}{3}-\sqrt{3}\right) d\theta = \frac{2}{3}(8-3\sqrt{3}\,)\pi$

29. $\displaystyle\int_{-\pi/2}^{\pi/2}\int_{0}^{2\cos\theta} 2r^2\cos\theta\,dr\,d\theta = \int_{-\pi/2}^{\pi/2}\left[\frac{2}{3}r^3\cos\theta\right]_0^{2\cos\theta}d\theta$

$$= \int_{-\pi/2}^{\pi/2}\frac{16}{3}\cos^4\theta\,d\theta = \frac{32}{3}\int_0^{\pi/2}\cos^4\theta\,d\theta = \frac{32}{3}\left(\frac{3}{16}\pi\right) = 2\pi$$

Ex. 46, Sect. 8.3

31. $\displaystyle\frac{b}{a}\int_0^{\pi}\int_0^{a\sin\theta} r\sqrt{a^2-r^2}\,dr\,d\theta = \frac{b}{a}\int_0^{\pi}\left[-\frac{1}{3}\left(a^2-r^2\right)^{3/2}\right]_0^{a\sin\theta}d\theta$

$$= \frac{1}{3}a^2 b\int_0^{\pi}\left(1-\cos^3\theta\right)d\theta = \frac{1}{3}\pi a^2 b$$

33. $\displaystyle A = 2\int_0^{\pi/4}\int_0^{2\cos 2\theta} r\,dr\,d\theta = 2\int_0^{\pi/4}\frac{1}{2}\left(2\cos 2\theta\right)^2 d\theta = 2\int_0^{\pi/4}\left(1+\cos 4\theta\right)d\theta = \frac{\pi}{2}$

SECTION 17.5

1. $\displaystyle M = \int_{-1}^{1}\int_0^1 x^2\,dy\,dx = \frac{2}{3}$

$\displaystyle x_M M = \int_{-1}^{1}\int_0^1 x^3\,dy\,dx = 0 \quad\Longrightarrow\quad x_M = 0$

$\displaystyle y_M M = \int_{1}^{1}\int_0^1 x^2 y\,dy\,dx = \int_{-1}^{1}\frac{1}{2}x^2\,dx = \frac{1}{3} \quad\Longrightarrow\quad y_M = \frac{1/3}{2/3} = \frac{1}{2}$

3. $\displaystyle M = \int_0^1\int_{x^2}^1 xy\,dy\,dx = \frac{1}{2}\int_0^1\left(x-x^5\right)dx = \frac{1}{6}$

$\displaystyle x_M M = \int_0^1\int_{x^2}^1 x^2 y\,dy\,dx = \frac{1}{2}\int_0^1\left(x^2-x^6\right)dx = \frac{2}{21} \quad\Longrightarrow\quad x_M = \frac{2/21}{1/6} = \frac{4}{7}$

$\displaystyle y_M M = \int_0^1\int_{x^2}^1 xy^2\,dy\,dx = \frac{1}{3}\int_0^1\left(x-x^7\right)dx = \frac{1}{8} \quad\Longrightarrow\quad y_M = \frac{1/8}{1/6} = \frac{3}{4}$

5. $\displaystyle M = \int_0^8\int_0^{x^{1/3}} y^2\,dy\,dx = \frac{1}{3}\int_0^8 x\,dx = \frac{32}{3}$

$\displaystyle x_M M = \int_0^8\int_0^{x^{1/3}} xy^2\,dy\,dx = \frac{1}{3}\int_0^8 x^2\,dx = \frac{512}{9} \quad\Longrightarrow\quad x_M = \frac{512/9}{32/3} = \frac{16}{3}$

$\displaystyle y_M M = \int_0^8\int_0^{x^{1/3}} y^3\,dy\,dx = \frac{1}{4}\int_0^8 x^{4/3}\,dx = \frac{96}{7} \quad\Longrightarrow\quad y_M = \frac{96/7}{32/3} = \frac{9}{7}$

7. $\displaystyle M = \int_0^1\int_{2x}^{3x} xy\,dy\,dx = \frac{5}{2}\int_0^1 x^3\,dx = \frac{5}{8}$

$\displaystyle x_M M = \int_0^1\int_{2x}^{3x} x^2 y\,dy\,dx = \frac{5}{2}\int_0^1 x^4\,dx = \frac{1}{2} \Longrightarrow\quad x_M = \frac{1/2}{5/8} = \frac{4}{5}$

$\displaystyle y_M M = \int_0^1\int_{2x}^{3x} xy^2\,dy\,dx = \frac{19}{3}\int_0^1 x^4\,dx = \frac{19}{15} \quad\Longrightarrow\quad y_M = \frac{19/15}{5/8} = \frac{152}{75}$

9. $M = \displaystyle\int_0^{2\pi} \int_0^{1+\cos\theta} r^2\, dr\, d\theta = \frac{1}{3} \int_0^{2\pi} (1 + 3\cos\theta + 3\cos^2\theta + \cos^3\theta)\, d\theta = \frac{5\pi}{3}$

$x_M\, M = \displaystyle\int_0^{2\pi} \int_0^{1+\cos\theta} r^3 \cos\theta\, dr\, d\theta = \frac{1}{4} \int_0^{2\pi} (1+\cos\theta)^4 \cos\theta\, d\theta$

$\qquad\qquad = \dfrac{1}{4} \displaystyle\int_0^{2\pi} \left[\cos\theta + 4\cos^2\theta + 6\cos^3\theta + 4\cos^4\theta + \cos^5\theta\right] d\theta$

$\qquad\qquad = \dfrac{7\pi}{4}$

Therefore, $x_M = \dfrac{7\pi/4}{5\pi/3} = \dfrac{21}{20}$.

$y_M\, M = \displaystyle\int_0^{2\pi} \int_0^{1+\cos\theta} r^3 \sin\theta\, dr\, d\theta = \frac{1}{4}\int_0^{2\pi} (1+\cos\theta)^4 \sin\theta\, d\theta = \frac{1}{4}\left[\frac{1}{5}(1+\cos\theta)^5\right]_0^{2\pi} = 0$

Therefore, $y_M = 0$.

11. $\Omega: \ -L/2 \le x \le L/2, \quad -W/2 \le y \le W/2$

$I_x = \displaystyle\iint_\Omega \frac{M}{LW} y^2\, dxdy = \frac{4M}{LW}\int_0^{W/2}\int_0^{L/2} y^2\, dx\, dy = \frac{1}{12} MW^2$

$\qquad\qquad\quad \text{symmetry}\ \underset{\big\uparrow}{\rule{0pt}{0pt}}$

$I_y = \displaystyle\iint_\Omega \frac{M}{LW} x^2\, dxdy = \frac{1}{12}ML^2, \quad I_z = \iint_\Omega \frac{M}{LW}\left(x^2 + y^2\right) dxdy = \frac{1}{12} M\left(L^2 + W^2\right)$

$K_x = \sqrt{I_x/M} = \dfrac{W\sqrt{3}}{6}, \quad K_y = \sqrt{I_y/M} = \dfrac{L\sqrt{3}}{6}$

$K_z = \sqrt{I_z/M} = \dfrac{\sqrt{3}\,\sqrt{L^2+W^2}}{6}$

13. $M = \displaystyle\iint_\Omega k\left(x + \frac{L}{2}\right) dxdy = \iint_\Omega \frac{1}{2} kL\, dxdy = \frac{1}{2} kL(\text{ area of } \Omega) = \frac{1}{2} kL^2 W$

$\qquad\qquad\qquad \text{symmetry}\ \underset{\big\uparrow}{\rule{0pt}{0pt}}$

$x_M\, M = \displaystyle\iint_\Omega x\left[k\left(x + \frac{L}{2}\right)\right] dxdy = \iint_\Omega \left(kx^2 + \frac{1}{2}Lx\right) dxdy$

$\qquad = \displaystyle\iint_\Omega kx^2\, dxdy = 4k\int_0^{W/2}\int_0^{L/2} x^2\, dx\, dy = \frac{1}{12} kWL^3$

$\qquad\quad \text{symmetry}\ \underset{\big\uparrow}{\rule{0pt}{0pt}} \qquad\qquad \text{symmetry}\ \underset{\big\uparrow}{\rule{0pt}{0pt}}$

$\qquad = \frac{1}{6}\left(\frac{1}{2}kL^2 W\right) L = \frac{1}{6} ML; \quad x_M = \frac{1}{6} L$

$y_M\, M = \displaystyle\iint_\Omega y\left[k\left(x + \frac{L}{2}\right)\right] dxdy = 0; \quad y_M = 0$

$\qquad\qquad \text{by symmetry}\ \underset{\big\uparrow}{\rule{0pt}{0pt}}$

15. $I_x = \iint\limits_{\Omega} \dfrac{4M}{\pi R^2} y^2 \, dx dy = \dfrac{4M}{\pi R^2} \displaystyle\int_0^{\pi/2} \int_0^R r^3 \sin^2 \theta \, dr \, d\theta$

$\qquad = \dfrac{4M}{\pi R^2} \left(\displaystyle\int_0^{\pi/2} \sin^2 \theta \, d\theta \right) \left(\displaystyle\int_0^R r^3 \, dr \right) = \dfrac{4M}{\pi R^2} \left(\dfrac{\pi}{4} \right) \left(\dfrac{1}{4} R^4 \right) = \dfrac{1}{4} MR^2$

$\qquad I_y = \tfrac{1}{4} MR^2, \quad I_z = \tfrac{1}{2} MR^2$

$\qquad K_x = K_y = \tfrac{1}{2} R, \quad K_z = R/\sqrt{2}$

17. I_M, the moment of inertia about the vertical line through the center of mass, is

$$\iint\limits_{\Omega} \dfrac{M}{\pi R^2} \left(x^2 + y^2 \right) dx dy$$

where Ω is the disc of radius R centered at the origin. Therefore

$$I_M = \dfrac{M}{\pi R^2} \int_0^{2\pi} \int_0^R r^3 \, dr \, d\theta = \dfrac{1}{2} MR^2.$$

We need $I_0 = \tfrac{1}{2} MR^2 + d^2 M$ where d is the distance from the center of the disc to the origin. Solving this equation for d, we have $d = \sqrt{I_0 - \tfrac{1}{2} MR^2} \Big/ \sqrt{M}$.

19. $\Omega : \ 0 \le x \le a, \quad 0 \le y \le b$

$\qquad I_x = \iint\limits_{\Omega} \dfrac{4M}{\pi ab} y^2 \, dx dy = \dfrac{4M}{\pi ab} \displaystyle\int_0^a \int_0^{\frac{b}{a}\sqrt{a^2 - x^2}} y^2 \, dy \, dx = \dfrac{1}{4} Mb^2$

$\qquad I_y = \iint\limits_{\Omega} \dfrac{4M}{\pi ab} x^2 \, dx dy = \dfrac{4M}{\pi ab} \displaystyle\int_0^a \int_0^{\frac{b}{a}\sqrt{a^2 - x^2}} x^2 \, dy \, dx = \dfrac{1}{4} Ma^2$

$\qquad I_z = \tfrac{1}{4} M \left(a^2 + b^2 \right)$

21. $I_x = \displaystyle\int_0^1 \int_{x^2}^1 xy^3 \, dy \, dx = \dfrac{1}{4} \int_0^1 (x - x^9) \, dx = \dfrac{1}{10}$

$\qquad I_y = \displaystyle\int_0^1 \int_{x^2}^1 x^3 y \, dy \, dx = \dfrac{1}{2} \int_0^1 (x^3 - x^7) \, dx = \dfrac{1}{16}$

$\qquad I_z = \displaystyle\int_0^1 \int_{x^2}^1 xy(x^2 + y^2) \, dy \, dx = I_x + I_y = \dfrac{13}{80}$

23. $I_x = \displaystyle\int_0^{2\pi} \int_0^{1+\cos\theta} r^4 \sin^2 \theta \, dr \, d\theta = \dfrac{1}{5} \int_0^{2\pi} (1 + \cos\theta)^5 \sin^2 \theta \, d\theta = \dfrac{33\pi}{40}$

$\qquad I_y = \displaystyle\int_0^{2\pi} \int_0^{1+\cos\theta} r^4 \cos^2 \theta \, dr \, d\theta = \dfrac{1}{5} \int_0^{2\pi} (1 + \cos\theta)^5 \cos^2 \theta \, d\theta = \dfrac{93\pi}{40}$

$\qquad I_z = \displaystyle\int_0^{2\pi} \int_0^{1+\cos\theta} r^4 \, dr \, d\theta = I_x + I_y = \dfrac{63\pi}{20}$

25. $\Omega: \ r_1^2 \leq x^2 + y^2 \leq r_2^2, \quad A = \pi\left(r_2^2 - r_1^2\right)$

(a) Place the diameter on the x-axis.

$$I_x = \iint_\Omega \frac{M}{A} y^2 \, dxdy = \frac{M}{A} \int_0^{2\pi} \int_{r_1}^{r_2} \left(r^2 \sin^2\theta\right) r \, dr \, d\theta = \frac{1}{4} M \left(r_2^2 + r_1^2\right)$$

(b) $\frac{1}{4} M \left(r_2^2 + r_1^2\right) + Mr_1^2 = \frac{1}{4} M \left(r_2^2 + 5r_1^2\right)$ (parallel axis theorem)

(c) $\frac{1}{4} M \left(r_2^2 + r_1^2\right) + Mr_2^2 = \frac{1}{4} M \left(5r_2^2 + r_1^2\right)$

27. $\Omega: \ r_1^2 \leq x^2 + y^2 \leq r_2^2, \quad A = \pi\left(r_2^2 - r_1^2\right)$

$$I = \iint_\Omega \frac{M}{A} \left(x^2 + y^2\right) dxdy = \frac{M}{A} \int_0^{2\pi} \int_{r_1}^{r_2} r^3 \, dr \, d\theta = \frac{1}{2} M (r_2^2 + r_1^2)$$

29. $M = \iint_\Omega k\left(R - \sqrt{x^2 + y^2}\right) dxdy = k \int_0^\pi \int_0^R \left(Rr - r^2\right) dr \, d\theta = \frac{1}{6} k\pi R^3$

$x_M = 0$ by symmetry

$$y_M M = \iint_\Omega y \left[k\left(R - \sqrt{x^2 + y^2}\right)\right] dxdy = k \int_0^\pi \int_0^R \left(Rr^2 - r^3\right) \sin\theta \, dr \, d\theta = \frac{1}{6} k R^4$$

$y_M = R/\pi$

31. Place P at the origin.

$$M = \iint_\Omega k\sqrt{x^2 + y^2} \, dxdy$$

$$= k \int_0^\pi \int_0^{2R\sin\theta} r^2 \, dr \, d\theta = \frac{32}{9} k R^3$$

$x_M = 0$ by symmetry

$$y_M M = \iint_\Omega y \left(k\sqrt{x^2 + y^2}\right) dxdy = k \int_0^\pi \int_0^{2R\sin\theta} r^3 \sin\theta \, dr \, d\theta = \frac{64}{15} k R^4$$

$y_M = 6R/5$

Answer: the center of mass lies on the diameter through P at a distance $6R/5$ from P.

33. Suppose Ω, a basic region of area A, is broken up into n basic regions $\Omega_1, \cdots, \Omega_n$ with areas $A_1, \cdots, A_n$. Then

$$\overline{x}A = \iint_\Omega x \, dxdy = \sum_{i=1}^n \left(\iint_{\Omega_i} x \, dxdy\right) = \sum_{i=1}^n \overline{x}_i A_i = \overline{x}_1 A_1 + \cdots + \overline{x}_n A_n.$$

The second formula can be derived in a similar manner.

SECTION 17.6

1. They are equal; they both give the volume of T.

3. $\displaystyle\iiint_\Pi \alpha\,dx\,dy\,dz = \alpha \iiint_\Pi dx\,dy\,dz = \alpha\,(\text{volume of }\Pi) = \alpha(a_2-a_1)(b_2-b_1)(c_2-c_1)$

5. Let $P_1 = \{x_0,\cdots,x_m\}$, $\;P_2=\{y_0,\cdots,y_n\}$, $\;P_3=\{z_0,\cdots,z_q\}$ be partitions of $[0,a]$, $[0,b]$, $[0,c]$ respectively and let $P = P_1 \times P_2 \times P_3$. Note that

$$x_{i-1}y_{j-1} \le \left(\frac{x_i+x_{i-1}}{2}\right)\left(\frac{y_j+y_{j-1}}{2}\right) \le x_i y_j$$

and therefore

$$x_{i-1}y_{j-1}\,\Delta x_i\,\Delta y_j\,\Delta z_k \le \tfrac14\left(x_i{}^2-x_{i-1}^2\right)\left(y_j{}^2-y_{j-1}^2\right)\Delta z_k \le x_i y_j\,\Delta x_i\,\Delta y_j \Delta z_k.$$

It follows that

$$L_f(P) \le \frac14 \sum_{i=1}^m\sum_{j=1}^n\sum_{k=1}^q \left(x_i{}^2-x_{i-1}^2\right)\left(y_j{}^2-y_{j-1}^2\right)\Delta z_k \le U_f(P).$$

The middle term can be written

$$\frac14\left(\sum_{i=1}^m x_i{}^2 - x_{i-1}^2\right)\left(\sum_{j=1}^n y_j{}^2 - y_{j-1}^2\right)\left(\sum_{k=1}^q \Delta z_k\right) = \frac14\,a^2 b^2 c.$$

7. $\overline{x}_1 = a,\;\;\overline{y}_1=b,\;\;\overline{z}_1=c;\;\;\;\;\overline{x}_0=A,\;\;\overline{y}_0=B,\;\;\overline{z}_0=C$

$$\overline{x}_1 V_1 + \overline{x}V = \overline{x}_0 V_0 \;\;\Longrightarrow\;\; a^2bc + (ABC-abc)\overline{x} = A^2BC$$

$$\Longrightarrow\;\; \overline{x} = \frac{A^2BC - a^2bc}{ABC-abc}$$

similarly

$$\overline{y} = \frac{AB^2C - ab^2c}{ABC-abc}, \;\;\;\; \overline{z} = \frac{ABC^2 - abc^2}{ABC-abc}$$

9. $\displaystyle M = \iiint_\Pi Kz\,dxdydz$

Let $P_1 = \{x_0,\cdots,x_m\}$, $\;P_2=\{y_0,\cdots,y_n\}$, $\;P_3=\{z_0,\cdots,z_q\}$ be partitions of $[0,a]$ and let $P = P_1 \times P_2 \times P_3$. Note that

$$z_{k-1} \le \tfrac12\left(z_k+z_{k-1}\right) \le z_k$$

and therefore

$$Kz_{k-1}\,\Delta x_i\,\Delta y_j\,\Delta z_k \le \tfrac12 K\,\Delta x_i\,\Delta y_j\left(z_k{}^2 - z_{k-1}^2\right) \le Kz_k\,\Delta x_i\,\Delta y_j \Delta z_k.$$

It follows that

$$L_f(P) \le \frac12 K \sum_{i=1}^m\sum_{j=1}^n\sum_{k=1}^q \Delta x_i\,\Delta y_j\left(z_k{}^2 - z_{k-1}^2\right) \le U_f(P).$$

The middle term can be written

$$\frac{1}{2} K \left(\sum_{i=1}^{m} \Delta x_i \right) \left(\sum_{j=1}^{n} \Delta y_j \right) \left(\sum_{k=1}^{q} z_k^2 - z_{k-1}^2 \right) = \frac{1}{2} K (a) (a) (a^2) = \frac{1}{2} K a^4.$$

$M = \frac{1}{2} K a^4$ where K is the constant of proportionality for the density function.

11.
$$I_z = \iiint_\Pi Kz \left(x^2 + y^2 \right) dx dy dz$$

$$= \underbrace{\iiint_\Pi Kx^2 z \, dx dy dz}_{I_1} + \underbrace{\iiint_\Pi Ky^2 z \, dx dy dz}_{I_2} \, .$$

We will calculate I_1 using the partitions we used in doing Exercise 9.   Note that

$$x_{i-1}^2 z_{k-1} \le \left(\frac{x_i^2 + x_i x_{i-1} + x_{i-1}^2}{3} \right) \left(\frac{z_k + z_{k-1}}{2} \right) \le x_i^2 z_k$$

and therefore

$$K x_{i-1}^2 z_{k-1} \, \Delta x_i \, \Delta y_j \, \Delta z_k \le \frac{1}{6} K \left(x_i^3 - x_{i-1}^3 \right) \Delta y_j \left(z_k^2 - z_{k-1}^2 \right) \le K x_i^2 z_k^2 \, \Delta x_i \, \Delta y_j \, \Delta z_k.$$

It follows that

$$L_f(P) \le \frac{1}{6} K \sum_{i=1}^{m} \sum_{j=1}^{n} \sum_{k=1}^{q} \left(x_i^3 - x_{i-1}^3 \right) \Delta y_j \left(z_k^2 - z_{k-1}^2 \right) \le U_f(P).$$

The middle term can be written

$$\frac{1}{6} K \left(\sum_{i=1}^{m} x_i^3 - x_{i-1}^3 \right) \left(\sum_{j=1}^{n} \Delta y_j \right) \left(\sum_{k=1}^{q} z_k^2 - z_{k-1}^2 \right) = \frac{1}{6} K a^3 (a)(a^2) = \frac{1}{6} K a^6.$$

Similarly $I_2 = \frac{1}{6} K a^6$ and therefore $I_z = \frac{1}{3} K a^6 = \frac{2}{3} \left(\frac{1}{2} K a^4 \right) a^2 = \frac{2}{3} M a^2.$

by Exercise 9⟍⟶

SECTION 17.7

1. $\displaystyle\int_0^a \int_0^b \int_0^c dx \, dy \, dz = \int_0^a \int_0^b c \, dy \, dz = \int_0^a bc \, dz = abc$

3. $\displaystyle\int_0^1 \int_1^{2y} \int_0^x (x + 2z) \, dz \, dx \, dy = \int_0^1 \int_1^{2y} \left[xz + z^2 \right]_0^x dx \, dy = \int_0^1 \int_1^{2y} 2x^2 \, dx \, dy$

$$= \int_0^1 \left[\frac{2}{3} x^3 \right]_1^{2y} dy = \int_0^1 \left(\frac{16}{3} y^3 - \frac{2}{3} \right) dy = \frac{2}{3}$$

5. $\displaystyle\int_0^2 \int_{-1}^1 \int_0^3 (z - xy) \, dz \, dy \, dx = \int_0^2 \int_{-1}^1 \left[\frac{1}{2} z^2 - xyz \right]_1^3 dy \, dx$

$$= \int_0^2 \int_{-1}^1 (4 - 2xy) \, dy \, dx = \int_0^2 \left[2y - xy^2 \right]_{-1}^1 dx = \int_0^2 8 \, dy = 16$$

7. $\displaystyle\int_0^{\pi/2}\int_0^1\int_0^{\sqrt{1-x^2}} x\cos z\,dy\,dx\,dz = \int_0^{\pi/2}\int_0^1 \left[xy\cos z\right]_0^{\sqrt{1-x^2}}\,dx\,dz$

$\displaystyle = \int_0^{\pi/2}\int_0^1 x\sqrt{1-x^2}\cos z\,dx\,dz = \int_0^{\pi/2}\left[-\frac{1}{3}(1-x^2)^{3/2}\cos z\right]_0^1\,dz = \frac{1}{3}\int_0^{\pi/2}\cos z\,dz = \frac{1}{3}$

9. $\displaystyle\int_1^2\int_y^{y^2}\int_0^{\ln x} ye^z\,dz\,dx\,dy = \int_1^2\int_y^{y^2}\left[ye^z\right]_0^{\ln x}\,dx\,dy$

$\displaystyle = \int_1^2\int_y^{y^2} y(x-1)\,dx\,dy = \int_1^2\left[\frac{1}{2}x^2 y - xy\right]_y^{y^2}\,dy = \int_1^2\left(\frac{1}{2}y^5 - \frac{3}{2}y^3 + y^2\right)dy = \frac{47}{24}$

11. $\displaystyle\iiint_\Pi f(x)g(y)h(z)\,dxdydz = \int_{c_1}^{c_2}\left[\int_{b_1}^{b_2}\left(\int_{a_1}^{a_2} f(x)g(y)h(z)\,dx\right)dy\right]dz$

$\displaystyle = \int_{c_1}^{c_2}\left[\int_{b_1}^{b_2} g(y)h(z)\left(\int_{a_1}^{a_2} f(x)\,dx\right)dy\right]dz$

$\displaystyle = \int_{c_1}^{c_2}\left[h(z)\left(\int_{a_1}^{a_2} f(x)\,dx\right)\left(\int_{b_1}^{b_2} g(y)\,dy\right)dz\right]$

$\displaystyle = \left(\int_{a_1}^{a_2} f(x)\,dx\right)\left(\int_{b_1}^{b_2} g(y)\,dy\right)\left(\int_{c_1}^{c_2} h(z)\,dz\right)$

13. $\displaystyle\left(\int_0^1 x^2\,dx\right)\left(\int_0^2 y^2\,dy\right)\left(\int_0^3 z^2\,dz\right) = \left(\frac{1}{3}\right)\left(\frac{8}{3}\right)\left(\frac{27}{3}\right) = 8$

15. $\displaystyle x_M M = \iiint_\Pi kx^2yz\,dxdydz = k\left(\int_0^a x^2\,dx\right)\left(\int_0^b y\,dy\right)\left(\int_0^c z\,dz\right)$

$\displaystyle = k\left(\tfrac{1}{3}a^3\right)\left(\tfrac{1}{2}b^2\right)\left(\tfrac{1}{2}c^2\right) = \tfrac{1}{12}ka^3 b^2 c^2.$

By Exercise 14, $M = \frac{1}{8}ka^2 b^2 c^2$. Therefore $\overline{x} = \frac{2}{3}a$. Similarly, $\overline{y} = \frac{2}{3}b$ and $\overline{z} = \frac{2}{3}c$.

17.

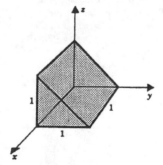

19. center of mass is the centroid

$$\bar{x} = \tfrac{1}{2} \quad \text{by symmetry}$$

$$\bar{y}V = \iiint\limits_{T} y\,dx\,dy\,dz = \int_0^1 \int_0^1 \int_0^{1-y} y\,dz\,dy\,dx = \int_0^1 \int_0^1 \left(y - y^2\right) dy\,dx$$

$$= \int_0^1 \left[\frac{1}{2}y^2 - \frac{1}{3}y^3\right]_0^1 dx = \int_0^1 \frac{1}{6}\,dx = \frac{1}{6}$$

$$\bar{z}V = \iiint\limits_{T} z\,dx\,dy\,dz = \int_0^1 \int_0^1 \int_0^{1-y} z\,dz\,dy\,dx = \int_0^1 \int_0^1 \frac{1}{2}(1-y)^2\,dy\,dx$$

$$= \frac{1}{2}\int_0^1 \int_0^1 \left(1 - 2y + y^2\right) dy\,dx = \frac{1}{2}\int_0^1 \left[y - y^2\frac{1}{3}y^3\right]_0^1 dx = \frac{1}{2}\int_0^1 \frac{1}{3}\,dx = \frac{1}{6}$$

$$V = \tfrac{1}{2} \text{ (by Exercise 18)}; \quad \bar{y} = \tfrac{1}{3}, \quad \bar{z} = \tfrac{1}{3}$$

21. $\displaystyle\int_{-r}^{r} \int_{-\phi(x)}^{\phi(x)} \int_{-\psi(x,y)}^{\psi(x,y)} k\left(r - \sqrt{x^2 + y^2 + z^2}\right) dz\,dy\,dx \quad \text{with} \quad \phi(x) = \sqrt{r^2 - x^2},$

$\psi(x,y) = \sqrt{r^2 - (x^2 + y^2)}, \quad k$ the constant of proportionality

23. $\displaystyle\int_0^1 \int_{-\sqrt{x-x^2}}^{\sqrt{x-x^2}} \int_{-2x-3y-10}^{1-y^2} dz\,dy\,dx$

25. $\displaystyle\int_{-1}^1 \int_{-2\sqrt{2-2x^2}}^{2\sqrt{2-2x^2}} \int_{3x^2+y^2/4}^{4-x^2-y^2/4} k\left(z - 3x^2 - \frac{1}{4}y^2\right) dz\,dy\,dx$

27. $\displaystyle\iiint\limits_{T} \left(x^2 z + y\right) dx\,dy\,dz = \int_0^2 \int_1^3 \int_0^1 \left(x^2 z + y\right) dx\,dy\,dz = \int_0^2 \int_1^3 \left[\frac{1}{3}x^3 z + xy\right]_0^1 dy\,dz$

$$= \int_0^2 \int_1^3 \left(\frac{1}{3}z + y\right) dy\,dz = \int_0^2 \left[\frac{1}{3}zy + \frac{1}{2}y^2\right]_1^3 dz = \int_0^2 \left(\frac{2}{3}z + 4\right) dz = \frac{28}{3}$$

29. $\displaystyle\iiint\limits_{T} x^2 y^2 z^2\,dx\,dy\,dz = \int_{-1}^0 \int_0^{y+1} \int_0^1 x^2 y^2 z^2\,dx\,dz\,dy + \int_0^1 \int_0^{1-y} \int_0^1 x^2 y^2 z^2\,dx\,dz\,dy$

$$= \int_{-1}^0 \int_0^{y+1} \left[\frac{1}{3}x^3 y^2 z^2\right]_0^1 dz\,dy + \int_0^1 \int_0^{1-y} \left[\frac{1}{3}x^3 y^2 z^2\right]_0^1 dz\,dy$$

$$= \frac{1}{3}\int_{-1}^0 \int_0^{y+1} y^2 z^2\,dz\,dy + \frac{1}{3}\int_0^1 \int_0^{1-y} \left[y^2 z^2\right]_0^1 dz\,dy$$

$$= \frac{1}{3}\int_{-1}^0 \left[\frac{1}{3}y^2 z^3\right]_0^{y+1} dy + \frac{1}{3}\int_0^1 \left[\frac{1}{3}y^2 z^3\right]_0^{1-y} dy$$

$$= \frac{1}{9}\int_{-1}^0 \left(y^5 + 3y^4 + 3y^3 + y^2\right) dy + \frac{1}{9}\int_0^1 \left(y^2 - 3y^3 + 3y^4 - y^5\right) dy = \frac{1}{270}$$

31. $\displaystyle\iiint\limits_{T} y^2 \, dx \, dy \, dz = \int_0^3 \int_0^{2-2x/3} \int_0^{6-2x-3y} y^2 \, dz \, dy \, dx = \int_0^3 \int_0^{2-2x/3} \left[y^2 z \right]_0^{6-2x-3y} dy \, dx$

$\displaystyle = \int_0^3 \int_0^{2-2x/3} \left(6y^2 - 2xy^2 - 3y^3 \right) dy \, dx$

$\displaystyle = \int_0^3 \left[2y^3 - \frac{2}{3}\,xy^3 - \frac{3}{4}\,y^4 \right]_0^{2-2x/3} dx$

$\displaystyle = \frac{1}{4} \int_0^3 \left(2 - \frac{2}{3}\,x \right) dx = \frac{12}{5}$

33. $\displaystyle V = \int_0^2 \int_{x^2}^{x+2} \int_0^x dz \, dy \, dx = \int_0^2 \int_{x^2}^{x+2} x \, dy \, dx = \int_0^2 \left(x^2 + 2x - x^3 \right) dx = \frac{8}{3}$

$\displaystyle \overline{x}V = \int_0^2 \int_{x^2}^{x+2} \int_0^x x \, dz \, dy \, dx = \int_0^2 \int_{x^2}^{x+2} x^2 \, dy \, dx = \int_0^2 \left(x^3 + 2x^2 - x^4 \right) dx = \frac{44}{15}$

$\displaystyle \overline{y}V = \int_0^2 \int_{x^2}^{x+2} \int_0^x y \, dz \, dy \, dx = \int_0^2 \int_{x^2}^{x+2} xy \, dy \, dx = \int_0^2 \frac{1}{2} \left(x^3 + 4x^2 + 4x - x^5 \right) dx = 6$

$\displaystyle \overline{z}V = \int_0^2 \int_{x^2}^{x+2} \int_0^x z \, dz \, dy \, dx = \int_0^2 \int_{x^2}^{x+2} \frac{1}{2} x^2 \, dy \, dx = \int_0^2 \frac{1}{2} \left(x^3 + 2x^2 - x^4 \right) dx = \frac{22}{15}$

$\displaystyle \overline{x} = \frac{11}{10}, \quad \overline{y} = \frac{9}{4}, \quad \overline{z} = \frac{11}{20}$

35. $\displaystyle V = \int_{-1}^2 \int_0^3 \int_{2-x}^{4-x^2} dz \, dy \, dx = \frac{27}{2}; \quad (\overline{x}, \overline{y}, \overline{z}) = \left(\frac{1}{2}, \frac{3}{2}, \frac{12}{5} \right)$

37. $\displaystyle V = \int_0^a \int_0^{\phi(x)} \int_0^{\psi(x,y)} dz \, dy \, dx = \frac{1}{6}\, abc \;\; \text{with} \;\; \phi(x) = b\left(1 - \frac{x}{a} \right), \quad \psi(x,y) = c\left(1 - \frac{x}{a} - \frac{y}{b} \right)$

$\displaystyle (\overline{x}, \overline{y}, \overline{z}) = \left(\tfrac{1}{4}\,a, \tfrac{1}{4}\,b, \tfrac{1}{4}\,c \right)$

39. $\Pi : 0 \leq x \leq a, \quad 0 \leq y \leq b, \quad 0 \leq z \leq c$

   (a) $\displaystyle I_z = \int_0^a \int_0^b \int_0^c \frac{M}{abc} \left(x^2 + y^2 \right) dz \, dy \, dx = \frac{1}{3}\, M \left(a^2 + b^2 \right)$

   (b) $\displaystyle I_M = I_z - d^2 M = \tfrac{1}{3} M \left(a^2 + b^2 \right) - \tfrac{1}{4} \left(a^2 + b^2 \right) M = \tfrac{1}{12} M \left(a^2 + b^2 \right)$

        $\uparrow$ parallel axis theorem (17.5.7)

   (c) $\displaystyle I = I_M + d^2 M = \tfrac{1}{12} M \left(a^2 + b^2 \right) + \tfrac{1}{4}\, a^2 M = \tfrac{1}{3}\, Ma^2 + \tfrac{1}{12}\, Mb^2$

        $\uparrow$ parallel axis theorem (17.5.7)

41. $M = \int_0^1 \int_0^1 \int_0^y k\left(x^2 + y^2 + z^2\right) dz\, dy\, dx = \int_0^1 \int_0^1 k\left(x^2 y + y^3 + \frac{1}{3}\, y^3\right) dy\, dx$

$$= \int_0^1 k\left(\frac{1}{2}\, x^2 + \frac{1}{3}\right) dx = \frac{1}{2}\, k$$

$(x_M, y_M, z_M) = \left(\frac{7}{12}, \frac{34}{45}, \frac{37}{90}\right)$

43. (a) 0 by symmetry

(b) $\iiint\limits_T (a_1 x + a_2 y + a_3 z + a_4)\, dx\, dy\, dz = \iiint\limits_T a_4\, dx\, dy\, dz = a_4 \text{ (volume of ball) } = \frac{4}{3}\, \pi a_4$

by symmetry $\nearrow$

45. $V = 8 \int_0^a \int_0^{\sqrt{a^2-x^2}} \int_0^{\sqrt{a^2-x^2-y^2}} dz\, dy\, dx = 8 \int_0^a \int_0^{\sqrt{a^2-x^2}} \sqrt{a^2 - x^2 - y^2}\, dy\, dx$

polar coordinates $\nearrow$

$$= 8 \int_0^{\pi/2} \int_0^a \sqrt{a^2 - r^2}\, r\, dr\, d\theta$$

$$= -4 \int_0^{\pi/2} \left[\frac{2}{3}(a^2 - r^2)^{3/2}\right]_0^a d\theta$$

$$= \frac{8}{3} \int_0^{\pi/2} d\theta = \frac{4}{3}\, \pi\, a^3$$

47. $M = \int_{-2}^2 \int_{-\sqrt{4-x^2}/2}^{\sqrt{4-x^2}/2} \int_{x^2+3y^2}^{4-y^2} k|x|\, dz\, dy\, dx = 4 \int_0^2 \int_0^{\sqrt{4-x^2}/2} \int_{x^2+3y^2}^{4-y^2} kx\, dz\, dy\, dx$

$$= 4k \int_0^2 \int_0^{\sqrt{4-x^2}/2} \left(4x - x^3 - 4xy^2\right) dy\, dx = \frac{4}{3}\, k \int_0^2 x\left(4 - x^2\right)^{3/2} dx = \frac{128}{15}\, k$$

49. $M = \int_{-1}^2 \int_0^3 \int_{2-x}^{4-x^2} k(1+y)\, dz\, dy\, dx = \frac{135}{4}\, k; \quad (x_M, y_M, z_M) = \left(\frac{1}{2}, \frac{9}{5}, \frac{12}{5}\right)$

51. (a) $V = \int_0^6 \int_{z/2}^3 \int_x^{6-x} dy\, dx\, dz$

(b) $V = \int_0^3 \int_0^{2x} \int_x^{6-x} dy\, dz\, dx$

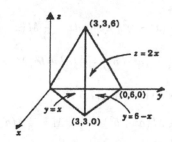

(c) $V = \int_0^6 \int_{z/2}^3 \int_{z/2}^y dx\, dy\, dz + \int_0^6 \int_3^{(12-z)/2} \int_{z/2}^{6-y} dx\, dy\, dz$

53.   (a) $V = \displaystyle\iint\limits_{\Omega_{xy}} 2y \, dy dz$ (b) $V = \displaystyle\iint\limits_{\Omega_{xy}} \left(\int_{-y}^{y} dx \right) dy dz$

      (c) $V = \displaystyle\int_{0}^{4} \int_{-\sqrt{4-y}}^{\sqrt{4-y}} \int_{-y}^{y} dx \, dz \, dy$ (d) $V = \displaystyle\int_{-2}^{2} \int_{0}^{4-z^2} \int_{-y}^{y} dx \, dy \, dz$

55.   (a) $\displaystyle\int_{2}^{4} \int_{3}^{5} \int_{1}^{2} \frac{\ln xy}{z} \, dz \, dy \, dx \cong 6.80703$ (b) $\displaystyle\int_{0}^{4} \int_{1}^{2} \int_{0}^{3} x\sqrt{yz} \, dz \, dy \, dx = \frac{16\sqrt{3}}{3} \left(4\sqrt{2} - 2 \right)$

SECTION 17.8

1. $r^2 + z^2 = 9$ **3.** $z = 2r$ **5.** $4r^2 = z^2$

7. $\displaystyle\int_{0}^{\pi/2} \int_{0}^{2} \int_{0}^{4-r^2} r \, dz \, dr \, d\theta$

     $= \displaystyle\int_{0}^{\pi/2} \int_{0}^{2} \left(4r - r^2 \right) dr \, d\theta$

     $= \displaystyle\int_{0}^{\pi/2} 4 \, d\theta = 2\pi$

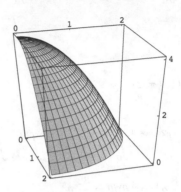

9. $\displaystyle\int_{0}^{2\pi} \int_{0}^{2} \int_{0}^{r^2} r \, dz \, dr \, d\theta$

     $= \displaystyle\int_{0}^{2\pi} \int_{0}^{2} r^3 \, dr \, d\theta$

     $= \displaystyle\int_{0}^{2\pi} 4 \, d\theta = 8\pi$

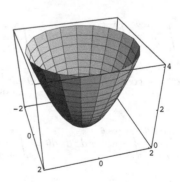

11. $\displaystyle\int_{0}^{1} \int_{0}^{\sqrt{1-x^2}} \int_{0}^{\sqrt{4-(x^2+y^2)}} dz \, dy \, dx = \int_{0}^{\pi/2} \int_{0}^{1} \int_{0}^{\sqrt{4-r^2}} r \, dz \, dr \, d\theta$

                 $= \displaystyle\int_{0}^{\pi/2} \int_{0}^{1} r\sqrt{4 - r^2} \, dr \, d\theta$

                 $= \displaystyle\int_{0}^{\pi/2} \left(\frac{8}{3} - \sqrt{3} \right) d\theta = \frac{1}{6} \left(8 - 3\sqrt{3} \right) \pi$

13. $\displaystyle\int_0^3 \int_0^{\sqrt{9-y^2}} \int_0^{\sqrt{9-x^2-y^2}} \frac{1}{\sqrt{x^2+y^2}}\, dz\, dx\, dy = \int_0^{\pi/2} \int_0^3 \int_0^{\sqrt{9-r^2}} \frac{1}{r}\cdot r\, dz\, dr\, d\theta$

$$= \int_0^{\pi/2} \int_0^3 \sqrt{9-r^2}\, dr\, d\theta$$

$$= \int_0^{\pi/2} \left[\frac{r}{2}\sqrt{9-r^2} + \frac{9}{2}\sin^{-1}\frac{r}{3}\right]_0^3 d\theta$$

$$= \frac{9\pi}{4}\int_0^{\pi/2} d\theta = \frac{9}{8}\pi^2$$

15. $\displaystyle\int_0^1 \int_0^{\sqrt{1-x^2}} \int_0^2 \sin(x^2+y^2)\, dz\, dy\, dx = \int_0^{\pi/2} \int_0^1 \int_0^2 \sin(r^2)r\, dz\, dr\, d\theta$

$$= \int_0^{\pi/2} \int_0^1 2r\sin(r^2)\, dr\, d\theta = \tfrac{1}{2}\pi(1-\cos 1) \cong 0.7221$$

17. $(0,1,2) \to (1,\tfrac{1}{2}\pi,2)$ **19.** $(0,-1,2) \to (1,\tfrac{3}{2}\pi,2)$

21. $\displaystyle V = \int_{-\pi/2}^{\pi/2} \int_0^{2a\cos\theta} \int_0^r r\, dz\, dr\, d\theta = \int_{-\pi/2}^{\pi/2} \int_0^{2a\cos\theta} r^2\, dr\, d\theta$

$$= \int_{-\pi/2}^{\pi/2} \frac{8}{3}a^3\cos^3\theta\, d\theta = \frac{32}{9}a^3$$

23. $\displaystyle V = \int_{-\pi/2}^{\pi/2} \int_0^{a\cos\theta} \int_0^{a-r} r\, dz\, dr\, d\theta = \int_{-\pi/2}^{\pi/2} \int_0^{a\cos\theta} r(a-r)\, dr\, d\theta$

$$= \int_{-\pi/2}^{\pi/2} a^3\left(\frac{1}{2}\cos^2\theta - \frac{1}{3}\cos^3\theta\right) d\theta = \frac{1}{36}a^3(9\pi - 16)$$

25. $\displaystyle V = \int_{-\pi/2}^{\pi/2} \int_0^{\cos\theta} \int_{r^2}^{r\cos\theta} r\, dz\, dr\, d\theta = \int_{-\pi/2}^{\pi/2} \int_0^{\cos\theta} \left(r^2\cos\theta - r^3\right) dr\, d\theta$

$$= \int_{-\pi/2}^{\pi/2} \frac{1}{12}\cos^4\theta\, d\theta = \frac{1}{32}\pi$$

27. $\displaystyle V = \int_0^{2\pi} \int_0^{1/2} \int_{r\sqrt{3}}^{\sqrt{1-r^2}} r\, dz\, dr\, d\theta = \int_0^{2\pi} \int_0^{1/2} \left(r\sqrt{1-r^2} - r^2\sqrt{3}\right) dr\, d\theta = \frac{1}{3}\pi\left(2-\sqrt{3}\right)$

29. $\displaystyle V = \int_0^{2\pi} \int_1^3 \int_0^{\sqrt{9-r^2}} r\, dz\, dr\, d\theta = \int_0^{2\pi} \int_1^3 r\sqrt{9-r^2}\, dr\, d\theta = \frac{32}{3}\pi\sqrt{2}$

31. Set the lower base of the cylinder on the xy-plane so that the axis of the cylinder coincides with the z-axis. Assume that the density varies directly as the distance from the lower base.

$$M = \int_0^{2\pi} \int_0^R \int_0^h kzr \, dz \, dr \, d\theta = \frac{1}{2} k\pi R^2 h^2$$

33. $I = I_z = k \int_0^{2\pi} \int_0^R \int_0^h zr^3 \, dr \, d\theta \, dz$

$$= \frac{1}{4} k\pi R^4 h^2 = \frac{1}{2} \left(\frac{1}{2} k\pi R^2 h^2 \right) R^2 = \frac{1}{2} MR^2$$

$\uparrow$
$\qquad\qquad$ from Exercise 31

35. Inverting the cone and placing the vertex at the origin, we have

$$V = \int_0^h \int_0^{2\pi} \int_0^{(R/h)z} r \, dr \, d\theta \, dz = \frac{1}{3} \pi R^2 h.$$

37. $I = \dfrac{M}{V} \displaystyle\int_0^h \int_0^{2\pi} \int_0^{(R/h)z} r^3 \, dr \, d\theta \, dz = \frac{3}{10} MR^2$

39. $V = \displaystyle\int_0^{2\pi} \int_0^1 \int_0^{1-r^2} r \, dz \, dr \, d\theta = \frac{1}{2} \pi$

41. $M = \displaystyle\int_0^{2\pi} \int_0^1 \int_0^{1-r^2} k(r^2 + z^2) r \, dz \, dr \, d\theta = \frac{1}{4} k\pi$

SECTION 17.9

1. $\left(\sqrt{3}, \; \frac{1}{4}\pi, \; \cos^{-1}\left[\frac{1}{3}\sqrt{3} \right] \right)$

3. $\left(\frac{3}{4}, \; \frac{3}{4}\sqrt{3}, \; \frac{3}{2}\sqrt{3} \right)$

5. $\rho = \sqrt{2^2 + 2^2 + (2\sqrt{6}/3)^2} = \dfrac{4\sqrt{6}}{3}$

$\phi = \cos^{-1}\left(\dfrac{2\sqrt{6}/3}{4\sqrt{6}/3} \right) = \cos^{-1}(1/2) = \dfrac{\pi}{3}$

$\theta = \tan^{-1}(1) = \dfrac{\pi}{4}$

$(\rho, \theta, \phi) = \left(\dfrac{4\sqrt{6}}{3}, \; \dfrac{\pi}{4}, \; \dfrac{\pi}{3} \right)$

7. $x = \rho \sin\phi \cos\theta = 3 \sin 0 \cos(\pi/2) = 0$

$z = \rho \cos\phi = 3 \cos 0 = 3$

$y = \rho \sin\phi \sin\theta = 3 \sin 0 \sin(\pi/2) = 0$

$(x, y, z) = (0, 0, 3)$

9. The circular cylinder $x^2 + y^2 = 1$;   the radius of the cylinder is 1 and the axis is the z-axis.

11. The lower nappe of the circular cone $z^2 = x^2 + y^2$.

13. Horizontal plane one unit above the xy-plane.

15. Sphere of radius 2 centered at the origin:

$$\int_0^{2\pi} \int_0^\pi \int_0^2 \rho^2 \sin\phi \, d\rho \, d\phi \, d\theta = \frac{8}{3} \int_0^{2\pi} \int_0^\pi \sin\phi \, d\phi \, d\theta = \frac{16}{3} \int_0^{2\pi} d\theta = \frac{32\pi}{3}$$

17. The first quadrant portion of the sphere that lies between the x,y-plane and the plane $z = \frac{3}{2}\sqrt{3}$.

$$\int_{\pi/6}^{\pi/2} \int_0^{\pi/2} \int_0^3 \rho^2 \sin\phi\, d\rho\, d\theta\, d\phi = 9 \int_{\pi/6}^{\pi/2} \int_0^{\pi/2} \sin\phi\, d\theta\, d\phi$$

$$= \tfrac{9}{2}\, \pi \int_{\pi/6}^{\pi/2} \sin\phi\, d\phi$$

$$= \tfrac{9}{2}\, \pi\, \left[-\cos\phi\right]_{\pi/6}^{\pi/2} = \tfrac{9}{4}\pi\sqrt{3}$$

19. $\displaystyle \int_0^1 \int_0^{\sqrt{1-x^2}} \int_{\sqrt{x^2+y^2}}^{\sqrt{2-x^2-y^2}} dz\, dy\, dx = \int_0^{\pi/4} \int_0^{\pi/2} \int_0^{\sqrt{2}} \rho^2 \sin\phi\, d\rho\, d\theta\, d\phi$

$$= \tfrac{2}{3}\sqrt{2} \int_0^{\pi/4} \int_0^{\pi/2} \sin\phi\, d\theta\, d\phi$$

$$= \tfrac{\sqrt{2}}{3}\, \pi \int_0^{\pi/4} \sin\phi\, d\phi = \frac{\sqrt{2}}{6}\, \pi\, (2 - \sqrt{2})$$

21. $\displaystyle \int_0^3 \int_0^{\sqrt{9-y^2}} \int_0^{\sqrt{9-x^2-y^2}} z\sqrt{x^2+y^2+x^2}\, dz\, dx\, dy$

$$= \int_0^{\pi/2} \int_0^{\pi/2} \int_0^3 \rho\cos\phi \cdot \rho \cdot \rho^2 \sin\phi\, d\rho\, d\theta\, d\phi$$

$$= \int_0^{\pi/2} \tfrac{1}{2}\sin 2\phi\, d\phi \int_0^{\pi/2} d\theta \int_0^3 \rho^4\, d\rho = \left[-\frac{1}{4}\cos 2\phi\right]_0^{\pi/2} \left(\frac{\pi}{2}\right) \left[\frac{1}{5}\rho^5\right]_0^3$$

$$= \frac{243\pi}{20}$$

23. $\displaystyle V = \int_0^{2\pi} \int_0^{\pi} \int_0^R \rho^2 \sin\phi\, d\rho\, d\phi\, d\theta = \frac{4}{3}\pi R^3$

25. $\displaystyle V = \int_0^{\alpha} \int_0^{\pi} \int_0^R \rho^2 \sin\phi\, d\rho\, d\phi\, d\theta = \frac{2}{3}\alpha R^3$

27. $\displaystyle M = \int_0^{2\pi} \int_0^{\tan^{-1}(r/h)} \int_0^{h\sec\phi} k\rho^3 \sin\phi\, d\rho\, d\phi\, d\theta$

$$= \int_0^{2\pi} \int_0^{\tan^{-1}(r/h)} \frac{kh^4}{4} \tan\phi\sec^3\phi\, d\phi\, d\theta$$

$$= \frac{kh^4}{4} \int_0^{2\pi} \frac{1}{3}\left[\sec^3\phi\right]_0^{\tan^{-1}(r/h)} d\theta = \frac{kh^4}{4} \int_0^{2\pi} \frac{1}{3}\left[\left(\frac{\sqrt{r^2+h^2}}{h}\right)^3 - 1\right] d\theta$$

$$= \frac{1}{6}\, k\pi h\, (r^2 + h^2)^{3/2} - h^3$$

29. center ball at origin;   density $= \dfrac{M}{V} = \dfrac{3M}{4\pi R^3}$

(a) $I = \dfrac{3M}{4\pi R^3} \displaystyle\int_0^{2\pi} \int_0^{\pi} \int_0^R \rho^4 \sin^3\phi\, d\rho\, d\phi\, d\theta = \frac{2}{5}MR^2$

(b) $I = \frac{2}{5}MR^2 + R^2M = \frac{7}{5}MR^2$ (parallel axis theorem)

31. center balls at origin;   density $= \dfrac{M}{V} = \dfrac{3M}{4\pi \left(R_2{}^3 - R_1{}^3\right)}$

(a) $I = \dfrac{3M}{4\pi \left(R_2{}^3 - R_1{}^3\right)} \displaystyle\int_0^{2\pi} \int_0^{\pi} \int_{R_1}^{R_2} \rho^4 \sin^3 \phi \, d\rho \, d\phi \, d\theta = \dfrac{2}{5} M \left(\dfrac{R_2{}^5 - R_1{}^5}{R_2{}^3 - R_1{}^3}\right)$

This result can be derived from Exercise 29 without further integration. View the solid as a ball of mass M_2 from which is cut out a core of mass M_1.

$$M_2 = \frac{M}{V} V_2 = \frac{3M}{4\pi \left(R_2{}^3 - R_1{}^3\right)} \left(\frac{4}{3} \pi R_2{}^3\right) = \frac{MR_2{}^3}{R_2{}^3 - R_1{}^3}; \quad \text{similarly} \quad M_1 = \frac{MR_1{}^3}{R_2{}^3 - R_1{}^3}.$$

Then

$$I = I_2 - I_1 = \tfrac{2}{5} M_2 R_2{}^2 - \tfrac{2}{5} M_1 R_1{}^2 = \frac{2}{5} \left(\frac{MR_2{}^3}{R_2{}^3 - R_1{}^3}\right) R_2{}^2 - \frac{2}{5} \left(\frac{MR_1{}^3}{R_2{}^3 - R_1{}^3}\right) R_1{}^2$$

$$= \frac{2}{5} M \left(\frac{R_2{}^5 - R_1{}^5}{R_2{}^3 - R_1{}^3}\right).$$

(b) Outer radius R and inner radius R_1 gives

$$\text{moment of inertia} = \frac{2}{5} M \left(\frac{R^5 - R_1{}^5}{R^3 - R_1{}^3}\right). \qquad \text{[part } (a)\text{]}$$

As $R_1 \to R$,

$$\frac{R^5 - R_1{}^5}{R^3 - R_1{}^3} = \frac{R^4 + R^3 R_1 + R^2 R_1{}^2 + R R_1{}^3 + R_1{}^4}{R^2 + R R_1 + R_1{}^2} \longrightarrow \frac{5R^4}{3R^2} = \frac{5}{3} R^2.$$

Thus the moment of inertia of spherical shell of radius R is

$$\frac{2}{5} M \left(\frac{5}{3} R^2\right) = \frac{2}{3} MR^2.$$

(c) $I = \tfrac{2}{3} MR^2 + R^2 M = \tfrac{5}{3} MR^2$ (parallel axis theorem)

33. $V = \displaystyle\int_0^{2\pi} \int_0^{\alpha} \int_0^{a} \rho^2 \sin \phi \, d\rho \, d\phi \, d\theta = \frac{2}{3} \pi \left(1 - \cos \alpha\right) a^3$

35. (a) Substituting $x = \rho \sin \phi \cos \theta, \quad y = \rho \sin \phi \sin \theta, \quad z = \rho \cos \phi$

into $x^2 + y^2 + (z - R)^2 = R^2$

we have $\rho^2 \sin^2 \phi + (\rho \cos \phi - R)^2 = R^2,$

which simplifies to $\rho = 2R \cos \phi.$

(b) $0 \le \theta \le 2\pi, \quad 0 \le \phi \le \pi/4, \quad R \sec \phi \le \rho \le 2R \cos \phi$

37.
$$V = \int_0^{2\pi} \int_0^{\pi/4} \int_0^{2} \rho^2 \sin \phi \, d\rho \, d\phi \, d\theta + \int_0^{2\pi} \int_{\pi/4}^{\pi/2} \int_0^{2\sqrt{2} \cos \phi} \rho^2 \sin \phi \, d\rho \, d\phi \, d\theta$$

$$= \frac{1}{3} \left(16 - 6\sqrt{2}\right) \pi$$

39. Encase T in a spherical wedge W. W has spherical coordinates in a box Π that contains S. Define f to be zero outside of T. Then
$$F(\rho, \theta, \phi) = f(\rho \sin \phi \cos \theta, \; \rho \sin \phi \sin \theta, \; \rho \cos \phi)$$
is zero outside of S and

$$\iiint_T f(x, y, z) \, dx dy dz = \iiint_W f(x, y, z) \, dx dy dz$$

$$= \iiint_\Pi F(\rho, \theta, \phi) \, \rho^2 \sin \phi \, d\rho d\theta d\phi$$

$$= \iiint_S F(\rho, \theta, \phi) \, \rho^2 \sin \phi \, d\rho d\theta d\phi.$$

41. T is the set of all (x, y, z) with spherical coordinates (ρ, θ, ϕ) in the set
$$S: \quad 0 \le \theta \le 2\pi, \quad 0 \le \phi \le \pi/4, \quad R \sec \phi \le \rho \le 2R \cos \phi.$$
T has volume $V = \frac{2}{3} \pi R^3$. By symmetry the $\mathbf{i}, \mathbf{j}$ components of force are zero and

$$\mathbf{F} = \left\{ \frac{3GmM}{2\pi R^3} \iiint_T \frac{z}{(x^2 + y^2 + z^2)^{3/2}} \, dx dy dz \right\} \mathbf{k}$$

$$= \left\{ \frac{3GmM}{2\pi R^3} \iiint_S \left(\frac{\rho \cos \phi}{\rho^3} \right) \rho^2 \sin \phi \, d\rho d\theta d\phi \right\} \mathbf{k}$$

$$= \left\{ \frac{3GmM}{2\pi R^3} \int_0^{2\pi} \int_0^{\pi/4} \int_{R \sec \phi}^{2R \cos \phi} \cos \phi \sin \phi \, d\rho \, d\phi \, d\theta \right\} \mathbf{k}$$

$$= \frac{GmM}{R^2} \left(\sqrt{2} - 1 \right) \mathbf{k}.$$

SECTION 17.10

1. $ad - bc$ **3.** $2(v^2 - u^2)$ **5.** $-3u^2 v^2$

7. abc **9.** $\rho^2 \sin \phi$

11. $J(\rho, \theta, \phi) = \begin{vmatrix} \sin \phi \cos \theta & \sin \phi \sin \theta & \cos \theta \\ -\rho \sin \phi \sin \theta & \rho \sin \phi \cos \theta & 0 \\ \rho \cos \phi \cos \theta & \rho \cos \phi \sin \theta & -\rho \sin \phi \end{vmatrix} = -\rho^2 \sin \phi; \quad |J(\rho, \theta, \phi)| = \rho^2 \sin \phi.$

13. Set $u = x + y$, $v = x - y$. Then
$$x = \frac{u+v}{2}, \quad y = \frac{u-v}{2} \quad \text{and} \quad J(u, v) = -\frac{1}{2}.$$
Ω is the set of all (x, y) with uv-coordinates in
$$\Gamma: \quad 0 \le u \le 1, \quad 0 \le v \le 2.$$

Then

$$\iint_\Omega (x^2 - y^2)\, dx\, dy = \iint_\Gamma \frac{1}{2}\, uv\, du\, dv = \frac{1}{2}\int_0^1 \int_0^2 uv\, dv\, du$$

$$= \frac{1}{2}\left(\int_0^1 u\, du\right)\left(\int_0^2 v\, dv\right) = \frac{1}{2}\left(\frac{1}{2}\right)(2) = \frac{1}{2}.$$

15. $\dfrac{1}{2}\displaystyle\int_0^1 \int_0^2 u\cos(\pi v)\, dv\, du = \dfrac{1}{2}\left(\int_0^1 u\, du\right)\left(\int_0^2 \cos(\pi v)\, dv\right) = \dfrac{1}{2}\left(\dfrac{1}{2}\right)(0) = 0$

17. Set $u = x - y$, $v = x + 2y$. Then

$$x = \frac{2u + v}{3}, \quad y = \frac{v - u}{3}, \quad \text{and} \quad J(u, v) = \frac{1}{3}.$$

Ω is the set of all (x, y) with uv-coordinates in the set

$$\Gamma: \ 0 \le u \le \pi, \quad 0 \le v \le \pi/2.$$

Therefore

$$\iint_\Omega \sin(x - y)\cos(x + 2y)\, dx\, dy = \iint_\Gamma \frac{1}{3}\sin u \cos v\, du\, dv = \frac{1}{3}\int_0^\pi \int_0^{\pi/2} \sin u\, \cos v\, dv\, du$$

$$= \frac{1}{3}\left(\int_0^\pi \sin u\, du\right)\left(\int_0^{\pi/2} \cos v\, dv\right) = \frac{1}{3}(2)(1) = \frac{2}{3}.$$

19. Set $u = xy$, $v = y$. Then

$$x = u/v, \quad y = v \quad \text{and} \quad J(u, v) = 1/v.$$

$$xy = 1, \quad xy = 4 \implies u = 1, \quad u = 4$$

$$y = x, \quad y = 4x \implies u/v = v, \quad 4u/v = v \implies v^2 = u, \quad v^2 = 4u$$

Ω is the set of all (x, y) with uv-coordinates in the set

$$\Gamma: \ 1 \le u \le 4, \quad \sqrt{u} \le v \le 2\sqrt{u}.$$

(a) $\quad A = \displaystyle\iint_\Gamma \frac{1}{v}\, du\, dv = \int_1^4 \int_{\sqrt{u}}^{2\sqrt{u}} \frac{1}{v}\, dv\, du = \int_1^4 \ln 2\, du = 3\ln 2$

(b) $\quad \overline{x}A = \displaystyle\int_1^4 \int_{\sqrt{u}}^{2\sqrt{u}} \frac{u}{v^2}\, dv\, du = \frac{7}{3}; \quad \overline{x} = \frac{7}{9\ln 2}$

$$\overline{y}A = \int_1^4 \int_{\sqrt{u}}^{2\sqrt{u}} dv\, du = \frac{14}{3}; \quad \overline{y} = \frac{14}{9\ln 2}$$

21. Set $u = x + y$, $v = 3x - 2y$. Then

$$x = \frac{2u + v}{5}, \quad y = \frac{3u - v}{5} \quad \text{and} \quad J(u, v) = -\frac{1}{5}.$$

With Γ: $\quad 0 \le u \le 1, \quad 0 \le v \le 2$

$$M = \int_0^1 \int_0^2 \frac{1}{5} \lambda \, dv \, du = \frac{2}{5} \lambda \quad \text{where} \quad \lambda \text{ is the density.}$$

Then

$$I_x = \int_0^1 \int_0^2 \left(\frac{3u-v}{5}\right)^2 \frac{1}{5} \lambda \, dv \, du = \frac{8\lambda}{375} = \frac{4}{75}\left(\frac{2}{5}\lambda\right) = \frac{4}{75} M,$$

$$I_y = \int_0^1 \int_0^2 \left(\frac{2u+v}{5}\right)^2 \frac{1}{5} \lambda \, dv \, du = \frac{28\lambda}{375} = \frac{14}{75}\left(\frac{2}{5}\lambda\right) = \frac{14}{75} M,$$

$$I_z = I_x + I_y = \frac{18}{75} M.$$

23. Set $u = x - 2y, \quad v = 2x + y$. Then

$$x = \frac{u+2v}{5}, \quad y = \frac{v-2u}{5} \quad \text{and} \quad J(u,v) = \frac{1}{5}.$$

Γ is the region between the parabola $v = u^2 - 1$ and the line $v = 2u + 2$. A sketch of the curves shows that

$$\Gamma: \quad -1 \le u \le 3, \quad u^2 - 1 \le v \le 2u + 2.$$

Then

$$A = \frac{1}{5}(\text{area of } \Gamma) = \frac{1}{5} \int_{-1}^3 \left[(2u+2) - (u^2 - 1) \right] du = \frac{32}{15}.$$

25. The choice $\theta = \pi/6$ reduces the equation to $13u^2 + 5v^2 = 1$. This is an ellipse in the uv-plane with area $\pi ab = \pi/\sqrt{65}$. Since $J(u,v) = 1$, the area of Ω is also $\pi/\sqrt{65}$.

27. $J = abc\rho^2 \sin\phi; \quad V = \int_0^{2\pi} \int_0^{\pi} \int_0^1 abc\rho^2 \sin\phi \, d\rho \, d\phi \, d\theta = \frac{4}{3}\pi abc$

29.
$$V = \frac{2}{3}\pi abc, \quad \lambda = \frac{M}{V} = \frac{3M}{2\pi abc}$$

$$I_x = \frac{3M}{2\pi abc} \int_0^{2\pi} \int_0^{\pi/2} \int_0^1 (b^2\rho^2 \sin^2\phi \sin^2\theta + c^2\rho^2 \cos^2\phi) abc\rho^2 \sin\phi \, d\rho \, d\phi \, d\theta$$

$$= \tfrac{1}{5} M(b^2 + c^2)$$

$$I_y = \tfrac{1}{5} M(a^2 + c^2), \quad I_z = \tfrac{1}{5} M(a^2 + b^2)$$

PROJECT 17.10

1. (a) $\quad \theta = \tan^{-1}\left[\left(\frac{ay}{bx}\right)^{1/\alpha}\right], \quad r = \left[\left(\frac{x}{a}\right)^{2/\alpha} + \left(\frac{y}{b}\right)^{2/\alpha}\right]^{\alpha/2}$

(b)
$$\left.\begin{array}{r} ar_1(\cos\theta_1)^\alpha = ar_2(\cos\theta_2)^\alpha \\ br_1(\sin\theta_1)^\alpha = br_2(\sin\theta_2)^\alpha \\ r_1 > 0, \quad 0 < \theta < \tfrac{1}{2}\pi \end{array}\right\} \implies r_1 = r_2, \quad \theta_1 = \theta_2$$

3. (a)

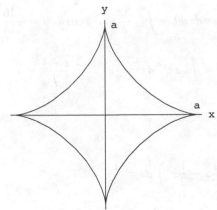

(b) $x = ar\cos^3\theta$, $y = ar\sin^3\theta$; $x^{\frac{2}{3}} + y^{\frac{2}{3}} = a^{\frac{2}{3}} \Longrightarrow r = 1$ and $x = a\cos^3\theta$, $y = a\sin^3\theta$

$$A = \int_{\frac{\pi}{2}}^{0} y(\theta)x'(\theta)\,d\theta = \int_{\frac{\pi}{2}}^{0} a\sin^3\theta(3a\cos^2\theta[-\sin\theta])\,d\theta$$

$$= 3a^2\int_0^{\frac{\pi}{2}} \sin^4\theta\cos^2\theta\,d\theta = 3a^2\int_0^{\frac{\pi}{2}}(\sin^4\theta - \sin^6\theta)\,d\theta$$

$$= 3a^2\left[\frac{3\cdot 1}{4\cdot 2}\frac{\pi}{2} - \frac{5\cdot 3\cdot 1}{6\cdot 4\cdot 2}\frac{\pi}{2}\right]\quad \text{(See Exercise 62(b) in 8.3)}$$

$$= \frac{3a^2\pi}{32}$$

(c) Entire area enclosed: $4\cdot\dfrac{3a^2\pi}{32} = \dfrac{3a^2\pi}{8}$

REVIEW EXERCISES

1. $\displaystyle\int_0^1\int_y^{\sqrt{y}} xy^2\,dx\,dy = \int_0^1\left[\frac{1}{2}x^2y^2\right]_y^{\sqrt{y}}dy = \int_0^1\left(\frac{1}{2}y^3 - \frac{1}{2}y^4\right)dy = \left[\frac{1}{8}y^4 - \frac{1}{10}y^5\right]_0^1 = \frac{1}{40}$

3. $\displaystyle\int_0^1\int_x^{3x} 2ye^{x^3}\,dy\,dx = \int_0^1\left[y^2e^{x^3}\right]_x^{3x}dx = \int_0^1(9x^2e^{x^3} - x^2e^{x^3})\,dx = \left[3e^{x^3} - \frac{1}{3}e^{x^3}\right]_0^1 = \frac{8}{3}e - \frac{8}{3}$

5. $\displaystyle\int_0^{\pi/4}\int_0^{2\sin\theta} r\cos\theta\,dr\,d\theta = \int_0^{\pi/4}\left[\frac{1}{2}r^2\cos\theta\right]_0^{2\sin\theta}d\theta = \int_0^{\pi/4} 2\sin^2\theta\cos\theta\,d\theta = \left[\frac{2}{3}\sin^3\theta\right]_0^{\pi/4} = \frac{\sqrt{2}}{6}$

7. $\displaystyle\int_0^2\int_0^{2-3x}\int_0^{x+y} x\,dz\,dy\,dx = \int_0^2\int_0^{2-3x}\left[xz\right]_0^{x+y}dy\,dx = \int_0^2\int_0^{2-3x}(x^2 + xy)\,dy\,dx$

$$= \int_0^2\left[x^2y + \frac{1}{2}xy^2\right]_0^{2-3x}dx = \int_0^2\left(\frac{3}{2}x^3 - 4x^2 + 2x\right)dx$$

$$= \left[\frac{3}{8}x^4 - \frac{4}{3}x^3 + x^2\right]_0^2 = -\frac{2}{3}$$

9. $\displaystyle\int_{-\frac{\pi}{2}}^{0}\int_{0}^{2\sin\theta}\int_{0}^{r^2} r^2\cos\theta\,dz\,dr\,d\theta = \int_{-\frac{\pi}{2}}^{0}\int_{0}^{2\sin\theta} r^4\cos\theta\,dr\,d\theta = \int_{-\frac{\pi}{2}}^{0}\frac{32}{5}\sin^5\theta\cos\theta\,d\theta = -\frac{16}{15}$

11. $\displaystyle\int_{0}^{1}\int_{y}^{1} e^{x^2}\,dx\,dy = \int_{0}^{1}\int_{0}^{x} e^{x^2}\,dy\,dx = \int_{0}^{1} e^{x^2}y\Big|_{0}^{x}\,dx = \int_{0}^{1} xe^{x^2}\,dx = \frac{1}{2}e^{x^2}\Big|_{0}^{1} = \frac{e-1}{2}$

13. $\displaystyle\int_{0}^{1}\int_{0}^{\sqrt{1-y^2}}\frac{1}{\sqrt{1-y^2}}dx\,dy = \int_{0}^{1} dy = 1$

15. $\displaystyle\int_{0}^{1}\int_{0}^{\sqrt{1-x^2}} xy\,dy\,dx = \int_{0}^{1}\left[\frac{1}{2}xy^2\right]_{0}^{\sqrt{1-x^2}}\,dx = \int_{0}^{1}\left(\frac{1}{2}x - \frac{1}{2}x^3\right)\,dx = \frac{1}{8}$

17. $\displaystyle\int_{0}^{2}\int_{x}^{3x-x^2}(x^2-xy)\,dy\,dx = \int_{0}^{2}\left[x^2y - \frac{1}{2}xy^2\right]_{x}^{3x-x^2}\,dx = \int_{0}^{2}\left(2x^4 - 2x^3 - \frac{1}{2}x^5\right)\,dx = -\frac{8}{15}$

19. $\displaystyle\int_{0}^{2}\int_{0}^{y}\int_{0}^{\sqrt{4-y^2}} 2xyz\,dz\,dx\,dy = \int_{0}^{2}\int_{x}^{2} xyz^2\Big|_{0}^{\sqrt{4-y^2}}\,dx\,dy = \int_{0}^{2}\int_{0}^{2} xy(4-y^2)\,dx\,dy$

$$= \int_{0}^{2}\left[\frac{1}{2}y^3(4-y^2)\right]\,dy = \frac{8}{3}$$

21. $\displaystyle\int_{0}^{2}\int_{0}^{\sqrt{4-x^2}}\int_{0}^{\sqrt{4-x^2-y^2}} xy\,dz\,dy\,dx = \int_{0}^{2}\int_{0}^{\sqrt{4-x^2}}\Big[xyz\Big]_{0}^{\sqrt{4-x^2-y^2}}\,dy\,dx$

$$= \int_{0}^{2}\left[-\frac{1}{3}x\sqrt{4-x^2-y^2}\right]_{0}^{\sqrt{4-x^2}}\,dx$$

$$= \int_{0}^{2}\frac{1}{3}x(4-x^2)^{\frac{3}{2}}\,dx = \frac{32}{15}$$

23. $\displaystyle\int_{0}^{2}\int_{0}^{\sqrt{4-y^2}} e^{\sqrt{x^2+y^2}}\,dx\,dy = \int_{0}^{\pi/2}\int_{0}^{2} e^r r\,dr\,d\theta = \frac{\pi}{2}\int_{0}^{2} re^r\,dr = \frac{\pi}{2}\Big[re^r - e^r\Big]_{0}^{2} = \frac{\pi}{2}(e^2+1)$

25. $\displaystyle V = \int_{0}^{3}\int_{0}^{2\pi}(9-r^2)r\,dr\,d\theta = 2\pi\int_{0}^{3}(9-r^2)r\,dr = \frac{81\pi}{2}$

27. $\displaystyle V = \int_{0}^{1}\int_{0}^{1-x}(x^2+y^2)\,dy\,dx = \int_{0}^{1}\left[x^2 - x^3 + \frac{1}{3}(1-x)^3\right]\,dx = \left[\frac{1}{3}x^3 - \frac{1}{4}x^4 - \frac{1}{12}(1-x)^4\right]_{0}^{1} = \frac{1}{6}$

29. $\displaystyle M = \int_{-\pi/2}^{\pi/2}\int_{0}^{\cos x} y\,dy\,dx = \int_{-\pi/2}^{\pi/2}\frac{1}{2}\cos^2 x\,dx = \frac{\pi}{4}$

$\displaystyle x_M M = \int_{-\pi/2}^{\pi/2}\int_{0}^{\cos x} xy\,dy\,dx = 0$ by symmetry

$\displaystyle y_M M = \int_{-\pi/2}^{\pi/2}\int_{0}^{\cos x} y^2\,dy\,dx = \int_{-\pi/2}^{\pi/2}\frac{1}{3}\cos^3 x\,dx = \frac{4}{9}$

The center of mass is: $\left(0, \frac{16}{9\pi}\right)$

31. $M = \displaystyle\int_0^{\pi/2} \int_r^R u^3 \, du \, d\theta = \dfrac{\pi}{8}(R^4 - r^4);$ (polar coordinates $[u, \theta]$)

By symmetry, $\bar{x} = \bar{y}$.

$x_M M = \displaystyle\int_0^{\pi/2} \int_r^R u^4 \cos\theta \, du \, d\theta = \dfrac{1}{5}(R^5 - r^5); \quad x_M = \dfrac{8(R^5 - r^5)}{5\pi(R^4 - r^4)}$

33.   Introduce a coordinate system as shown in the figure.

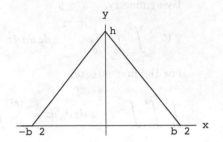

(a) $A = \frac{1}{2}bh;$ by symmetry, $\bar{x} = 0$

$\bar{y}\, A = \displaystyle\int_{-b/2}^0 \int_0^{\frac{2h}{b}(x+\frac{b}{2})} y \, dy \, dx + \int_0^{\frac{b}{2}} \int_0^{-\frac{2h}{b}(x-\frac{b}{2})} y \, dy \, dx$

$= \dfrac{bh^2}{6} \quad \Longrightarrow \quad \bar{y} = \dfrac{h}{3}$

(b) $I = \displaystyle\int_{-b/2}^0 \int_0^{\frac{2h}{b}(x+\frac{b}{2})} \lambda y^2 \, dy \, dx + \int_0^{b/2} \int_0^{-\frac{2h}{b}(x-\frac{b}{2})} \lambda y^2 \, dy \, dx = \dfrac{\lambda b h^3}{12} = \frac{1}{6} M h^2$

(c) $I = 2 \displaystyle\int_0^{b/2} \int_0^{-\frac{2h}{b}(x-\frac{b}{2})} \lambda x^2 \, dx \, dy = \dfrac{1}{48}\lambda h b^3 = \frac{1}{24} M b^2$

35. $V = \displaystyle\int_0^2 \int_0^x \int_0^{2x+2y+1} dz \, dy \, dx = \int_0^2 \int_0^x (2x + 2y + 1) \, dy \, dx = 10$

37.   The curve of intersection of the two surfaces is the circle: $x^2 + y^2 = 4, \; x = 3$

$V = \displaystyle\int_{-2}^2 \int_{-\sqrt{4-x^2}}^{\sqrt{4-x^2}} \int_{2x^2+y^2}^{12-x^2-2y^2} dz \, dy \, dx = \int_{-2}^2 \int_{-\sqrt{4-x^2}}^{\sqrt{4-x^2}} 3\left(4 - x^2 - y^2\right) dy \, dx$

$= 3 \displaystyle\int_0^{2\pi} \int_0^2 \left(4 - r^2\right) r \, dr \, d\theta$

$= 3 \displaystyle\int_0^{2\pi} \left[2r^2 - \tfrac{1}{4} r^4\right]_0^2 d\theta = 12 \int_0^{2\pi} = 24\pi$

39. $V = \displaystyle\int_0^{2\pi} \int_0^{\pi/3} \int_{\sec\phi}^2 \rho^2 \sin\phi \, d\rho \, d\phi \, d\theta = \int_0^{2\pi} \int_0^{\pi/3} \left[\tfrac{1}{3}\rho^3\right]_{\sec\phi}^2 d\phi \, d\theta$

$= \dfrac{1}{3} \displaystyle\int_0^{2\pi} \int_0^{\pi/3} \left(8 - \sec^3\phi\right) \sin\phi \, d\phi \, d\theta$

$= \dfrac{1}{3} \displaystyle\int_0^{2\pi} \left[-8\cos\phi - \tfrac{1}{2}\sec^2\phi\right]_0^{\pi/3} d\theta$

$= \dfrac{1}{3}\left(\dfrac{5}{2}\right)(2\pi) = \dfrac{5\pi}{3}$

41. $V = \displaystyle\int_0^{2\pi} \int_{\pi/4}^{\pi/2} \int_0^1 \rho^2 \sin\phi \, d\rho \, d\phi \, d\theta = \int_0^{2\pi} \int_{\pi/4}^{\pi/2} \dfrac{1}{3}\sin\phi \, d\phi \, d\theta = \dfrac{\sqrt{2}\pi}{3}$

43. (a) $V = \displaystyle\int_0^1 \int_0^x \int_0^{\sqrt{1-x^2}} dz\,dy\,dx + \int_0^1 \int_0^y \int_0^{\sqrt{1-y^2}} dz\,dx\,dy$

$$= 2\int_0^1 \int_0^x \sqrt{1-x^2}\,dy\,dx = 2\int_0^1 x\sqrt{1-x^2}\,dx = \frac{2}{3}$$

By symmetry, $\quad \overline{x} = \overline{y}.$

$$\overline{x}\,V = \int_0^1 \int_0^x \int_0^{\sqrt{1-x^2}} x\,dz\,dy\,dx + \int_0^1 \int_0^y \int_0^{\sqrt{1-y^2}} x\,dz\,dx\,dy$$

For the first integral:

$$\int_0^1 \int_0^x \int_0^{\sqrt{1-x^2}} x\,dz\,dy\,dx = \int_0^1 \int_0^x x\sqrt{1-x^2}\,dy\,dx$$

$$= \int_0^1 x^2\sqrt{1-x^2}\,dx = \int_0^{\pi/2} \sin^2 u\,\cos^2 u\,du = \frac{\pi}{16}$$

$$x = \sin u \uparrow$$

For the second integral:

$$\int_0^1 \int_0^y \int_0^{\sqrt{1-y^2}} x\,dz\,dx\,dy = \int_0^1 \int_0^y x\sqrt{1-y^2}\,dx\,dy = \int_0^1 \frac{1}{2}y^2\sqrt{1-y^2}\,dy = \frac{\pi}{32}$$

Thus, $\ \overline{x}\,V = \dfrac{3\pi}{32} \ \implies \ \overline{x} = \overline{y} = \dfrac{9\pi}{64}$

Now calculate $\overline{z}$:

$$\overline{z}\,V = \int_0^1 \int_0^x \int_0^{\sqrt{1-x^2}} z\,dz\,dy\,dx + \int_0^1 \int_0^y \int_0^{\sqrt{1-y^2}} z\,dz\,dx\,dy;$$

$$\int_0^1 \int_0^x \int_0^{\sqrt{1-x^2}} z\,dz\,dy\,dx = \int_0^1 \int_0^x \frac{1}{2}(1-x^2)\,dy\,dx = \frac{1}{2}\int_0^1 \left(x - x^3\right)\,dx = \frac{1}{8}$$

and similarly,

$$\int_0^1 \int_0^y \int_0^{\sqrt{1-y^2}} z\,dz\,dy\,dx = \frac{1}{8}.$$

Therefore, $\ \overline{z}\,V = \dfrac{1}{4} \ \implies \ \overline{z} = \dfrac{3}{8}$

(b) $I_z = \displaystyle\int_0^1 \int_0^x \int_0^{\sqrt{1-x^2}} \lambda\left(\sqrt{x^2+y^2}\right)^2 dz\,dy\,dx + \int_0^1 \int_0^y \int_0^{\sqrt{1-y^2}} \lambda\left(\sqrt{x^2+y^2}\right)^2 dz\,dx\,dy;$

$$\int_0^1 \int_0^x \int_0^{\sqrt{1-x^2}} \lambda\left(\sqrt{x^2+y^2}\right)^2 dz\,dy\,dx = \int_0^{\pi/4} \int_0^{\sec\theta} \int_0^{r\sin\theta} \lambda r^3\,dz\,dr\,d\theta = \frac{3}{20}\lambda$$

and $\displaystyle\int_0^1 \int_0^y \int_0^{\sqrt{1-y^2}} \lambda(\sqrt{x^2+y^2})^2 dz\,dx\,dy = \frac{3}{20}\lambda \ \implies \ I_z = \frac{3}{10}\lambda$

45. Denote polar coordinates by $[u, \theta]$.

(a) $M = \displaystyle\int_0^{2\pi} \int_0^r \int_0^h u^3 \, dz \, du \, d\theta = 2\pi h \int_0^r u^3 \, du = \dfrac{\pi h r^4}{2}$

(b) By symmetry, $x_M = y_M = 0$

(c) $z_M M = \displaystyle\int_0^{2\pi} \int_0^r \int_0^h u^3 z \, dz \, du \, d\theta = \dfrac{\pi h^2 r^4}{4} \Longrightarrow z_M = h/2$

47. (a) $M = \displaystyle\int_0^1 \int_0^{2\pi} \int_r^1 r^2 \, dz \, d\theta \, dr = 2\pi \int_0^1 \int_r^1 r^2 \, dz \, dr = 2\pi \int_0^1 r^2(1 - r) \, dr = \dfrac{\pi}{6}$

(b) By symmetry, $x_M = y_M = 0$

$z_M M = \displaystyle\int_0^1 \int_0^{2\pi} \int_r^1 r^2 z \, dz \, d\theta \, dr = \pi \int_0^1 r^2 (1 - r^2) \, dr = \dfrac{2\pi}{15} \Longrightarrow z_M = \dfrac{4}{5}$

(c) $I_z = \displaystyle\int_0^1 \int_0^{2\pi} \int_r^1 r^4 \, dz \, d\theta \, dr = \dfrac{\pi}{15}$

49. $J(u, v) = \begin{vmatrix} e^u \cos v & e^u \sin v \\ -e^u \sin v & e^u \cos v \end{vmatrix} = e^{2u}$

51. Set $\quad x = \dfrac{v - u}{2}, \quad y = \dfrac{v + u}{2} \Longrightarrow u = y - x, \quad v = y + x, \quad 1 \le v \le 2, \quad J = -\dfrac{1}{2}$

at $x = 0$, $y = u$, $y = v \Longrightarrow u = v$

at $y = 0$, $-x = u$, $x = v \Longrightarrow u = -v$

$\displaystyle\iint_\Omega \cos\left(\dfrac{y - x}{y + x}\right) dx \, dy = \int_1^2 \int_{-v}^v \dfrac{1}{2} \cos\left(\dfrac{u}{v}\right) du \, dv = \int_i^2 v \sin 1 \, dv = \dfrac{3}{2} \sin 1$

CHAPTER 18

SECTION 18.1

1. (a) $\mathbf{h}(x,y) = y\,\mathbf{i} + x\,\mathbf{j};$ $\mathbf{r}(u) = u\,\mathbf{i} + u^2\,\mathbf{j},$ $u \in [0,1]$

$x(u) = u,$ $y(u) = u^2;$ $x'(u) = 1,$ $y'(u) = 2u$

$\mathbf{h}(\mathbf{r}(u)) \cdot \mathbf{r}'(u) = y(u)\,x'(u) + x(u)\,y'(u) = u^2(1) + u(2u) = 3u^2$

$\displaystyle \int_C \mathbf{h}(\mathbf{r}) \cdot d\mathbf{r} = \int_0^1 3u^2\,du = 1$

(b) $h(x,y) = y\,\mathbf{i} + x\,\mathbf{j};$ $\mathbf{r}(u) = u^3\,\mathbf{i} - 2u\,\mathbf{j},$ $u \in [0,1]$

$x(u) = u^3,$ $y(u) = -2u;$ $x'(u) = 3u^2,$ $y'(u) = -2$

$\mathbf{h}(\mathbf{r}(u)) \cdot \mathbf{r}'(u) = y(u)\,x'(u) + x(u)\,y'(u) = (-2u)(3u^2) + u^3(-2) = -8u^3$

$\displaystyle \int_C \mathbf{h}(\mathbf{r}) \cdot d\mathbf{r} = \int_0^1 -8u^3\,du = -2$

3. $h(x,y) = y\,\mathbf{i} + x\,\mathbf{j};$ $\mathbf{r}(u) = \cos u\,\mathbf{i} - \sin u\,\mathbf{j},$ $u \in [0, 2\pi]$

$x(u) = \cos u,$ $y(u) = -\sin u;$ $x'(u) = -\sin u,$ $y'(u) = -\cos u$

$\mathbf{h}(\mathbf{r}(u)) \cdot \mathbf{r}'(u) = y(u)\,x'(u) + x(u)\,y'(u) = \sin^2 u - \cos^2 u$

$\displaystyle \int_C \mathbf{h}(\mathbf{r}) \cdot d\mathbf{r} = \int_0^{2\pi} (\sin^2 u - \cos^2 u)\,du = 0$

5. (a) $\mathbf{r}(u) = (2-u)\,\mathbf{i} + (3-u)\,\mathbf{j},$ $u \in [0,1]$

$\displaystyle \int_C \mathbf{h}(\mathbf{r}) \cdot d\mathbf{r} = \int_0^1 (-5 + 5u - u^2)\,du = -\frac{17}{6}$

(b) $\mathbf{r}(u) = (1+u)\,\mathbf{i} + (2+u)\,\mathbf{j},$ $u \in [0,1]$

$\displaystyle \int_C \mathbf{h}(\mathbf{r}) \cdot d\mathbf{r} = \int_0^1 (1 + 3u + u^2)\,du = \frac{17}{6}$

7. $C = C_1 \cup C_2 \cup C_3$ where,

$C_1 : \mathbf{r}(u) = (1-u)(-2\,\mathbf{i}) + u(2\,\mathbf{i}) = (4u - 2)\,\mathbf{i},$ $u \in [0,1]$

$C_2 : \mathbf{r}(u) = (1-u)(2\,\mathbf{i}) + u(2\,\mathbf{j}) = (2-2u)\,\mathbf{i} + 2u\,\mathbf{j},$ $u \in [0,1]$

$C_3 : \mathbf{r}(u) = (1-u)(2\,\mathbf{j}) + u(-2\,\mathbf{i}) = -2u\,\mathbf{i} + (2-2u)\,\mathbf{j},$ $u \in [0,1]$

$\displaystyle \int_C = \int_{C_1} + \int_{C_2} + \int_{C_3} = 0 + (-4) + (-4) = -8$

9.

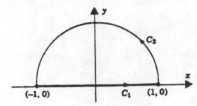

$$C_1 : \mathbf{r}(u) = (-1 + 2u)\,\mathbf{i}, \quad u \in [0, 1]$$

$$C_2 : \mathbf{r}(u) = \cos u\,\mathbf{i} + \sin u\,\mathbf{j}, \quad u \in [0, \pi]$$

$$\int_C = \int_{C_1} + \int_{C_2} = 0 + (-\pi) = -\pi$$

11. (a) $\mathbf{r}(u) = u\,\mathbf{i} + u\,\mathbf{j} + u\,\mathbf{k}, \quad u \in [0, 1]$

$$\int_C \mathbf{h}(\mathbf{r}) \cdot d\mathbf{r} = \int_0^1 3u^2\, du = 1$$

  (b) $\displaystyle \int_C \mathbf{h}(\mathbf{r}) \cdot d\mathbf{r} = \int_0^1 (2u^3 + u^5 + 3u^6)\, du = \frac{23}{21}$

13. (a) $\mathbf{r}(u) = 2u\,\mathbf{i} + 3u\,\mathbf{j} - u\,\mathbf{k}, \quad u \in [0, 1]$

$$\int_C \mathbf{h}(\mathbf{r}) \cdot d\mathbf{r} = \int_0^1 (2\cos 2u + 3\sin 3u + 3u^2)\, du = \left[\sin 2u - \cos 3u + u^3\right]_0^1 = 2 + \sin 2 - \cos 3$$

  (b) $\displaystyle \int_C \mathbf{h}(\mathbf{r}) \cdot d\mathbf{r} = \int_0^1 \left(2u\cos u^2 + 3u^2 \sin u^3 - u^4\right) du = \left[\sin u^2 - \cos u^3 - \frac{1}{5}u^5\right]_0^1 = \frac{4}{5} + \sin 1 - \cos 1$

15. $\mathbf{r}(u) = u\,\mathbf{i} + u^2\,\mathbf{j}, \quad u \in [0, 2]$

$$\int_C \mathbf{F}(\mathbf{r}) \cdot d\mathbf{r} = \int_0^2 \left[(u + 2u^2) + (2u + u^2)2u\right] du = \int_0^2 (2u^3 + 6u^2 + u)\, du = 26$$

17. $\mathbf{r}(u) = (1 - u)(\mathbf{j} + 4\,\mathbf{k}) + u(\mathbf{i} - 4\,\mathbf{k})$

$$= u\,\mathbf{i} + (1 - u)\,\mathbf{j} + (4 - 8u)\,\mathbf{k}, \quad u \in [0, 1]$$

$$\int_C \mathbf{F}(\mathbf{r}) \cdot d\mathbf{r} = \int_0^1 (-32u + 97u^2 - 64u^3)\, du = \frac{1}{3}$$

19. $\mathbf{r}(u) = \cos u\,\mathbf{i} + \sin u\,\mathbf{j} + u\,\mathbf{k}, \quad u \in [0, 2\pi]$

$$\int_C \mathbf{F}(\mathbf{r}) \cdot d\mathbf{r} = \int_0^{2\pi} \left[-\cos^2 u \sin u + \cos^2 u \sin u + u^2\right] du = \int_0^{2\pi} u^2\, du = \frac{8\pi^3}{3}$$

21.

$$\int_C \mathbf{q} \cdot d\mathbf{r} = \int_a^b [\mathbf{q} \cdot \mathbf{r}'(u)]\, du = \int_a^b \frac{d}{du}[\mathbf{q} \cdot \mathbf{r}(u)]\, du$$

$$= [\mathbf{q} \cdot \mathbf{r}(b)] - [\mathbf{q} \cdot \mathbf{r}(a)]$$

$$= \mathbf{q} \cdot [\mathbf{r}(b) - \mathbf{r}(a)]$$

$$\int_C \mathbf{r} \cdot d\mathbf{r} = \int_a^b [\mathbf{r}(u) \cdot \mathbf{r}'(u)]\, du$$

$$= \frac{1}{2} \int_a^b \|\mathbf{r}\|\, d\|\mathbf{r}\| \quad \text{(see Exercise 57, Section 14.1)}$$

$$= \frac{1}{2} \left(\|\mathbf{r}(b)\|^2 - \|\mathbf{r}(a)\|^2 \right)$$

23. $\displaystyle\int_C \mathbf{f}(\mathbf{r}) \cdot d\mathbf{r} = \int_a^b [\mathbf{f}(\mathbf{r}(u)) \cdot \mathbf{r}'(u)]\, du = \int_a^b [f(u)\,\mathbf{i} \cdot \mathbf{i}]\, du = \int_a^b f(u)\, du$

25. $E : \mathbf{r}(u) = a\cos u\,\mathbf{i} + b\sin u\,\mathbf{j}, \quad u \in [0, 2\pi]$

$$W = \int_0^{2\pi} \left[\left(-\frac{1}{2}b\sin u \right)(-a\sin u) + \left(\frac{1}{2}a\cos u \right)(b\cos u) \right] du = \frac{1}{2} \int_0^{2\pi} ab\, du = \pi ab$$

If the ellipse is traversed in the opposite direction, then $W = -\pi ab$. In both cases $|W| = \pi ab =$ area of the ellipse.

27. $\mathbf{r}(t) = \alpha t\,\mathbf{i} + \beta t^2\,\mathbf{j} + \gamma t^3\,\mathbf{k}$

$\mathbf{r}'(t) = \alpha\,\mathbf{i} + 2\beta t\,\mathbf{j} + 3\gamma t^2\,\mathbf{k}$

force at time $t = m\mathbf{r}''(t) = m(2\beta\,\mathbf{j} + 6\gamma t\mathbf{k})$

$$W = \int_0^1 [m(2\beta\,\mathbf{j} + 6\gamma t\,\mathbf{k}) \cdot (\alpha\,\mathbf{i} + 2\beta t\mathbf{j} + 3\gamma t^2\,\mathbf{k})]\, dt$$

$$= m \int_0^1 (4\beta^2 t + 18\gamma^2 t^3)\, dt = \left(2\beta^2 + \frac{9}{2}\gamma^2 \right) m$$

29.   Take $C : \mathbf{r}(t) = r\cos t\,\mathbf{i} + r\sin t\,\mathbf{j}, \quad t \in [0, 2\pi]$

$$\int_C \mathbf{v}(\mathbf{r}) \cdot d\mathbf{r} = \int_0^{2\pi} [\mathbf{v}(\mathbf{r}(t)) \cdot \mathbf{r}'(t)]\, dt$$

$$= \int_0^{2\pi} [f(x(t), y(t))\,\mathbf{r}(t) \cdot \mathbf{r}'(t)]\, dt$$

$$= \int_0^{2\pi} f(x(t), y(t))\, [\mathbf{r}(t) \cdot \mathbf{r}'(t)]\, dt = 0$$

since for the circle $\mathbf{r}(t) \cdot \mathbf{r}'(t) = 0$ identically. The circulation is zero.

31. (a) $\mathbf{r}(u) = (1-u)(\mathbf{i}+2\mathbf{k}) + u(\mathbf{i}+3\mathbf{j}+2\mathbf{k}) = \mathbf{i}+3u\mathbf{j}+2\mathbf{k}, \quad u \in [0,1].$

$$\int_C \mathbf{F}(\mathbf{r}) \cdot d\mathbf{r} = \int_0^1 \frac{9uk}{(5+9u^2)^{3/2}} \, du = \left[\frac{-k}{\sqrt{5+9u^2}} \right]_0^1 = \frac{k}{\sqrt{5}} - \frac{k}{\sqrt{14}}$$

(b) Let C be an arc on the sphere $\|\mathbf{r}\| = r = 5.$

$$\int_C \mathbf{F}(\mathbf{r}) \cdot d\mathbf{r} = \int_{C_2} \frac{k\mathbf{r}}{\|\mathbf{r}\|^3} \cdot d\mathbf{r}$$

$$= \frac{k}{5^3} \int_{C_2} \mathbf{r} \cdot d\mathbf{r} = \frac{k}{5^3} \int_{C_2} \|\mathbf{r}\| \, d\|\mathbf{r}\| \quad \text{(see Exercise 57, Section 14.1)}$$

$$= \frac{k}{5^3} \left[\frac{1}{2} \|\mathbf{r}\|^2 \right]_{(3,4,0)}^{(0,4,3)} = 0$$

33. $\mathbf{r}(u) = u\mathbf{i} + \alpha u(1-u)\mathbf{j}, \quad \mathbf{r}'(u) = \mathbf{i} + \alpha(1-2u)\mathbf{j}, \quad u \in [0,1]$

$$W(\alpha) = \int_C \mathbf{F}(\mathbf{r}) \cdot d\mathbf{r} = \int_0^1 \left[(\alpha^2 u^2 (1-u)^2 + 1) + [u + \alpha u(1-u)]\alpha(1-2u) \right] dx$$

$$= \int_0^1 \left[1 + (\alpha + \alpha^2)u - (2\alpha + 2\alpha^2)u^2 + \alpha^2 u^4 \right] du = 1 - \frac{1}{6}\alpha + \frac{1}{30}\alpha^2$$

$$W'(\alpha) = -\frac{1}{6} + \frac{1}{15}\alpha \quad \Longrightarrow \quad \alpha = \frac{15}{6} = \frac{5}{2}$$

The work done by $\mathbf{F}$ is a minimum when $\alpha = 5/2.$

SECTION 18.2

1. $\mathbf{h}(x,y) = \nabla f(x,y) \quad \text{where} \quad f(x,y) = \frac{1}{2}(x^2 + y^2)$

C is closed $\implies \int_C \mathbf{h}(\mathbf{r}) \cdot d\mathbf{r} = 0$

3. $\mathbf{h}(x,y) = \nabla f(x,y) \quad \text{where} \quad f(x,y) = x\cos \pi y; \quad \mathbf{r}(0) = \mathbf{0}, \quad \mathbf{r}(1) = \mathbf{i} - \mathbf{j}$

$$\int_C \mathbf{h}(\mathbf{r}) \cdot d\mathbf{r} = \int_C \nabla f(\mathbf{r}) \cdot d\mathbf{r} = f(\mathbf{r}(1)) - f(\mathbf{r}(0)) = f(1,-1) - f(0,0) = -1$$

5. $\mathbf{h}(x,y) = \nabla f(x,y) \quad \text{where} \quad f(x,y) = \frac{1}{2}x^2 y^2; \quad \mathbf{r}(0) = \mathbf{j}, \quad \mathbf{r}(1) = -\mathbf{j}$

$$\int_C \mathbf{h}(\mathbf{r}) \cdot d\mathbf{r} = \int_C \nabla f(\mathbf{r}) \cdot d\mathbf{r} = f(\mathbf{r}(1)) - f(\mathbf{r}(0)) = f(0,-1) - f(0,1) = 0 - 0 = 0$$

7. $\mathbf{h}(x,y) = \nabla f(x,y) \quad \text{where} \quad f(x,y) = x^2 y - xy^2; \quad \mathbf{r}(0) = \mathbf{i}, \; \mathbf{r}(\pi) = -\mathbf{i}$

$$\int_C \mathbf{h}(\mathbf{r}) \cdot d\mathbf{r} = \int_C \nabla f(\mathbf{r}) \cdot d\mathbf{r} = f(\mathbf{r}(\pi)) - f(\mathbf{r}(0)) = f(-1,0) - f(1,0) = 0 - 0 = 0$$

9. $\mathbf{h}(x,y) = \nabla f(x,y)$ where $f(x,y) = (x^2 + y^4)^{3/2}$

$$\int_C \mathbf{h}(\mathbf{r}) \cdot d\mathbf{r} = \int_C \nabla f(\mathbf{r}) \cdot d\mathbf{r} = f(1,0) - f(-1,0) = 1 - 1 = 0$$

11. $\mathbf{h}(x,y)$ is not a gradient, but part of it,

$$2x \cosh y\, \mathbf{i} + (x^2 \sinh y - y)\mathbf{j},$$

is a gradient. Since we are integrating over a closed curve, the contribution of the gradient part is 0. Thus

$$\int_C \mathbf{h}(\mathbf{r}) \cdot d\mathbf{r} = \int_C (-y\mathbf{i}) \cdot d\mathbf{r}.$$

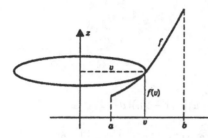

$C_1 : \mathbf{r}(u) = \mathbf{i} + (-1 + 2u)\mathbf{j}, \quad u \in [0,1]$

$C_2 : \mathbf{r}(u) = (1 - 2u)\mathbf{i} + \mathbf{j}, \quad u \in [0,1]$

$C_3 : \mathbf{r}(u) = -\mathbf{i} + (1 - 2u)\mathbf{j}, \quad u \in [0,1]$

$C_4 : \mathbf{r}(u) = (-1 + 2u)\mathbf{i} - \mathbf{j}, \quad u \in [0,1]$

$$\int_C \mathbf{h}(\mathbf{r}) \cdot d\mathbf{r} = \int_{C_1} (-y\,\mathbf{i}) \cdot d\mathbf{r} + \int_{C_2} (-y\,\mathbf{i}) \cdot d\mathbf{r} + \int_{C_3} (-y\,\mathbf{i}) \cdot d\mathbf{r} + \int_{C_4} (-y\,\mathbf{i}) \cdot d\mathbf{r}$$

$$= \quad 0 \quad + \int_0^1 -\mathbf{i} \cdot (-2\,\mathbf{i})\, du + \quad 0 \quad + \int_0^1 \mathbf{i} \cdot (2\,\mathbf{i})\, du$$

$$= \quad 0 \quad + \int_0^1 2\, du \quad + \quad 0 \quad + \int_0^1 2\, du$$

$$= \quad 4$$

13. $\mathbf{h}(x,y) = (3x^2 y^3 + 2x)\,\mathbf{i} + (3x^3 y^2 - 4y)\,\mathbf{j}; \quad \dfrac{\partial P}{\partial y} = 9x^2 y^2 = \dfrac{\partial Q}{\partial x}.$ Thus $\mathbf{h}$ is a gradient.

(a) $\mathbf{r}(u) = u\mathbf{i} + e^u\,\mathbf{j}, \quad \mathbf{r}'(u) = \mathbf{i} + e^u\,\mathbf{j}, \quad u \in [0,1]$

$$\int_C \mathbf{h}(\mathbf{r}) \cdot d\mathbf{r} = \int_0^1 \left[(3u^2 e^{3u} + 2u) + 3u^3 e^{3u} - 4e^{2u} \right] du = \left[u^3 e^{3u} + u^2 - 2e^{2u} \right]_0^1 = e^3 - 2e^2 + 3$$

(b) $\dfrac{\partial f}{\partial x} = 3x^2 y^3 + 2x \implies f(x,y) = x^3 y^3 + x^2 + g(y);$

$\dfrac{\partial f}{\partial y} = 3x^3 y^2 + g'(y) = 3x^3 - 4y \implies g'(y) = -4y \implies g(y) = -2y^2$

Therefore, $f(x,y) = x^3 y^3 + x^2 - 2y^2.$

Now, at $u = 0$, $r(0) = 0\mathbf{i} + \mathbf{j} = (0,1)$; at $u = 1$, $r(1) = \mathbf{i} + e\mathbf{j} = (1,e)$ and

$$\int_C \mathbf{h}(\mathbf{r}) \cdot d\mathbf{r} = \left[x^3 y^3 + x^2 - 2y^2 \right]_{(0,1)}^{(1,e)} = e^3 - 2e^2 + 3$$

15. $\mathbf{h}(x,y) = (e^{2y} - 2xy)\,\mathbf{i} + (2xe^{2y} - x^2 + 1)\,\mathbf{j};\quad \dfrac{\partial P}{\partial y} = 2e^{2y} - 2x = \dfrac{\partial Q}{\partial x}.$ Thus $\mathbf{h}$ is a gradient.

(a) $\mathbf{r}(u) = ue^u\,\mathbf{i} + (1+u)\,\mathbf{j},\quad \mathbf{r}'(u) = (1+u)e^u\,\mathbf{i} + \mathbf{j},\quad u \in [0,1]$

$$\int_C \mathbf{h}(\mathbf{r})\cdot d\mathbf{r} = \int_0^1 \left[e^2(3ue^{3u} + e^{3u} - 2u^3e^{2u} - 5u^2e^{2u} - 2ue^{2u} + 1 \right]\,du$$
$$= \left[e^2 ue^{3u} - u^3e^{2u} - u^2e^{2u} + u \right]_0^1 = e^5 - 2e^2 + 1$$

(b) $\dfrac{\partial f}{\partial x} = e^{2y} - 2xy \implies f(x,y) = xe^{2y} - x^2y + g(y).$

$$\dfrac{\partial f}{\partial y} = 2xe^{2y} - x^2 + g'(y) = 3x^3 - 4y \implies g'(y) = 1 \implies g(y) = y$$

Therefore, $f(x,y) = xe^{2y} - x^2y + y.$

Now, at $u = 0$, $r(0) = 0\,\mathbf{i} + \mathbf{j} = (0,1)$; at $u = 1$, $r(1) = e\,\mathbf{i} + 2\,\mathbf{j} = (e,2)$ and

$$\int_C \mathbf{h}(\mathbf{r})\cdot d\mathbf{r} = \left[xe^{2y} - x^2y + y \right]_{(0,1)}^{(e,2)} = e^5 - 2e^2 + 1$$

17. $\mathbf{h}(x,y,z) = (2xz + \sin y)\,\mathbf{i} + x\cos y\,\mathbf{j} + x^2\,\mathbf{k};$

$$\dfrac{\partial P}{\partial y} = \cos y = \dfrac{\partial Q}{\partial x},\quad \dfrac{\partial P}{\partial z} = 2x = \dfrac{\partial R}{\partial x},\quad \dfrac{\partial Q}{\partial z} = 0 = \dfrac{\partial R}{\partial y}.$$ Thus $\mathbf{h}$ is a gradient.

$$\dfrac{\partial f}{\partial x} = 2xz + \sin y, \implies f(x,y,z) = x^2z + x\sin y + g(y,z)$$

$$\dfrac{\partial f}{\partial y} = x\cos y + \dfrac{\partial g}{\partial y} = x\cos y, \implies g(y,z) = h(z) \implies f(x,y,z) = x^2z + x\sin y + h(z)$$

$$\dfrac{\partial f}{\partial z} = x^2 + h'(z) = x^2 \implies h'(z) = 0 \implies h(z) = C$$

Therefore, $f(x,y,z) = x^2z + x\sin y$ (take $C = 0$)

$$\int_C \mathbf{h}(\mathbf{r})\cdot d\mathbf{r} = \int_C \nabla f \cdot d\mathbf{r} = \left[x^2z + x\sin y \right]_{\mathbf{r}(0)}^{\mathbf{r}(2\pi)} = \left[x^2z + x\sin y \right]_{(1,0,0)}^{(1,0,2\pi)} = 2\pi$$

19. $\mathbf{h}(x,y,z) = (2xy + z^2)\,\mathbf{i} + x^2\,\mathbf{j} + 2xz\,\mathbf{k};$

$$\dfrac{\partial P}{\partial y} = 2x = \dfrac{\partial Q}{\partial x},\quad \dfrac{\partial P}{\partial z} = 2z = \dfrac{\partial R}{\partial x},\quad \dfrac{\partial Q}{\partial z} = 0 = \dfrac{\partial R}{\partial y}.$$ Thus $\mathbf{h}$ is a gradient.

$$\dfrac{\partial f}{\partial x} = 2xy + z^2 \implies f(x,y,z) = x^2y + xz^2 + g(y,z)$$

$$\dfrac{\partial f}{\partial y} = x^2 + \dfrac{\partial g}{\partial y} = x^2 \implies g(y,z) = h(z) \implies f(x,y,z) = x^2y + xz^2 + h(z)$$

$$\dfrac{\partial f}{\partial z} = 2xz + h'(z) = 2xz \implies h'(z) = 0 \implies h(z) = C$$

Therefore, $f(x,y,z) = x^2y + xz^2$ (take $C = 0$)

$$\int_C \mathbf{h}(\mathbf{r}) \cdot d\mathbf{r} = \int_C \nabla f \cdot d\mathbf{r} = \left[x^2y + xz^2 \right]_{\mathbf{r}(0)}^{\mathbf{r}(1)} = \left[x^2y + xz^2 \right]_{(0,2,0)}^{(2,3,-1)} = 14$$

21. $\mathbf{F}(x,y) = (x + e^{2y})\,\mathbf{i} + (2y + 2xe^{2y})\,\mathbf{j}; \quad \dfrac{\partial P}{\partial y} = 2e^{2y} = \dfrac{\partial Q}{\partial x}.$ Thus $\mathbf{F}$ is a gradient.

$$\frac{\partial f}{\partial x} = x + e^{2y} \implies f(x,y) = \frac{1}{2}x^2 + xe^{2y} + g(y);$$

$$\frac{\partial f}{\partial y} = 2xe^{2y} + g'(y) = 2y + 2xe^{2y} \implies g'(y) = 2y \implies g(y) = y^2 \text{ (take } C = 0)$$

Therefore, $f(x,y) = \dfrac{1}{2}x^2 + xe^{2y} + y^2.$

$$\int_C \mathbf{F}(\mathbf{r}) \cdot d\mathbf{r} = \int_C \nabla f \cdot d\mathbf{r} = \left[\frac{1}{2}x^2 + xe^{2y} + y^2 \right]_{\mathbf{r}(0)}^{\mathbf{r}(2\pi)} = \left[\frac{1}{2}x^2 + xe^{2y} + y^2 \right]_{(3,0)}^{(3,0)} = 0$$

23. Set $f(x,y,z) = g(x)$ and $C : \mathbf{r}(u) = u\,\mathbf{i}, \quad u \in [a,b].$

In this case

$$\nabla f(\mathbf{r}(u)) = g'(x(u))\mathbf{i} = g'(u)\mathbf{i} \quad \text{and} \quad \mathbf{r}'(u) = \mathbf{i},$$

so that

$$\int_C \nabla f(\mathbf{r}) \cdot d\mathbf{r} = \int_a^b \left[\nabla f(\mathbf{r}(u)) \cdot \mathbf{r}'(u) \right] du = \int_a^b g'(u)\, du.$$

Since $\quad f(\mathbf{r}(b)) - f(\mathbf{r}(a)) = g(b) - g(a),$

$$\int_C \nabla f(\mathbf{r}) \cdot d\mathbf{r} = f(\mathbf{r}(b)) - f(\mathbf{r}(a)) \quad \text{gives} \quad \int_a^b g'(u)\, du = g(b) - g(a).$$

25. $\mathbf{F}(\mathbf{r}) = kr\,\mathbf{r} = k\sqrt{x^2+y^2+z^2}\,(x\,\mathbf{i} + y\,\mathbf{j} + z\,\mathbf{k}), \quad k > 0$ constant.

$$\frac{\partial P}{\partial y} = \frac{kxy}{\sqrt{x^2+y^2+z^2}} = \frac{\partial Q}{\partial x}, \quad \frac{\partial P}{\partial z} = \frac{kxz}{\sqrt{x^2+y^2+z^2}} = \frac{\partial R}{\partial x}, \quad \frac{\partial Q}{\partial z} = \frac{kyz}{\sqrt{x^2+y^2+z^2}} = \frac{\partial R}{\partial y}$$

Therefore, $\mathbf{F}$ is a gradient field.

$$\frac{\partial f}{\partial x} = kx\sqrt{x^2+y^2+z^2} \implies f(x,y,z) = \frac{k}{3}\left(x^2+y^2+z^2\right)^{3/2} + g(y,z).$$

$$\frac{\partial f}{\partial y} = ky\sqrt{x^2+y^2+z^2} + \frac{\partial g}{\partial y} = ky\sqrt{x^2+y^2+z^2} \implies f(x,y,z) = \frac{k}{3}\left(x^2+y^2+z^2\right)^{3/2} + h(z)$$

$$\frac{\partial f}{\partial z} = kz\sqrt{x^2+y^2+z^2} + h'(z) = kz\sqrt{x^2+y^2+z^2} \implies h(z) = C, \text{ constant}$$

Therefore, $f(x,y,z) = \frac{k}{3}\left(x^2+y^2+z^2\right)^{3/2} + C.$

27. $\mathbf{F}(\mathbf{r}) = \nabla\left(\dfrac{mG}{r}\right)$; $\quad W = \displaystyle\int_C \mathbf{F}(\mathbf{r})\cdot d\mathbf{r} = mG\left(\dfrac{1}{r_2} - \dfrac{1}{r_1}\right)$

29. $\mathbf{F}(x,y,z) = 0\,\mathbf{i} + 0\,\mathbf{j} + \dfrac{-mGr_0^2}{(r_0+z)^2}\,\mathbf{k}$; $\quad \dfrac{\partial P}{\partial y} = 0 = \dfrac{\partial Q}{\partial x}, \quad \dfrac{\partial P}{\partial z} = 0 = \dfrac{\partial R}{\partial x}, \quad \dfrac{\partial Q}{\partial z} = 0 = \dfrac{\partial R}{\partial y}.$

Therefore, $\mathbf{F}(x,y,z)$ is a gradient.

$\dfrac{\partial f}{\partial x} = 0 \implies f(x,y,z) = g(y,z); \quad \dfrac{\partial f}{\partial y} = \dfrac{\partial g}{\partial y} = 0 \implies g(y,z) = h(z).$

Therefore $f(x,y,z) = h(z)$.

Now $\dfrac{\partial f}{\partial z} = h'(z) = \dfrac{-mGr_0^2}{(r_0+z)^2} \implies f(x,y,z) = h(z) = \dfrac{mGr_0^2}{r_0+z}$

SECTION 18.3

1. If f is continuous, then $-f$ is continuous and has antiderivatives u. The scalar fields $U(x,y,z) = u(x)$ are potential functions for $\mathbf{F}$:

$$\nabla U = \dfrac{\partial U}{\partial x}\,\mathbf{i} + \dfrac{\partial U}{\partial y}\,\mathbf{j} + \dfrac{\partial U}{\partial z}\,\mathbf{k} = \dfrac{du}{dx}\,\mathbf{i} = -f\,\mathbf{i} = -\mathbf{F}.$$

3. The scalar field $U(x,y,z) = \alpha z + d$ is a potential energy function for $\mathbf{F}$. We know that the total mechanical energy remains constant. Thus, for any times t_1 and t_2,

$$\tfrac{1}{2}m[v(t_1)]^2 + U(\mathbf{r}(t_1)) = \tfrac{1}{2}m[v(t_2)]^2 + U(\mathbf{r}(t_2)).$$

This gives

$$\tfrac{1}{2}m[v(t_1)]^2 + \alpha z(t_1) + d = \tfrac{1}{2}m[v(t_2)]^2 + \alpha z(t_2) + d.$$

Solve this equation for $v(t_2)$ and you have the desired formula.

5. (a) We know that $-\nabla U$ points in the direction of maximum decrease of U. Thus $\mathbf{F} = -\nabla U$ attempts to drive objects toward a region where U has lower values.

(b) At a point where u has a minimum, $\nabla U = \mathbf{0}$ and therefore $\mathbf{F} = \mathbf{0}$.

7. (a) By conservation of energy $\tfrac{1}{2}mv^2 + U = E$. Since E is constant and U is constant, v is constant.

(b) ∇U is perpendicular to any surface where U is constant. Obviously so is $\mathbf{F} = -\nabla U$.

9. $f(x,y,z) = -\dfrac{k}{\sqrt{x^2+y^2+z^2}}$ is a potential function for $\mathbf{F}$. The work done by $\mathbf{F}$ moving an object along C is:

$$W = \int_C \mathbf{F}(\mathbf{r})\cdot d\mathbf{r} = \int_a^b \nabla f \cdot d\mathbf{r} = f[\mathbf{r}(b)] - f[\mathbf{r}(a)].$$

Since $\mathbf{r}(a) = (x_0, y_0, z_0)$ and $\mathbf{r}(b) = (x_1, y_1, z_1)$ are points on the unit sphere,

$$f[\mathbf{r}(b)] = f[\mathbf{r}(a)] = -k \quad \text{and so} \quad W = 0$$

SECTION 18.4

1. $\mathbf{r}(u) = u\,\mathbf{i} + 2u\,\mathbf{j}, \quad u \in [0,1]$

$$\int_C (x-2y)\,dx + 2x\,dy = \int_0^1 \{[x(u)-2y(u)]x'(u) + 2x(u)\,y'(u)\}\,du = \int_0^1 u\,du = \frac{1}{2}$$

3. $C = C_1 \cup C_2$

$C_1: \mathbf{r}(u) = u\,\mathbf{i}, \quad u \in [0,1]; \qquad C_2: \mathbf{r}(u) = \mathbf{i} + 2u\,\mathbf{j}, \quad u \in [0,1]$

$$\int_{C_1} (x-2y)\,dx + 2x\,dy = \int_{C_1} x\,dx = \int_0^1 x(u)\,x'(u)\,du = \int_0^1 u\,du = \frac{1}{2}$$

$$\int_{C_2} (x-2y)\,dx + 2x\,dy = \int_{C_2} 2x\,dy = \int_0^1 4\,du = 4$$

$$\int_C = \int_{C_1} + \int_{C_2} = \frac{9}{2}$$

5. $\mathbf{r}(u) = 2u^2\,\mathbf{i} + u\,\mathbf{j}, \quad u \in [0,1]$

$$\int_C y\,dx + xy\,dy = \int_0^1 [y(u)\,x'(u) + x(u)\,y(u)\,y'(u)]\,du = \int_0^1 (4u^2 + 2u^3)\,du = \frac{11}{6}$$

7. $C = C_1 \cup C_2 \qquad C_1: \mathbf{r}(u) = u\,\mathbf{j}, \quad u \in [0,1]; \qquad C_2: \mathbf{r}(u) = 2u\,\mathbf{i} + \mathbf{j}, \quad u \in [0,1]$

$$\int_{C_1} y\,dx + xy\,dy = 0$$

$$\int_{C_2} y\,dx + xy\,dy = \int_{C_2} y\,dx = \int_0^1 y(u)\,x'(u)\,du = \int_0^1 2\,du = 2$$

$$\int_C = \int_{C_1} + \int_{C_2} = 2$$

9. $\mathbf{r}(u) = 2u\,\mathbf{i} + 4u\,\mathbf{j}, \quad u \in [0,1]$

$$\int_C y^2\,dx + (xy - x^2)\,dy = \int_0^1 \{y^2(u)x'(u) + [x(u)y(u) - x^2(u)]\,y'(u)\}\,du$$

$$= \int_0^1 \left[(4u)^2(2) + (8u^2 - 4u^2)(4)\right]\,du = \int_0^1 48u^2\,du = 16$$

11. $\mathbf{r}(u) = \dfrac{1}{8}u^2\,\mathbf{i} + u\,\mathbf{j}, \quad u \in [0,4]$

$$\int_C y^2\,dx + (xy - x^2)\,dy = \int_0^4 \{y^2(u)x'(u) + [x(u)y(u) - x^2(u)]\,y'(u)\}\,du$$

$$= \int_0^4 \left[u^2\left(\frac{u}{4}\right) + \left(\frac{u^2}{8}(u) - \left(\frac{u^2}{8}\right)^2(1)\right)\right]\,du$$

$$= \int_0^4 \left[\frac{3}{8}u^3 - \frac{1}{64}u^4\right]\,du = \frac{104}{5}$$

13. $\mathbf{r}(u) = u\,\mathbf{i} + u\,\mathbf{j}, \quad u \in [0,1]$

$$\int_C (y^2 + 2x + 1)\,dx + (2xy + 4y - 1)\,dy$$

$$= \int_0^1 \left\{ [y^2(u) + 2x(u) + 1]x'(u) + [2x(u)y(u) + 4y(u) - 1]y'(u) \right\}\,du$$

$$\int_0^1 \left[(u^2 + 2u + 1) + (2u^2 + 4u - 1) \right]\,du = \int_0^1 (3u^2 + 6u)\,du = 4$$

15. $\mathbf{r}(u) = u\,\mathbf{i} + u^3\,\mathbf{j}, \quad u \in [0,1]$

$$\int_C (y^2 + 2x + 1)\,dx + (2xy + 4y - 1)\,dy$$

$$= \int_0^1 \left\{ [y^2(u) + 2x(u) + 1]x'(u) + [2x(u)y(u) + 4y(u) - 1]y'(u) \right\}\,du$$

$$= \int_0^1 \left[(u^6 + 2u + 1) + (2u^4 + 4u^3 - 1)3u^2 \right]\,du = \int_0^1 \left(7u^6 + 12u^5 - 3u^2 + 2u + 1 \right)\,du = 4$$

17. $\mathbf{r}(u) = u\,\mathbf{i} + u\,\mathbf{j} + u\,\mathbf{k}, \quad u \in [0,1]$

$$\int_C y\,dx + 2z\,dy + x\,dz = \int_0^1 [y(u)\,x'(u) + 2z(u)\,y'(u) + x(u)\,z'(u)]\,du = \int_0^1 4u\,du = 2$$

19. $C = C_1 \cup C_2 \cup C_3$

$C_1 : \mathbf{r}(u) = u\,\mathbf{k}, \ u \in [0,1]; \quad C_2 : \mathbf{r}(u) = u\,\mathbf{j} + \mathbf{k}, \ u \in [0,1]; \quad C_3 : \mathbf{r}(u) = u\,\mathbf{i} + \mathbf{j} + \mathbf{k}, \ u \in [0,1]$

$$\int_{C_1} y\,dx + 2z\,dy + x\,dz = 0$$

$$\int_{C_2} y\,dx + 2z\,dy + x\,dz = \int_{C_2} 2z\,dy = \int_0^1 2z(u)\,y'(u)\,du = \int_0^1 2\,du = 2$$

$$\int_{C_3} y\,dx + 2z\,dy + x\,dz = \int_{C_3} y\,dx = \int_0^1 y(u)\,x'(u)\,du = \int_0^1 du = 1$$

$$\int_C = \int_{C_1} + \int_{C_2} + \int_{C_3} = 3$$

21. $\mathbf{r}(u) = 2u\,\mathbf{i} + 2u\,\mathbf{j} + 8u\,\mathbf{k}, \quad u \in [0,1]$

$$\int_C xy\,dx + 2z\,dy + (y + z)\,dz$$

$$= \int_0^1 \left\{ x(u)y(u)x'(u) + 2z(u)y'(u) + [y(u) + z(u)]z'(u) \right\}\,du$$

$$= \int_0^1 \left[(2u)(2u)(2) + 2(8u)(2) + (2u + 8u)(8) \right]\,du$$

$$= \int_0^1 \left(8u^2 + 112u \right)\,du = \frac{176}{3}$$

23. $\mathbf{r}(u) = u\,\mathbf{i} + u\,\mathbf{j} + 2u^2\,\mathbf{k}, \quad u \in [0, 2]$

$$\int_C xy\,dx + 2z\,dy + (y + z)\,dz$$

$$= \int_0^2 \{x(u)y(u)x'(u) + 2z(u)y'(u) + [y(u) + z(u)]z'(u)\}\,du$$

$$= \int_0^2 \left[(u)(u)(1) + 2(2u^2)(1) + (u + 2u^2)(4u)\right]\,du$$

$$= \int_0^2 \left(8u^3 + 9u^2\right)\,du = 56$$

25. $\mathbf{r}(u) = (u - 1)\,\mathbf{i} + (1 + 2u^2)\,\mathbf{j} + u\,\mathbf{k}, \quad u \in [1, 2]$

$$\int_C x^2 y\,dx + y\,dy + xz\,dz$$

$$= \int_1^2 \left[x^2(u)y(u)x'(u) + y(u)y'(u) + x(u)z(u)z'(u)\right]\,du$$

$$= \int_1^2 \left[(u - 1)^2(1 + 2u^2)(1) + (1 + 2u^2)(4u) + (u - 1)u\right]\,du$$

$$= \int_1^2 \left(2u^4 + 4u^3 + 4u^2 + u + 1\right)\,du = \frac{1177}{30}$$

27. (a) $\dfrac{\partial P}{\partial y} = 6x - 4y = \dfrac{\partial Q}{\partial x}$

$$\frac{\partial f}{\partial x} = x^2 + 6xy - 2y^2 \quad\Longrightarrow\quad f(x, y) = \frac{1}{3}x^3 + 3x^2 y - 2xy^2 + g(y)$$

$$\frac{\partial f}{\partial y} = 3x^2 - 4xy + g'(y) = 3x^2 - 4xy + 2y \quad\Longrightarrow\quad g'(y) = 2y \quad\Longrightarrow\quad g(y) = y^2 + C$$

Therefore, $f(x, y) = \dfrac{1}{3}x^3 + 3x^2 y - 2xy^2 + y^2$ (take $C = 0$)

(b) (i) $\displaystyle\int_C (x^2 + 6xy - 2y^2)\,dx + (3x^2 - 4xy + 2y)\,dy = [f(x, y)]_{(3,0)}^{(0,4)} = 7$

(ii) $\displaystyle\int_C' (x^2 + 6xy - 2y^2)\,dx + (3x^2 - 4xy + 2y)\,dy = [f(x, y)]_{(4,0)}^{(0,3)} = -\frac{37}{3}$

29. $s'(u) = \sqrt{[x'(u)]^2 + [y'(u)]^2} = a$

(a) $M = \displaystyle\int_C k(x+y)\,ds = k\int_0^{\pi/2} [x(u)+y(u)]\,s'(u)\,du = ka^2\int_0^{\pi/2} (\cos u + \sin u)\,du = 2ka^2$

$$x_M M = \int_C kx(x+y)\,ds = k\int_0^{\pi/2} x(u)\,[x(u)+y(u)]\,s'(u)\,du$$

$$= ka^3\int_0^{\pi/2} (\cos^2 u + \cos u \sin u)\,du = \frac{1}{4}ka^3(\pi+2)$$

$$y_M M = \int_C ky(x+y)\,ds = k\int_0^{\pi/2} y(u)\,[x(u)+y(u)]\,s'(u)\,du$$

$$= ka^3\int_0^{\pi/2} (\sin u \cos u + \sin^2 u)\,du = \frac{1}{4}ka^3(\pi+2)$$

$x_M = y_M = \frac{1}{8}a(\pi+2)$

(b)

$$I = \int_C k(x+y)y^2\,ds = k\int_0^{\pi/2} \left[x(u)y^2(u) + y^3(u)\right] s'(u)\,du$$

$$= ka^4\int_0^{\pi/2} \left[\sin^2 u \cos u + \sin^3 u\right] du$$

$$= ka^4\int_0^{\pi/2} \left[\sin^2 u \cos u + (1 - \cos^2 u)\sin u\right] du$$

$$= ka^4\left[\frac{1}{3}\sin^3 u - \cos u + \frac{1}{3}\cos^3 u\right]_0^{\pi/2} = ka^4$$

$I = \frac{1}{2}a^2 M.$

31. (a) $I_z = \displaystyle\int_C k(x+y)a^2\,ds = a^2\int_C k(x+y)\,ds = a^2 M = Ma^2$

(b) The distance from a point (x^*, y^*) to the line $y = x$ is $|y^* - x^*|/\sqrt{2}$.   Therefore

$$I = \int_C k(x+y)\left[\frac{1}{2}(y-x)^2\right]ds = \frac{1}{2}k\int_0^{\pi/2} (a\cos u + a\sin u)(a\sin u - a\cos u)^2 a\,du$$

$$= \frac{1}{2}ka^4\int_0^{\pi/2} (\sin u - \cos u)^2 \frac{d}{du}(\sin u - \cos u)\,du$$

$$= \frac{1}{2}ka^4\left[\frac{1}{3}(\sin u - \cos u)^3\right]_0^{\pi/2} = \frac{1}{3}ka^4.$$

From Exercise 29, $M = 2ka^2$.   Therefore

$$I = \frac{1}{6}(2ka^2)a^2 = \frac{1}{6}Ma^2.$$

33. (a) $s'(u) = \sqrt{a^2 + b^2}$

$$L = \int_C ds = \int_0^{2\pi} \sqrt{a^2 + b^2}\,du = 2\pi\sqrt{a^2 + b^2}$$

(b) $x_M = 0$, $y_M = 0$ (by symmetry)

$$z_M = \frac{1}{L}\int_C z\,ds = \frac{1}{2\pi\sqrt{a^2+b^2}}\int_0^{2\pi} bu\sqrt{a^2+b^2}\,du = b\pi$$

(c) $I_x = \displaystyle\int_C \frac{M}{L}(y^2+z^2)\,ds = \frac{M}{2\pi}\int_0^{2\pi}(a^2\sin^2 u + b^2 u^2)\,du = \frac{1}{6}M(3a^2+8b^2\pi^2)$

$I_y = \frac{1}{6}M(3a^2+8b^2\pi^2)$ similarly

$I_z = Ma^2$ (all the mass is at distance a from the z-axis)

35.
$$M = \int_C k(x^2+y^2+z^2)\,ds$$

$$= k\sqrt{a^2+b^2}\int_0^{2\pi}(a^2+b^2u^2)\,du = \frac{2}{3}\pi k\sqrt{a^2+b^2}\,(3a^2+4\pi^2 b^2)$$

SECTION 18.5

1. (a) $\displaystyle\oint_C xy\,dx + x^2\,dy = \int_{C_1} xy\,dx + x^2\,dy + \int_{C_2} xy\,dx + x^2\,dy + \int_{C_3} xy\,dx + x^2\,dy$,   where

$C_1: \mathbf{r}(u) = u\mathbf{i} + u\mathbf{j}$, $u \in [0,1]$; $C_2: \mathbf{r}(u) = (1-u)\mathbf{i} + \mathbf{j}$, $u \in [0,1]$

$C_3: \mathbf{r}(u) = (1-u)\mathbf{j}$, $u \in [0,1]$.

$$\int_{C_1} xy\,dx + x^2\,dy = \int_0^1 (u^2 + u^2)\,du = \frac{2}{3}$$

$$\int_{C_2} xy\,dx + x^2\,dy = \int_0^1 -(1-u)\,du = -\frac{1}{2}$$

$$\int_{C_3} xy\,dx + x^2\,dy = \int_0^1 0^2(-1)\,du = 0$$

Therefore, $\displaystyle\oint_C xy\,dx + x^2\,dy = \frac{2}{3} - \frac{1}{2} = \frac{1}{6}$.

(b) $\displaystyle\oint_C xy\,dx + x^2\,dy = \iint_\Omega x\,dx\,dy = \int_0^1\int_0^y x\,dx\,dy = \int_0^1\left[\frac{1}{2}x^2\right]_0^y du = \frac{1}{2}\int_0^1 y^2\,dy = \frac{1}{6}$

3. (a) $C: \mathbf{r}(u) = 2\cos u\,\mathbf{i} + 3\sin u\,\mathbf{j}$, $u \in [0, 2\pi]$

$$\oint_C (3x^2+y)\,dx + (2x+y^3)\,dy$$

$$= \int_0^{2\pi}\left[(12\cos^2 u + 3\sin u)(-2\sin u) + (4\cos u + 27\sin^3 u)3\cos u\right]du$$

$$= \int_0^{2\pi}\left[-24\cos^2 u\,\sin u - 6\sin^2 u + 12\cos^2 u + 81\sin^3 u\,\cos u\right]du$$

$$= \left[8\cos^3 u - 3u + \frac{3}{2}\sin 2u + 6u + 3\sin 2u + \frac{81}{4}\sin^4 u\right]_0^{2\pi} = 6\pi$$

(b) $\displaystyle\oint_C (3x^2+y)\,dx + (2x+y^3)\,dy = \iint_\Omega 1\,dx\,dy = $ area of ellipse $\Omega = 6\pi$

5. $\oint_C 3y\,dx + 5x\,dy = \iint_\Omega (5-3)\,dxdy = 2A = 2\pi$

7. $\oint_C x^2\,dy = \iint_\Omega 2x\,dxdy = 2\overline{x}A = 2\left(\dfrac{a}{2}\right)(ab) = a^2 b$

9.
$$\oint_C (3xy + y^2)\,dx + (2xy + 5x^2)\,dy = \iint_\Omega [(2y + 10x) - (3x + 2y)]\,dxdy$$

$$= \iint_\Omega 7x\,dxdy = 7\,\overline{x}A = 7(1)(\pi) = 7\pi$$

11. $\oint_C (2x^2 + xy - y^2)\,dx + (3x^2 - xy + 2y^2)\,dy = \iint_\Omega [(6x - y) - (x - 2y)]\,dxdy$

$$= \iint_\Omega (5x + y)\,dxdy = (5\overline{x} + \overline{y})A = (5a + 0)(\pi r^2) = 5a\pi r^2$$

13. $\oint_C e^x \sin y\,dx + e^x \cos y\,dy = \iint_\Omega [e^x \cos y - e^x \cos y]\,dxdy = 0$

15. $\oint_C 2xy\,dx + x^2\,dy = \iint_\Omega [2x - 2x]\,dxdy = 0$

17. $C: \mathbf{r}(u) = a\cos u\,\mathbf{i} + a\sin u\,\mathbf{j}; \quad u \in [0, 2\pi]$

$$A = \oint_C -y\,dx = \int_0^{2\pi} (-a\sin u)(-a\sin u)\,du = a^2 \int_0^{2\pi} \sin^2 u\,du = a^2 \left[\dfrac{1}{2}u - \dfrac{1}{4}\sin 2u\right]_0^{2\pi} = \pi a^2$$

19. $A = \oint_C x\,dy,$ where $C = C_1 \cup C_2;$

$$C_1 : \mathbf{r}(u) = u\,\mathbf{i} + \dfrac{4}{u}\,\mathbf{j},\ 1 \le u \le 4; \quad C_2 : \mathbf{r}(u) = (4 - 3u)\,\mathbf{i} + (1 + 3u)\,\mathbf{j},\ 0 \le u \le 1.$$

$$\oint_{C_1} x\,dy = \int_1^4 u\left(\dfrac{-4}{u^2}\right)du = -4\int_1^4 \dfrac{1}{u}\,du = -4\ln 4;$$

$$\oint_{C_2} x\,dy = \int_0^1 (4 - 3u)3\,du = \int_0^1 (12 - 9u)\,du = \dfrac{15}{2}.$$

Therefore, $A = \frac{15}{2} - 4\ln 4.$

21. $\oint_C (ay + b)\,dx + (cx + d)\,dy = \iint_\Omega (c - a)\,dxdy = (c - a)A$

23. We take the arch from $x = 0$ to $x = 2\pi R$. (Figure 9.11.1)   Let C_1 be the line segment from $(0,0)$ to $(2\pi R, 0)$ and let C_2 be the cycloidal arch from $(2\pi R, 0)$ back to $(0,0)$. Letting $C = C_1 \cup C_2$, we have

$$A = \oint_C x\, dy = \int_{C_1} x\, dy + \int_{C_2} x\, dy = 0 + \int_{C_2} x\, dy$$

$$= \int_{2\pi}^{0} R(\theta - \sin\theta)(R\sin\theta)\, d\theta$$

$$= R^2 \int_{0}^{2\pi} (\sin^2\theta - \theta\sin\theta)\, d\theta$$

$$= R^2 \left[\frac{\theta}{2} - \frac{\sin 2\theta}{4} + \theta\cos\theta - \sin\theta \right]_{0}^{2\pi} = 3\pi R^2.$$

25. Taking Ω to be of type II (see Figure 18.5.2), we have

$$\iint_\Omega \frac{\partial Q}{\partial x}(x, y)\, dxdy = \int_c^d \int_{\psi_1(y)}^{\psi_2(y)} \frac{\partial Q}{\partial x}(x, y)\, dx\, dy$$

$$= \int_c^d \{ Q[\psi_2(y), y] - Q[\psi_1(y), y] \}\, dy$$

$$(*) = \int_c^d Q[\psi_2(y), y]\, dy - \int_c^d Q[\psi_1(y), y]\, dy.$$

The graph of $x = \psi_2(y)$ from $x = c$ to $x = d$ is the curve

$$C_4 : \mathbf{r}_4(u) = \psi_2(u)\,\mathbf{i} + u\,\mathbf{j}, \qquad u \in [c, d].$$

The graph of $x = \psi_1(y)$ from $x = c$ to $x = d$ is the curve

$$C_3 : \mathbf{r}_3(u) = \psi_1(u)\,\mathbf{i} + u\,\mathbf{j}, \qquad u \in [c, d].$$

Then

$$\oint_C Q(x, y)\, dy = \int_{C_4} Q(x, y)\, dy - \int_{C_3} Q(x, y)\, dy$$

$$= \int_c^d Q[\psi_2(u), u]\, du - \int_c^d Q[\psi_1(u), u]\, du.$$

Since u is a dummy variable, it can be replaced by y. Comparison with $(*)$ gives the result.

27. Suppose that f is harmonic. By Green's theorem,

$$\int_C \frac{\partial f}{\partial y}\, dx - \frac{\partial f}{\partial x}\, dy = \iint_\Omega \left(-\frac{\partial^2 f}{\partial^2 x} - \frac{\partial^2 f}{\partial^2 y} \right)\, dxdy = \iint_\Omega 0\, dxdy = 0.$$

29. $\displaystyle \oint_{C_1} = \oint_{C_2} + \oint_{C_3}$

31. $\displaystyle \frac{\partial P}{\partial y} = \frac{-2xy}{(x^2 + y^2)^2} = \frac{\partial Q}{\partial x}$ except at $(0,0)$

(a) If C does not enclose the origin, and Ω is the region enclosed by C, then

$$\oint_C \frac{x}{x^2+y^2}\,dx + \frac{y}{x^2+y^2}\,dy = \iint_\Omega 0\,dxdy = 0.$$

(b) If C does enclose the origin, then

$$\oint_C = \oint_{C_a}$$

where $C_a : \mathbf{r}(u) = a\cos u\,\mathbf{i} + a\sin u\,\mathbf{j}, \quad u \in [\,0, 2\pi\,]$ is a small circle in the inner region of C.
In this case

$$\oint_C = \int_0^{2\pi}\left[\frac{a\cos u}{a^2}(-a\sin u) + \frac{a\sin u}{a^2}(a\cos u)\right]du = \int_0^{2\pi} 0\,du = 0.$$

The integral is still 0.

33. If Ω is the region enclosed by C, then

$$\oint_C \mathbf{v}\cdot d\mathbf{r} = \oint_C \frac{\partial\phi}{\partial x}\,dx + \frac{\partial\phi}{\partial y}\,dy = \iint_\Omega \left\{\frac{\partial}{\partial x}\left(\frac{\partial\phi}{\partial y}\right) - \frac{\partial}{\partial y}\left(\frac{\partial\phi}{\partial x}\right)\right\}dxdy$$

$$= \iint_\Omega 0\,dxdy = 0.$$

equality of mixed partials

35. $A = \dfrac{1}{2}\oint_C(-y\,dx + x\,dy)$

$$= \left[\int_{C_1} + \int_{C_2} + \cdots \int_{C_n}\right]$$

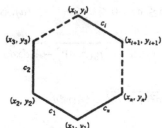

Now

$$\int_{C_i}(-y\,dx + x\,dy) = \int_0^1 \{[y_i + u(y_{i+1}-y_i)]\,(x_{i+1}-x_i) + [x_i + u(x_{i+1}-x_i)]\,(y_{i+1}-y_i)\}\,du$$

$$= x_iy_{i+1} - x_{i+1}y_i, \quad i = 1,2,\ldots,n; \ x_{n+1}=x_1,\ y_{n+1}=y_1$$

Thus, $A = \dfrac{1}{2}\,[(x_1y_2 - x_2y_1) + (x_2y_3 - x_3y_2) + \cdots + (x_ny_1 - x_1y_n)]$

SECTION 18.6

1. $4[(u^2-v^2)\mathbf{i} - (u^2+v^2)\mathbf{j} + 2uv\,\mathbf{k}]$ **3.** $2(\mathbf{j}-\mathbf{i})$

5. $\mathbf{r}(u,v) = 3\cos u\,\cos v\,\mathbf{i} + 2\sin u\,\cos v\,\mathbf{j} + 6\sin v\,\mathbf{k}, \quad u\in[\,0,2\pi\,],\ v\in[\,0,\pi/2\,]$

7. $\mathbf{r}(u,v) = 2\cos u\,\cos v\,\mathbf{i} + 2\sin u\,\cos v\,\mathbf{j} + 2\sin v\,\mathbf{k}, \quad u\in[\,0,2\pi\,],\ v\in(\,\pi/4,\pi/2\,]$

9. The surface consists of all points of the form $(x, g(x,z), z)$ with $(x,z) \in \Omega$. This set of points is given by

$$\mathbf{r}(u,v) = u\,\mathbf{i} + g(u,v)\,\mathbf{j} + v\,\mathbf{k}, \quad (u,v) \in \Omega.$$

11. $x^2/a^2 + y^2/b^2 + z^2/c^2 = 1$;   ellipsoid

13. $x^2/a^2 - y^2/b^2 = z$;   hyperbolic paraboloid

15. For each $v \in [a, b]$, the points on the surface at level $z = f(v)$ form a circle of radius v.
That circle can be parametrized:
$$\mathbf{R}(u) = v \cos u\,\mathbf{i} + v \sin u\,\mathbf{j} + f(v)\mathbf{k}, \quad u \in [0, 2\pi].$$

Letting v range over $[a,b]$, we obtain the entire surface:
$$\mathbf{r}(u,v) = v \cos u\,\mathbf{i} + v \sin u\,\mathbf{j} + f(v)\mathbf{k}; \quad 0 \le u \le 2\pi, \quad a \le v \le b.$$

17. Since γ is the angle between p and the xy-plane, γ is the angle between the upper normal to p and $\mathbf{k}$. (Draw a figure.) Therefore, by 18.6.5,

$$\text{area of } \Gamma = \iint_\Omega \sec\gamma\,dxdy = (\sec\gamma)A_\Omega = A_\Omega \quad \sec\gamma.$$

$$\gamma \text{ is constant}$$

19. The surface is the graph of the function
$$f(x,y) = c\left(1 - \frac{x}{a} - \frac{y}{b}\right) = \frac{c}{ab}(ab - bx - ay)$$
defined over the triangle $\Omega : 0 \le x \le a, \quad 0 \le y \le b(1 - x/a)$. Note that Ω has area $\frac{1}{2}ab$.

$$A = \iint_\Omega \sqrt{[f_x'(x,y)]^2 + [f_y'(x,y)]^2 + 1}\;dxdy$$

$$= \iint_\Omega \sqrt{c^2/a^2 + c^2/b^2 + 1}\;dxdy$$

$$= \frac{1}{ab}\sqrt{a^2b^2 + a^2c^2 + b^2c^2}\iint_\Omega dx\,dy = \frac{1}{2}\sqrt{a^2b^2 + a^2c^2 + b^2c^2}.$$

21. $f(x,y) = x^2 + y^2, \quad \Omega : 0 \le x^2 + y^2 \le 4$

$$A = \iint_\Omega \sqrt{4x^2 + 4y^2 + 1}\;dx\,dy \qquad [\text{change to polar coordinates}]$$

$$= \int_0^{2\pi}\int_0^2 \sqrt{4r^2 + 1}\,r\,dr\,d\theta$$

$$= 2\pi\left[\tfrac{1}{12}(4r^2 + 1)^{3/2}\right]_0^2 = \tfrac{1}{6}\pi(17\sqrt{17} - 1)$$

23. $f(x,y) = a^2 - (x^2 + y^2), \quad \Omega : \frac{1}{4}a^2 \le x^2 + y^2 \le a^2$

$$A = \iint_\Omega \sqrt{4x^2 + 4y^2 + 1}\, dxdy \qquad [\text{change to polar coordinates}]$$

$$= \int_0^{2\pi} \int_{a/2}^a r\sqrt{4r^2 + 1}\, dr\, d\theta = 2\pi \left[\frac{1}{12}(4r^2 + 1)^{3/2} \right]_{a/2}^a$$

$$= \frac{\pi}{6} \left[(4a^2 + 1)^{3/2} - (a^2 + 1)^{3/2} \right]$$

25. $f(x,y) = \frac{1}{3}(x^{3/2} + y^{3/2}), \quad \Omega : 0 \le x \le 1, \quad 0 \le y \le x$

$$A = \iint_\Omega \frac{1}{2}\sqrt{x + y + 4}\, dxdy$$

$$= \int_0^1 \int_0^x \frac{1}{2}\sqrt{x + y + 4}\, dy\, dx = \int_0^1 \left[\frac{1}{3}(x + y + 4)^{3/2} \right]_0^x dx$$

$$= \int_0^1 \frac{1}{3}\left[(2x + 4)^{3/2} - (x + 4)^{3/2} \right] dx = \frac{1}{3}\left[\frac{1}{5}(2x + 4)^{5/2} - \frac{2}{5}(x + 4)^{5/2} \right]_0^1$$

$$= \frac{1}{15}(36\sqrt{6} - 50\sqrt{5} + 32)$$

27. The surface $x^2 + y^2 + z^2 - 4z = 0$ is a sphere of radius 2 centered at $(0,0,2)$:

$$x^2 + y^2 + z^2 - 4z = 0 \iff x^2 + y^2 + (z - 2)^2 = 4.$$

The quadric cone $z^2 = 3(x^2 + y^2)$ intersects the sphere at height $z = 3$:

$$\left. \begin{array}{r} x^2 + y^2 + z^2 - 4z = 0 \\ z^2 = 3(x^2 + y^2) \end{array} \right\} \implies \begin{array}{c} 3(x^2 + y^2) + 3z^2 - 12z = 0 \\ 4z^2 - 12z = 0 \\ z = 3. \quad (\text{since } z \ge 2) \end{array}$$

The surface of which we are asked to find the area is a spherical segment of width 1 (from $z = 3$ to $z = 4$) in a sphere of radius 2. The area of the segment is 4π. (Exercise 27, Section 9.9.)

A more conventional solution. The spherical segment is the graph of the function

$$f(x,y) = 2 + \sqrt{4 - (x^2 + y^2)}, \quad \Omega : 0 \le x^2 + y^2 \le 3.$$

Therefore

$$A = \iint_\Omega \sqrt{\left(\frac{-x}{\sqrt{4 - x^2 - y^2}} \right)^2 + \left(\frac{-y}{\sqrt{4 - x^2 - y^2}} \right)^2 + 1}\, dxdy$$

$$= \iint_\Omega \frac{2}{\sqrt{4 - (x^2 + y^2)}}\, dxdy$$

$$= \int_0^{2\pi} \int_0^{\sqrt{3}} \frac{2r}{\sqrt{4 - r^2}}\, dr\, d\theta \qquad [\text{changed to polar coordinates}]$$

$$= 2\pi \left[-2\sqrt{4 - r^2} \right]_0^{\sqrt{3}} = 4\pi$$

29. (a) $$\iint_\Omega \sqrt{\left[\frac{\partial g}{\partial y}(y,z)\right]^2 + \left[\frac{\partial g}{\partial z}(y,z)\right]^2 + 1}\; dydz = \iint_\Omega \sec\left[\alpha(y,z)\right] dydz$$

where α is the angle between the unit normal with positive $\mathbf{i}$ component and the positive x-axis

(b) $$\iint_\Omega \sqrt{\left[\frac{\partial h}{\partial x}(x,z)\right]^2 + \left[\frac{\partial h}{\partial z}(x,z)\right]^2 + 1}\; dxdz = \iint_\Omega \sec\left[\beta(x,z)\right] dxdz$$

where β is the angle between the unit normal with positive $\mathbf{j}$ component and the positive y-axis

31. (a) $\mathbf{N}(u,v) = v\cos u \sin\alpha\cos\alpha\,\mathbf{i} + v\sin u\sin\alpha\cos\alpha\,\mathbf{j} - v\sin^2\alpha\,\mathbf{k}$

(b) $$A = \iint_\Omega \|\mathbf{N}(u,v)\|\,dudv = \iint_\Omega v\sin\alpha\,dudv$$
$$= \int_0^{2\pi}\int_0^s v\sin\alpha\,dv\,du = \pi s^2\sin\alpha$$

33. (a) Set $x = a\cos u\,\cosh v,\quad y = b\sin u\,\cosh v,\quad z = c\sinh v$. Then,
$$\frac{x^2}{a^2} + \frac{y^2}{b^2} - \frac{z^2}{c^2} = 1.$$

(b)

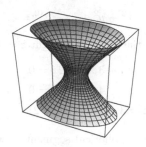

(c) $$A = \iint_\Omega \|\mathbf{N}(u,v)\|\,dv\,du$$
$$= \int_0^{2\pi}\int_{-\ln 2}^{\ln 2} \sqrt{64\cos^2 u\,\cosh^2 v + 144\sin^2 u\,\cosh^2 v + 36\cosh^2 v\,\sinh^2 v}\; dv\,du$$

35. $A = \sqrt{A_1{}^2 + A_2{}^2 + A_3{}^2};$ the unit normal to the plane of Ω is a vector of the form
$$\cos\gamma_1\,\mathbf{i} + \cos\gamma_2\,\mathbf{j} + \cos\gamma_3\,\mathbf{k}.$$
Note that
$$A_1 = A\cos\gamma_1,\quad A_2 = A\cos\gamma_2,\quad A_3 = A\cos\gamma_3.$$
Therefore
$$A_1{}^2 + A_2{}^2 + A_3{}^2 = A^2[\cos^2\gamma_1 + \cos^2\gamma_2 + \cos^2\gamma_3] = A^2.$$

37. (a) (We use Exercise 36.) $f(r,\theta) = r + \theta;$ $\Omega : 0 \le r \le 1,$ $0 \le \theta \pi$

$$A = \iint_{\Omega} \sqrt{r^2\left[f_r'(r,\theta)\right]^2 + \left[f_\theta'(r,\theta)\right]^2 + r^2}\, dr d\theta = \iint_{\Omega} \sqrt{2r^2 + 1}\, dr d\theta$$

$$= \int_0^\pi \int_0^1 \sqrt{2r^2 + 1}\, dr\, d\theta = \frac{1}{4}\sqrt{2}\pi\left[\sqrt{6} + \ln\left(\sqrt{2} + \sqrt{3}\right)\right]$$

(b) $f(r,\theta) = re^\theta;$ $\Omega : 0 \le r \le a,$ $0 \le \theta \le 2\pi$

$$A = \iint_{\Omega} r\sqrt{2e^{2\theta} + 1}\, dr d\theta = \left(\int_0^{2\pi} \sqrt{2e^{2\theta} + 1}\, d\theta\right)\left(\int_0^a r\, dr\right)$$

$$= \tfrac{1}{2}a^2\left[\sqrt{2e^{4\pi} + 1} - \sqrt{3} + \ln\left(1 + \sqrt{3}\right) - \ln\left(1 + \sqrt{2e^{4\pi} + 1}\right)\right]$$

SECTION 18.7

For Exercises 1–6 we have $\sec\left[\gamma(x,y)\right] = \sqrt{y^2 + 1}.$ $\mathbf{N}(x,y) = -y\mathbf{j} + \mathbf{k},$ so $\|N(x,y)\| = \sqrt{y^2 + 1}.$

1. $\displaystyle\iint_S d\sigma = \int_0^1 \int_0^1 \sqrt{y^2 + 1}\, dx\, dy = \int_0^1 \sqrt{y^2 + 1}\, dy = \frac{1}{2}[\sqrt{2} + \ln(1 + \sqrt{2})]$

3. $\displaystyle\iint_S 3y\, d\sigma = \int_0^1 \int_0^1 3y\sqrt{y^2 + 1}\, dy\, dx = \int_0^1 3y\sqrt{y^2 + 1}\, dy = \left[(y^2 + 1)^{3/2}\right]_0^1 = 2\sqrt{2} - 1$

5. $\displaystyle\iint_S \sqrt{2}z\, d\sigma = \iint_S y\, d\sigma = \frac{1}{3}(2\sqrt{2} - 1)$ (Exercise 3)

7. $\displaystyle\iint_S xy\, d\sigma;$ $S : \mathbf{r}(u,v) = (6 - 2u - 3v)\mathbf{i} + u\mathbf{j} + v\mathbf{k},$ $0 \le u \le 3 - \frac{3}{2}v,$ $0 \le v \le 2$

$$\|\mathbf{N}(u,v)\| = \|(-2\mathbf{i} + \mathbf{j}) \times (-3\mathbf{i} + \mathbf{k})\| = \sqrt{14}$$

$$\iint_S xy\, d\sigma = \sqrt{14}\iint_{\Omega} x(u,v)y(u,v)\, du\, dv$$

$$= \sqrt{14}\iint_{\Omega} (6 - 2u - 3v)u\, du\, dv$$

$$= \sqrt{14}\int_0^2 \int_0^{3-3v/2} (6u - 2u^2 - 3uv)\, du\, dv$$

$$= \sqrt{14}\left[3\left(3 - \tfrac{3}{2}v\right)^2 - \tfrac{2}{3}\left(3 - \tfrac{3}{2}v\right)^3 - \tfrac{3}{2}v\left(3 - \tfrac{3}{2}v\right)^2\right] dv = \frac{9}{2}\sqrt{14}$$

9. $\displaystyle\iint_S x^2 z \, d\sigma; \quad S : \mathbf{r}(u,v) = (\cos u \, \mathbf{i} + v \, \mathbf{j} + \sin u \, \mathbf{k}, \quad 0 \le u \le \pi, \quad 0 \le v \le 2.$

$$\mathbf{N}(u,v) = \begin{vmatrix} \mathbf{i} & \mathbf{j} & \mathbf{k} \\ -\sin u & 0 & \cos u \\ 0 & 1 & 0 \end{vmatrix} = -\cos u \, \mathbf{i} - \sin u \, \mathbf{k} \quad \text{and} \quad \|\mathbf{N}(u,v)\| = 1.$$

$$\iint_S x^2 z \, d\sigma = \iint_\Omega \cos^2 u \, \sin u \, du \, dv = \int_0^2 \int_0^\pi \cos^2 u \, \sin u \, du \, dv = \frac{4}{3}$$

11. $\displaystyle\iint_S (x^2 + y^2) \, d\sigma; \quad S : \mathbf{r}(u,v) = \cos u \cos v \, \mathbf{i} + \cos u \sin v \, \mathbf{j} + \sin u \, \mathbf{k}, \quad 0 \le u \le \pi/2, \quad 0 \le v \le 2\pi.$

$$\mathbf{N}(u,v) = \begin{vmatrix} \mathbf{i} & \mathbf{j} & \mathbf{k} \\ -\sin u \cos v & -\sin u \sin v & \cos u \\ -\cos u \sin v & \cos u \cos v & 0 \end{vmatrix} = -\cos^2 u \cos v \, \mathbf{i} + \cos^2 u \sin v \, \mathbf{j} - \sin u \cos u \, \mathbf{k};$$

$\|\mathbf{N}(u,v)\| = \cos u.$

$$\iint_S (x^2 + y^2) \, d\sigma = \iint_\Omega \cos^2 u \cos u \, du \, dv = \int_0^{2\pi} \int_0^{\pi/2} \cos^3 u \, du \, dv = \frac{4}{3}\pi$$

13. $\displaystyle M = \iint_S \lambda(x,y,x) \, d\sigma = \int_0^a \int_0^{a-x} k\sqrt{3} \, dy \, dx = \int_0^a k\sqrt{3}\,(a-x) \, dx = \frac{1}{2}a^2 k\sqrt{3}$

15. $\displaystyle M = \iint_S \lambda(x,y,z) \, d\sigma = \int_0^a \int_0^{a-x} kx^2 \sqrt{3} \, dy \, dx = \int_0^a k\sqrt{3}x^2(a-x) \, dx = \frac{1}{12}a^4 k\sqrt{3}$

17. $S : \mathbf{r}(u,v) = a\cos u \cos v \, \mathbf{i} + a\sin u \cos v \, \mathbf{j} + a\sin v \, \mathbf{k}$ with $0 \le u \le 2\pi, \quad 0 \le v \le \frac{1}{2}\pi.$ By a previous calculation $\|\mathbf{N}(u,v)\| = a^2 \cos v.$

$\overline{x} = 0, \quad \overline{y} = 0 \quad$ (by symmetry)

$$\overline{z}A = \iint_S z \, d\sigma = \iint_\Omega z(u,v) \, \|\mathbf{N}(u,v)\| \, du\,dv = \int_0^{2\pi} \int_0^{\pi/2} a^3 \sin v \cos v \, dv \, du = \pi a^3$$

$\overline{z} = \frac{1}{2}a \quad \text{since} \quad A = 2\pi a^2$

19. $\mathbf{N}(u,v) = (\mathbf{i} + \mathbf{j} + 2\mathbf{k}) \cdot (\mathbf{i} - \mathbf{j}) = 2\mathbf{i} + 2\mathbf{j} - 2\mathbf{k}$

flux in the direction of $\mathbf{N} = \iint_S \left(\mathbf{v} \cdot \dfrac{\mathbf{N}}{\|\mathbf{N}\|} \right) d\sigma = \iint_\Omega [\mathbf{v}(x(u), y(u), z(u)) \cdot \mathbf{N}(u,v)] \, du\, dv$

$$= \iint_\Omega [(u+v)\mathbf{i} - (u-v)\mathbf{j}] \cdot [2\mathbf{i} + 2\mathbf{j} - 2\mathbf{k}] \, du\, dv.$$

$$= \iint_\Omega 4v \, du\, dv = 4 \int_0^1 \int_0^1 v \, dv\, du = 2$$

21. With $\quad \mathbf{v} = z\,\mathbf{k}$

$$\text{flux} = \iint_S (\mathbf{v} \cdot \mathbf{n})\, d\sigma = \frac{1}{a} \iint_S z^2 \, d\sigma = \frac{1}{a} \iint_\Omega (a^2 \sin^2 v)(a^2 \cos v)\, du\, dv$$

$$= a^3 \int_0^{2\pi} \int_{-\pi/2}^{\pi/2} (\sin^2 v \cos v)\, du\, dv = \frac{4}{3}\pi a^3$$

23. With $\quad \mathbf{v} = y\,\mathbf{i} - x\,\mathbf{j}$

$$\text{flux} = \iint_S (\mathbf{v} \cdot \mathbf{n})\, d\sigma = \frac{1}{a} \iint_S \underbrace{(yx - xy)}_{0} \, d\sigma = 0$$

25. With $\quad \mathbf{v} = x\,\mathbf{i} + y\,\mathbf{j} + z\,\mathbf{k}$

$$\text{flux} = \iint_S (\mathbf{v} \cdot \mathbf{n})\, d\sigma = \iint_\Omega (-v_1 f'_x - v_2 f'_y + v_3)\, dx\, dy$$

$$= \iint_\Omega [-x(-1) - y(-1) + (a - x - y)]\, dx\, dy = a \iint_\Omega dx\, dy = aA = \frac{1}{2}\sqrt{3}\, a^3$$

27. With $\quad \mathbf{v} = x^2\,\mathbf{i} - y^2\,\mathbf{j}$

$$\text{flux} = \iint_S (\mathbf{v} \cdot \mathbf{n})\, d\sigma = \iint_\Omega (-v_1 f'_x - v_2 f'_y + v_3)\, dx\, dy$$

$$= \iint_\Omega [-x^2(-1) - (-y^2)(-1) + 0]\, dx\, dy = \int_0^a \int_0^{a-x} (x^2 - y^2)\, dy\, dx$$

$$= \int_0^a \left[ax^2 - x^3 - \frac{1}{3}(a-x)^3 \right] dx = \left[\frac{1}{3}ax^3 - \frac{1}{4}x^4 + \frac{1}{12}(a-x)^4 \right]_0^a = 0$$

29. With $\quad \mathbf{v} = xz\,\mathbf{j} - xy\,\mathbf{k}$

$$\text{flux} = \iint_S (\mathbf{v} \cdot \mathbf{n})\, d\sigma = \iint_\Omega (-v_1 f'_x - v_2 f'_y + v_3)\, dx\, dy$$

$$= \iint_\Omega (-x^3 y - xy)\, dx\, dy = \int_0^1 \int_0^2 (-x^3 y - xy)\, dy\, dx$$

$$= \int_0^1 -2(x^3 + x)\, dx = -\frac{3}{2}$$

31. $\mathbf{n} = \dfrac{1}{a}(x\,\mathbf{i} + y\,\mathbf{j})$

$$\text{flux} = \iint_S (\mathbf{v} \cdot \mathbf{n})\, d\sigma = \frac{1}{a} \iint_S [(x\,\mathbf{i} + y\,\mathbf{j} + z\,\mathbf{k}) \cdot (x\,\mathbf{i} + y\,\mathbf{j})]\, d\sigma$$

$$= \frac{1}{a} \iint_S (x^2 + y^2)\, d\sigma = a \iint_S d\sigma = a\,(\text{area of } S) = a\,(2\pi a l) = 2\pi a^2 l$$

33.   With $\mathbf{v} = x\,\mathbf{i} - y\,\mathbf{j} + \frac{3}{2}\,z\,\mathbf{k}$

$$\text{flux} = \iint_S (\mathbf{v} \cdot \mathbf{n})\, d\sigma = \iint_\Omega (-v_1 f'_x - v_2 f'_y + v_3)\, dxdy = \iint_\Omega 2y^{3/2}\, dxdy$$

$$= \int_0^1 \int_0^{1-x} 2y^{3/2}\, dy\, dx = \int_0^1 \frac{4}{5}(1-x)^{5/2}\, dx = \frac{8}{35}$$

35.   With $\mathbf{v} = y^2\,\mathbf{j}$

$$\text{flux} = \iint_S (\mathbf{v} \cdot \mathbf{n})\, d\sigma = \iint_\Omega (-v_1 f'_x - v_2 f'_y + v_3)\, dxdy = \iint_\Omega -y^{5/2}\, d\sigma$$

$$= \int_0^1 \int_0^{1-x} -y^{5/2}\, dy\, dx = \int_0^1 -\frac{2}{7}(1-x)^{7/2}\, dx = -\frac{4}{63}$$

37. $\overline{x} = 0,\quad \overline{y} = 0$ by symmetry.   You can verify that $\|\mathbf{N}(u,v)\| = v \sin \alpha$.

$$\overline{z}A = \iint_S z\, d\sigma = \iint_\Omega (s \cos \alpha)(v \sin \alpha)\, du\, dv = \sin \alpha \cos \alpha \int_0^{2\pi} \int_0^s v^2\, dv\, du = \tfrac{2}{3}\pi \sin \alpha \cos \alpha\, s^3$$

$$\overline{z} = \tfrac{2}{3}s \cos \alpha \quad \text{since} \quad A = \pi s^2 \sin \alpha$$

39. $f(x,y) = \sqrt{x^2 + y^2}$ on $\Omega : 0 \le x^2 + y^2 \le 1;\quad \lambda(x,y,z) = k\sqrt{x^2 + y^2}$

$x_M = 0,\quad y_M = 0$ (by symmetry)

$$z_M M = \iint_S z\lambda(x,y,z)\, d\sigma = \iint_\Omega k(x^2 + y^2) \sec\left[\gamma(x,y)\right]\, dxdy$$

$$= k\sqrt{2} \iint_\Omega (x^2 + y^2)\, dxdy$$

$$= k\sqrt{2} \int_0^{2\pi} \int_0^1 r^3\, dr\, d\theta = \frac{1}{2}\sqrt{2}\pi k$$

$z_M = \tfrac{3}{4}$ since $M = \tfrac{2}{3}\sqrt{2}\pi k$ (Exercise 38)

41.   no answer required

43. $x_M M = \displaystyle\iint\limits_{S} x\lambda\,(x,y,z)\,d\sigma = \iint\limits_{S} kx(y^2+z^2)\,d\sigma$

$= 2\sqrt{3}\,k \displaystyle\iint\limits_{\Omega} (u+v)\left[(u-v)^2+4u^2\right]\,du\,dv$

$= 2\sqrt{3}\,k \displaystyle\int_0^1 \int_0^1 (5u^3-2u^2v+uv^2+5u^2v-2uv^2+v^3)\,dv\,du$

$= 2\sqrt{3}\,k \displaystyle\int_0^1 \left(5u^3-u^2+\frac{1}{3}u+\frac{5}{2}u^2-\frac{2}{3}u+\frac{1}{4}\right)\,du = \frac{11}{3}\sqrt{3}k$

$x_M = \dfrac{11}{9}$ since $M = 3\sqrt{3}k$ (Exercise 42)

45. Total flux out of the solid is 0. It is clear from a diagram that the outer unit normal to the cylindrical side of the solid is given by $\mathbf{n} = x\,\mathbf{i}+y\,\mathbf{j}$ in which case $\mathbf{v}\cdot\mathbf{n}=0$. The outer unit normals to the top and bottom of the solid are $\mathbf{k}$ and $-\mathbf{k}$ respectively. So, here as well, $\mathbf{v}\cdot\mathbf{n}=0$ and the total flux is 0.

47. The surface $z = \sqrt{2-(x^2+y^2)}$ is the upper half of the sphere $x^2+y^2+z^2=2$. The surface intersects the surface $z=x^2+y^2$ in a circle of radius 1 at height $z=1$. Thus the upper boundary of the solid, call it S_1, is a segment of width $\sqrt{2}-1$ on a sphere of radius $\sqrt{2}$. The area of S_1 is therefore $2\pi\sqrt{2}(\sqrt{2}-1)$. (Exercise 27, Section 9.9). The upper unit normal to S_1 is the vector

$$\mathbf{n} = \frac{1}{\sqrt{2}}(x\,\mathbf{i}+y\,\mathbf{j}+z\,\mathbf{k}).$$

Therefore

$$\text{flux through } S_1 = \iint\limits_{S_1} (\mathbf{v}\cdot\mathbf{n})\,d\sigma = \frac{1}{\sqrt{2}}\iint\limits_{S_1} \overbrace{(x^2+y^2+z^2)}^{2}\,d\sigma$$

$$= \sqrt{2}\iint\limits_{S_1} d\sigma = \sqrt{2}\,(\text{area of } S_1) = 4\pi(\sqrt{2}-1).$$

The lower boundary of the solid, call it S_2, is the graph of the function

$$f(x,y) = x^2+y^2 \quad \text{on} \quad \Omega : 0 \le x^2+y^2 \le 1.$$

Taking $\mathbf{n}$ as the lower unit normal, we have

$$\text{flux through } S_2 = \iint\limits_{S_2} (\mathbf{v}\cdot\mathbf{n})\,d\sigma = \iint\limits_{\Omega} \left(v_1 f_x' + v_2 f_y' - v_3\right)\,dx\,dy$$

$$= \iint\limits_{\Omega} (x^2+y^2)\,dx\,dy = \int_0^{2\pi}\int_0^1 r^3\,dr\,d\theta = \frac{1}{2}\pi.$$

The total flux out of the solid is $4\pi(\sqrt{2}-1)+\dfrac{1}{2}\pi = (4\sqrt{2}-\dfrac{7}{2})\pi.$

SECTION 18.8

1. $\nabla \cdot \mathbf{v} = 2$, $\quad \nabla \times \mathbf{v} = \mathbf{0}$ 3. $\quad \nabla \cdot \mathbf{v} = 0$, $\quad \nabla \times \mathbf{v} = \mathbf{0}$ 5. $\quad \nabla \cdot \mathbf{v} = 6$, $\quad \nabla \times \mathbf{v} = \mathbf{0}$

7. $\nabla \cdot \mathbf{v} = yz + 1$, $\quad \nabla \times \mathbf{v} = -x\,\mathbf{i} + xy\,\mathbf{j} + (1-x)z\,\mathbf{k}$

9. $\nabla \cdot \mathbf{v} = 1/r^2$, $\quad \nabla \times \mathbf{v} = \mathbf{0}$

11. $\nabla \cdot \mathbf{v} = 2(x+y+z)e^{r^2}$, $\qquad \nabla \times \mathbf{v} = 2e^{r^2}\left[(y-z)\mathbf{i} - (x-z)\mathbf{j} + (x-y)\,\mathbf{k}\right]$

13. $\nabla \cdot \mathbf{v} = f'(x)$, $\quad \nabla \times \mathbf{v} = \mathbf{0}$ 15.   use components.

17. $\nabla \cdot \mathbf{v} = \dfrac{\partial P}{\partial x} + \dfrac{\partial Q}{\partial y} + \dfrac{\partial R}{\partial z} = 2 + 4 - 6 = 0$

19. $\nabla \times \mathbf{F} = \begin{vmatrix} \mathbf{i} & \mathbf{j} & \mathbf{k} \\ \dfrac{\partial}{\partial x} & \dfrac{\partial}{\partial y} & \dfrac{\partial}{\partial z} \\ x & y & -2z \end{vmatrix} = \mathbf{0}$

21. $\nabla^2 f = 12(x^2 + y^2 + z^2)$ 23. $\quad \nabla^2 f = 2y^3 z^4 + 6x^2 yz^4 + 12x^2 y^3 z^2$

25. $\nabla^2 f = e^r(1 + 2r^{-1})$ 27.   (a) $2r^2$ (b) $-1/r$

29. $\nabla^2 f = \nabla^2 g(r) = \nabla \cdot (\nabla g(r)) = \nabla \cdot \left(g'(r)r^{-1}\mathbf{r}\right)$

$$= \left[(\nabla g'(r)) \cdot r^{-1}\mathbf{r}\right] + g'(r)\left(\nabla \cdot r^{-1}\mathbf{r}\right)$$

$$= \left\{\left[g''(r)r^{-1}\mathbf{r}\right] \cdot r^{-1}\mathbf{r}\right\} + g'(r)(2r^{-1})$$

$$= g''(r) + 2r^{-1}g'(r)$$

31. $\dfrac{\partial f}{\partial x} = 2x + y + 2z$, $\quad \dfrac{\partial^2 f}{\partial x^2} = 2$; $\quad \dfrac{\partial f}{\partial y} = 4y + x - 3z$, $\quad \dfrac{\partial^2 f}{\partial y^2} = 4$;

$$\dfrac{\partial f}{\partial z} = -6z + 2x - 3y, \quad \dfrac{\partial^2 f}{\partial z^2} = -6;$$

$$\dfrac{\partial^2 f}{\partial x^2} + \dfrac{\partial^2 f}{\partial y^2} + \dfrac{\partial^2 f}{\partial z^2} = 2 + 4 - 6 = 0$$

33. $n = -1$

SECTION 18.9

1. $\displaystyle \iint\limits_{S} (\mathbf{v}\cdot\mathbf{n})\,d\sigma = \iiint\limits_{T} (\boldsymbol{\nabla}\cdot\mathbf{v})\,dxdydz = \iiint\limits_{T} 3\,dxdydz = 3V = 4\pi$

3. $\displaystyle \iint\limits_{S} (\mathbf{v}\cdot\mathbf{n})\,d\sigma = \iiint\limits_{T} (\boldsymbol{\nabla}\cdot\mathbf{v})\,dxdydz = \iiint\limits_{T} 2(x+y+z)\,dxdydz.$

The flux is zero since the function $f(x,y,z) = 2(x+y+z)$ satisfies the relation $f(-x,-y,-z) = -f(x,y,z)$ and T is symmetric about the origin.

5.

face	$\mathbf{n}$	$\mathbf{v}\cdot\mathbf{n}$	flux	
$x=0$	$-\mathbf{i}$	0	0	
$x=1$	$\mathbf{i}$	1	1	
$y=0$	$-\mathbf{j}$	0	0	total flux $= 3$
$y=1$	$\mathbf{j}$	1	1	
$z=0$	$-\mathbf{k}$	0	0	
$z=1$	$\mathbf{k}$	1	1	

$$\iiint\limits_{T} (\boldsymbol{\nabla}\cdot\mathbf{v})\,dxdydz = \iiint\limits_{T} 3\,dxdydz = 3V = 3$$

7.

face	$\mathbf{n}$	$\mathbf{v}\cdot\mathbf{n}$	flux
$x=0$	$-\mathbf{i}$	0	0
$x=1$	$\mathbf{i}$	1	1
$y=0$	$-\mathbf{j}$	xz	
$y=1$	$\mathbf{j}$	$-xz$	
$z=0$	$-\mathbf{k}$	0	0
$z=1$	$\mathbf{k}$	1	1

fluxes add up to 0   total flux $= 2$

$$\iiint\limits_{T} (\boldsymbol{\nabla}\cdot\mathbf{v})\,dxdydz = \iiint\limits_{T} 2\,(x+z)\,dxdydz = 2\,(\bar{x}+\bar{z})V = 2\left(\tfrac{1}{2}+\tfrac{1}{2}\right)1 = 2$$

9. $\text{flux} = \iiint\limits_{T} (1 + 4y + 6z) \, dxdydz = (1 + 4\bar{y} + 6\bar{z})V = (1 + 0 + 3)\, 9\pi = 36\pi$

11.
$$\text{flux} = \iiint\limits_{T} (2x + x - 2x) \, dxdydz \iiint\limits_{T} x \, dxdydz$$

$$= \int_0^1 \int_0^{1-x} \int_0^{1-x-y} x \, dz \, dy \, dx$$

$$= \int_0^1 \int_0^{1-x} (x - x^2 - xy) \, dy \, dx$$

$$= \int_0^1 \left[xy - x^2 y - \frac{1}{2} xy^2 \right]_0^{1-x} dx$$

$$= \int_0^1 \left(\frac{1}{2} x - x^2 + \frac{1}{2} x^3 \right) dx = \frac{1}{24}$$

13. $\text{flux} = \iiint\limits_{T} 2(x + y + z) \, dxdydz = \int_0^4 \int_0^2 \int_0^{2\pi} 2(r \cos\theta + r \sin\theta + z) r \, d\theta \, dr \, dz$

$$= \int_0^4 \int_0^2 4\pi \, rz \, dr \, dz$$

$$= \int_0^4 8\pi \, z \, dz = 64\pi$$

15. $\text{flux} = \iiint\limits_{T} (2y + 2y + 3y) \, dxdydz = 7\bar{y}V = 0$

17. $\text{flux} = \iiint\limits_{T} (A + B + C) \, dxdydz = (A + B + C)V$

19. Let T be the solid enclosed by S and set $\mathbf{n} = n_1 \mathbf{i} + n_2 \mathbf{j} + n_3 \mathbf{k}$.

$$\iint\limits_{S} n_1 \, d\sigma = \iint\limits_{S} (\mathbf{i} \cdot \mathbf{n}) \, d\sigma = \iiint\limits_{T} (\boldsymbol{\nabla} \cdot \mathbf{i}) \, dxdydz = \iiint\limits_{T} 0 \, dxdydz = 0.$$

Similarly

$$\iint\limits_{S} n_2 \, d\sigma = 0 \quad \text{and} \quad \iint\limits_{S} n_3 \, d\sigma = 0.$$

21. A routine computation shows that $\boldsymbol{\nabla} \cdot (\boldsymbol{\nabla} f \times \boldsymbol{\nabla} g) = 0.$ Therefore

$$\iint\limits_{S} [(\boldsymbol{\nabla} f \times \boldsymbol{\nabla} g) \cdot \mathbf{n}] \, d\sigma = \iiint\limits_{T} [\boldsymbol{\nabla} \cdot (\boldsymbol{\nabla} f \times \boldsymbol{\nabla} g)] \, dxdydz = 0.$$

23. Set $\mathbf{F} = F_1\mathbf{i} + F_2\mathbf{j} + F_3\mathbf{k}$.

$$F_1 = \iint\limits_{S} [\rho(z-c)\mathbf{i}\cdot\mathbf{n}]\,d\sigma = \iiint\limits_{T} [\boldsymbol{\nabla}\cdot\rho(z-c)\mathbf{i}]\,dxdydz$$

$$= \iiint\limits_{T} \underbrace{\frac{\partial}{\partial x}[\rho(z-c)]}\,dxdydz = 0.$$

Similarly $F_2 = 0$.

$$F_3 = \iint\limits_{S} [\rho(z-c)\mathbf{k}\cdot\mathbf{n}]\,d\sigma = \iiint\limits_{T} [\boldsymbol{\nabla}\cdot\rho(z-c)\mathbf{k}]\,dxdydz$$

$$= \iiint\limits_{T} \frac{\partial}{\partial z}[\rho(z-c)]\,dxdydz$$

$$= \iiint\limits_{T} \rho\,dxdydz = W.$$

PROJECT 18.9

1. For $\mathbf{r} \neq \mathbf{0}$, $\quad \boldsymbol{\nabla}\cdot\mathbf{E} = \boldsymbol{\nabla}\cdot qr^{-3}\mathbf{r} = q(-3+3)r^{-3} = 0$ by (17.8.8)

3. On $S_a, \mathbf{n} = \dfrac{\mathbf{r}}{r}$, and thus $\mathbf{E}\cdot\mathbf{n} = q\dfrac{\mathbf{r}}{r^3}\cdot\dfrac{\mathbf{r}}{r} = \dfrac{q}{r^2} = \dfrac{q}{a^2}$

Thus flux of $\mathbf{E}$ out of $S_a = \iint\limits_{S_a} (\mathbf{E}\cdot\mathbf{n})\,d\sigma = \iint\limits_{S_a} \dfrac{q}{a^2}\,d\sigma = \dfrac{q}{a^2}(\text{area of } S_a) = \dfrac{q}{a^2}(4\pi a^2) = 4\pi q.$

SECTION 18.10

For Exercises 1–4: $\quad \mathbf{n} = x\mathbf{i} + y\mathbf{j} + z\mathbf{k} \quad$ and $\quad C: \mathbf{r}(u) = \cos u\,\mathbf{i} + \sin u\,\mathbf{j}, \quad u \in [0, 2\pi]$.

1. (a) $\displaystyle\iint\limits_{S} [(\boldsymbol{\nabla}\times\mathbf{v})\cdot\mathbf{n}]\,d\sigma = \iint\limits_{S} (\mathbf{0}\cdot\mathbf{n})\,d\sigma = 0$

    (b) S is bounded by the unit circle $\quad C: \mathbf{r}(u) = \cos u\,\mathbf{i} + \sin u\,\mathbf{j}, \quad u \in [0, 2\pi]$.

$$\oint_C \mathbf{v}(\mathbf{r})\cdot d\mathbf{r} = 0 \quad \text{since } \mathbf{v} \text{ is a gradient.}$$

3. (a) $\displaystyle\iint_S [(\nabla \times \mathbf{v}) \cdot \mathbf{n}]\, d\sigma = \iint_S [(-3y^2\mathbf{i} + 2z\mathbf{j} + 2\mathbf{k}) \cdot \mathbf{n}]\, d\sigma$

$$= \iint_S (-3xy^2 + 2yz + 2z)\, d\sigma$$

$$= \underbrace{\iint_S (-3xy^2)\, d\sigma}_{0} + \underbrace{\iint_S 2yz\, d\sigma}_{0} + 2\iint_S z\, d\sigma = 2\bar{z}V = 2(\tfrac{1}{2})2\pi = 2\pi$$

Exercise 17, Section 17.7

(b) $\displaystyle\oint_C \mathbf{v}(\mathbf{r}) \cdot d\mathbf{r} = \oint_C z^2\, dx + 2x\, dy = \oint_C 2x\, dy = \int_0^{2\pi} 2\cos^2 u\, du = 2\pi$

5. (a) $\displaystyle\iint_S [(\nabla \times \mathbf{v}) \cdot \mathbf{n}]\, d\sigma = \iint_S \tfrac{1}{3}\sqrt{3}\, d\sigma = \tfrac{1}{3}\sqrt{3}A = 2$

(b) $\displaystyle\oint_C \mathbf{v}(\mathbf{r}) \cdot d\mathbf{r} = \left(\int_{C_1} + \int_{C_2} + \int_{C_3}\right) \mathbf{v}(\mathbf{r}) \cdot d\mathbf{r} = -2 + 2 + 2 = 2$

7. (a) $\displaystyle\iint_S [(\nabla \times \mathbf{v}) \cdot \mathbf{n}]\, d\sigma = \iint_S (y\mathbf{k} \cdot \mathbf{n})\, d\sigma = \tfrac{1}{3}\sqrt{3}\iint_S y\, d\sigma = \tfrac{1}{3}\sqrt{3}\,\bar{y}A = \tfrac{4}{3}$

(b) $\displaystyle\oint_C \mathbf{v}(\mathbf{r}) \cdot d\mathbf{r} = \left(\int_{C_1} + \int_{C_2} + \int_{C_3}\right) \mathbf{v}(\mathbf{r}) \cdot d\mathbf{r} = \left(\tfrac{4}{3} - \tfrac{32}{5}\right) + \tfrac{32}{5} + 0 = \tfrac{4}{3}$

9. The bounding curve is the set of all (x, y, z) with

$$x^2 + y^2 = 4 \quad\text{and}\quad z = 4.$$

Traversed in the positive sense with respect to $\mathbf{n}$, it is the curve $-C$ where

$$C : \mathbf{r}(u) = 2\cos u\,\mathbf{i} + 2\sin u\,\mathbf{j} + 4\mathbf{k}, \qquad u \in [0, 2\pi].$$

By Stokes's theorem the flux we want is

$$-\int_C \mathbf{v}(\mathbf{r}) \cdot d\mathbf{r} = -\int_C y\, dx + z\, dy + x^2z^2\, dz$$

$$= -\int_0^{2\pi} (-4\sin^2 u + 8\cos u)\, du = 4\pi.$$

11. The bounding curve C for S is the bounding curve of the elliptical region $\Omega : \tfrac{1}{4}x^2 + \tfrac{1}{9}y^2 = 1$. Since

$$\nabla \times \mathbf{v} = 2x^2yz^2\mathbf{i} - 2xy^2z^2\mathbf{j}$$

is zero on the xy-plane, the flux of $\nabla \times \mathbf{v}$ through Ω is zero, the circulation of $\mathbf{v}$ about C is zero, and therefore the flux of $\nabla \times \mathbf{v}$ through S is zero.

13. C bounds the surface

$$S: z = \sqrt{1 - \tfrac{1}{2}(x^2 + y^2)}, \qquad (x,y) \in \Omega$$

with $\Omega : x^2 + (y - \tfrac{1}{2})^2 \leq \tfrac{1}{4}$. Routine calculation shows that $\nabla \times \mathbf{v} = y\mathbf{k}$. The circulation of $\mathbf{v}$ with respect to the upper unit normal $\mathbf{n}$ is given by

$$\iint\limits_{S} (y\mathbf{k} \cdot \mathbf{n}) \, d\sigma = \iint\limits_{\Omega} y \, dxdy = \overline{y}A = \frac{1}{2}\left(\frac{\pi}{4}\right) = \frac{1}{8}\pi.$$

$$(18.7.9)$$

If $-\mathbf{n}$ is used, the circulation is $-\tfrac{1}{8}\pi$.   Answer: $\pm\tfrac{1}{8}\pi$.

15. $\nabla \times \mathbf{v} = \mathbf{i} + 2\mathbf{j} + \mathbf{k}$. The paraboloid intersects the plane in a curve C that bounds a flat surface S that projects onto the disc $x^2 + (y - \tfrac{1}{2})^2 = \tfrac{1}{4}$ in the xy-plane. The upper unit normal to S is the vector $\mathbf{n} = \tfrac{1}{2}\sqrt{2}\,(-\mathbf{j} + \mathbf{k})$. The area of the base disc is $\tfrac{1}{4}\pi$. Letting γ be the angle between $\mathbf{n}$ and $\mathbf{k}$, we have $\cos\gamma = \mathbf{n} \cdot \mathbf{k} = \tfrac{1}{2}\sqrt{2}$ and $\sec\gamma = \sqrt{2}$. Therefore the area of S is $\tfrac{1}{4}\sqrt{2}\pi$. The circulation of $\mathbf{v}$ with respect to $\mathbf{n}$ is given by

$$\iint\limits_{S} [(\nabla \times \mathbf{v}) \cdot \mathbf{n}] \, d\sigma = \iint\limits_{S} -\frac{1}{2}\sqrt{2} \, d\sigma = \left(-\frac{1}{2}\sqrt{2}\right)(\text{area of } S) = -\frac{1}{4}\pi.$$

If $-\mathbf{n}$ is used, the circulation is $\tfrac{1}{4}\pi$.   Answer: $\pm\tfrac{1}{4}\pi$.

17. Straightforward calculation shows that

$$\nabla \times (\mathbf{a} \times \mathbf{r}) = \nabla \times [(a_2 z - a_3 y)\,\mathbf{i} + (a_3 x - a_1 z)\,\mathbf{j} + (a_1 y - a_2 x)\mathbf{k}] = 2\mathbf{a}.$$

19. In the plane of C, the curve C bounds some Jordan region that we call Ω. The surface $S \cup \Omega$ is a piecewise–smooth surface that bounds a solid T. Note that $\nabla \times \mathbf{v}$ is continuously differentiable on T.

Thus, by the divergence theorem,

$$\iiint\limits_{T} [\nabla \cdot (\nabla \times \mathbf{v})] \, dxdydz = \iint\limits_{S \cup \Omega} [(\nabla \times \mathbf{v}) \cdot \mathbf{n}] \, d\sigma$$

where $\mathbf{n}$ is the outer unit normal. Since the divergence of a curl is identically zero, we have

$$\int\limits_{S \cup \Omega}\!\!\int [(\nabla \times \mathbf{v}) \cdot \mathbf{n}] \, d\sigma = 0.$$

Now $\mathbf{n}$ is $\mathbf{n}_1$ on S and $\mathbf{n}_2$ on Ω. Thus

$$\iint\limits_{S} [(\nabla \times \mathbf{v}) \cdot \mathbf{n}_1] \, d\sigma + \iint\limits_{\Omega} [(\nabla \times \mathbf{v}) \cdot \mathbf{n}_2] \, d\sigma = 0.$$

This gives

$$\iint_S [(\boldsymbol{\nabla} \times \mathbf{v}) \cdot \mathbf{n}_1] \, d\sigma = \iint_\Omega [(\boldsymbol{\nabla} \times \mathbf{v}) \cdot (-\mathbf{n}_2)] \, d\sigma = \oint_C \mathbf{v}(\mathbf{r}) \cdot d\mathbf{r}$$

where C is traversed in a positive sense with respect to $-\mathbf{n}_2$ and therefore in a positive sense with respect to $\mathbf{n}_1$. ($-\mathbf{n}_2$ points toward S.)

REVIEW EXERCISES

1. (a) $\mathbf{r}(u) = u\mathbf{i} + u\mathbf{j}, \quad 0 \le u \le 1; \quad \displaystyle\int_C \mathbf{h} \cdot d\mathbf{r} = \int_0^1 (u^3 - u^2) \, du = -\frac{1}{12}$

    (b) $\displaystyle\int_C \mathbf{h} \cdot d\mathbf{r} = \int_0^1 (2u^8 - 3u^7) \, du = -\frac{11}{72}$

3. Since $\mathbf{h}(x,y) = \nabla f$ where $f(x,y) = x^2 y^2 + \frac{1}{2}x^2 - y$,

$$\int_C \mathbf{h}(\mathbf{r}) \cdot d\mathbf{r} = f(2,4) - f(-1,2) = \frac{119}{2}$$

for *any* curve C beginning at $(-1, 2)$ and ending at $(2, 4)$.

5. $\mathbf{h}(x,y,z) = \sin y \, \mathbf{i} + xe^{xy} \, \mathbf{j} + \sin z \, \mathbf{k}; \quad \mathbf{r}(u) = u^2 \mathbf{i} + u \mathbf{j} + u^3 \mathbf{k}, \quad u \in [0,3]$

$x(u) = u^2 \quad y(u) = u \quad z(u) = u^3, \quad x'(u) = 2u, \quad y'(u) = 1, \quad z'(u) = 3u^2$

$\mathbf{h}(\mathbf{r}(u)) \cdot \mathbf{r}'(u) = 2u \sin u + u^2 e^{u^3} + 3u^2 \sin u^3$

$$\int_C \mathbf{h}(\mathbf{r}) \cdot d\mathbf{r} = \int_0^3 \left(2u \sin u + u^2 e^{u^3} + 3u^2 \sin u^3 \right) du$$

$$= \left[-2u \cos u + 2 \sin u + \tfrac{1}{3} e^{u^3} - \cos u^3 \right]_0^3$$

$$= \tfrac{2}{3} - 6 \cos 3 + 2 \sin 3 + \tfrac{1}{3} e^{27} - \cos 27$$

7. $\mathbf{F}(x,y,z) = xy \, \mathbf{i} + yz \, \mathbf{j} + xz \, \mathbf{k}; \quad \mathbf{r}(u) = u \mathbf{i} + u^2 \mathbf{j} + u^3 \mathbf{k}.$

$\mathbf{F}(\mathbf{r}(u)) \cdot \mathbf{r}' = u^3 + 5u^6; \quad W = \displaystyle\int_{-1}^2 (u^3 + 5u^6) \, du = \left[\frac{1}{4} u^4 + \frac{5}{7} u^7 \right]_{-1}^2 = \frac{2685}{28}$

9. A vector equation for the line segment is: $\mathbf{r}(u) = (1 + 2u) \, \mathbf{i} + 4u \, \mathbf{k}, \quad u \in [0,1].$

$\mathbf{F}(\mathbf{r}(u)) \cdot \mathbf{r}' = C \dfrac{2 + 20u}{\sqrt{1 + 4u + 20u^2}}; \quad \displaystyle\int_C \mathbf{F} \cdot d\mathbf{r} = C \int_0^1 \frac{(20u + 2)}{\sqrt{1 + 4u + 20u^2}} \, du = 4C$

11. $\dfrac{\partial(ye^{xy}+2x)}{\partial y}=e^{xy}+xye^{xy}=\dfrac{\partial(xe^{xy}-2y)}{\partial x}\implies \mathbf{h}$ is a gradient.

(a) $\mathbf{h}(\mathbf{r}(u))\cdot\mathbf{r}'=3u^2e^{u^3}-4u^3+2u;\qquad \displaystyle\int_C \mathbf{h}\cdot d\mathbf{r}=\int_0^2\left(3u^2e^{u^3}-4u^3+2u\right)du=e^8-13$

(b) Let $f(x,y)=e^{xy}+x^2-y^2$. Then $\nabla f=\mathbf{h}$ and $\displaystyle\int_C \mathbf{h}\cdot d\mathbf{r}=f(2,4)-f(0,0)=e^8-13$

13. $\mathbf{h}(x,y,z)=\nabla f$ where $f(x,y,z)=x^4y^3z^2$.

(a) $\mathbf{h}(\mathbf{r}(u))=4u^{15}\,\mathbf{i}+3u^{14}\,\mathbf{j}+2u^{13}\,\mathbf{k};\quad \mathbf{r}'(u)=\mathbf{i}+2u\,\mathbf{j}+3u^2\,\mathbf{k}$

$$\int_C \mathbf{h}(\mathbf{r})\cdot d\mathbf{r}=\int_0^1 16\,u^{15}\,du=1.$$

(b) $\displaystyle\int_C \mathbf{h}(\mathbf{r})\cdot d\mathbf{r}=f(\mathbf{r}(1))-f(\mathbf{r}(0))=f(1,1,1)-f(0,0,0)=1.$

15. (a) $\mathbf{r}(u)=(1-u)\mathbf{i}+u\mathbf{j},\quad 0\le u\le 1.$

$$\int_C 2xy^{1/2}\,dx+yx^{1/2}\,dy=\int_0^1\left[2(1-u)u^{1/2}(-1)+u(1-u)^{1/2}\right]du$$

$$=-2\int_0^1(1-u)u^{1/2}\,du+\int_0^1 u(1-u)^{1/2}\,du$$

$$=-\int_0^1(1-u)u^{1/2}\,du=-\frac{4}{15}$$

(b) $\mathbf{r}_1=\mathbf{i}+u\mathbf{j},\quad 0\le u\le 1;\quad \mathbf{r}_2=(1-u)\mathbf{i}+\mathbf{j}$

$$\int_C 2xy^{1/2}dx+yx^{1/2}dy=\int_0^1 u\,du+\int_0^1 -2(1-u)\,du=-\frac{1}{2}$$

(c) $\mathbf{r}=\cos u\,\mathbf{i}+\sin u\,\mathbf{j},\quad 0\le u\le \pi/2$

$$\int_C 2xy^{1/2}dx+yx^{1/2}dy=\int_0^{\pi/2}\left(-2\sin^{3/2}u\,\cos u+\cos^{3/2}u\,\sin u\right)du=-\frac{2}{5}$$

17.
$$\int_C ye^{xy}\,dx+\cos x\,dy+\left(\frac{xy}{z}\right)dz=\int_0^2\left(u^2e^{u^3}+2u\cos u+3u^2\right)du$$

$$=\left[\tfrac{1}{3}e^{u^3}+2u\sin u+2\cos u+u^3\right]_0^2$$

$$=\frac{1}{3}e^8+\frac{17}{3}+4\sin 2+2\cos 2$$

19. (a) Set $C_1 : \mathbf{r}(u) = u\,\mathbf{i} + u^2\,\mathbf{j}, \quad 0 \le u \le 1; \quad C_2 : \mathbf{r}(u) = (1-u)\,\mathbf{i} + \sqrt{1-u}\,\mathbf{j}, \quad 0 \le u \le 1.$

Then, $C = C_1 + C_2.$

$$\oint_C xy^2\,dx - x^2 y\,dy = \int_{C_1} xy^2\,dx - x^2 y\,dy + \int_{C_2} xy^2\,dx - x^2 y\,dy$$

$$= \int_0^1 (u^5 - 2u^5)\,du + \int_0^1 \left[-(1-u)^2 + \tfrac{1}{2}(1-u)^2 \right]\,du$$

$$= \int_0^1 (-u^5)\,du - \tfrac{1}{2}\int_0^1 (1-u)^2\,du = \left[-\tfrac{1}{6}u^6 + \tfrac{1}{6}(1-u)^3 \right]_0^1 = -\tfrac{1}{3}$$

(b) $P = xy^2; \quad Q = -x^2 y$

$$\oint_C xy^2\,dx - x^2 y\,dy = \int_0^1 \int_{x^2}^{\sqrt{x}} (-4xy)\,dy\,dx = \int_0^1 (2x^2 - 2x^5)\,dx = -\frac{1}{3}$$

21. $P = x - 2y^2; \quad Q = 2xy$

$$\oint_C (x - 2y^2)\,dx + 2xy\,dy = \int_0^2 \int_0^1 6y\,dy\,dx = 6$$

23. $P = \ln(x^2 + y^2); \quad Q = \ln(x^2 + y^2); \quad \dfrac{\partial Q}{\partial x} - \dfrac{\partial P}{\partial y} = \dfrac{2x - 2y}{x^2 + y^2}$

$$\oint_C \ln(x^2 + y^2)\,dx + \ln(x^2 + y^2)\,dy = \iint_\Omega \frac{2x - 2y}{x^2 + y^2}\,dx\,dy$$

$$= \int_0^\pi \int_1^2 \frac{2r\cos\theta - 2r\sin\theta}{r^2}\,r\,dr\,d\theta$$

$$= 2\int_0^\pi \int_1^2 (\cos\theta - \sin\theta)\,dr\,d\theta = -4$$

25. $\displaystyle \oint y^2\,dx = \iint_\Omega -2y\,dx\,dy = \int_0^{2\pi} \int_0^{1+\sin\theta} -2r^2\sin\theta\,dr\,d\theta = \int_0^{2\pi} (-\tfrac{2}{3})(1+\sin\theta)^3\sin\theta\,d\theta = -\frac{5\pi}{2}$

27. $C_1 : \mathbf{r}(u) = -u\,\mathbf{i} + (4 - u^2)\,\mathbf{j}, \quad -2 \le u \le 2; \quad C_2 : \mathbf{r}(u) = u\,\mathbf{i}, \quad -2 \le u \le 2; \quad C = C_1 \cup C_2$

$$A = \frac{1}{2}\int_C (-y\,dx + x\,dy) = \frac{1}{2}\int_{C_1} (-y\,dx + x\,dy) + \frac{1}{2}\int_{C_2} (-y\,dx + x\,dy)$$

$$= \frac{1}{2}\int_{-2}^2 -(4 - u^2)(-1)\,du - u(-2u)\,du + \frac{1}{2}\int_{-2}^2 0\,du$$

$$= \int_{-2}^2 (4 + u^2)\,du = \frac{32}{3}$$

29. By symmetry, it is sufficient to consider the upper part of the sphere: $z = \sqrt{4 - x^2 - y^2}$

$$\frac{\partial z}{\partial x} = \frac{-x}{\sqrt{4 - x^2 - y^2}}, \qquad \frac{\partial z}{\partial y} = \frac{-y}{\sqrt{4 - x^2 - y^2}}$$

Let Ω be the projection of the sphere onto the xy plane, then

$$S = 2 \iint_\Omega \sqrt{(z_x)^2 + (z_y)^2 + 1}\, dx\, dy = 2 \iint_\Omega \frac{2}{\sqrt{4 - x^2 - y^2}}\, dx\, dy$$

$$= 4 \int_{-\pi/2}^{\pi/2} \int_0^{2\cos\theta} \frac{1}{\sqrt{4 - r^2}}\, r\, dr\, d\theta$$

$$= 4 \int_{-\pi/2}^{\pi/2} \left(2 - 2\sqrt{1 - \cos^2\theta}\right) d\theta = 8(\pi - 2)$$

31. $\dfrac{\partial z}{\partial x} = \dfrac{x}{\sqrt{x^2 + y^2}}, \quad \dfrac{\partial z}{\partial y} = \dfrac{y}{\sqrt{x^2 + y^2}}.$

The projection Ω of the surface onto the xy plane is the disk $x^2 + y^2 \le 9.$

$$S = \iint_\Omega \sqrt{(z_x)^2 + (z_y)^2 + 1}\, dx\, dy = \iint_\Omega \sqrt{2}\, dx\, dy = \int_0^{2\pi} \int_0^3 \sqrt{2}\, r\, dr\, d\theta = 9\,\pi\,\sqrt{2}$$

33. $\displaystyle \iint_S yz\, d\sigma = \sqrt{2} \int_0^{2\pi} \int_0^1 r^2 \sin\theta (r\sin\theta + 4)\, dr\, d\theta = \frac{\sqrt{2}\,\pi}{4}$

35. The cylindrical surface S_1 is parametrized by: $x = u,\ y = 2\cos v,\ z = 2\sin v,\ 0 \le u \le 2,\ 0 \le v \le 2\pi.$

$$\mathbf{N}(u, v) = -2\cos v\, \mathbf{i} - 2\sin v\, \mathbf{j}, \quad \|\mathbf{N}(u, v)\| = 2$$

$$\iint_{S_1} \left(x^2 + y^2 + z^2\right) d\sigma = \int_0^2 \int_0^{2\pi} \left(u^2 + 4\right) 2\, dv\, du = \frac{128\pi}{3}$$

The disc $S_2 : x = 0,\ y^2 + z^2$ is parametrized by: $x = 0,\ y = u\cos v,\ z = u\sin v,\ 0 \le u \le 2,$

$0 \le v \le 2\pi.$

$$\mathbf{N} = u\, \mathbf{i}, \quad \|\mathbf{N}(u, v)\| = u; \qquad \iint_{S_2} \left(x^2 + y^2 + z^2\right) d\sigma = \int_0^2 \int_0^{2\pi} \left(0 + u^2\right) u\, dv\, du = 8\pi$$

The disc $S_3 : x = 2,\ y^2 + z^2$ is parametrized by: $x = 2,\ y = u\cos v,\ z = u\sin v,\ 0 \le u \le 2,$

$0 \le v \le 2\pi.$

$$\mathbf{N} = u\, \mathbf{i}, \quad \|\mathbf{N}(u, v)\| = u; \qquad \iint_{S_2} \left(x^2 + y^2 + z^2\right) d\sigma = \int_0^2 \int_0^{2\pi} \left(4 + u^2\right) u\, dv\, du = 24\pi$$

Thus, $\displaystyle \iint_S \left(x^2 + y^2 + z^2\right) d\sigma = \frac{128\pi}{3} + 8\pi + 24\pi = \frac{224\pi}{3}$

37. $\nabla \cdot \mathbf{v} = 4x,$ $\nabla \times \mathbf{v} = 2y\mathbf{k}$

39. $\nabla \cdot \mathbf{v} = 1 + xy,$ $\nabla \times \mathbf{v} = (xz - x)\mathbf{i} - yz\mathbf{j} + z\mathbf{k}$

41. (a) $\nabla \cdot \mathbf{v} = z - x + y$

$$\int_0^1 \int_0^1 \int_0^1 (z - x + y)\,dz\,dy\,dx = \frac{1}{2}$$

(b) at $x = 0,$ $\mathbf{n} = -\mathbf{i}, \mathbf{v} \cdot \mathbf{n} = 0, \int_0^1 \int_0^1 0\,dy\,dz = 0$

at $x = 1,$ $\mathbf{n} = \mathbf{i}, \mathbf{v} \cdot \mathbf{n} = z, \int_0^1 \int_0^1 z\,dy\,dz = 1/2$

at $y = 0,$ $\mathbf{n} = -\mathbf{j}, \mathbf{v} \cdot \mathbf{n} = xy = 0, \int_0^1 \int_0^1 0\,dx\,dz = 0$

at $y = 1,$ $\mathbf{n} = \mathbf{j}, \mathbf{v} \cdot \mathbf{n} = -xy = -x, \int_0^1 \int_0^1 -x\,dx\,dz = -1/2$

at $z = 0,$ $\mathbf{n} = -\mathbf{k}, \mathbf{v} \cdot \mathbf{n} = 0, \int_0^1 \int_0^1 0\,dy\,dx = 0$

at $z = 1,$ $\mathbf{n} = \mathbf{k}, \mathbf{v} \cdot \mathbf{n} = yz, \int_0^1 \int_0^1 y\,dy\,dx = 1/2$

The sum is $1/2$

43. The projection of S onto the xy-plane is: $\Omega : x^2 + y^2 \le 9.$

$$\iint_S \mathbf{v} \cdot \mathbf{n}\,d\sigma = \iint_\Omega \left(4x^2 + 2xyz + z^2\right) dx\,dy$$

$$= \iint_\Omega \left(4x^2 + 2xy\left[9 - x^2 - y^2\right] + \left[9 - x^2 - y^2\right]^2\right) dx\,dy$$

$$= \int_0^{2\pi} \int_0^3 \left[4r^2 \cos^2\theta + r^2(9 - r^2)\sin 2\theta + (9 - r^2)^2\right] r\,dr\,d\theta = 324\pi$$

45. (a) $(\nabla \times \mathbf{v}) \cdot \mathbf{n} = (\mathbf{i} + \mathbf{j} + \mathbf{k}) \cdot \left(-\frac{1}{2}x\mathbf{i} - \frac{1}{2}y\mathbf{j} + \frac{\sqrt{4 - x^2 - y^2}}{2}\mathbf{k}\right) = -\frac{1}{2}x - \frac{1}{2}y + \frac{\sqrt{4 - x^2 - y^2}}{2}$

$$\iint_S \left(-\frac{1}{2}x - \frac{1}{2}y + \frac{\sqrt{4-x^2-y^2}}{2}\right) d\sigma = \iint_S \left(-\frac{1}{2}x - \frac{1}{2}y + \frac{\sqrt{4-x^2-y^2}}{2}\right) \frac{2}{\sqrt{4-x^2-y^2}} \, dx \, dy$$

$$= \iint \left(\frac{-x}{\sqrt{4-x^2-y^2}} - \frac{-y}{\sqrt{4-x^2-y^2}} + 1\right) dx \, dy$$

$$= \int_0^{2\pi} \int_0^2 \left(-\frac{r\cos\theta}{\sqrt{4-r^2}} - \frac{r\sin\theta}{\sqrt{4-r^2}} + 1\right) r \, dr \, d\theta = 4\pi$$

(b) $\mathbf{r}(\theta) = 2\cos\theta\,\mathbf{i} + 2\sin\theta\,\mathbf{j}, \quad 0 \le \theta \le 2\pi$

$$\iint_S [(\boldsymbol{\nabla} \times \mathbf{v}) \cdot \mathbf{n}] d\sigma = \oint_C \mathbf{v}(\mathbf{r}) \cdot d\mathbf{r} = \int_0^{2\pi} 4\cos^2\theta \, d\theta = 4\pi$$

CHAPTER 19

SECTION 19.1

1. $y' + xy = xy^3 \implies y^{-3}y' + xy^{-2} = x.$ Let $v = y^{-2}, \quad v' = -2y^{-3}y'.$

$$-\frac{1}{2}v' + xv = x$$
$$v' - 2xv = -2x$$
$$e^{-x^2}v' - 2xe^{-x^2}v = -2xe^{-x^2}$$
$$e^{-x^2}v = e^{-x^2} + C$$
$$v = 1 + Ce^{x^2}$$
$$y^2 = \frac{1}{1 + Ce^{x^2}}.$$

3. $y' - 4y = 2e^x y^{\frac{1}{2}} \implies y^{-\frac{1}{2}}y' - 4y^{\frac{1}{2}} = 2e^x.$ Let $v = y^{\frac{1}{2}}, \quad v' = \frac{1}{2}y^{-\frac{1}{2}}y'.$

$$2v' - 4v = 2e^x$$
$$v' - 2v = e^x$$
$$e^{-2x}v' - 2e^{-2x}v = e^{-x}$$
$$e^{-2x}v = -e^{-x} + C$$
$$v = -e^x + Ce^{2x}$$
$$y = (Ce^{2x} - e^x)^2.$$

5. $(x-2)y' + y = 5(x-2)^2 y^{\frac{1}{2}} \implies y^{-\frac{1}{2}}y' + \frac{1}{x-2}y^{\frac{1}{2}} = 5(x-2).$ Let $v = y^{\frac{1}{2}}, \quad v' = \frac{1}{2}y^{-\frac{1}{2}}y'.$

$$2v' + \frac{1}{x-2}v = 5(x-2)$$
$$v' + \frac{1}{2(x-2)}v = \frac{5}{2}(x-2)$$
$$\sqrt{x-2}\,v' + \frac{1}{2\sqrt{x-2}}v = \frac{5}{2}(x-2)^{\frac{3}{2}}$$
$$\sqrt{x-2}\,v = (x-2)^{\frac{5}{2}} + C$$
$$v = (x-2)^2 + \frac{C}{\sqrt{x-2}}$$
$$y = \left[(x-2)^2 + \frac{C}{\sqrt{x-2}}\right]^2.$$

7. $y' + xy = y^3 e^{x^2}$ $\implies$ $y^{-3}y' + xy^{-2} = e^{x^2}$.   Let $v = y^{-2}$, $v' = -2y^{-3}y'$.

$$-\frac{1}{2}v' + xv = e^{x^2}$$

$$v' - 2xv = -2e^{x^2}$$

$$e^{-x^2}v' - 2xe^{-x^2}v = -2$$

$$e^{-x^2}v = -2x + C$$

$$v = -2xe^{x^2} + Ce^{x^2}$$

$$y^{-2} = Ce^{x^2} - 2xe^{x^2}.$$

$C = 4$ $\implies$ $y^{-2} = 4e^{x^2} - 2xe^{x^2}.$

9. $2x^3 y' - 3x^2 y = y^3$ $\implies$ $y^{-3}y' - \frac{3}{2x}y^{-2} = \frac{1}{2x^3}$.   Let $v = y^{-2}$, $v' = -2y^{-3}y'$.

$$-\frac{1}{2}v' - \frac{3}{2x}v = \frac{1}{2x^3}$$

$$v' + \frac{3}{x}v = -\frac{1}{x^3}$$

$$x^3 v' + 3x^2 v = -1$$

$$x^3 v = -x + C$$

$$v = \frac{C - x}{x^3}$$

$$y^2 = \frac{x^3}{C - x}$$

$1 = \dfrac{1}{C - x}$ $\implies$ $C = 2$ $\implies$ $y^2 = \dfrac{x^3}{2 - x}.$

11. $y' - \frac{y}{x}\ln y = xy$ $\implies$ $\frac{y'}{y} - \frac{1}{x}\ln y = x$.   Let $u = \ln y$, $u' = \frac{y'}{y}$.

$$u' - \frac{1}{x}u = x$$

$$\frac{1}{x}u' - \frac{1}{x^2}u = 1$$

$$\frac{1}{x}u = x + C$$

$$u = x^2 + Cx$$

$$\ln y = x^2 + Cx.$$

13. $f(x,y) = \dfrac{x^2 + y^2}{2xy}$; $f(tx, ty) = \dfrac{(tx)^2 + (ty)^2}{2(tx)(ty)} = \dfrac{t^2(x^2 + y^2)}{t^2(2xy)} = \dfrac{x^2 + y^2}{2xy} = f(x,y)$

Set $vx = y$. Then, $v + xv' = y'$ and

$$v + xv' = \frac{x^2 + v^2 x^2}{2vx^2} = \frac{1 + v^2}{2v}$$

$$v - \frac{1 + v^2}{2v} + xv' = 0$$

$$v^2 - 1 + 2xvv' = 0$$

$$\frac{1}{x}\,dx + \frac{2v}{v^2 - 1}\,dv = 0$$

$$\int \frac{1}{x}\,dx + \int \frac{2v}{v^2 - 1}\,dv = C$$

$$\ln|x| + \ln|v^2 - 1| = K \quad \text{or} \quad x(v^2 - 1) = C$$

Replacing v by y/x, we get

$$x\left(\frac{y^2}{x^2} - 1\right) = C \quad \text{or} \quad y^2 - x^2 = Cx$$

15. $f(x,y) = \dfrac{x - y}{x + y}$; $f(tx, ty) = \dfrac{(tx) - (ty)}{tx + ty} = \dfrac{t(x - y)}{t(x + y)} = \dfrac{x - y}{x + y} = f(x,y)$

Set $vx = y$. Then, $v + xv' = y'$ and

$$v + xv' = \frac{x - vx}{x + vx} = \frac{1 - v}{1 + v}$$

$$v^2 + 2v - 1 + x(1 + v)v' = 0$$

$$\frac{1}{x}\,dx + \frac{1 + v}{v^2 + 2v - 1}\,dv = 0$$

$$\int \frac{1}{x}\,dx + \int \frac{1 + v}{v^2 + 2v - 1}\,dv = C$$

$$\ln|x| + \tfrac{1}{2}\ln|v^2 + 2v - 1| = K \quad \text{or} \quad x\sqrt{v^2 + 2v - 1} = C$$

Replacing v by y/x, we get

$$x\sqrt{\frac{y^2}{x^2} + 2\frac{y}{x} - 1} = C \quad \text{or} \quad y^2 + 2xy - x^2 = C$$

17. $f(x,y) = \dfrac{x^2 e^{y/x} + y^2}{xy}$; $f(tx, ty) = \dfrac{(tx)^2 - e^{(ty)/(tx)} + (ty)^2}{(tx)(ty)} = \dfrac{t^2 \left(x^2 e^{y/x} + y^2\right)}{t^2(xy)} = f(x,y)$

Set $vx = y$. Then, $v + xv' = y'$ and

$$v + xv' = \frac{x^2 e^v + v^2 x^2}{vx^2} = \frac{e^v + v^2}{v}$$

$$v^2 + xvv' = e^v + v^2$$

$$-e^v + xvv' = 0$$

$$\frac{1}{x}\, dx = ve^{-v}\, dv$$

$$\int \frac{1}{x}\, dx = \int ve^{-v}\, dv$$

$$\ln|x| = -ve^{-v} - e^{-v} + C$$

Replacing v by y/x, and simplifying, we get

$$y + x = xe^{y/x}(C - \ln|x|)$$

19. $f(x,y) = \dfrac{y}{x} + \sin(y/x)$; $f(tx, ty) = \dfrac{(ty)}{tx} + \sin[(ty/tx)] = \dfrac{y}{x} + \sin(y/x) = f(x,y)$

Set $vx = y$. Then, $v + xv' = y'$ and

$$v + xv' = \frac{vx}{x} + \sin[(vx)/x] = v + \sin v$$

$$xv' = \sin v$$

$$\csc v\, dv = \frac{1}{x}\, dx$$

$$\int \csc v\, dv = \int \frac{1}{x}\, dx$$

$$\ln|\csc v - \cot v| = \ln|x| + K \qquad \text{or} \qquad \csc v - \cot v = Cx$$

Replacing v by y/x, and simplifying, we get

$$1 - \cos(y/x) = Cx\, \sin(y/x)$$

21. The differential equation is homogeneous since

$$f(x,y) = \frac{y^3 - x^3}{xy^2}; \qquad f(tx, ty) = \frac{(ty)^3 - (tx)^3}{(tx)(ty)^2} = \frac{t^3(y^3 - x^3)}{t^3(xy^2)} = \frac{y^3 - x^3}{xy^2} = f(x,y)$$

Set $vx = y$. Then, $v + xv' = y'$ and

$$v + xv' = \frac{(vx)^3 - x^3}{v^2 x^3} = \frac{v^3 - 1}{v^2}$$

$$1 + xv^2 v' = 0$$

$$\frac{1}{x}\, dx + v^2\, dv = 0$$

$$\int \frac{1}{x}\, dx + \int v^2\, dv = 0$$

$$\ln|x| + \frac{1}{3} v^3 = C$$

Replacing v by y/x, we get

$$y^3 + 3x^3 \ln|x| = Cx^3$$

Applying the side condition $y(1) = 2$, we have

$$8 + 3\ln 1 = C \quad \Longrightarrow \quad C = 8 \quad \text{and} \quad y^3 + 3x^3 \ln|x| = 8x^3$$

SECTION 19.2

1. $\dfrac{\partial P}{\partial y} = 2xy - 1 = \dfrac{\partial Q}{\partial x};$ the equation is exact on the whole plane.

$$\frac{\partial f}{\partial x} = xy^2 - y \quad \Longrightarrow \quad f(x,y) = \tfrac{1}{2}x^2y^2 - xy + \varphi(y)$$

$$\frac{\partial f}{\partial y} = x^2y - x + \varphi'(y) = x^2y - x \quad \Longrightarrow \quad \varphi'(y) = 0 \quad \Longrightarrow \quad \varphi(y) = 0 \ \text{(omit the constant)}*$$

Therefore $f(x,y) = \tfrac{1}{2}x^2y^2 - xy,$ and a one-parameter family of solutions is:

$$\tfrac{1}{2}x^2y^2 - xy = C$$

* We will omit the constant at this step throughout this section.

3. $\dfrac{\partial P}{\partial y} = e^y - e^x = \dfrac{\partial Q}{\partial x};$ the equation is exact on the whole plane.

$$\frac{\partial f}{\partial x} = e^y - ye^x \quad \Longrightarrow \quad f(x,y) = xe^y - ye^x + \varphi(y)$$

$$\frac{\partial f}{\partial y} = xe^y - e^x + \varphi'(y) = xe^y - e^x \quad \Longrightarrow \quad \varphi'(y) = 0 \quad \Longrightarrow \quad \varphi(y) = 0$$

Therefore $f(x,y) = xe^y - ye^x,$ and a one-parameter family of solutions is:

$$xe^y - ye^x = C$$

5. $\dfrac{\partial P}{\partial y} = \dfrac{1}{y} + 2x = \dfrac{\partial Q}{\partial x};$ the equation is exact on the upper half plane.

$$\frac{\partial f}{\partial x} = \ln y + 2xy \quad \Longrightarrow \quad f(x,y) = x\ln y + x^2y + \varphi(y)$$

$$\frac{\partial f}{\partial y} = \frac{x}{y} + x^2 + \varphi'(y) = \frac{x}{y} + x^2 \quad \Longrightarrow \quad \varphi'(y) = 0 \quad \Longrightarrow \quad \varphi(y) = 0$$

Therefore $f(x,y) = x\ln y + x^2y,$ and a one-parameter family of solutions is:

$$x\ln y + x^2y = C$$

7. $\dfrac{\partial P}{\partial y} = \dfrac{1}{x} = \dfrac{\partial Q}{\partial x};$ the equation is exact on the right half plane.

$$\frac{\partial f}{\partial x} = \frac{y}{x} + 6x \quad \Longrightarrow \quad f(x,y) = y\ln x + 3x^2 + \varphi(y)$$

$$\frac{\partial f}{\partial y} = \ln x + \varphi'(y) = \ln x - 2 \implies \varphi'(y) = -2 \implies \varphi(y) = -2y$$

Therefore $f(x,y) = y \ln x + 3x^2 - 2y$,   and a one-parameter family of solutions is:

$$y \ln x + 3x^2 - 2y = C$$

9. $\dfrac{\partial P}{\partial y} = 3y^2 - 2y \sin x = \dfrac{\partial Q}{\partial x};$ the equation is exact on the whole plane.

$$\frac{\partial f}{\partial x} = y^3 - y^2 \sin x - x \implies f(x,y) = xy^3 + y^2 \cos x - \tfrac{1}{2} x^2 + \varphi(y)$$

$$\frac{\partial f}{\partial y} = 3xy^2 + 2y \cos x + \varphi'(y) = 3xy^2 + 2y \cos x + e^{2y} \implies \varphi'(y) = e^{2y} \implies \varphi(y) = \tfrac{1}{2} e^{2y}$$

Therefore $f(x,y) = xy^3 + y^2 \cos x - \tfrac{1}{2} x^2 + \tfrac{1}{2} e^{2y}$,   and a one-parameter family of solutions is:

$$xy^3 + y^2 \cos x - \tfrac{1}{2} x^2 + \tfrac{1}{2} e^{2y} = C$$

11.   (a)   Yes: $\dfrac{\partial}{\partial y}[p(x)] = 0 = \dfrac{\partial}{\partial x}[q(y)].$

     (b)   For all x, y such that $p(y)q(x) \neq 0$, $\dfrac{1}{p(y)q(x)}$ is an integrating factor.

        Multiplying the differential equation by $\dfrac{1}{p(y)q(x)},$ we get

$$\frac{1}{q(x)} + \frac{1}{p(y)} y' = 0$$

        which has the form of the differential equation in part (a).

13. $\dfrac{\partial P}{\partial y} = e^{y-x} - 1$ and $\dfrac{\partial Q}{\partial x} = e^{y-x} - xe^{y-x};$ the equation is not exact.

     Since $\dfrac{1}{Q}\left(\dfrac{\partial P}{\partial y} - \dfrac{\partial Q}{\partial x}\right) = \dfrac{1}{xe^{y-x}-1}\left(xe^{y-x} - 1\right) = 1, \quad \mu(x) = e^{\int dx} = e^x$ is

     an integrating factor. Multiplying the given equation by e^x,   we get

$$(e^y - ye^x) + (xe^y - e^x)\, y' = 0$$

     This is the equation given in Exercise 3. A one-parameter family of solutions is:

$$xe^y - ye^x = C$$

15. $\dfrac{\partial P}{\partial y} = 6x^2y + e^y = \dfrac{\partial Q}{\partial x};$ the equation is exact.

$$\frac{\partial f}{\partial x} = 3x^2y^2 + x + e^y \implies f(x,y) = x^3y^2 + \tfrac{1}{2} x^2 + xe^y + \varphi(y)$$

$$\frac{\partial f}{\partial y} = 2x^3y + xe^y + \varphi'(y) = 2x^3y + y + xe^y \implies \varphi'(y) = y \implies \varphi(y) = \tfrac{1}{2} y^2$$

Therefore $f(x,y) = x^3y^2 + \tfrac{1}{2} x^2 + xe^y + \tfrac{1}{2} y^2$,   and a one-parameter family of solutions is:

$$x^3y^2 + \tfrac{1}{2} x^2 + xe^y + \tfrac{1}{2} y^2 = C$$

17. $\dfrac{\partial P}{\partial y} = 3y^2$ and $\dfrac{\partial Q}{\partial x} = 0;$ the equation is not exact.

Since $\dfrac{1}{Q}\left(\dfrac{\partial P}{\partial y} - \dfrac{\partial Q}{\partial x}\right) = \dfrac{1}{3y^2}(3y^2) = 1,$ $\mu(x) = e^{\int dx} = e^x$ is an

an integrating factor. Multiplying the given equation by $e^x,$ we get

$$(y^3 e^x + x e^x + e^x) + (3y^2 e^x)\, y' = 0$$

$\dfrac{\partial f}{\partial x} = y^3 e^x + x e^x + e^x$ $\implies$ $f(x,y) = y^3 e^x + x e^x + \varphi(y)$

$\dfrac{\partial f}{\partial y} = 3y^2 e^x + \varphi'(y) = 3y^2 e^x$ $\implies$ $\varphi'(y) = 0$ $\implies$ $\varphi(y) = 0$

Therefore $f(x,y) = y^3 e^x + x e^x,$ and a one-parameter family of solutions is:

$$y^3 e^x + x e^x = C$$

19. $\dfrac{\partial P}{\partial y} = 1 = \dfrac{\partial Q}{\partial x};$ the equation is exact.

$\dfrac{\partial f}{\partial x} = x^2 + y$ $\implies$ $f(x,y) = \tfrac{1}{3}x^3 + xy + \varphi(y)$

$\dfrac{\partial f}{\partial y} = x + \varphi'(y) = x + e^y$ $\implies$ $\varphi'(y) = e^y$ $\implies$ $\varphi(y) = e^y$

Therefore $f(x,y) = \tfrac{1}{3}x^3 + xy + e^y,$ and a one-parameter family of solutions is:

$$\tfrac{1}{3}x^3 + xy + e^y = C$$

Setting $x = 1,\ y = 0,$ we get $C = \tfrac{4}{3}$ and

$$\tfrac{1}{3}x^3 + xy + e^y = \tfrac{4}{3}   \quad \text{or} \quad   x^3 + 3xy + 3e^y = 4$$

21. $\dfrac{\partial P}{\partial y} = 4y$ and $\dfrac{\partial Q}{\partial x} = 2y;$ the equation is not exact.

Since $\dfrac{1}{Q}\left(\dfrac{\partial P}{\partial y} - \dfrac{\partial Q}{\partial x}\right) = \dfrac{1}{2xy}(2y) = \dfrac{1}{x},$ $\mu(x) = e^{\int (1/x)\, dx} = e^{\ln x} = x$ is an

integrating factor. Multiplying the given equation by $x,$ we get

$$(2xy^2 + x^3 + 2x) + (2x^2 y)\, y' = 0$$

$\dfrac{\partial f}{\partial y} = 2x^2 y$ $\implies$ $f(x,y) = x^2 y^2 + \varphi(x)$

$\dfrac{\partial f}{\partial x} = 2xy^2 + \varphi'(x) = 2xy^2 + x^3 + 2x$ $\implies$ $\varphi'(x) = x^3 + 2x$ $\implies$ $\varphi = \tfrac{1}{4}x^4 + x^2$

Therefore $f(x,y) = x^2 y^2 + \frac{1}{4} x^4 + x^2$,   and a one-parameter family of solutions is:

$$x^2 y^2 + \tfrac{1}{4} x^4 + x^2 = C$$

Setting $x = 1$, $y = 0$, we get $C = \frac{5}{4}$ and

$$x^2 y^2 + \tfrac{1}{4} x^4 + x^2 = \tfrac{5}{4} \qquad \text{or} \qquad 4x^2 y^2 + x^4 + 4x^2 = 5$$

23. $\dfrac{\partial P}{\partial y} = 3y^2$ and $\dfrac{\partial Q}{\partial x} = y^2$;   the equation is not exact.

Since $\dfrac{1}{P}\left(\dfrac{\partial P}{\partial y} - \dfrac{\partial Q}{\partial x}\right) = \dfrac{1}{y^3}(2y^2) = \dfrac{2}{y}$, $w(y) = e^{-\int (2/y)\, dy} = e^{-2\ln y} = y^{-2}$ is an

integrating factor. Multiplying the given equation by y^{-2}, we get

$$y + \left(y^{-2} + x\right) y' = 0$$

$\dfrac{\partial f}{\partial x} = y \implies f(x,y) = xy + \varphi(y)$

$\dfrac{\partial f}{\partial y} = x + \varphi'(y) = y^{-2} + x \implies \varphi'(y) = y^{-2} \implies \varphi(y) = -\dfrac{1}{y}$

Therefore $f(x,y) = xy - \dfrac{1}{y}$, and a one-parameter family of solutions is: $xy - \dfrac{1}{y} = C$

Setting $x = -2$, $y = -1$, we get $C = 3$ and the solution $xy - \dfrac{1}{y} = 3$.

25. $\dfrac{\partial P}{\partial y} = -2y \sinh(x - y^2) = \dfrac{\partial Q}{\partial x}$;   the equation is exact.

$\dfrac{\partial f}{\partial x} = \cosh(x - 2y^2) + e^{2x} \implies f(x,y) = \sinh(x - y^2) + \tfrac{1}{2} e^{2x} + \varphi(y)$

$\dfrac{\partial f}{\partial y} = -2y \cosh(x - y^2) + \varphi'(y) = y - 2y \cosh(x - y^2) \implies \varphi'(y) = y \implies \varphi(y) = \tfrac{1}{2} y^2$

Therefore $f(x,y) = \sinh(x - y^2) + \tfrac{1}{2} e^{2x} + \tfrac{1}{2} y^2$,   and a one-parameter family of solutions is:

$$\sinh(x - y^2) + \tfrac{1}{2} e^{2x} + \tfrac{1}{2} y^2 = C$$

Setting $x = 2$, $y = \sqrt{2}$, we get $C = \tfrac{1}{2} e^4 + 1$ and the solution

$$\sinh(x - y^2) + \tfrac{1}{2} e^{2x} + \tfrac{1}{2} y^2 = \tfrac{1}{2} e^4 + 1$$

27. (a) $\dfrac{\partial P}{\partial y} = 2xy + kx^2$ and $\dfrac{\partial Q}{\partial x} = 2xy + 3x^2 \implies k = 3$.

(b) $\dfrac{\partial P}{\partial y} = e^{2xy} + 2xye^{2xy}$ and $\dfrac{\partial Q}{\partial x} = ke^{2xy} + 2kxye^{2xy} \implies k = 1$.

29. $y' = y^2 x^3$;   the equation is separable.

$$y^{-2}\, dy = x^3\, dx \implies -\dfrac{1}{y} = \tfrac{1}{4} x^4 + C \implies y = \dfrac{-4}{x^4 + C}$$

31. $y' + \dfrac{4}{x} y = x^4$;   the equation is linear.

$$H(x) = \int (4/x)\, dx = 4 \ln x = \ln x^4, \quad \text{integrating factor:} \quad e^{\ln x^4} = x^4$$

$$x^4 y' + 4x^3 y = x^8$$

$$\tfrac{d}{dx}\left[x^4 y \right] = x^8$$

$$x^4 y = \tfrac{1}{9} x^9 + C$$

$$y = \tfrac{1}{9} x^5 + C x^{-4}$$

33. $\dfrac{\partial P}{\partial y} = e^{xy} + xy e^{xy} = \dfrac{\partial Q}{\partial x}$;   the equation is exact.

$$\dfrac{\partial f}{\partial x} = y e^{xy} - 2x \quad \Longrightarrow \quad f(x,y) = e^{xy} - x^2 + \varphi(y)$$

$$\dfrac{\partial f}{\partial y} = x e^{xy} + \varphi'(y) = \dfrac{2}{y} + x e^{xy} \quad \Longrightarrow \quad \varphi'(y) = \dfrac{2}{y} \quad \Longrightarrow \quad \varphi(y) = 2 \ln |y|$$

Therefore $f(x,y) = e^{xy} - x^2 + 2 \ln |y|$,   and a one-parameter family of solutions is:

$$e^{xy} - x^2 + 2 \ln |y| = C$$

SECTION 19.3

1. $y' = y \quad \Longrightarrow \quad y = C e^x$.   Also, $y(0) = 1 \quad \Longrightarrow \quad C = 1$
Thus $y = e^x$ and $y(1) = 2.71828$
(a) 2.48832, relative error= 8.46%.
(b) 2.71825, relative error= 0.001%.

3. (a) 2.59374, relative error= 4.58%.
(b) 2.71828, relative error= 0%.

5. $y' = 2x \quad \Longrightarrow \quad y = x^2 + C$.   Also, $y(2) = 5 \quad \Longrightarrow \quad C = 1$
Thus $y = x^2 + 1$ and $y(1) = 2$.
(a) 1.9, relative error= 5.0%.
(b) 2.0, relative error= 0%.

7. $y' = \dfrac{1}{2y}$
Thus $y = \sqrt{x}$ and $y(2) = \sqrt{2} \simeq 1.41421$.
(a) 1.42052, relative error= −0.45%.
(b) 1.41421, relative error= 0%.

9. (a) 2.65330, relative error= 2.39%.
(b) 2.71828, relative error= 0%.

PROJECT 19.3

1. (a) and (b)

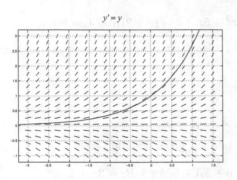

$$y' = y$$

(c) $y - y' = 0$ $H(x) = \int -dx = -x$;   integrating factor: e^{-x}

$e^{-x}y' - e^{-x}y = 0$

$\dfrac{d}{dx}(e^{-x}y) = 0$

$e^{-x}y = C$

$y = Ce^x$

$y(0) = 1 \implies C = 1.$ Thus $y = e^x$.

3. (a) and (b)

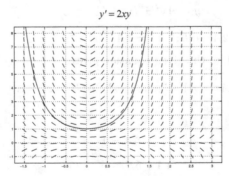

$$y' = 2xy$$

(c) $y' - 2xy = 0$ $H(x) = \int -2x\,dx = -x^2$;   integrating factor: e^{-x^2}

$e^{-x^2}y' - 2xe^{x^2}y = 0$

$\dfrac{d}{dx}(e^{-x^2}y) = 0$

$e^{-x^2}y = C$

$y = Ce^{x^2}$

$y(0) = 1 \implies C = 1.$ Thus $y = e^{x^2}$.

SECTION 19.4

1. First consider the reduced equation. The characteristic equation is:

$$r^2 + 5r + 6 = (r+2)(r+3) = 0$$

and $u_1(x) = e^{-2x}$, $u_2(x) = e^{-3x}$ are fundamental solutions. A particular solution of the given equation has the form

$$y = Ax + B.$$

The derivatives of y are: $y' = A$, $y'' = 0$.

Substituting y and its derivatives into the given equation gives

$$0 + 5A + 6(Ax + B) = 3x + 4.$$

Thus,

$$6A = 3$$

$$5A + 6B = 4$$

The solution of this pair of equations is: $A = \frac{1}{2}$, $B = \frac{1}{4}$,   and $y = \frac{1}{2}x + \frac{1}{4}$.

3. First consider the reduced equation. The characteristic equation is:

$$r^2 + 2r + 5 = 0$$

and $u_1(x) = e^{-x}\cos 2x$, $u_2(x) = e^{-x}\sin 2x$ are fundamental solutions. A particular solution of the given equation has the form

$$y = Ax^2 + Bx + C.$$

The derivatives of y are: $y' = 2Ax + B$, $y'' = 2A$.

Substituting y and its derivatives into the given equation gives

$$2A + 2(2Ax + B) + 5(Ax^2 + Bx + C) = x^2 - 1.$$

Thus,

$$5A = 1$$

$$4A + 5B = 0$$

$$2A + 2B + 5C = -1$$

The solution of this system of equations is: $A = \frac{1}{5}$, $B = -\frac{4}{25}$, $C = -\frac{27}{125}$,   and

$$y = \frac{1}{5}x^2 - \frac{4}{25}x - \frac{27}{125}.$$

5. First consider the reduced equation. The characteristic equation is:

$$r^2 + 6r + 9 = (r+3)^2 = 0$$

and $u_1(x) = e^{-3x}$, $u_2(x) = xe^{-3x}$ are fundamental solutions. A particular solution of the given equation has the form

$$y = Ae^{3x}.$$

The derivatives of y are: $y' = 3Ae^{3x}$, $y'' = 9Ae^{3x}$.

Substituting y and its derivatives into the given equation gives

$$9Ae^{3x} + 18Ae^{3x} + 9Ae^{3x} = e^{3x}.$$

Thus, $36A = 1 \implies A = \dfrac{1}{36}$,   and $y = \frac{1}{36} e^{3x}$.

7. First consider the reduced equation. The characteristic equation is:

$$r^2 + 2r + 2 = 0$$

and $u_1(x) = e^{-x}\cos x$, $u_2(x) = e^{-x}\sin x$ are fundamental solutions. A particular solution of the given equation has the form

$$y = Ae^x.$$

The derivatives of y are: $y' = Ae^x$, $y'' = Ae^x$.

Substituting y and its derivatives into the given equation gives

$$Ae^x + 2Ae^x + 2Ae^x = e^x.$$

Thus, $5A = 1 \implies A = \frac{1}{5}$ and $y = \frac{1}{5} e^x$.

9. First consider the reduced equation. The characteristic equation is:

$$r^2 - r - 12 = (r-4)(r+3) = 0$$

and $u_1(x) = e^{4x}$, $u_2(x) = e^{-3x}$ are fundamental solutions. A particular solution of the given equation has the form

$$y = A\cos x + B\sin x.$$

The derivatives of y are: $y' = -A\sin x + B\cos x$, $y'' = -A\cos x - B\sin x$.

Substituting y and its derivatives into the given equation gives

$$-A\cos x - B\sin x - (-A\sin x + B\cos x) - 12(A\cos x + B\sin x) = \cos x.$$

Thus,

$$-13A - B = 1$$
$$A - 13B = 0$$

The solution of this system of equations is: $A = -\frac{13}{170}$, $B = -\frac{1}{170}$,   and

$$y = -\frac{13}{170}\cos x - \frac{1}{170}\sin x.$$

is a particular solution of the complete equation.

11. First consider the reduced equation. The characteristic equation is:

$$r^2 + 7r + 6 = (r + 6)(r + 1) = 0$$

and $u_1(x) = e^{-6x}$, $u_2(x) = e^{-x}$ are fundamental solutions. A particular solution of the given equation has the form

$$y = A \cos 2x + B \sin 2x.$$

The derivatives of y are: $y' = -2A \sin 2x + 2B \cos 2x$, $y'' = -4A \cos 2x - 4B \sin 2x$.

Substituting y and its derivatives into the given equation gives

$$-4A \cos 2x - 4B \sin 2x + 7(-2A \sin 2x + 2B \cos 2x) + 6(A \cos 2x + B \sin 2x) = 3 \cos 2x.$$

Thus,

$$2A + 14B = 3$$
$$-14A + 2B = 0$$

The solution of this system of equations is: $A = \frac{3}{100}$, $B = \frac{21}{100}$ and

$$y = \frac{3}{10} \cos 2x + \frac{21}{100} \sin 2x.$$

13. First consider the reduced equation. The characteristic equation is:

$$r^2 - 2r + 5 = 0$$

and $u_1(x) = e^x \cos 2x$, $u_2(x) = e^x \sin 2x$ are fundamental solutions. A particular solution of the given equation has the form

$$y = Ae^{-x} \cos 2x + Be^{-x} \sin 2x$$

The derivatives of y are: $y' = -Ae^{-x} \cos 2x - 2Ae^{-x} \sin 2x - Be^{-x} \sin 2x + 2Be^{-x} \cos 2x$,

$y'' = 4Ae^{-x} \sin 2x - 3Ae^{-x} \cos 2x - 4Be^{-x} \cos 2x - 3Be^{-x} \sin 2x$.

Substituting y and its derivatives into the given equation gives

$$4Ae^{-x} \sin 2x - 3Ae^{-x} \cos 2x - 4Be^{-x} \cos 2x - 3Be^{-x} \sin 2x-$$

$$2\left(-Ae^{-x} \cos 2x - 2Ae^{-x} \sin 2x - Be^{-x} \sin 2x + 2Be^{-x} \cos 2x\right) +$$

$$5\left(Ae^{-x} \cos 2x + Be^{-x} \sin 2x\right) = e^{-x} \sin 2x.$$

Equating the coefficients of $e^{-x} \cos 2x$ and $e^{-x} \sin 2x$ we get,

$$8A + 4B = 1$$
$$4A - 8B = 0$$

The solution of this system of equations is: $A = \frac{1}{10}$, $B = \frac{1}{20}$ and

$$y = \frac{1}{10} e^{-x} \cos 2x + \frac{1}{20} e^{-x} \sin 2x.$$

15. First consider the reduced equation. The characteristic equation is:

$$r^2 + 6r + 8 = (r + 4)(r + 2) = 0$$

and $u_1(x) = e^{-4x}$, $u_2(x) = e^{-2x}$ are fundamental solutions. A particular solution of the given equation has the form

$$y = Axe^{-2x}.$$

The derivatives of y are: $y' = Ae^{-2x} - 2Axe^{-2x}$, $y'' = -4Ae^{-2x} + 4Axe^{-2x}$.

Substituting y and its derivatives into the given equation gives

$$-4Ae^{-2x} + 4Axe^{-2x} + 6\left(Ae^{-2x} - 2Axe^{-2x}\right) + 8Axe^{-2x} = 3e^{-2x}$$

Thus, $2A = 3$ $\implies$ $A = \frac{3}{2}$ and $y = \frac{3}{2}xe^{-2x}$.

17. First consider the reduced equation: $y'' + y = 0$.   The characteristic equation is:

$$r^2 + 1 = 0$$

and $u_1(x) = \cos x$, $u_2(x) = \sin x$ are fundamental solutions. A particular solution of the given equation has the form

$$y = Ae^x.$$

The derivatives of y are: $y' = y'' = Ae^x$.

Substitute y and its derivatives into the given equation:

$$Ae^x + Ae^x = e^x \implies A = \frac{1}{2} \quad \text{and} \quad y = \tfrac{1}{2}e^x.$$

The general solution of the given equation is: $y = C_1 \cos x + C_2 \sin x + \frac{1}{2}e^x$.

19. First consider the reduced equation: $y'' - 3y' - 10y = 0$.   The characteristic equation is:

$$r^2 - 3r - 10 = (r-5)(r+2) = 0$$

and $u_1(x) = e^{5x}$, $u_2(x) = e^{-2x}$ are fundamental solutions. A particular solution of the given equation has the form

$$y = Ax + B.$$

The derivatives of y are: $y' = A$, $y'' = 0$.

Substitute y and its derivatives into the given equation:

$$-3A - 10(Ax + B) = -x - 1 \implies A = \tfrac{1}{10}, \quad B = \tfrac{7}{100} \quad \text{and} \quad y = \tfrac{1}{10}x + \tfrac{7}{100}$$

The general solution of the given equation is:

$$y = C_1 e^{5x} + C_2 e^{-2x} + \tfrac{1}{10}x + \tfrac{7}{100}$$

21. First consider the reduced equation: $y'' + 3y' - 4y = 0$.   The characteristic equation is:

$$r^2 + 3r - 4 = (r+4)(r-1) = 0$$

and $u_1(x) = e^x$, $u_2(x) = e^{-4x}$ are fundamental solutions. A particular solution of the given equation has the form

$$y = Axe^{-4x}.$$

The derivatives of y are: $\quad y' = Ae^{-4x} - 4Axe^{-4x}, \quad y'' = -8Ae^{-4x} + 16Axe^{-4x}.$

Substitute y and its derivatives into the given equation:

$$-8Ae^{-4x} + 16Axe^{-4x} + 3\left(Ae^{-4x} - 4Axe^{-4x}\right) - 4Axe^{-4x} = e^{-4x}.$$

This implies $\quad -5A = 1, \quad$ so $\quad A = -\frac{1}{5} \quad$ and $\quad y = -\frac{1}{5}xe^{-4x}.$

The general solution of the given equation is: $\quad y = C_1 e^x + C_2 e^{-4x} - \frac{1}{5}xe^{-4x}.$

23. First consider the reduced equation: $\quad y'' + y' - 2y = 0.$ The characteristic equation is:

$$r^2 + r - 2 = (r + 2)(r - 1) = 0$$

and $\quad u_1(x) = e^{-2x}$, $u_2(x) = e^x \quad$ are fundamental solutions. A particular solution of the given equation has the form

$$y = x(A + Bx)e^x.$$

The derivatives of y are:

$$y' = (A + (2B + A)x + Bx^2)e^x, \quad y'' = (2A + 2B + (4B + A)x + Bx^2)e^x.$$

Substitute y and its derivatives into the given equation:

$$(2A + 2B + (4B + A)x + Bx^2 + A + (2B + A)x + Bx^2 - 2Ax - 2Bx^2)e^x = 3xe^x.$$

This implies $\quad A = -\frac{1}{3} \quad, B = \frac{1}{2} \quad$ so $\quad y = x(-\frac{1}{3} + \frac{1}{2}x)e^x.$

The general solution of the given equation is: $\quad y = C_1 e^{-2x} + C_2 e^x - \frac{1}{3}xe^x + \frac{1}{2}x^2 e^x.$

25. Let $y_1(x)$ be a solution of $\quad y'' + ay' + by = \phi_1(x), \quad$ let $y_2(x)$ be a solution of $\quad y'' + ay' + by = \phi_2(x),$ and let $\quad z = y_1 + y_2.$ Then

$$z'' + az' + bz = (y_1'' + y_2'') + a(y_1' + y_2') + b(y_1 + y_2)$$
$$= (y_1'' + ay_1' + by_1) + (y_2'' + ay_2' + y_2) = \phi_1 + \phi_2.$$

27. First consider the reduced equation: $\quad y'' + 4y' + 3y = 0.$ The characteristic equation is:

$$r^2 + 4r + 3 = (r + 3)(r + 1) = 0$$

and $\quad u_1(x) = e^{-3x}$, $u_2(x) = e^{-x} \quad$ are fundamental solutions. Since $\quad \cosh x = \frac{1}{2}\left(e^x + e^{-x}\right),\quad$ a particular solution of the given equation has the form

$$y = Ae^x + Bxe^{-x}$$

The derivatives of y are: $\quad y' = Ae^x + Be^{-x} - Bxe^{-x} \quad y'' = Ae^x - 2Be^{-x} + Bxe^{-x}.$

Substitute y and its derivatives into the given equation:

$$Ae^x - 2Be^{-x} + Bxe^{-x} + 4\left(Ae^x + Be^{-x} - Bxe^{-x}\right) + 3\left(Ae^x + Bxe^{-x}\right) = \tfrac{1}{2}\left(e^x + e^{-x}\right).$$

Equating coefficients, we get $\quad A = \tfrac{1}{16}, \quad B = \tfrac{1}{4}, \quad$ and so $\quad y = \tfrac{1}{16}e^x + \tfrac{1}{4}xe^{-x}.$

The general solution of the given equation is: $\quad y = C_1 e^{-3x} + C_2 e^{-x} + \tfrac{1}{16}e^x + \tfrac{1}{4}xe^{-x}.$

29. First consider the reduced equation $\quad y'' - 2y' + y = 0.\quad$ The characteristic equation is:

$$r^2 - 2r + 1 = (r-1)^2 = 0$$

and $\quad u_1(x) = e^x, \quad u_2(x) = xe^x \quad$ are fundamental solutions. Their Wronskian is given by

$$W = u_1 u_2' - u_2 u_1' = e^x(e^x + xe^x) - xe^x(e^x) = e^{2x}$$

Using variation of parameters, a particular solution of the given equation will have the form

$$y = u_1 z_1 + u_2 z_2,$$

where

$$z_1 = -\int \frac{xe^x(xe^x \cos x)}{e^{2x}}\, dx = -\int x^2 \cos x\, dx = -x^2 \sin x - 2x\cos x + 2\sin x,$$

$$z_2 = \int \frac{e^x(xe^x \cos x)}{e^{2x}}\, dx = \int x\cos x\, dx = x\sin x + \cos x$$

Therefore,

$$y = e^x\left(-x^2 \sin x - 2x\cos x + 2\sin x\right) + xe^x\left(x\sin x + \cos x\right) = 2e^x \sin x - xe^x \cos x.$$

31. First consider the reduced equation $\quad y'' - 4y' + 4y = 0.\quad$ The characteristic equation is:

$$r^2 - 4r + 4 = (r-2)^2 = 0$$

and $\quad u_1(x) = e^{2x}, \quad u_2(x) = xe^{2x} \quad$ are fundamental solutions. Their Wronskian is given by

$$W = u_1 u_2' - u_2 u_1' = e^{2x}\left(e^{2x} + 2xe^{2x}\right) - xe^{2x}(2e^{2x}) = e^{4x}.$$

Using variation of parameters, a particular solution of the given equation will have the form

$$y = u_1 z_1 + u_2 z_2,$$

where

$$z_1 = -\int \frac{xe^{2x}\left(\tfrac{1}{3}x^{-1}e^{2x}\right)}{e^{4x}}\, dx = -\frac{1}{3}\int dx = -\tfrac{1}{3}x,$$

$$z_2 = \int \frac{e^{2x}\left(\tfrac{1}{3}x^{-1}e^{2x}\right)}{e^{4x}}\, dx = \frac{1}{3}\int \frac{1}{x}\, dx = \tfrac{1}{3}\ln|x|.$$

Therefore,

$$y = e^{2x}\left(-\tfrac{1}{3}x\right) + xe^{2x}\left(\tfrac{1}{3}\ln|x|\right) = -\tfrac{1}{3}xe^{2x} + \tfrac{1}{3}x\ln|x|\,e^{2x}.$$

Note: Since $\quad u = -\frac{1}{3}xe^{2x}\quad$ is a solution of the reduced equation,

$$y = \frac{1}{3}x\,\ln|x|\,e^{2x}$$

is also a particular solution of the given equation.

33. First consider the reduced equation $\quad y'' + 4y' + 4y = 0$. The characteristic equation is:

$$r^2 + 4r + 4 = (r+2)^2 = 0$$

and $\quad u_1(x) = e^{-2x}, \quad u_2(x) = xe^{-2x}\quad$ are fundamental solutions. Their Wronskian is given by

$$W = u_1 u_2' - u_2 u_1' = e^{-2x}\left(e^{-2x} - 2xe^{2x}\right) - xe^{-2x}(-2e^{-2x}) = e^{-4x}.$$

Using variation of parameters, a particular solution of the given equation will have the form

$$y = u_1 z_1 + u_2 z_2,$$

where

$$z_1 = -\int \frac{xe^{-2x}\left(x^{-2}e^{-2x}\right)}{e^{-4x}}\,dx = -\int \frac{1}{x}\,dx = -\ln|x|$$

$$z_2 = \int \frac{e^{-2x}\left(x^{-2}e^{-2x}\right)}{e^{-4x}}\,dx = \int \frac{1}{x^2}\,dx = -\frac{1}{x}$$

Therefore,

$$y = e^{-2x}\left(-\ln|x|\right) + xe^{-2x}\left(-\frac{1}{x}\right) = -e^{-2x}\ln|x| - e^{-2x}.$$

Note: Since $\quad u = -e^{-2x}\quad$ is a solution of the reduced equation, we can take

$$y = -\ln|x|e^{2x}.$$

35. First consider the reduced equation $\quad y'' - 2y' + 2y = 0$. The characteristic equation is:

$$r^2 - 2r + 2 = 0$$

and $\quad u_1(x) = e^x \cos x, \quad u_2(x) = e^x \sin x\quad$ are fundamental solutions. Their Wronskian is given by

$$W = e^x \cos x\left[e^x \sin x + e^x \cos x\right] - e^x \sin x\left[e^x \cos x - e^x \sin x\right] = e^{2x}$$

Using variation of parameters, a particular solution of the given equation will have the form

$$y = u_1 z_1 + u_2 z_2,$$

where

$$z_1 = -\int \frac{e^x \sin x \cdot e^x \sec x}{e^{2x}}\,dx = -\int \tan x\,dx = -\ln|\sec x| = \ln|\cos x|$$

$$z_2 = \int \frac{e^x \cos x \cdot e^x \sec x}{e^{2x}}\,dx = \int dx = x$$

Therefore,

$$y = e^x \cos x\left(\ln|\cos x|\right) + e^x \sin x(x) = e^x \cos x\,\ln|\cos x| + xe^x \sin x.$$

37. Assume that the forcing function $F(t) = F_0$ (constant). Then the differential equation has a particular solution of the form $i = A$. The derivatives of i are: $i' = i'' = 0$. Substituting i and its derivatives into the equation, we get

$$\frac{1}{C}A = F_0 \implies A = CF_0 \implies i = CF_0.$$

The characteristic equation for the reduced equation is:

$$Lr^2 + Rr + \frac{1}{C} = 0 \implies r_1, r_2 = \frac{-R \pm \sqrt{R^2 - 4L/C}}{2L} = \frac{-R\sqrt{C} \pm \sqrt{CR^2 - 4L}}{2L\sqrt{C}}$$

(a) If $CR^2 = 4L$, then the characteristic equation has only one root: $r = -R/2L$, and $u_1 = e^{-(R/2L)t}$, $u_2 = te^{-(R/2L)t}$ are fundamental solutions.

The general solution of the given equation is:

$$i(t) = C_1 e^{-(R/2L)t} + C_2 t e^{-(R/2L)t} + CF_0$$

and its derivative is:

$$i'(t) = -C_1(R/2L)e^{-(R/2L)t} + C_2 e^{-(R/2L)t} - C_2(R/2L)t e^{-(R/2L)t}.$$

Applying the side conditions $i(0) = 0$, $i'(0) = F_0/L$, we get

$$C_1 + CF_0 = 0$$

$$(-R/2L)C_1 + C_2 = F_0/L$$

The solution is $C_1 = -CF_0$, $C_2 = \dfrac{F_0}{2L}(2 - RC)$.

The current in this case is:

$$i(t) = -CF_0 e^{-(R/2L)t} + \frac{F_0}{2L}(2 - RC)\,t\,e^{-(R/2L)t} + CF_0.$$

(b) If $CR^2 - 4L < 0$ then the characteristic equation has complex roots:

$$r_1 = -R/2L \pm i\beta, \quad \text{where} \quad \beta = \sqrt{\frac{4L - CR^2}{4CL^2}} \quad \text{(here } i^2 = -1\text{)}$$

and fundamental solutions are: $u_1 = e^{-(R/2L)t}\cos\beta t$, $u_2 = e^{-(R/2L)t}\sin\beta t$.

The general solution of the given differential equation is:

$$i(t) = e^{-(R/2L)t}\left(C_1 \cos\beta t + C_2 \sin\beta t\right) + CF_0$$

and its derivative is:

$$i'(t) = (-R/2L)e^{-(R/2L)t}\left(C_1 \cos\beta t + C_2 \sin\beta t\right) + \beta e^{-(R/2L)t}\left(-C_1 \sin\beta t + C_2 \cos\beta t\right).$$

Applying the side conditions $i(0) = 0$, $i'(0) = F_0/L$, we get

$$C_1 + CF_0 = 0$$

$$(-R/2L)C_1 + \beta C_2 = F_0/L$$

The solution is $C_1 = -CF_0$, $C_2 = \dfrac{F_0}{2L\beta}(2 - RC)$.

The current in this case is:

$$i(t) = e^{-(R/2L)t}\left(\frac{F_0}{2L\beta}(2 - RC)\sin\beta t - CF_0 \cos\beta t\right) + CF_0.$$

39. (a) Let $y_1(x) = \sin(\ln x^2)$. Then

$$y_1' = \left(\frac{2}{x}\right)\cos(\ln x^2) \quad \text{and} \quad y_1'' = -\left(\frac{4}{x^2}\right)\sin(\ln x^2) - \left(\frac{2}{x^2}\right)\cos(\ln x^2)$$

Substituting y_1 and its derivatives into the differential equation, we have

$$x^2\left[-\left(\frac{4}{x^2}\right)\sin(\ln x^2) - \left(\frac{2}{x^2}\right)\cos(\ln x^2)\right] + x\left[\left(\frac{2}{x}\right)\cos(\ln x^2)\right] + 4\sin(\ln x^2) = 0$$

The verification that y_2 is a solution is done in exactly the same way.

The Wronskian of y_1 and y_2 is:

$$W(x) = y_1 y_2' - y_2 y_1'$$

$$= \sin(\ln x^2)\left[-\left(\frac{2}{x}\right)\sin(\ln x^2)\right] - \cos(\ln x^2)\left[\left(\frac{2}{x}\right)\cos(\ln x^2)\right]$$

$$= -\tfrac{2}{x} \neq 0 \text{ on } (0, \infty)$$

(b) To use the method of variation of parameters as described in the text, we first re-write the equation in the form

$$y'' + x^{-1}y' + 4x^{-2}y = x^{-2}\sin(\ln x).$$

Then, a particular solution of the equation will have the form $y = y_1 z_1 + y_2 z_2$, where

$$z_1 = -\int \frac{\cos(\ln x^2)x^{-2}\sin(\ln x)}{-2/x}\,dx$$

$$= \tfrac{1}{2}\int \cos(2\ln x)x^{-1}\sin(\ln x)\,dx$$

$$= \tfrac{1}{2}\int \cos 2u \sin u\,du \qquad (u = \ln x)$$

$$= \tfrac{1}{2}\int (2\cos^2 u - 1)\sin u\,du$$

$$= -\tfrac{1}{3}\cos^3 u + \tfrac{1}{2}\sin u$$

and

$$z_2 = \int \frac{\sin(\ln x^2)x^{-2}\sin(\ln x)}{-2/x}\,dx$$

$$= -\tfrac{1}{2}\int \sin(2\ln x)x^{-1}\sin(\ln x)\,dx$$

$$= -\tfrac{1}{2}\int \sin 2u \sin u\,du \qquad (u = \ln x)$$

$$= -\int \sin^2 u \cos u\,du$$

$$= -\tfrac{1}{3}\sin^3 u$$

Thus, $y = \sin 2u\left(-\tfrac{1}{3}\cos^3 u + \tfrac{1}{2}\sin u\right) - \cos 2u\left(\tfrac{1}{3}\sin^3 u\right)$ which simplifies to:

$$y = \tfrac{1}{3}\sin u = \tfrac{1}{3}\sin(\ln x).$$

SECTION 19.5

1. The equation of motion is of the form

$$x(t) = A \sin(\omega t + \phi_0).$$

The period is $T = 2\pi/\omega = \pi/4$. Therefore $\omega = 8$. Thus

$$x(t) = A \sin(8t + \phi_0) \quad \text{and} \quad v(t) = 8A \cos(8t + \phi_0).$$

Since $x(0) = 1$ and $v(0) = 0$, we have

$$1 = A \sin \phi_0 \quad \text{and} \quad 0 = 8A \cos \phi_0.$$

These equations are satisfied by taking $A = 1$ and $\phi_0 = \pi/2$.

Therefore the equation of motion reads

$$x(t) = \sin\left(8t + \tfrac{1}{2}\pi\right).$$

The amplitude is 1 and the frequency is $8/2\pi = 4/\pi$.

3. We can write the equation of motion as

$$x(t) = A \sin\left(\frac{2\pi}{T}t\right).$$

Differentiation gives

$$v(t) = \frac{2\pi A}{T} \cos\left(\frac{2\pi}{T}t\right).$$

The object passes through the origin whenever $\sin[(2\pi/T)] = 0$.
Then $\cos[(2\pi/T)\,t] = \pm 1$ and $v = \pm 2\pi A/T$.

5. In this case $\phi_0 = 0$ and, measuring t in seconds, $T = 6$.

Therefore $\omega = 2\pi/6 = \pi/3$ and we have

$$x(t) = A \sin\left(\frac{\pi}{3}t\right), \quad v(t) = \frac{\pi A}{3} \cos\left(\frac{\pi}{3}t\right).$$

Since $v(0) = 5$, we have $\pi A/3 = 5$ and therefore $A = 15/\pi$.

The equation of motion can be written

$$x(t) = (15/\pi) \sin\left(\tfrac{1}{3}\pi t\right)$$

7. $x(t) = \dot{x}_0 \sin\left(\sqrt{k/m}\,t + \tfrac{1}{2}\pi\right)$

9. The equation of motion for the bob reads

$$x(t) = x_0 \sin\left(t\sqrt{k/m} + \tfrac{1}{2}\pi\right). \qquad \text{(Exercise 7)}$$

Since $v(t) = \sqrt{k/m}\,x_0 \cos\left(\sqrt{k/m}\,t + \tfrac{1}{2}\pi\right)$, the maximum speed is $\sqrt{k/m}\,x_0$.

The bob takes on half of that speed where $\left| \cos \left(\sqrt{k/m}\, t + \tfrac{1}{2}\pi \right) \right| = \tfrac{1}{2}$. Therefore

$$\left| \sin \left(\sqrt{k/m}\, t + \tfrac{1}{2}\pi \right) \right| = \sqrt{1 - \tfrac{1}{4}} = \tfrac{1}{2}\sqrt{3} \quad \text{and} \quad x(t) = \pm \tfrac{1}{2}\sqrt{3}\, x_0.$$

11. $\text{KE} = \tfrac{1}{2}m[v(t)]^2 = \tfrac{1}{2}m(k/m)x_0{}^2 \cos^2 \left(\sqrt{k/m}\, t + \tfrac{1}{2}\pi \right)$

$$= \tfrac{1}{4}kx_0{}^2 \left[1 + \cos \left(2\sqrt{k/m}\, t + \pi \right) \right].$$

$$\text{Average KE} = \frac{1}{2\pi\sqrt{m/k}} \int_0^{2\pi\sqrt{m/k}} \tfrac{1}{4}kx_0{}^2 \left[1 + \cos \left(2\sqrt{k/m}\, t + \pi \right) \right]\, dt$$

$$= \tfrac{1}{4}kx_0{}^2.$$

13. Setting $y(t) = x(t) - 2$, we can write $x''(t) = 8 - 4x(t)$ as $y''(t) + 4y(t) = 0$.

This is simple harmonic motion about the point $y = 0$; that is, about the point $x = 2$. The equation of motion is of the form

$$y(t) = A \sin (2t + \phi_0).$$

The condition $x(0) = 0$ implies $y(0) = -2$ and thus

(∗) $A \sin \phi_0 = -2$

Since $y'(t) = x'(t)$ and $y'(t) = 2A \cos(2t + \phi_0)$, the condition $x'(0) = 0$ gives $y'(0) = 0$, and thus

(∗∗) $2A \cos \phi_0 = 0.$

Equations (∗) and (∗∗) are satisfied by $A = 2$, $\phi_0 = \tfrac{3}{2}\pi$. The equation of motion can therefore be written

$$y(t) = 2 \sin \left(2t + \frac{3}{2}\pi \right).$$

The amplitude is 2 and the period is π.

15. (a) Take the downward direction as positive. We begin by analyzing the forces on the buoy at a general position x cm beyond equilibrium. First there is the weight of the buoy: $F_1 = mg$. This is a downward force. Next there is the buoyancy force equal to the weight of the fluid displaced; this force is in the opposite direction: $F_2 = -\pi r^2 (L + x)\rho$. We are neglecting friction so the total force is

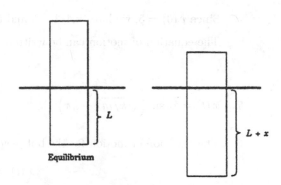

Equilibrium

$$F = F_1 + F_2 = mg - \pi r^2 (L + x)\rho = (mg - \pi r^2 L\rho) - \pi r^2 x\rho.$$

We are assuming at the equilibrium point that the forces (weight of buoy and buoyant force of fluid) are in balance:

$$mg - \pi r^2 L \rho = 0.$$

Thus,

$$F = -\pi r^2 x \rho.$$

By Newton's

$$F = ma \qquad (\text{force} = \text{mass} \times \text{acceleration})$$

we have

$$ma = -\pi r^2 x \rho \qquad \text{and thus} \qquad a + \frac{\pi r^2 \rho}{m} x = 0.$$

Thus, at each time t,

$$x''(t) + \frac{\pi r^2 \rho}{m} x(t) = 0.$$

(b) The usual procedure shows that

$$x(t) = x_0 \sin\left(r\sqrt{\pi \rho / m}\, t + \tfrac{1}{2}\pi\right).$$

The amplitude A is x_0 and the period T is $(2/r)\sqrt{m\pi/\rho}$.

17. From (19.5.4), we have

$$x(t) = Ae^{(-c/2m)t} \sin(\omega t + \phi_0) = \frac{A}{e^{(c/2m)t}} \sin(\omega t + \phi_0) \quad \text{where} \quad \omega = \frac{\sqrt{4km - \omega^2}}{2m}.$$

If c increases, then both the amplitude, $\left| \dfrac{A}{e^{(c/2m)t}} \right|$ and the frequency $\dfrac{\omega}{2\pi}$ decrease.

19. Set $x(t) = 0$ in (19.5.6). The result is:

$$C_1 e^{(-c/2m)t} + C_2 t e^{(-c/2m)t} = 0 \quad \Longrightarrow \quad C_1 + C_2 t = 0 \quad \Longrightarrow \quad t = -C_1/C_2$$

Thus, there is at most one value of t at which $x(t) = 0$.

The motion changes directions when $x'(t) = 0$:

$$x'(t) = -C_1(c/2m)e^{(-c/2m)t} + C_2 e^{(-c/2m)t} - C_2(c/2m)t e^{(-c/2m)t}.$$

Now,

$$x'(t) = 0 \quad \Longrightarrow \quad -C_1(c/2m) + C_2 - C_2 t(c/2m) = 0 \quad \Longrightarrow \quad t = \frac{C_2 - C_1(c/2m)}{C_2(c/2m)}$$

and again we conclude that there is at most one value of t at which $x'(t) = 0$.

21. $x(t) = A\sin(\omega t + \phi_0) + \dfrac{F_0/m}{\omega^2 - \gamma^2} \cos(\gamma t)$

If $\omega/\gamma = m/n$ is rational, then $2\pi m/\omega = 2\pi n/\gamma$ is a period.

23. The characteristic equation is

$$r^2 + 2\alpha r + \omega^2 = 0; \qquad \text{the roots are} \quad r_1, r_2 = -\alpha \pm \sqrt{\alpha^2 - \omega^2}$$

Since $0 < \alpha < \omega$, $\alpha^2 < \omega^2$ and the roots are complex. Thus, $u_1(t) = e^{-\alpha t} \cos \beta t$, $u_2(t) = e^{-\alpha t} \sin \beta t$, where $\beta = \sqrt{\omega^2 - \alpha^2}$ are fundamental solutions, and the general solution is:

$$x(t) = e^{-\alpha t}(C_1 \cos \beta t + C_2 \sin \beta t); \quad \beta = \sqrt{\alpha^2 - \omega^2}$$

25. Set $\omega = \gamma$ in the particular solution x_p given in Exercise 24. Then we have

$$x_p = \frac{F_0}{2\alpha\gamma m} \sin \gamma t$$

As $c = 2\alpha m \to 0^+$, the amplitude $\left| \frac{F_0}{2\alpha\gamma m} \right| \to \infty$

27. $\left(\omega^2 - \gamma^2\right)^2 + 4\alpha^2\gamma^2 = \omega^4 + \gamma^4 + 2\gamma^2(2\alpha^2 - \omega^2)$ increases as γ increases.

REVIEW EXERCISES

1. The equation is linear: $\quad H(x) = \int 1 dx = x \Longrightarrow e^{H(x)} = e^x$

$$\frac{d}{dx}(e^x y) = 2e^{-x} \Longrightarrow e^x y = -2e^{-x} + C; \quad \text{the solution is:} \quad y = -2e^{-2x} + Ce^{-x}$$

3. The equation is separable:

$$\frac{y}{y^2 + 1} \, dy = \frac{1}{\cos^2 x} \, dx = \sec^2 x \, dx$$

$$\frac{1}{2} \ln(y^2 + 1) = \tan x + C$$

The solution is: $\quad \ln(1 + y^2) = 2 \tan x + C$

5. The equation can be written $\quad y' - \frac{2}{x} y = \frac{1}{x^2} y^2 \quad$ a Bernoulli equation.

$$y^{-2} y' - \frac{2}{x} y^{-1} = \frac{1}{x^2}$$

Let $v = y^{-1}$. Then $v' = -y^{-2} y'$, and we get the linear equation

$$v' + \frac{2}{x} v = -\frac{1}{x^2}.$$

Integrating factor: $\quad H(x) = \int (2/x) \, dx = \ln x^2$ and $e^{H(x)} = x^2$.

$$x^2 v' + 2xv = -1$$

$$x^2 v = -x + C$$

$$v = -\frac{1}{x} + \frac{C}{x^2} = \frac{C - x}{x^2}$$

The solution for the original equation is $y = \dfrac{x^2}{C - x}$

7. Since $\dfrac{\partial(y\sin x + xy\cos x)}{\partial y} = \sin x + x\cos x = \dfrac{\partial(x\sin x + y^2)}{\partial x}$,   the equation is exact.

$$f(x,y) = \int (y\sin x + xy\cos x)\,dx = xy\sin x + \phi(y).$$

$\frac{\partial f}{\partial y} = x\sin x + \phi'(y) = y^2 + x\sin x \Longrightarrow \phi'(y) = y^2 \Longrightarrow \phi(y) = \frac{1}{3}y^3$

The solution is $x\sin x + \frac{1}{3}y^3 = C$

9. The equation is separable:

$$\frac{1+y}{y}\,dy = (x^2 - 1)\,dx$$

$$\ln|y| + y = \frac{1}{3}x^3 - x + C$$

11. The equation can be written as $y' + \dfrac{2}{x}y = x^2$,   a linear equation.

$H(x) = \ln x^2; \quad e^{H(x)} = x^2.$

$$x^2 y' + 2xy = x^4$$

$$x^2 y = \frac{1}{5}x^5 + C$$

$$y = \frac{1}{5}x^3 + Cx^{-2}$$

13. The differential equation is homogeneous.

Set $v = y/x$. Then $y = vx$ and $\dfrac{dy}{dx} = v + x\dfrac{dv}{dx}$.

$$v + x\frac{dv}{dx} = \frac{x^2 + x^2 v^2}{2x^2 v} = \frac{1+v^2}{2v}$$

$$x\frac{dv}{dx} = \frac{1+v^2}{2v} - v = \frac{1-v^2}{2v}$$

$$\frac{2v}{1-v^2}\,dv = \frac{1}{x}\,dx$$

$$-\ln|1-v^2| = \ln|x| + C$$

$$1 - v^2 = \frac{C}{x}$$

Replacing v by y/x, we get $x^2 - y^2 = Cx$.

Applying the initial condition $y(1) = 2$ gives $C = -3$. The solution of the initial-value problem is: $x^2 + 3x - y^2 = 0$.

15. Since $\dfrac{\partial(x+y)^2}{\partial y} = 2x + 2y = \dfrac{\partial(2xy + x^2 - 1)}{\partial x}$,   the equation is exact.

$$f(x,y) = \int (x+y)^2\,dx = \int (x^2 + 2xy + y^2)\,dx = \frac{1}{3}x^3 + x^2 y + xy^2 + \phi(y).$$

$\dfrac{\partial f}{\partial y} = x^2 + 2xy + \phi'(y) = 2xy + x^2 - 1 \Longrightarrow \phi'(y) = -1 \quad \Longrightarrow \quad \phi(y) = -y$

The general solution is: $\frac{1}{3}x^3 + x^2 y + xy^2 - y = C$

Applying the initial condition $y(1) = 1$ gives $C = 4/3$. The solution of the initial-value problem is:

$\frac{1}{3}x^3 + x^2 y + xy^2 - y = 4/3$

17. The equation is a Bernoulli equation; rewrite it as: $y^{-2}y' + xy^{-1} = x$.

Set $v = y^{-1}$. Then $v' = -y^{-2}y'$, and we have

$$v' - xv = -x,$$

a linear equation. $H(x) = \int(-x)\,dx = -\frac{1}{2}x^2$ and $e^{H(x)} = e^{-x^2/2}$

$$e^{-x^2/2}v' - xe^{-x^2/2}v = -xe^{-x^2/2}$$

$$e^{-x^2/2}v = e^{-x^2/2} + C$$

$$v = 1 + Ce^{x^2/2}$$

$$y = \frac{1}{1 + Ce^{x^2/2}}$$

Applying the initial condition $y(0) = 2$ gives $C = -1/2$. The solution of the initial-value problem is: $y = \dfrac{2}{2 - e^{x^2/2}}$.

19. The characteristic equation is: $r^2 - 2r + 2 = 0$. The roots are: $r_1,\ r_2 = 1 \pm i$.

The general solution is:

$$e^x(C_1 \cos x + C_2 \sin x).$$

21. The characteristic equation for the reduced equation is: $r^2 - r - 2 = 0$. The roots are: $r = 2, -1$.

Use undetermined coefficients to find a particular solution of the nonhomogeneous equation:

$$z = A\cos 2x + B\sin 2x$$
$$z' = -2A\sin 2x + 2B\cos 2x$$
$$z'' = -4A\cos 2x - 4B\sin 2x$$

Substituting $z,\ z',\ z''$ into the differential equation yields the pair of equations:

$$-6A - 2B = 0,\ 2A - 6B = 1 \quad \Longrightarrow \quad A = \frac{1}{20},\ B = -\frac{3}{20}.$$

The general solution is: $y = C_1 e^{2x} + C_2 e^{-x} + \frac{1}{20}\cos 2x - \frac{3}{20}\sin 2x$

23. The characteristic equation for the reduced equation is: $r^2 - 6r + 9 = 0$. The roots are: $r_1 = r_2 = 3$.

Use undetermined coefficients to find a particular solution of the nonhomogeneous equation.

Since $z = e^{3x}$ and $z = xe^{3x}$ are solutions of the reduced equation, set $z = Ax^2 e^{3x}$.

$$z = Ax^2 e^{3x}$$
$$z' = 2Axe^{3x} + 3Ax^2 e^{3x}$$
$$z'' = 2Ae^{3x} + 12Axe^{3x} + 9Ax^2 e3x$$

Substituting z, z', z'' into the differential equation gives:

$$2A = 3 \implies A = \frac{3}{2}.$$

The general solution is: $y = C_1 e^{3x} + C_2 xe^{3x} + \frac{3}{2} x^2 e^{3x}$.

25. The characteristic equation for the reduced equation is: $r^2 - 2r + 1 = 0$. The roots are: $r_1 = r_2 = 1$.

Use variation of parameters to find a particular solution of the nonhomogeneous equation.

Set $u_1 = e^x$ and $u_2 = xe^x$. Then their Wronskian is $W(x) = e^{2x}$.

$$z_1 = -\int \frac{xe^x(1/x)e^x}{e^{2x}} \, dx = -\int dx = -x \qquad z_2 = \int \frac{e^x(1/x)e^x}{e^{2x}} \, dx = \int (1/x) \, dx = \ln x$$
$$y_p = -xe^x + xe^x \ln x$$

The general solution of the equation is: $y = C_1 e^x + C_2 xe^x + xe^x \ln x, \; x > 0$

27. The characteristic equation for the reduced equation is: $r^2 + 4r + 4 = 0$. The roots are: $r_1 = r_2 = -2$.

Use undetermined coefficients to find a particular solution of the nonhomogeneous equation:

$$z = Ax^2 e^{-2x} + Be^{2x}$$
$$z' = 2Axe^{-2x} - 2Ax^2 e^{-2x} + 2Be^{2x}$$
$$z'' = 2Ae^{-2x} - 8Axe^{-2x} + 4Ax^2 e^{-2x} + 4Be^{2x}$$

Substituting z, z', z'' into the differential equation yields the equations:

$$2A = 4, \; 16B = 2 \implies A = 2, \; B = \frac{1}{8}.$$

The general solution is: $y = C_1 e^{-2x} + C_2 xe^{-2x} + \frac{1}{8} e^{2x} + 2x^2 e^{-2x}$

29. First find the general solution of the differential equation.

The characteristic equation for the reduced equation is: $r^2 + r = 0$. The roots are: $r_1 = -1, \; r_2 = 0$.

Use undetermined coefficients to find a particular solution of the nonhomogeneous equation:

Set $z = Ax^2 + Bx$

$$z = Ax^2 + Bx$$
$$z' = 2Ax + B$$
$$z'' = 2A$$

Substituting z, z', z'' into the differential equation yields the equations:

$$2A = 1, \; A + B = 0 \implies A = 1/2, \; B = -1.$$

The general solution of the differential equation is: $y = C_1 e^{-x} + C_2 + \frac{1}{2}x^2 - x$

Applying the initial conditions $y(0) = 1$, $y'(0) = 0$, we get the pair of equations

$$C_1 + C_2 = 1, \quad -C_1 - 1 = 0, \quad \implies \quad C_1 = -1, \ C_2 = 2.$$

The solution of the initial-value problem is: $\quad y = 2 - e^{-x} + \frac{1}{2}x^2 - x$

31. First find the general solution of the differential equation.

The characteristic equation for the reduced equation is: $r^2 - 5r + 6 = 0$. The roots are: $r_1 = 2$, $r_2 = 3$.

Use undetermined coefficients to find a particular solution of the nonhomogeneous equation:

Set $z = Axe^{2x}$

$$z = Axe^{2x}$$
$$z' = Ae^{2x} + 2Axe^{2x}$$
$$z'' = 4Ae^{2x} + 4Axe^{2x}$$

Substituting z, z', z'' into the differential equation gives: $-A = 10$, $A = -10$.

The general solution of the differential equation is: $y = C_1 e^{2x} + C_2 e^{3x} - 10xe^{2x}$

Applying the initial conditions $y(0) = 1$, $y'(0) = 1$, we get $C_1 = -8$, $C_2 = 9$.

The solution of the initial-value problem is: $\quad y = 9e^{3x} - 8e2x - 10xe^{2x}$

33. Assume $x(t) = A\sin(wt + \phi_0)$.

From $T = 2\pi/\omega = \pi/2$, $\omega = 4$ and $x(t) = A\sin(4t + \phi_0)$

$x(0) = 2 \implies A\sin(\phi)0) = 2; \quad x'(0) = 0 \implies 4A\cos(\phi_0) = 0 \implies \phi_0 = \dfrac{\pi}{2}$ and $A = 2$.

Therefore,

$$x(t) = 2\sin(4t + \pi/2); \quad \text{amplitude } A = 2; \quad \text{frequency } 2/\pi.$$

35. Assume that the downward direction is positive. Then

$$4x''(t) = -64x(t) + 8\sin 4t, \quad x(0) = -\frac{1}{2}, \quad x'(0) = 0$$

This equation can be written as

$$x'' + 16x = 2\sin 4t$$

The characteristic equation for the reduced equation is: $r^2 + 16 = 0$ and the roots are $r = \pm 4i$.

Use undetermined coefficients to find a particular solution of the nonhomogeneous equation:

Set $z = At\cos 4t + Bt\sin 4t$

$$z = At\cos 4t + Bt\sin 4t$$
$$z' = A\cos 4t - 4At\sin 4t + B\sin 4t + 4Bt\cos 4t$$
$$z'' = -8A\sin 4t - 16At\cos 4t + 8B\cos 4t - 16Bt\sin 4t$$

Substituting z, z', z'' into the differential equation yields the equations, we get $A = -\frac{1}{4}$, $B = 0$.

The general solution of the differential equation is:

$$x(t) = C_1 \cos 4t + C_2 \sin 4t - \frac{1}{4}t \cos 4t.$$

Applying the initial conditions $x(0) = -1/2$, $x'(0) = 0$, we get $C_1 = -1/2$, $C_2 = 1/16$.

The equation of motion is:

$$x(t) = -\frac{1}{2} \cos 4t - \frac{1}{4}t \cos 4t + \frac{1}{16} \sin 4t$$

NOTES

NOTES

NOTES

NOTES

NOTES

NOTES

NOTES

NOTES

NOTES

NOTES

NOTES

NOTES

NOTES

NOTES

NOTES

NOTES

NOTES

NOTES

NOTES

NOTES

NOTES

NOTES

NOTES

NOTES

NOTES

NOTES

NOTES

NOTES

NOTES

NOTES

NOTES

NOTES

NOTES